Organic Farming, Prototype for Sustainable Agricultures

Stéphane Bellon • Servane Penvern
Editors

Organic Farming, Prototype for Sustainable Agricultures

Springer

Editors
Stéphane Bellon
Unité Ecodéveloppement
INRA
Avignon Cedex 09
France

Servane Penvern
Unité Ecodéveloppement
INRA
Avignon Cedex 09
France

ISBN 978-94-007-7926-6 ISBN 978-94-007-7927-3 (eBook)
DOI 10.1007/978-94-007-7927-3
Springer Dordrecht Heidelberg New York London

Library of Congress Control Number: 2013957940

Printed on acid-free paper

Springer is part of Springer Science+Business Media (www.springer.com)

Preface

Organic Farming–A Role Model for Productive and Ecologically-Sustainable Farming Systems

Europe has been the pacesetter for organic farming for 40 years. The fact that between 10 and 20 % of the farms and the agricultural land area have become certified organic in a few leading countries has attracted the attention of the scientific community and of policy makers. Scientific studies on public goods delivered by organic farms have become more numerous and encompass topical aspects such as soil fertility building, carbon sequestration, biodiversity at the plant, animal, and microorganism levels, and eutrophication of semi-natural and natural ecosystems, etc. Support schemes for farmers have compensated for the delivery of public goods.

The steady economic growth of the global organic food market has further fueled the public interest in organic agriculture. Is it a viable strategy that reduces the trade-offs between food and feed production on the one hand, while maintaining the regulating and supporting ecosystem services and landscape quality on the other? "Yes, but…" is the most often heard answer. "Yes" for the fact that organic farms are likely to reduce detrimental impacts on the environment and to maintain the quality of ecosystems. "But" because crop and livestock yields are, on average, less on organic farms. Without any changes to the wasteful way in which society handles, uses, and consumes food, a large-scale transformation of high-yielding farmland to organic cultivation might accelerate deforestation and (re)cultivation of ecologically-sensitive land.

The state-of-the art of scientific data on productivity is divergent and controversial. While the crop productivity of organic farms appears to be 0.7–0.8 of that of intensive farms in temperate zones, the yield ratio in marginal regions of Africa where subsistence farming is still widely spread, has been found to be in favour of organic farms Hence, in resource- and income-poor countries, organic farming seems to offer an appropriate and low-cost way to increase productivity and to improve farm livelihood.

Despite its success in Europe and for specific cash crops on the world market, organic farming is still a niche, with only 1 % of agricultural land under organic cultivation worldwide. Organic agriculture is challenged to unlock its potential: both as a role model and a real pathway to sustainability in agriculture and food systems. As a farming system, it is knowledge-intensive and resistant to overspecialisation. This is a challenge for scientists, farm advisors and farmers, and needs to be addressed by improving education and by enabling participation and inter-disciplinary research.

The concept of eco-functional intensification goes far beyond the restrictive use of fertilisers and pesticides. It requires a fundamental redesign of farms and fields, and entails more co-operation within the organic sector. Accordingly, livestock needs to be integrated into the nutrient and organic matter circuits in order to improve the robustness and resilience of both crops and animals, with the selection of well-adapted varieties and breeds. Finally, development pathways in organic agriculture also challenge agricultural sciences. While the basic principles of organic agriculture are persuasive and dynamic agroecological approaches, existing standards for the certification of farms and foods have become outdated. Creative research work and out-of-the-box thinking are needed to unleash social, ecological, and technological innovation in organic agriculture.

This book gives an outstanding analysis of what has been achieved, as well as an insight into what the future avenues for organic farming will be.

Director of the Research
Institute of Organic
Agriculture (FiBL)
Switzerland

Professor of Research
Management, University
of Kassel-Witzenhausen
Germany

Urs Niggli

Acknowledgments

We, as editors, would like to thank all the contributors who accepted to write chapters and who made this book possible. We are particularly indebted to them for their flexibility in adapting their needs to the overall demands of such a project. Through their involvement in organic research projects and communities, they were able to contribute state-of the-art research results and provide new avenues for research. We also warmly thank the invisible colleagues who, despite their busy schedules and heavy workloads, accepted and took the time to review the manuscripts of all of the chapters: Claudine Basset-Mens, Patrick Brachet, Yvan Capowiez, Marianne Cerf, Peter Cornish, Enrique Dapena, Ika Darnhofer, Jean-Philippe Deguine, Dominique Desclaux, Jean-Jacques Drevon, Philippe Fleury, Ken Giller, Stuart Hill, Louise Jackson, Claire Lamine, Edith Lammerts van Bueren, Derek Lynch, Rod MacRae, Ian A Merwin, Michel Meuret, Naoufel Mzoughi, Astrid D. Oberson, Guillaume Ollivier, Susanne Padel, Daniel Plénet, Alain Ratnadass, Michael Raviv, Ewa Rembialowska, Paul Robin, Natacha Sautereau, Otto Schmid, Eva Schmidtner, Bernard Seguin, Sylvaine Simon, Marc Tchamitchian, Isabelle Vagneron, Patrick Veysset and Christoph Winckler.

We also thank the Internal Committee for Organic Agriculture (CIAB), comprising representatives from various scientific disciplines and responsible for the implementation of INRA's research programme, for the book rational preparation, reviewing and support: Joel Abecassis, Didier Andrivon, Marc Benoit, Jacques Cabaret, Philippe Debaeke, Sophie Prache and Isabelle Savini.

The structure and content of the book is also the fruit of the many interactions we had with various colleagues: from our research unit, Ecodevelopment (INRA, Avignon, France); and during meetings of the Scientific Council of Organic Agriculture (CSAB) and the various activities of the Mixed Thematic Network dedicated to OF&F development (RMT DévAB). Many exchanges with various stakeholders in the French organic sector also encouraged us to move forward with this initiative. Mentioning all of them would be too long, but organic farmers and their representative organisations were indeed a major source of inspiration to address current issues and dynamics. At the international level, we highly benefited from the relationships established with people involved in research projects, conferences and events. In addition to allowing us to compare our ideas with those of other communities, it also enabled us to improve our synthesis of the existing knowledge, to discover and appreciate other books, and to contribute to new projects. A pioneer scientist, Bertil Sylvander, requires special mention: he opened the way for research in OF&F at INRA, both with the CIAB and through international commitments. Concerning the editing of this book, we are grateful to Maryse Walsh (from Springer) who allowed us to see this book project through to the end.

We would also like to express our gratitude to Isabelle Savini and Alice Vettoretti, both deeply involved in getting this book into print. Their constructive comments and assistance made it possible to complete the project. We also thank Gail Wagman for her contribution in revising and improving the final English versions of the various chapters.

Contents

1 Organic Food and Farming as a *Prototype* for Sustainable Agricultures 1
Stéphane Bellon and Servane Penvern

Part I Systems' Functionning

2 Soil Phosphorus Management in Organic Cropping Systems: From Current Practices to Avenues for a More Efficient Use of P Resources 23
Thomas Nesme, Bruno Colomb, Philippe Hinsinger and Christine A. Watson

3 Eco-Functional Intensification by Cereal-Grain Legume Intercropping in Organic Farming Systems for Increased Yields, Reduced Weeds and Improved Grain Protein Concentration 47
Laurent Bedoussac, Étienne-Pascal Journet, Henrik Hauggaard-Nielsen, Christophe Naudin, Guénaëlle Corre-Hellou, Loïc Prieur, Erik Steen Jensen and Eric Justes

4 Regulatory Framework for Plant Protection in Organic Farming 65
Bernhard Speiser, Lucius Tamm and Franco P. Weibel

5 Conservation Biocontrol: Principles and Implementation in Organic Farming 83
Sylvaine Simon, Adrien Rusch, Eric Wyss and Jean-Pierre Sarthou

6 Agroecological Crop Protection in Organic Farming: Relevance and Limits 107
Jean-Philippe Deguine and Servane Penvern

7 **Adapting Apple Ideotypes to Low-Input Fruit Production Agro-Ecosystems** ... 131
Luciana Parisi, Laurent Jamar, Marc Lateur, François Laurens and Pierre-Eric Lauri

8 **Alternatives to Synthetic Chemical Antiparasitic Drugs in Organic Livestock Farming in Europe** ... 149
Hervé Hoste, Smaragda Sotiraki, Helena Mejer, Felix Heckendorn, Veronika Maurer and Stig Thamsborg

9 **Animal Healthcare Strategies in Organic and Conventional Farming** ... 171
Christian Nicourt and Jacques Cabaret

10 **Optimisation of Breeding Systems and Land use to Maximise Feed Self-Sufficiency and Economic Outcomes in Organic Sheep-for-Meat Production** ... 181
Marc Benoit and Gabriel Laignel

11 **Experiencing Organic Mixed Crop Dairy Systems: A Step-by-Step Design Centred on a Long-term Experiment** ... 201
Xavier Coquil, Jean-Louis Fiorelli, André Blouet and Catherine Mignolet

Part II Organic Performances

12 **Using Life Cycle Analysis to Analyse the Environmental Performances of Organic and Non-organic Apple Orchards** ... 221
Aude Alaphilippe, Sylvaine Simon and Frank Hayer

13 **The Potential of Organic Agriculture to Mitigate the Influence of Agriculture on Global Warming—A Review** ... 239
Adrian Muller and Claude Aubert

14 **The Freedoms and Capabilities of Farm Animals: How Can Organic Husbandry Fulfill Them?** ... 261
Jacques Cabaret, Caroline Chylinski and Mette Vaarst

15 **Breaking with the Animal Production Paradigm: A Major Issue for Organic Husbandry** ... 279
Jocelyne Porcher

16 Food Quality and Possible Positive Health Effects of Organic Products 295
Denis Lairon and Machteld Huber

17 Advances, Issues and Challenges in Organic Lamb Meat Quality 313
Sophie Prache

18 Is Organic Farming Unsustainable? Analysis of the Debate About the Conventionalisation of the Organic Label 325
Geneviève Teil

Part III Development Dynamics

19 Framing the Social, Ecological and Economic Goods and Services Derived from Organic Agriculture in the Canadian Context 347
Derek H. Lynch, Jennifer Sumner and Ralph C. Martin

20 Does the Seed Sector Offer Meet the Needs of Organic Cropping Diversity? Challenges for Organic Crop Varieties 367
Dominique Desclaux and Jean-Marie Nolot

21 Seeds for Organic Agriculture: Development of Participatory Plant Breeding and Farmers' Networks in France 383
Véronique Chable, Julie Dawson, Riccardo Bocci and Isabelle Goldringer

22 Considerations for Enabling the Ecological Redesign of Organic and Conventional Agriculture: A Social Ecology and Psychosocial Perspective 401
Stuart B. Hill

23 Transitions Towards Organic Farming at the Farm and at the Local Scales: The Role of Innovative Production and Organisational Modes and Networks 423
Claire Lamine, Mireille Navarrete and Aurélie Cardona

24 Contributing to a Transition to Sustainability of Agri-Food Systems: Potentials and Pitfalls for Organic Farming 439
Ika Darnhofer

25 Multi-Scale Integrated Assessment of Regional Conversion to Organic Farming (OF) ... 453
Santiago Lopez-Ridaura, Sylvestre Delmotte, Christophe Le Page, Laure Le Quéré, Gaël Goulevant, Philippe Chauvelon, Alain Sandoz and Jean-Claude Mouret

Index ... 479

Contributors

Aude Alaphilippe INRA, UE695 UERI Unité Expérimentale Recherches Intégrées, Saint Marcel les Valence, France

Claude Aubert Lempdes, France

Laurent Bedoussac ENFA-INRA, UMR1248 AGIR, Castanet-Tolosan Cedex, France

Stéphane Bellon INRA, UR0767 Écodéveloppement, Domaine Saint Paul, Site Agroparc, Avignon Cedex 9, France

Marc Benoit INRA, UR1213 Herbivores Saint-Genès-Champanelle, France

André Blouet INRA, UR055 SAD ASTER, Mirecourt, France

Riccardo Bocci AIAB, Florence, Italy

Jacques Cabaret INRA-Université de Tours, UMR1282 Infectiologie et Santé Publique, Nouzilly, France

Université Paris I—Sorbonne, Paris, France

Aurélie Cardona INRA, UR0767 Écodéveloppement, Domaine Saint Paul, Site Agroparc, Avignon Cedex 9, France

Véronique Chable INRA, UR980 SAD-Paysage, Rennes, France

Philippe Chauvelon Tour du Valat, Research Centre for the Conservation of Mediterranean Wetlands, Arles, France

Caroline Chylinski INRA-Université de Tours, UMR1282 Infectiologie et Santé Publique, Nouzilly, France

Bruno Colomb INRA-INP Toulouse, UMR1248 AGIR, Castanet-Tolosan, France

Xavier Coquil INRA, UR055 SAD ASTER, Mirecourt, France

Guénaëlle Corre-Hellou LUNAM Université, Groupe ESA, UR LEVA, SFR 4207 QUASAV, Angers Cedex 01, France

Ika Darnhofer Department of Economic and Social Sciences, University of Natural Resources and Life Sciences, Vienna, Austria

Julie Dawson Department of Horticulture, University of Wisconsin-Madison, Madison, WI, USA

INRA-CNRS-INAPG-Université de Paris 11, UMR0320 Génétique Végétale, Gif-sur-Yvette, France

Jean-Philippe Deguine CIRAD, UMR PVBMT, Saint-Pierre, La Réunion, France

Sylvestre Delmotte INRA-SupAgro-CIRAD, UMR0951 Innovation, Montpellier, France

Dominique Desclaux INRA, UE398 Diascope, Mauguio, France

Jean-Louis Fiorelli INRA, UR055 SAD ASTER, Mirecourt, France

Isabelle Goldringer INRA-CNRS-INAPG-Université de Paris 11, UMR0320 Génétique Végétale, Gif-sur-Yvette, France

Gaël Goulevant INRA-SupAgro-CIRAD, UMR0951 Innovation, Montpellier, France

Henrik Hauggaard-Nielsen Department of Chemical and Biochemical Engineering, Technical University of Denmark, Lyngby, Denmark

Frank Hayer Agroscope Reckenholz Tanikon ART, Zurich, Switzerland

Felix Heckendorn Department of Livestock Science, Research Institute of Organic Agriculture (FiBL), Frick, Switzerland

Stuart B. Hill School of Education, University of Western Sydney (Kingswood Campus), Locked Bag 1797, Penrith, Australia

Philippe Hinsinger INRA, UMR1222 Eco&Sols, Montpellier, France

Hervé Hoste INRA-Université de Toulouse-ENVT, UMR1225 IHAP Interactions Hôte-Agents Pathogènes, Toulouse, France

Machteld Huber Department of Healthcare & Nutrition, Louis Bolk Institute, Driebergen, The Netherlands

Laurent Jamar Department of Life Sciences, Unit Breeding & Biodiversity, CRA-W, Gembloux, Belgium

Erik Steen Jensen Department of Biosystems and Technology, Swedish University of Agricultural Sciences, Alnarp, Sweden

Étienne-Pascal Journet CNRS, UMR2594 LIPM, INRA, UMR1248 AGIR, Castanet-Tolosan, France

Eric Justes INRA-Université de Toulouse-INPT, UMR1248 AGIR, Castanet-Tolosan, France

Gabriel Laignel INRA, UR 1213 Herbivores, Saint-Genès-Champanelle, France

Denis Lairon Univ d'Aix Marseille-INSERM-INRA, UMR1062 NORT Nutrition, obésité et risque thrombotique, Marseille, France

Claire Lamine INRA, UR0767 Écodéveloppement, Domaine Saint Paul, Site Agroparc, Avignon Cedex 9, France

Marc Lateur Department of Life Sciences, Unit Breeding & Biodiversity, CRA-W, Gembloux, Belgium

François Laurens INRA, UMR IRHS, Beaucouzé Cedex, France

Pierre-Eric Lauri INRA, UMR AGAP, Montpellier, France

Christophe Le Page CIRAD, UPR Green, Montpellier, France

Laure Le Quéré INRA-SupAgro-CIRAD, UMR0951 Innovation, Montpellier, France

Santiago Lopez-Ridaura INRA-SupAgro-CIRAD, UMR0951 Innovation, Montpellier, France

Derek H. Lynch Department of Plant and Animal Sciences, Dalhousie University, Truro, Nova Scotia Canada

Ralph C. Martin Department of Plant Agriculture, University of Guelph, Guelph, Ontario, Canada

Veronika Maurer Department of Livestock Science, Research Institute of Organic Agriculture (FiBL), Frick, Switzerland

Helena Mejer Department of Veterinary Disease Biology, Faculty of Health and Medical Sciences, University of Copenhagen, Frederiksberg, Denmark

Catherine Mignolet INRA, UR055 SAD ASTER, Mirecourt, France

Jean-Claude Mouret INRA-SupAgro-CIRAD, UMR0951 Innovation, Montpellier, France

Adrian Muller Research Institute of Organic Agriculture (FiBL), Frick, Switzerland

Christophe Naudin LUNAM Université, Groupe ESA, UR LEVA, SFR 4207 QUASAV, Angers Cedex 01, France

Mireille Navarrete INRA, UR0767 Écodéveloppement, Domaine Saint Paul, Site Agroparc, Avignon Cedex 9, France

Thomas Nesme BORDEAUX SCIENCES AGRO-INRA, UMR1220 TCEM Transfert Sol-Plante et Cycle des Éléments Minéraux dans les Ecosystèmes Cultivés, Gradignan Cedex, France

Christian Nicourt INRA, UR1216 RiTME, Ivry, France

Jean-Marie Nolot INRA-Université de Toulouse-INPT, UMR1248 AGIR, Castanet-Tolosan, France

Luciana Parisi INRA, UR407 Pathologie Végétale, Avignon Cedex 9, France

Servane Penvern INRA, UR0767 Écodéveloppement, Domaine Saint Paul, Site Agroparc, Avignon Cedex 9, France

Jocelyne Porcher INRA-SupAgro-CIRAD, UMR951 Innovation, Montpellier, France

Sophie Prache INRA, UR1213 Herbivores, Saint-Genès-Champanelle, France

Loïc Prieur CREAB Midi-Pyrénées, LEGTA Auch-Beaulieu, Auch, France

Adrien Rusch INRA-Université de Bordeaux, UMR1065 Santé et Agroécologie du Vignoble, Villenave d'Ornon, France

INRA-Université de Bordeaux, UMR1065 Santé et Agroécologie du Vignoble, Villenave d'Ornon, France

Alain Sandoz Tour du Valat, Research Centre for the Conservation of Mediterranean Wetlands, Arles, France

Jean-Pierre Sarthou Université de Toulouse-INP-ENSAT-INRA, UMR1248 AGIR, Castanet Tolosan, France

Sylvaine Simon INRA, UE695 UERI Unité Expérimentale Recherches Intégrées, Saint-Marcel-lès-Valence, France

Smaragda Sotiraki Veterinary Research Institute (VRI), Hellenic Agricultural Organisation "Demeter" (former NAGREF), Thermi, Thessaloniki, Greece

Bernhard Speiser Research Institute of Organic Agriculture (FiBL), Frick Switzerland

Jennifer Sumner Department of Leadership, Higher and Adult Education, Ontario Institute for Studies in Education, University of Toronto, Toronto, Ontario, Canada

Lucius Tamm Research Institute of Organic Agriculture (FiBL), Frick, Switzerland

Geneviève Teil INRA-AgroParisTech, UMR1048 SADAPT, Paris, France

Stig Thamsborg Department of Veterinary Disease Biology, Faculty of Health and Medical Sciences, University of Copenhagen, Frederiksberg, Denmark

Mette Vaarst Department of Animal Science, Aarhus University, Tjele, Denmark

Christine A. Watson Scottish Agricultural College, Aberdeen, UK

Franco P. Weibel Research Institute of Organic Agriculture (FiBL), Frick, Switzerland

Eric Wyss Research Institute of Organic Agriculture (FiBL), Frick, Switzerland

GLOBE Switzerland, c/o Federal Office for the Environment, Berne, Switzerland

Chapter 1
Organic Food and Farming as a *Prototype* for Sustainable Agricultures

Stéphane Bellon and Servane Penvern

Abstract Many agricultural models claim to serve as a foundation towards sustainability. This introductory chapter examines how research results in organic food and farming (OF&F) may contribute to meaningful innovations and transitions for sustainable agricultures. To support this, we refer to three different interpretations of the concept of prototype. Each of them is developed in the three sections of the book. First, prototype theory is used as a mode of graded categorisation in cognitive sciences where categories are relative and boundaries may be fuzzy, making it possible to confront OF&F to other agricultures. The first section addresses production, protection and agro-ecological processes with the aim of increasing self-sufficiency. It addresses the validity domain of research findings for other agricultures. The second interpretation of OF&F as a prototype refers to its ability to outperform existing agricultures. This could also serve as a basis for outcome-based OF&F, which is currently mean-based. Three main challenges are developed in the second section: environmental issues, animal welfare and the quality of organic products. The third interpretation refers to OF&F development pathways. OF&F internal dynamics can be seen as enabling transformations. The third section combines two implications: renewal of an organic framework open to other stakeholders and identification of transition pathways for OF&F systems, including the territorial level. The prototype concept is useful for tackling the multiple challenges of the dynamic relationships of OF&F with other forms of agriculture. If OF&F is more than a niche, shifting from a *prototype* to a generalisable model still remains an issue.

Keywords Alternative agricultural model · Innovation · Redesign · Transition · Prototyping · Performance · Development pathways · Research

---

S. Bellon (✉) · S. Penvern
INRA, UR0767 Écodéveloppement, 228 route de l'Aérodrome, Domaine Saint Paul, Site Agroparc, CS 40509, 84914 Avignon Cedex 9, France
e-mail: stephane.bellon@avignon.inra.fr

S. Bellon, S. Penvern (eds.), *Organic Farming, Prototype for Sustainable Agricultures*, DOI 10.1007/978-94-007-7927-3_1,

1.1 Introduction

Many agricultural models claim to serve as a foundation towards sustainability (Koohafkan et al. 2011). Organic food and farming (OF&F) is one of these candidate models, and probably the most acknowledged worldwide. It is recognised in the scientific and political arenas (McIntyre et al. 2009; National Research Council 2010), as well as by society as a whole. In spite of the limited share it occupies on the global food market, everybody has an opinion about OF&F practices or can identify its products. The legitimacy of OF&F is also due to its history and evolution in terms of practices, principles and regulations (Besson 2009; Francis 2009; Kristiansen et al. 2006; Lockeretz 2007). Several books and many articles about OF&F are published every year in various languages. They concern OF&F's ability to address agricultural and societal challenges: how to feed humanity, how to alleviate the impacts of climate change, how to enhance ecosystem services, etc. As the number of outcomes expected from agriculture multiplies, we are increasingly aware that there is no ready-made solution to address complex issues. Moreover, it is expected that research will contribute to the design of such solutions, together with stakeholders who have created and implemented promising initiatives and transition pathways. Among the possible solutions to be explored, we can mention the shift from intensive or high external-input agricultural systems to "knowledge intensive" or "ecologically intensive" agricultures.

In this book, our purpose is to identify to what extent and under what conditions research results in OF&F may contribute to meaningful innovations and transitions for sustainable agriculture. Its ambition is to present and critically review major biotechnical and socio-economic aspects of organic agriculture that can also be relevant to other agricultures. In this perspective, French scientists who contributed to a national organic congress, DinABio[1], were identified as potential authors, together with scientists from other countries and continents. These research scientists from Europe, North America and Australasia have all made important contributions and are all still active in organic research projects. They represent a wide range of scientific disciplines and use most of the available research methods. Subsequently, the topics addressed in this book combine different cultures and realities of organics.

OF&F is considered as a *prototype* both in its own dynamics and in its relationships with other forms of agriculture. Beyond the abandonment of chemical fertilisers and pesticides, OF&F is studied on the basis of its own acquisitions and diversity: process-based more than product-based; reconnecting agriculture with its ecological origins; broader than the traditional focus on soil fertility and dynamics in that its rules and practices are constantly evolving; and building on the combination of modern science and farmers' own experiences, references and

[1] The first DinABio congress took place in May 2008 at the INRA Centre of Montpellier (SupAgro). The proceedings were published in the online journal "Innovations Agronomiques" (available at: http://www6.inra.fr/ciag/Revue/Volume-4-Janvier-2009). The second congress took place on the 13–14th of November 2013 in Tours (France). Proceedings available at: http://www6.inra.fr/ciag/Revue/Volume-32-Novembre-2013.

knowledge of organic terrain. OF&F is also related and sometimes overlaps with other agricultural regimes (e.g., Ollivier and Bellon 2013). Such relationships are a premise of this book. Reference to OF&F as a *prototype* was introduced in the first French action plan for OF&F development (1998–2003) by its coordinator (Riquois 1999). His main argument was that due to its principles and strong constraints, OF&F is a laboratory for the development of sustainable agriculture and food production. Without chemical crutches or safety nets, organic farmers must imagine alternative methods that can be relevant to other situations. This position puts stress on the modernity of organic agriculture, while strongly reconnecting it with other forms of agriculture instead of relegating it to a ghetto or a niche. The concept referred to in this book is also shared in McIntyre et al. (2009) and Halberg and Muller (2013), who both focus on the contribution of organic agriculture to sustainability.

In the first section of this introductory chapter, we incorporate the concept of *prototype* and expand it in three directions. Each of them is then developed in three subsequent sections, briefly presenting a group of chapters that support a specific facet of the *prototype*.

1.2 Three Facets of the Concept of *Prototype*

The concept of *prototype* frames to what extent OF&F is a good representative of sustainable agricultures. Prototype theory is a mode of graded categorisation in cognitive science that stresses the fact that category membership is not homogenous and that some members are better representatives of a category than others (Qi et al. 2006; Rosch and Mervis 1975). Whereas the apple is a good representative of fruits, the penguin is not the best one for birds. However, every categorisation is relative, and boundaries are sometimes fuzzy between production models. A given farming situation can thus also be characterised in terms of distance between several categories of agricultural regimes. For example, OF&F and agroecology share some commonalities in their principles and practices (Bellon et al. 2011a). Both OF&F and low-input farming have been addressed jointly in research projects (e.g., Quality Low Input Food project: http://www.qlif.org/). Analogies can be made with the way the concept of prototype is used in the industrial sector, e.g., in car design. As an example, safety belts and other equipment such as ABS were tested and used in racing cars—where constraints are high—before being commonly used in all cars. There is, however, a difference between designing a solar-powered car and working on reducing fuel consumption on a car engine to minimise the use of non-renewable resources. The same comparison can be applied to agriculture, with its biotechnical and ecological processes. This is a first interpretation of *prototype*.

The second interpretation of OF&F as a *prototype* refers to its ability to achieve a set of performances. For example, increasing productivity or closing yield yaps can be done at the expense of closing nutrient cycles and achieving a health-enhancing food system. According to various authors (Azadi et al. 2011; Dima and Odero

1997; Lampkin et al. 2006; Leifeld 2012; Lockie et al. 2006), OF&F appears as a promising and innovative means of tackling the challenges facing agriculture and food production with respect to sustainability (climate change, food security and safety, biodiversity and enhancement of ecosystem services, endogenous rural development). Its outcomes in terms of environmental stewardship are also acknowledged. There is nevertheless a disagreement on the exact nature of the relationships between OF&F and sustainable agriculture (Bergström et al. 2008; Leifeld 2012; Rigby and Cáceres 2001). Hence, OF&F is either advocated or questioned, like two sides of the same coin. In the search for more sustainable agricultures, several issues must be addressed in parallel. For example, combining food security and environmental preservation can be tackled at the expense of social justice for farm workers. Alternative agricultural models will have to combine a wider range of performances. Given the uncertainties of future developments, previous concepts that guided research—such as stability, income maximisation, technical fine-tuning and biological optimisation—need to be balanced with system properties—such as adaptability, resilience and flexibility. Maintaining or strengthening such properties will also orient and determine both system performances and the criteria to evaluate them. The issue of assessing and combining OF&F multiple performances is still pending. Such debates are an important premise of this book. Although there is probably no definite answer, readers will find elements based on research findings from scholars involved in OF&F research, acquainted with the history and reality of the organic movements.

The third perspective of prototyping refers to OF&F development pathways. OF&F internal dynamics and diversity can be approached as enabling transitions or transformations. This is closely linked to the idea of sustainable agriculture considered as a programme, not as a steady state. Organic agriculture is also involved in a progress loop (Rahmann et al. 2009). However, there are no steps or linear timelines in transitions. Subsequently, the Efficiency-Substitution-Redesign (ESR) model presented by key authors in agroecology (Gliessman 2010; Hill 1985; Rosset and Altieri 1997) as levels of conversion to sustainable agriculture in a stepwise process reflects neither the complexity of farmers' trajectories nor the dynamics of the organic sector. Interestingly, Gliessman (2007) refers to a fourth level of agroecological transitions that enables "a more direct connection between those who grow the food and those who consume it, with a goal of re-establishing a culture of sustainability that takes into account the interactions between all the components of the food system". This specific link with the food system is an integral part of OF&F premises and is treated in the third section of this book.

These three facets of prototype are in fact interrelated, in as much as production processes are also valuated within the scope of expected new performances such as OF&F ecosystem services (Sandhu et al. 2010), whether separately or in combination, as a potential development pathway (Fleury 2011). Beyond the classical version of sustainability as a "3-legged stool" based on people, planet and profit, we also suggest that these three other "P"s are relevant in OF&F achievements and challenges: processes, performances and pathways. They are addressed as a series of integrated chapters in the three sections of this book.

1.3 *OF&F as a Prototype* That Interacts With Other Agricultures

OF&F lies at the heart of alternative agricultures and is an extreme case in terms of constraints in a continuum that begins with "conventional" agriculture and that also includes integrated production, which can be a step towards OF&F. Even if constraints are relative and not always synonymous with innovation, they contribute to identifying a target for change. Goulet and Vinck (2012) analyse no-till agriculture—a shift to farming techniques that have eliminated plowing—as an innovation through subtraction, i.e., innovation founded on reducing a practice or ceasing to use (subtracting, detaching) a given artefact. As for OF&F, farmers shifting to conservation agriculture practices also speak about conversion, referring to the implementation of another type of soil functioning (Sangor and Abrol 2004). Likewise, Gliessman (2007) also uses conversion to OF&F to exemplify agroecological transitions. Conversion to organics can thus be approached as a more general figure of transition in agriculture. Some commonalities actually appear among ecologically-based agricultures for which natural processes and regulations assume an important role once again in system functioning and higher self-sufficiency.

Organic farming may have different forms of interactions, in space and in time, with other types of agriculture. Such interactions can be as follows: coexistence, integration of practices from one system to another one, mixity within the same system, and evolution from one production system to another one:

- With *coexistence*, interactions are usually limited (e.g., planting edges to limit pesticide drift from neighbouring fields), and sometimes negative (e.g., in the case of contamination with GMOs). However, organic farmers not only belong to dedicated organisations, but are also part of professional networks or unions that are made up of conventional farmers (Ruault 2000). Environmental issues also lead farmers to act more collectively. Environmental performances increasingly depend both on the diversity of organics and on the other neighbouring systems. This is the case for water quality in watersheds (Thieu et al. 2011) and biodiversity conservation in a landscape matrix (Gabriel et al. 2010). Beyond the local effects of organic practices, organic fields are embedded in a landscape composed of different production systems that condition organic performances. The proportion and dispersion of organic farms will thus influence the organic capacity to preserve water quality and biodiversity.
- *Integration* refers to practices used in organic farming as well as in 'conventional' farms (Gosling and Shepherd 2005). These connections make it possible to spread innovations (such as mechanical weeding, composting, etc.) in the wider field of agriculture. Likewise, some technologies adapted to integrated production or protection can also be applied to OF&F (such as mate disruption or alternative animal therapies presented in the first section of this book) (Watson et al. 2008).
- *Mixity* occurs at various levels (farm, food chain, etc.), as well as in institutions (extension, training, research). It is controversial from the point of view of

production since the definition and principles of organic agriculture encourage farmers to convert the whole farm. Lamine et al. (Chap. 23) studied these relationships at the farm level and in the context of territorial agri-food systems.
- *Evolution* is related to farmers' trajectories and linked to the previous, since mixity can be temporary. On the production side, farmers who shifted their farms to organics provide concrete examples of changes in practices and systems. They are involved in another form of agriculture through transition processes that encompass longer time spans than the formal 2–3 year "conversion" period.

In OF&F, input substitution is somehow a prerequisite. Beyond the abandonment of chemical fertilisers and pesticides, alternative agricultural methods can be implemented following principles that can serve as a framework to guide practices. However, innovations at the farm level can be either incremental or radical. The umbrella organisation, IFOAM, gives an overall definition of organic agriculture (IFOAM 2005) and suggests four guiding principles (Health, Ecology, Fairness and Care), which represent the basic aims that OF&F systems must strive to achieve (Luttikholt 2007), especially in a system redesign perspective. A fifth principle, the one of Reality, could be added to address how the implementation of the four previous ones can be orchestrated as time goes by. However, increasing the efficiency and effectiveness of organic practices is often considered as a priority for many stakeholders. This is also the case within European Commission (EC) regulations and expert groups that focus on inputs, whereas the organic EU regulation (EC 834/2007) also stipulates that "organic practices and inputs should be based on risk assessment, and the use of precautionary and preventative measures, when appropriate". The first sections of the EU regulations (EC 834/2007) indicate the general directions in which the future development of OF&F can be oriented, but such standards can also be interpreted as a list of eligible inputs. To some extent, a genuine conversion to OF&F, i.e., one committed to an agroecosystem redesign strategy, can be considered as a system innovation (Elzen and Wieczorek 2005; Padel 2001). The difficulty to implement redesign approaches at the practical or experimental level has already been introduced (Bellon et al. 2010), especially in the areas of crop protection and animal healthcare. Various papers in this book will tackle this issue, from a system perspective.

The first section of this book addresses production, protection and processes in organic farming with the aim of increasing self-sufficiency. It is based on systemic approaches at various levels and enhances both agronomic and ecological knowledge. It draws on examples in crop, livestock and mixed crop-livestock situations. It combines two viewpoints: (i) addressing an issue (definition of plant ideotypes, animal healthcare) relevant for OF&F, while linking it with other agricultural regimes, candidates for sustainable agricultures (agroecology, low-input, etc.), (ii) focusing on OF&F and addressing the validity domain of findings for other agricultures. This section includes ten chapters.

Two chapters deal with "landcare" and management of soil fertility, considered as a joint effort in environmental stewardship. Chapter 2 deals with phosphorus, which is likely to become a bottleneck for productivity, and suggests ways to close nutrient cycles. Chapter 3 shows the interest of intercropping as a way to improve Nitrogen Use Efficiency in organic systems when combining species in space.

Four chapters are then dedicated to crop protection, i.e., one of the limiting factors to the development of ecologically-based agricultures, especially in organic fruit production, which allows for a long-term design approach and for a combination of methods. Chapter 4 introduces the regulations and approaches relevant to crop protection, including those specific to OF&F and those that apply to agriculture as a whole. It exemplifies new research avenues for orchard management and self-regulation. Chapter 5 completes the previous one while focusing on conservation biocontrol at various scales in space and time. It also takes a look at how organic management practices can affect pests and their natural enemies, both in annual and perennial crops. Following the first facet of *prototype*, Chap. 6 draws on agroecological principles and experiences to define the basis for alternative crop protection strategies in OF&F. It provides a general framework to manage weeds, pests and diseases. Interactions among cropping methods are also an important asset in organic fruit production and protection. A last chapter 7 considers the role of cultivars, with the identification of ideotypes adapted to integrated or organic cropping systems, with their site or regional specificities.

The four next chapters are dedicated to livestock and health management. Chapter 8 proposes a set of four control principles for parasitic infections in various livestock production systems. Both the combination and implementation of such principles provide a range of technical options that can alleviate animal parasitism, particularly important when access to the outdoors is advocated. The next chapter 9 is based on an interdisciplinary approach of healthcare strategies implemented by livestock farmers, enabling a holistic and long-term vision of a set of diseases (pathocoenosis), instead of tackling each disease separately. Chapter 10 is also based on a network of commercial farms and addresses sheep farmers' livestock management strategies. Various land use and breeding patterns are feasible, but self-sufficiency objectives or technical and economic outcomes restrict the range of sustainable options. The last Chap. 11 of this section is based on a long-term experiment of two organic mixed crop dairy systems. Transitions towards self-sufficient and adaptive systems are supported with a stepwise redesign approach.

1.4 *OF&F* Performances: A Prototype Between Societal Expectations and Scientific Evidence

The second interpretation of *prototype* focuses on OF&F performances and their consequences in terms of research methods or agendas. OF&F development in research and practice shows the significant role of organic agriculture in relation to the future of sustainable agriculture and food production. This is reflected in topics addressed within the scientific arena and by the internal dynamics in OF&F at higher levels of organisation than individual farms.

In terms of scientific production, the knowledge base in organic food and farming is increasing, following exponential growth. The topics addressed are also becoming more diversified, whereas much attention was given to soil fertility and performances in the 1980s (Drinkwater 2009; Ollivier et al. 2011). As the number of publications has grown, meta-analyses and reviews have appeared on several issues: productivity (Badgley and Perfecto 2007; de Ponti et al. 2012; Seufert et al. 2012), profitability (Nemes 2009), conversion (Lamine and Bellon 2009), environmental impacts (Bengtsson et al. 2005; Blanchart et al. 2005; Hole et al. 2005; Mondelaers et al. 2009; Tuomisto et al. 2012), and the nutritional quality of organic food (Dangour et al. 2010; Lairon 2009). Consensus can be found in the literature as to the fact that organic farming delivers some substantial benefits over other production systems, in particular, in terms of resource conservation and multicriteria assessment (Bengtsson et al. 2005; Cavigelli et al. 2008; Reganold et al. 2010; Tuomisto et al. 2012). Some other expectations are still unfulfilled and have been faced with more intensive questioning, e.g., on actual yields and economic profitability, or on the quality of its products (Bergström et al. 2008; Kristiansen et al. 2006; Reganold and Dobermann 2012; Seufert et al. 2012; Trewavas 2001). However, the literature is dominated by comparisons of organic and other forms of agriculture, generally "conventional", although in many cases, such comparisons are questionable. Three drawbacks can be mentioned: (i) the difficulty of a rigorous pairwise matching when comparing organic and conventional agricultures (Mayen et al. 2010; Nemes 2009; Offermann and Lampkin 2006); (ii) in meta-analyses or reviews referring to the same topic (e.g., productivity, biodiversity or food quality in OF&F) and published the same year, the overlapping among sets of references is often below 50%. Conclusions may differ as a result of varying selection criteria for articles and the limited number of co-citations; (iii) comparisons within organics can also be biased when spatial and temporal scales are considered. The implementation of organic standards differs among countries, and specific labels exist (e.g., in biodynamic agriculture). Scale effects should also be considered when addressing the impact of organic agriculture on biodiversity (Gabriel et al. 2010). As for temporal scales, both evolutions in the organic standards and in the maturity of organic systems (e.g., number of years after conversion or position in the ESR model) are usually not dealt with. These elements are in favour of robust and extended databases, as well as long-term experiments (e.g., system trials) or monitoring (such as in Long-Term Ecological Research). A continuous process of monitoring and re-evaluation (Rigby and Cáceres 2001) is particularly at stake when dealing with sustainability, since the identification of technologies seen as sustainable today is based on different assumptions regarding the sustainable management of natural resources (Hubert 2002), maintaining their productive capacity over time. So far, OF&F appears as a good compromise among multiple performances. However, it is closer to a decathlon than a sprint in the competitive universe of alternative agricultures. Recognising that a specific performance is not at its maximum also provides room for improvements. More generally, sustainability and other system properties such as self-sufficiency can open or orient R&D agendas.

Prototyping is also a stepwise approach for designing alternative management systems (Blazy et al. 2009; Sterk et al. 2007; Vereijken 1997), usually in cooperation with commercial farmers and sometimes on experimental stations or farms. Prototyping is mainly used in transformative agendas where new production systems have to fulfil multi-objectives and address sustainability issues. This is relevant and has been applied to OF&F (Kabourakis 1996). However, it does not fully account for transition processes and redesign perspectives.

The multidimensionality of transitions to organics has been shown (Lamine and Bellon 2009). Transition trajectories entail new relationships with nature, with techniques, with consumers and with other stakeholders (peers, neighbours, advisers, certifying bodies, etc.). Research methods must be adapted accordingly, valuing farmers' knowledge in partnerships but also implementing interdisciplinary and systemic approaches. Organic farmers were pioneers in experiencing and designing farming methods or systems when research in OF&F was still marginal. Exchanges among farmers and farm or field visits ("campesino a campesino") remain an important component in the conversion process. Many scientists report the role of farmers' knowledge and experiences for the development of organic farming as independent of traditional scientific institutions (Aeberhard and Rist 2009; Baars 2011; Sayre 2011). In OF&F, there is no optimal, universal and immutable system but, instead, multiple methods that must be combined and activated depending on a set of varying situations and contexts. Sharing standards does not entail standardisation of practices, and OF&F appears as an asset for breakthroughs in designing agricultural systems. This is probably why it has been used in various agricultural expertises and scenarii (e.g., Butault et al. 2010; Sørensen et al. 2005). Both within and outside OF&F, many farmers have successfully dealt with complex problems and eventually developed systemic solutions to issues that have not yet been addressed by researchers, advisors or policy makers (Lichtfouse et al. 2009). Morgan and Murdoch (2000) even argue that farmers become "knowing agents" again when shifting to organic agriculture. Innovative agricultural systems are knowledge-intensive and can benefit from a closer collaboration between farmers and scientists. This approach is consistent with the Agricultural Knowledge Systems (AKS) paradigm, where organisational and institutional changes have a large part to play, beyond the classical innovation pattern that gives priority to technological change (Elzen and Barbier 2012).

As for scientists, mixity also exists when they do not work exclusively within OF&F. This questions whether organic research defines specific objects or is included in a wider research area (Sylvander and Bellon 2003). Various national assessments (Denmark, France, the US, etc.) and papers (Watson et al. 2008) support the first position, based on the scientific outputs (mostly publications) resulting from the integration of organic research into universities and mainstream institutes. It can also be argued that both specific site and system interactions occur in organic agriculture where the range of variation of environmental and production factors is high (Hokazono and Hayashi 2012; Lyon et al. 2011). A combination of analytical and systemic approaches is relevant to tackle the variety of questions in OF&F. Whereas analytic research can contribute to our understanding of biological processes and solve specific problems, systemic approaches can address the complexity of real situations, a prerequisite for performance assessment, and enable the formulation of hypotheses to be tested analytically. Transdisciplinary research and partnerships also have an important role to play in understanding the complexities of the ecological approach to agriculture typified by organic farming (Bellon et al. 2011b; Cabaret et al. 2003; Sylvander and Bellon 2003).

The second section of the book tackles major challenges in terms of organic performances. Evaluation is an important dimension for farmers, advisers and scientists. Apart from its normative perspective, it can be seen as a valuation opportunity or as a deliberation on values, including those that are not as yet recognised on the market. This could also serve as a basis for outcome-based OF&F, whereas it is currently mean-based. Three main challenges are considered: environmental issues, animal welfare and the quality of organic products. As for environment, Chap. 12 uses Life Cycle Analysis, a normalised method increasingly used in agriculture. Its application to organic fruit production shows a lower impact than comparable conventional systems when the results are expressed in ha/year. It also points out key steps of the production process and suggests some improvements in terms of methods and practices. At a global level, Chap. 13 consists in a review of the contribution of organic agriculture to mitigate climate change, enhancing OF&F systemic and multifunctional attributes. Whereas the application of high animal welfare standards is part of the EU regulations (EC 834/2007), its implementation is still controversial. Chapter 14 examines the potential of combining a capabilities approach with other tools to meet a wider range of animal requirements. OF&F is also associated with a paradigm shift both in agriculture and science, thus calling for renewed scientific approaches and methods. Chapter 15 suggests such a shift, from a focus on animal production to another paradigm that leaves more room for animals and the workers who care for them. It is based on comprehensive fieldwork in various countries. Finally, three papers deal with the quality of organic products. Chapter 16 reviews the main characteristics of organic foods in terms of their nutritional, safety and health aspects. It also provides methodological perspectives for further investigations. Chapter 17 completes the previous one and contributes to a knowledge gap in many reviews regarding the quality of animal products. It focuses on the quality of organic lamb. The last chap. 18 of the second section of the book refers to OF&F as a quality sign. It examines how the conventionalisation thesis can also affect organic standards and labelling. The co-existence of regimes of action, with different recommendation domains (input-based vs. process-based or product-oriented) is considered as an asset for OF&F development.

1.5 *OF&F* as a Prototype with its Own Dynamics and Diversity: Transition Pathways

A third interpretation of prototype refers to OF&F dynamics and internal diversity. It is also related to a design perspective in as much as (re)design is an endless and iterative process based on knowledge and practical acquisitions. In spite of its meaningful history (some 100 years), OF&F is still immature compared to the beginning of agriculture (10,000 years BCE). Various trends can be identified in recent OF&F dynamics, including internationalisation, globalisation, institutionalisation, specialisation and conventionalisation.

A steady demand exists for organic products, together with a growing recognition of the impacts of OF&F on the environment and rural development. Moreover, OF&F is still growing in many countries, both in terms of area, number of farmers and range of products and services provided (Willer and Kilcher 2012). Such an *internationalisation* of OF&F can be viewed as an opportunity for many farmers, livelihoods and operators willing to change their practices or rationale. It also potentially generates competition in the market arena and makes it necessary to reflect on the integration of newcomers in the organic agriculture sector. Adequation between supply and demand in organic products indeed varies among countries, but signs of decline in demand are appearing in the leading European countries. An imbalance between both terms entails risks of fluctuations in farmer income and product price. Efficient marketing and logistics also depend on critical mass and organisations. The structuring of commodity or value chains is at stake, including imports and exports, and some national plans or agencies include support for market organisation. Policies that encourage organic public procurement of food also contribute to fostering demand, whereas support payments vary among EU countries (Schwarz et al. 2010).

The *globalisation* of the organic sector has already been presented in two books (Halberg et al. 2006; Kristiansen et al. 2006), showing how new stakeholders can contribute to its growth and development without compromising OF&F anchorage in an "ecology of contexts" that provides a useful device for understanding agroecosystems (Bland and Bell 2007). However, a persistent debate in the literature on agroecological farming and on the impact of agricultural research in general has been the question of scaling out (broad adoption over wide areas and by many farmers) and scaling up (institutionalising supportive policies for alternatives) successful experiences (Pachicho and Fujisaka 2004), cited by (Rosset and Martínez-Torres 2012). Scaling out relates to a classical lifecycle that considers innovation as a diffusion process depicted with an S-curve (Rogers 1995). Padel (2001) criticised its relevance for conversion, based on a large number of studies of organic farmers in several countries over a 20-year time span. Among others, she observed that few studies have attempted to carry out a rigorous comparison of earlier and later adopters in terms of farm, market and personal characteristics. In fact, this S-curve depicts the growth of a technology more than its development or adaptations over time. Likewise, growth targets for the organic sector are usually expressed in quantitative terms: market share, number of organic farmers, % of agricultural area (e.g. MacRae et al. 2009). Scaling up relates to the development of OF&F. At the European level, environment is a driving factor that justifies support for organic farming and is therefore still a key component of OF&F development (Fleury 2011; Guyomard 2009). Development also includes targets supported by national or regional action plans, combining a broader mix of measures than the recognition of environmental and other benefits of organic farming (Lampkin and Stolze 2006). Such measures include research, training, advice, consumer promotion and market organisation. They contribute to extending an earlier trend, known as the *institutionalisation* of organic agriculture (Piriou 2002), which leads to a closer relationship between organic farmers and the agricultural profession, sometimes at the expense of alliances with other parties (consumers, civil society, etc.). Two other major trends have been identified in OF&F dynamics.

Some authors describe the increasing *specialisation* of organic farms (Allard et al. 2000), at odds with the classical mixed crop-livestock organic model. Its consequences would be an accentuation of technical problems such as weed control and fertilisation management and the parallel specialisation of research, development and extension, leading to a focus on a commodity approach to organics, as in the case of livestock production (Hovi and Garcia Trujillo 2000; Roderick et al. 2004). Other scientists instead put the accent on the process of *conventionalisation*, whereby OF&F evolves just like within a conventional agriculture framework. It can be briefly characterised by the concentration of capital among fewer and larger growers and intermediaries who are better equipped to deal with retailers, the erosion of organic standards, the generalisation of substitution with prescribed or eligible inputs, and a greater dependence of farmers on input suppliers and supermarkets (Buck et al. 1997; Darnhofer et al. 2010; Hall and Mogyorody 2001). This evolution would eventually lead to a bifurcation into two distinct organic sub-sectors: a conventionalised one and a resistant one (Holt and Reed 2006).

An accepted definition and common principles provide an "organic" framework. However, differences appear in the implementation of principles and agricultural methods. There are also on-going debates on the level of requirements among countries and regions. This internal diversity has probably existed since the inception of OF&F. It entails composing with multiple and successive definitions and interpretations: from Sir Albert Howard in the 1940's to IFOAM's principles and missions that consist in leading, initiating and assisting the organic movements in their full diversity (IFOAM 2005). While acknowledging the diversity of organic farming and its subsequent development models (Sylvander et al. 2006), three issues are at stake. Firstly, the internal adaptive or evolution capability of organic farmers leads to continuous progress or innovation in existing organic systems. A second challenge consists in providing perspectives both to newcomers, with the combination of mixed or hybrid forms of production (partly organic) and chains (short, medium, long), and to the early converters. Finally, one remaining question is the balance to be maintained among farmers or farming system categories both in terms of development models and pathways for the organic sector.

The organic sector demonstrated its capability to address these trends and issues, usually at its own initiative:

- New conventions and economies (FNAB 2012) are alternative strategies to the conventionalisation thesis.
- Food security and climate change have been dealt with in dedicated conferences (FAO 2007[2]; Enita 2008[3]), combining scientific inputs and field experiences.
- A Technology Platform (TP "Organics") has prepared a Vision Research Agenda (Niggli et al. 2008) and a Strategic Research Agenda (Schmid et al. 2009).

[2] International Conference on Organic Agriculture and Food Security, FAO, Rome 3–5 May 2007, Proceedings obtainable under: ftp://ftp.fao.org/paia/organicag/ofs/02-Edwards.pdf

[3] Agriculture biologique et changement climatique: colloque International, ENITA Clermont-Ferrand, 17–18 avril 2008, Clermont-Ferrand, France.

More broadly, cultivating an ecological basis is also a challenge for organic agriculture if it is to remain an ecological system (Francis 2009) or a sustainable farming system that advocates "strong ecological modernisation" and agroecological principles (Horlings and Marsden 2011). Several of the chapters in the book question the above-mentioned trends and present research results that support or counteract them, including at regional levels.

This introduces the last section of the book, dedicated to OF&F development pathways. It includes two main sections. The first one links the need to extend performance criteria used in OF&F (beyond productivity as a major value) with a renewed organic framework that leaves room for other stakeholders. Indeed, societal expectations in terms of agricultural performances will determine public and policy measures for the promotion and extension of organics. The second one focuses on transition pathways in organic farming and food systems, including at territorial levels. In the first section, Chap. 19 reviews the variety of goods and services derived from OF&F, based on the Canadian context. It suggests both practical and theoretical applications, in the form of a regional pilot initiative and a "civil commons" framework. The following two chapters focus on the institutional framework of plant breeding methods in OF&F. Chapter 20 presents how the diversity of organic farming styles also entails different requirements in terms of varieties and, subsequently, of research needs. This also questions the breeding system and its evaluation procedures. Chapter 21 proposes an innovative breeding scheme that responds to organic principles and the specific needs of farmers, based on participatory approaches and farmers' networks. The last section of the book includes four papers that combine retrospective and prospective visions. The two first chapters provide information about the previously mentioned ESR model. Its promoter authored Chap. 22, which includes his position for redesigning agriculture. Beyond an opposition between deep to shallow organics, this chapter suggests that human beings and the sciences play a central role in transforming agricultural and food systems. Based on interdisciplinary work and case studies, Chap. 23 shows the interest of combining a diversity of agricultural systems and actors to foster transitions. Combinations occur both in space and time, including in the context of territorial and agri-food system dynamics. The next chap. 24 goes further. On the one hand, it questions the internal evolution capability of the organic sector. On the other, it shows its transformative potential with regard to the global agro-food system. In the case of the Camargue region (France), a protected area also known for its rice production, the last chap. 25 shows how the regional extension of organic production can be considered, integrating three models for up-scaling OF&F in scenario analyses.

1.6 Outlook

OF&F is more than a niche, considering its different meanings: (i) a niche market restricted to a segment for consumers willing to pay for qualities attached to certified products or environmentally-friendly production methods; (ii) an ecological niche favouring a focus on some "hot spots" such as water catchments where conversion would be encouraged; (iii) a technological niche (Elzen et al. 2012) that serves as a breeding ground where radical innovations that cannot initially fit into the main socio-technical regime (or agri-food system) can emerge. The combination of these three meanings is part of the idea of *prototype*, considered as an agri-food system of innovation (Allaire and Sylvander 1997), in the search for a new model of agricultural production and consumption. Compared with other alternatives that focus on environmental (low-input agriculture, precision farming), quality (PDO, PGI) or social (fair-trade) dimensions, OF&F embraces multiple dimensions that inextricably link food, farming and environment. OF&F thus contributes to reconnecting food production and consumption, as well as to rethinking the relationships between agriculture and nature, while enhancing the natural tendencies of regulations to reduce the use of pesticides. It also considers environment as a resource and takes relationships between science and society into account, while encouraging scientists to be sensitive to societal expectations. Shifting from a *prototype* to a generalisable model is still an issue. It would probably inspire and stimulate the entire agricultural and agri-food sector, master its own dynamics and co-evolve with other forms of agriculture.

References

Aeberhard A, Rist S (2009) Transdisciplinary knowledge production in organic agriculture. Agrarforschung 16:414–419

Allaire G, Sylvander B (1997) Qualité spécifique et innovation territoriale. Cahiers d'économie et sociologie rurales 24:29–59

Allard G, David C, Henning JC (2000) L'agriculture biologique face à son développement: les enjeux futurs. Les colloques, INRA, 396 pages Editions Quae

Azadi H, Schoonbeek S, Mahmoudi H, Derudder B, De Maeyer P, Witlox F (2011) Organic agriculture and sustainable food production system: Main potentials. Agric Ecosys & Environ 144:92–94. DOI: 10.1016/j.agee.2011.08.001

Baars T (2011) Experiential science; towards an integration of implicit and reflected practitioner-expert knowledge in the scientific development of organic farming. J Agric Environ Ethics 24:601–628. DOI: 10.1007/s10806-010-9281-3

Badgley C, Perfecto I (2007) Can organic agriculture feed the world? Renew Agric Food Sys 22:80–86

Bellon S, Desclaux D, Le Pichon V (2010) Innovation and research in organic farming: A multi-level approach to facilitate cooperation among stakeholders, 9th European IFSA Symposium, Vienna, Austria. pp. 703–717

Bellon S, Lamine C, Olivier G, de Abreu LS (2011a) The relationships between organic farming and agroecology, in: D. Neuhoff, et al. (Eds.), Third Scientific Conference of ISOFAR, ISOFAR, Namyangju, Republic of Korea. pp. 235–238

Bellon S, Penvern S, Ollivier G, Debaeke P, Cabaret J (2011b) Promouvoir des travaux sur l'AB dans un institut de recherche (INRA): enseignements d'un collectif interdisciplinaire, Les transversalités de l'agriculture biologique, Strasbourg, France

Bengtsson J, Ahnstrom J, Weibull AC (2005) The effects of organic agriculture on biodiversity and abundance: a meta-analysis. J Appl Ecol 42:261–269

Bergström L, Kirchmann H, Thorvaldsson G (2008) Widespread Opinions About Organic Agriculture-Are They Supported by Scientific Evidence? In: Kirchmann H, Bergström L (eds.) Organic Crop Production-Ambitions and Limitations. Springer, Netherlands, pp. 1–11

Besson Y (2009) Une histoire d'exigences: philosophie et agrobiologie. L'actualité de la pensée des fondateurs de l'agriculture biologique pour son développement contemporain. Innovations Agronomiques 4:329–362

Blanchart E, Cabidoche Y-M, Gautronneau Y, Moreau R (2005) Les aspects spatiaux et environnementaux de l'agriculture biologique, in: M. François, et al. (Eds.), Agriculture biologique en Martinique: quelles perspectives de développement?, IRD Editions, Paris (FRA), pp. 325–379

Bland WL, Bell MM (2007) A holon approach to agroecology. Intern J Agric Sustain 5:280–294. DOI: 10.1080/14735903.2007.9684828

Blazy J-M, Ozier-Lafontaine H, Doré T, Thomas A, Wery J (2009) A methodological framework that accounts for farm diversity in the prototyping of crop management systems. Application to banana-based systems in Guadeloupe. Agric Sys 101:30–41. DOI: http://dx.doi.org/10.1016/j.agsy.2009.02.004

Buck D, Getz C, Guthman J (1997) From farm to table: the organic vegetable commodity chain of northern California. Sociologia Ruralis 37:3–20

Butault JP, Dedryver CA, Gary C, Guichard L, Jacquet F, Meynard JM, Nicot P, Pitrat M, Reau R, Sauphanor B, Savini I, Volay T (2010) Ecophyto R & D. Quelles voies pour réduire l'usage des pesticides? Synthèse du rapport d'étude., INRA Editeur (France), 90 pages

Cabaret J, Bellon S, Gautronneau Y (2003) Quelle recherche pour l'agriculture biologique. Pour, 178:116–126

Cavigelli MA, Teasdale JR, Conklin AE (2008) Long-term agronomic performance of organic and conventional field crops in the mid-Atlantic region. Agron J 100:785–794

Dangour AD, Lock K, Hayter A, Aikenhead A, Allen E, Uauy R (2010) Nutrition-related health effects of organic foods: a systematic review. Am J Clin Nutr 92:203–210. DOI: 10.3945/ajcn.2010.29269

Darnhofer I, Lindenthal T, Bartel-Kratochvil R, Zollitsch W (2010) Conventionalisation of organic farming practices: from structural criteria towards an assessment based on organic principles. A review. Agron Sustain Dev 30:67–81. DOI: http://dx.doi.org/10.1051/agro/2009011

Dima SJ, Odero AN (1997) Organic farming for sustainable agricultural production: a brief theoretical review and preliminary empirical evidence. Environ Resour Econ 10:177–188

Drinkwater LE (2009) Ecological knowledge: foundation for sustainable organic agriculture, in: C. Francis (ed.), Organic farming: the ecological system. American Society of Agronomy, Inc., Crop Science Society of America, Inc., Soil Science Society of America, Inc.: Madison. pp. 19–47

Elzen B, Wieczorek A (2005) Transitions towards sustainability through system innovation. Technological forecasting and social change 72:651–661

Elzen B, Barbier M (2012) System Innovations, Knowledge Regimes, and Design Practices towards Transitions for Sustainable Agriculture Inra-Science for Action and Development

Elzen B, Barbier M, Cerf M, Grin J (2012) Stimulating transitions towards sustainable farming systems, Farming Systems Research into the 21st century: The new dynamic, Springer. pp. 431–455

Fleury P (2011) Agriculture biologique et environnement. Des Enjeux Convergents Educagri éditions/ACTA

FNAB (2012) La nouvelle économie bio-Témoignages et expériences. pp. 31

Francis C (2009) (ed.) Organic Farming: the ecological system. American Society of Agronomy, Inc., Crop Science Society of America, Inc., Soil Science Society of America, Inc.: Madison, 353 pages

Gabriel D, Sait SM, Hodgson JA, Schmutz U, Kunin WE, Benton TG (2010) Scale matters: the impact of organic farming on biodiversity at different spatial scales. Ecol Lett 13:858–869. DOI: 10.1111/j.1461-0248.2010.01481.x

Gliessman S (2010) The framework for conversion. The conversion to sustainable agriculture. CRC Press, Boca Raton, 3–14

Gliessman SR (2007) (Ed.) Agroecology: the ecology of sustainable food systems, CRC Press, Taylor & Francis, New York, USA, 384 pages

Gosling P, Shepherd M (2005) Long-term changes in soil fertility in organic arable farming systems in England, with particular reference to phosphorus and potassium. Agric. Ecosys & Environ 105:425–432

Goulet F, Vinck D (2012) L'innovation par retrait. Contribution à une sociologie du détachement. Revue française de sociologie 53:195–224

Guyomard H (2009) Politiques publiques et agriculture biologique. Innovations Agronomiques, 4:499–511

Halberg N, Muller A (2013) (ed.) Organic agriculture for sustainable livelihoods. Routledge, London and New York, 296 pages

Halberg N, Alroe HF, Knudsen MT, Kristensen ES (2006) Global development of organic agriculture: challenges and prospects CABI Publishing, Wallingford, 391 pages

Hall A, Mogyorody V (2001) Organic farmers in Ontario: An examination of the conventionalization argument. Sociologia Ruralis 41:399–422

Hill SB (1985) Redesigning the food system for sustainability. Alternatives 12:32–36

Hokazono S, Hayashi K (2012) Variability in environmental impacts during conversion from conventional to organic farming: a comparison among three rice production systems in Japan. J Clean Prod 28:101–112. DOI: 10.1016/j.jclepro.2011.12.005

Hole DG, Perkins AJ, Wilson JD, Alexander IH, Grice PV, Evans AD (2005) Does organic farming benefit biodiversity? Biol Conserv 122:113–130

Holt GC, Reed MJ (2006) Sociological perspectives of organic research: to policy and beyond, in: GCRM. Holt (ed.), Sociological perspectives of organic agriculture: from pioneer to policy, Cabi, Wallingford, pp. 284–304

Horlings LG, Marsden TK (2011) Towards the real green revolution? Exploring the conceptual dimensions of a new ecological modernisation of agriculture that could 'feed the world'. Global Environ Chan 21:441–452. DOI: http://dx.doi.org/10.1016/j.gloenvcha.2011.01.004

Hovi M, Garcia Trujillo R (2000) Diversity of livestock systems and definition of animal welfare. Proceedings of the second NAHWOA Workshop, Cordoba, Spain, 8–11 January 2000, University of Reading Library (RUL), 165 pages

Hubert B (2002) Agricultures and sustainable development: The stakes of knowledge and research attitudes. Think forward and act now. Dossiers de l'Environnement de l'INRA 22:41–56

IFOAM (2005) Principles of Organic Agriculture, IFOAM, Bonn. pp. 4, http://www.ifoam.org/en/organic-landmarks/principles-organic-agriculture

Kabourakis E (1996) Prototyping and dissemination of ecological olive production systems: a methodology for designing and a first step towards validation and dissemination of prototype Ecological Olive Productions Systems (EOPS) in Crete, Wageningen University, The Netherlands, 181 pages

Koohafkan P, Altieri MA, Gimenez EH (2011) Green agriculture: foundations for bio-diverse, resilient and productive agricultural systems. Intern J Agric Sus 10:61–75. DOI: 10.1080/14735903.2011.610206

Kristiansen P, Taji A, Reganold J (2006) Organic agriculture: opportunities and challenges, in: PTARJ Kristiansen (ed.), Organic agriculture: a global perspective, Cabi, Wallingford, pp. 421–441

Lairon D (2009) Nutritional quality and safety of organic food. A review. Agron Sust Develop. DOI: 10.1051/agro/2009019

Lamine C, Bellon S (2009) Conversion to organic farming: a multidimensional research object at the crossroads of agricultural and social sciences. A review. Agron Sust Develop 29:97–112. DOI: 10.1051/agro:2008007

Lampkin N, Stolze M (2006) European action plan for organic food and farming. Law, Science and. Policy 3:59–73

Lampkin N, Fowler SM, Jackson A, Jeffreys I, Lobley M, Measures M, Padel S, Reed M, Roderick S, Woodward L (2006) Sustainability assessment for organic farming-integrating financial, environmental, social and animal welfare benchmarking. Asp Appl Biol, 79:9–13

Leifeld J (2012) How sustainable is organic farming? Agric Ecosys Environ 150:121–122. DOI: 10.1016/j.agee.2012.01.020

Lichtfouse E, Navarrete M, Debaeke P, Souchere V, Alberola C, Menassieu J (2009) Agronomy for sustainable agriculture. A review. Agron Sust Develop 29:1–6. DOI: 10.1051/agro:2008054
Lockeretz W (2007) Organic farming: an international history. Cabi Publishing, Willingford
Lockie S, Lyons K, Lawrence G, Halpin D (2006) Going organic: mobilizing networks for environmentally responsible food production. Cabi Publishing, Willingford, 208 pages
Luttikholt LWM (2007) Principles of organic agriculture as formulated by the International Federation of Organic Agriculture Movements. NJAS-Wageningen J Life Sci 54:347–357
Lyon A, Bell MM, Gratton C, Jackson R (2011) Farming without a recipe: Wisconsin graziers and new directions for agricultural science. J Rural Stud 27:384–393. DOI: http://dx.doi.org/10.1016/j.jrurstud.2011.04.002
MacRae R, Martin RC, Juhasz M, Langer J (2009) Ten percent organic within 15 years: Policy and program initiatives to advance organic food and farming in Ontario, Canada. Renew Agr Food Syst 24:120–136. DOI: 10.1017/s1742170509002531
Mayen CD, Balagtas JV, Alexander CE (2010) Technology Adoption and Technical Efficiency: Organic and Conventional Dairy Farms in the United States. Amer J Agr Econ 92:181–195. DOI: 10.1093/ajae/aap018
McIntyre BD, Herren HR, Wakhungu J, Watson RT (2009) International Assessment of Agricultural Knowledge, Science and Technology for Development (IAASTD): Global Report. Island Press, Washington, DC, 106 pages
Mondelaers K, Aertsens J, Van Huylenbroeck G (2009) A meta-analysis of the differences in environmental impacts between organic and conventional farming. Br Food J 111:1098–1119. DOI: 10.1108/00070700910992925
Morgan K, Murdoch J (2000) Organic vs. conventional agriculture: Knowledge, power and innovation in the food chain. Geoforum 31:159–173
National Research Council (2010) Toward sustainable agricultural systems in the 21st century. The National Academies, Washington, DC
Nemes N (2009) Comparative Analysis of Organic and Non-Organic Farming Systems: A Critical Assessment of Farm Profitability. In: Natural Resources Management and Environment Department (Ed.) Department (Ed.), FAO, Rome. 33 pages
Niggli U, Slabe A, Schmid O, Halberg N, Schluter M (2008) Vision for an Organic Food and Farming Research Agenda to 2025. Published by IFOAM-EU and FiBL, 48 pages
Offermann F, Lampkin N (2006) Organic farming in FADNs-comparison issues and analysis. In: Rippin, M., Willer, H., Lampkin, N., Vaughan, A. (Ed), Towards a European framework for organic market information. Proceedings of the second EISFOM European Seminar, Brussels, 10–11 November, 2005, Research Institute of Organic Agriculture (FiBL), Frick Switzerland. pp. 106–111
Ollivier G, Bellon S (2013) Dynamiques paradigmatiques des agricultures écologisées dans les communautés scientifiques internationales. Nat Sci Soc, 21(2):166–181. DOI: 10.1051/nss/2013093
Ollivier G, Bellon S, Penvern S (2011) Dynamic of thematic and citation patterns in Organic Food & Farming research. 3rd Scientific Conference of ISOFAR, september 28th-october 1st 2011, Namyangju, Republic of Korea, 4 pages
Pachicho D, Fujisaka S (2004) (Ed.) Scaling up and out: achieving widespread impact through agricultural research, Centro Internacional de Agricultura Tropical, CIAT Publications, Cali, Colombia, 293 pages
Padel S (2001) Conversion to organic farming: a typical example of the diffusion of an innovation? Sociologia Ruralis 41:40–61
Piriou S (2002) L'institutionnalisation de l'agriculture biologique (1980-2000), École Nationale Supérieure Agronomique de Rennes, Ph. D. Thesis, 451 pages
de Ponti T, Rijk B, van Ittersum MK (2012) The crop yield gap between organic and conventional agriculture. Agric Syst 108:1–9. DOI: 10.1016/j.agsy.2011.12.004
Qi F, Zhu AX, Harrower M, Burt JE (2006) Fuzzy soil mapping based on prototype category theory. Geoderma 136:774–787. DOI: 10.1016/j.geoderma.2006.06.001
Rahmann G, Oppermann R, Paulsen HM, Weissmann F (2009) Good, but not good enough? Research and development needs in Organic Farming. Landbauforschung Volkenrode 59:29–40
Reganold JP, Dobermann A (2012) Agriculture: Comparing apples with oranges. Nature 485:176–177

Reganold JP, Andrews PK, Reeve JR, Carpenter-Boggs L, Schadt CW, Alldredge JR, Ross CF, Davies NM, Zhou J (2010) Fruit and soil quality of organic and conventional strawberry agro-ecosystems. PLoS ONE 5:e12346

Rigby D, Cáceres D (2001) Organic farming and the sustainability of agricultural systems. Agric Syst 68:21–40. DOI: 10.1016/S0308-521X(00)00060-3

Riquois A (1999) L'agriculture biologique: un prototype au service de l'agriculture conventionnelle pour un développement durable. Aménagement et Nature, 132:49–61

Roderick S, Henriksen B, Trujillo RG, Bestman M, Walkenhorst M (2004) The diversity of organic livestock systems in Europe. In: Vaarst MRSLVLW (Ed.) Animal health and welfare in organic agriculture. CABI Publishing, Wallingford, pp 29–56

Rogers ME (1995) Diffusion of innovation (4th edition). Free Press, New York

Rosch E, Mervis CB (1975) Family resemblances: Studies in the internal structure of categories. Cognitive Psychol 7:573–605. DOI: http://dx.doi.org/10.1016/0010-0285(75)90024-9

Rosset PM, Altieri MA (1997) Agroecology versus input substitution: a fundamental contradiction of sustainable agriculture. Soc Natu Resour 10:283–295

Rosset PM, Martínez-Torres ME (2012) Rural Social Movements and Agroecology: Context, Theory, and Process. Ecology and Society 17. DOI: 10.5751/es-05000-170317

Ruault C (2000) Évolution des réseaux professionnels des agriculteurs et formes de conseil en agriculture biologique: quels enjeux pour le développement? Le cas de la Bretagne. In: G. Allard, C. David, J. Henning (Eds.), L'agriculture biologique face à son développement. Les enjeux futurs, Editions INRA, France. pp. 289–311

Sandhu HS, Wratten SD, Cullen R (2010) Organic agriculture and ecosystem services. Environ Sci Pol 13:1–7

Sangor S, Abrol IP (2004) Conservation agriculture for transition to sustainable agriculture. Current Science 88:686–687

Sayre L (2011) Des livres anciens aux vieux agriculteurs, in: P. Béguin, B. Dedieu, E. Sabourin, (Eds.), Travail en agriculture, Edition l'Harmattan, Paris

Schmid O, Padel S, Halberg N, Huber M, Darnhofer I, Micheloni C, Koopmans C, Bügel S, Stopes C, Willer H (2009) Strategic research agenda for organic food and farming, Technology Platform Organics. IFOAM EU Group, Brussels. pp. 116

Schwarz G, Nieberg H, Sanders J (2010) (Ed.) Organic Farming Support Payments in the EU, German National Library, Germany

Seufert V, Ramankutty N, Foley JA (2012) Comparing the yields of organic and conventional agriculture. Nature 485:229–232

Sørensen CG, Madsen NA, Jacobsen BH (2005) Organic farming scenarios: Operational analysis and costs of implementing innovative technologies. Biosys Engineer 91:127–137

Sterk B, van Ittersum MK, Leeuwis C, Wijnands FG (2007) Prototyping and farm system modelling-Partners on the road towards more sustainable farm systems? Eur J Agron 26:401–409. DOI: http://dx.doi.org/10.1016/j.eja.2006.12.006

Sylvander B, Bellon S (2003) INRA and organic farming: towards a research programme, in: Oecd (Ed.), Organic agriculture: sustainability, markets and policies. OECD workshop on organic agriculture, Washington, D.C., USA, 23-26 September 2002, CABI Publishing, Wallingford, pp. 383–392

Sylvander B, Bellon S, Benoit M (2006) Facing the organic reality: the diversity of development models and their consequences on research policies, in: CB Anreasen et al. (eds.), European Joint Organic Congress: Organic Farming and European Rural Development, Odense (DK), pp. 58–61

Thieu V, Billen G, Garnier J, Benoit M (2011) Nitrogen cycling in a hypothetical scenario of generalised organic agriculture in the Seine, Somme and Scheldt watersheds. Reg Environ Change 11:359–370. DOI: 10.1007/s10113-010-0142-4

Trewavas A (2001) Urban myths of organic farming. Nature 410:409–410

Tuomisto HL, Hodge ID, Riordan P, Macdonald DW (2012) Does organic farming reduce environmental impacts?-A meta-analysis of European research. J Env Man 112:309–320. DOI: 10.1016/j.jenvman.2012.08.018

Vereijken P (1997) A methodical way of prototyping integrated and ecological arable farming systems (I/EAFS) in interaction with pilot farms, in: M.K. van Ittersum and S.C. van de Geijn (Eds.), Developments in Crop Science, Elsevier, pp. 293–308

Watson CA, Walker RL, Stockdale EA (2008) Research in organic production systems-past, present and future. J Agric Sci 146:1–19. DOI: 10.1017/s0021859607007460

Willer H, Kilcher L (2012) (Ed.) The World of Organic Agriculture. Statistics and Emerging Trends 2012

Part I
Systems' Functioning

Chapter 2
Soil Phosphorus Management in Organic Cropping Systems: From Current Practices to Avenues for a More Efficient Use of P Resources

Thomas Nesme, Bruno Colomb, Philippe Hinsinger and Christine A. Watson

Abstract Phosphorus (P) is a major nutrient for all living organisms and a key production factor in agriculture. In crop production, it is usually supplied to soils through fertilisers or recycled manure and compost. Organic production guidelines ban the use of highly soluble, manufactured P fertilisers and, thus, recommend recycling P from livestock manure and compost. In this chapter, after an overview of P dynamics in soils, we explore the consequences of such guidelines in terms of field- and farm-gate P budget, soil P availability and crop productivity. Moreover, we propose some avenues for the more effective use of P resources, ranging from rhizosphere-based processes (e.g., soil microorganism manipulation), genotype selection and cropping practices (e.g., intercropping), to farming system design (e.g., a combination of crops and animals at the farm scale). Finally, the potential benefits of these options are compared with respect to soil P status, field- and farm-P budgets.

Keywords Farm inflow and outflow · Farm-gate budget · Field budget · Genotype · Mixed farm · Phosphorus · Rhizosphere · Roots · Stockless farm

T. Nesme (✉)
BORDEAUX SCIENCES AGRO-INRA, UMR1220 TCEM Transfert Sol-Plante et Cycle des Éléments Minéraux dans les Ecosystèmes Cultivés, CS 40201, 33175 Gradignan, France
e-mail: thomas.nesme@agro-bordeaux.fr

B. Colomb
INRA-INP Toulouse, UMR1248 AGIR, 31326 Castanet-Tolosan, France

P. Hinsinger
INRA, UMR1222 Eco&Sols, 34060 Montpellier, France

C. A. Watson
Scottish Agricultural College, Craibstone Estate, Aberdeen, AB21 9YA UK

S. Bellon, S. Penvern (eds.), *Organic Farming, Prototype for Sustainable Agricultures*,
DOI 10.1007/978-94-007-7927-3_2,

2.1 Introduction

Phosphorus (P) is a major nutrient for all living organisms and is a key production factor in agriculture. Its scarcity in soils results in P being a limiting factor for crop production in many soils (Cordell et al. 2009). Crop production results in substantial off-take of P, making it necessary to replace P outputs by P inputs over the long term in order to avoid depleting the soil P reserve (except in soils with high P reserves; see Section 4).

IFOAM principles state that organic agricultural production is to be based on ecological processes and recycling. Inputs to organic farms should be reduced by reuse, recycling and efficient management of materials in order to maintain and improve environmental quality and conserve resources. Therefore, organic agriculture should strive to attain ecological balance through the design of farming systems, the establishment of habitats and the maintenance of genetic and agricultural diversity like in natural ecosystems (www.ifoam.org). This is particularly true for nutrients such as P since the available P reserve in many soils is not large (even if notable exceptions to this exist), and P is not renewable in the same way as nitrogen since there is no notable atmospheric P reservoir. Moreover, manufactured, chemical P-fertilisers are banned in organic production guidelines. Only some types of P-containing products can thus be used. For example, European organic production regulations allow only two types of products: rock phosphates and P-containing organic materials (Council Regulation (EC) No. 834/2007).

Virtually all P-containing organic materials are derived directly or indirectly from rock phosphates. They are generally extracted from sedimentary deposits that contain apatite-like calcium phosphate minerals and are mainly located in North Africa, China and the USA (Cordell et al. 2009; Jasinski 2011). However, rock phosphate reserves are facing over-exploitation, dissipation and poor recycling. Their depletion is projected over the next 50–100 years, depending on food and feed demand (Van Vuuren et al. 2010), but this is still subject to much debate. Therefore, in the coming years, rock phosphate prices are likely to rise.

These issues raise questions about the sustainability of P management in organic cropping systems. First, what are the consequences of organic cropping and farming systems for soil P status and crop yields? Second, can we identify some avenues for the better use of soil P reserves by taking advantage of the functional diversity of plants and soil organisms in the rhizosphere (i.e., the soil close to roots)?

In this chapter, we will introduce some basics about the fate of P in soils and P management. We will then focus on the options for better use of soil P reserves through an understanding of the fate of P in low-input soils, particularly considering rhizosphere dynamics. Finally, we will discuss the consequences of current cropping and farming practices for P management in organic systems and will identify options for better P management at both the field and farm levels.

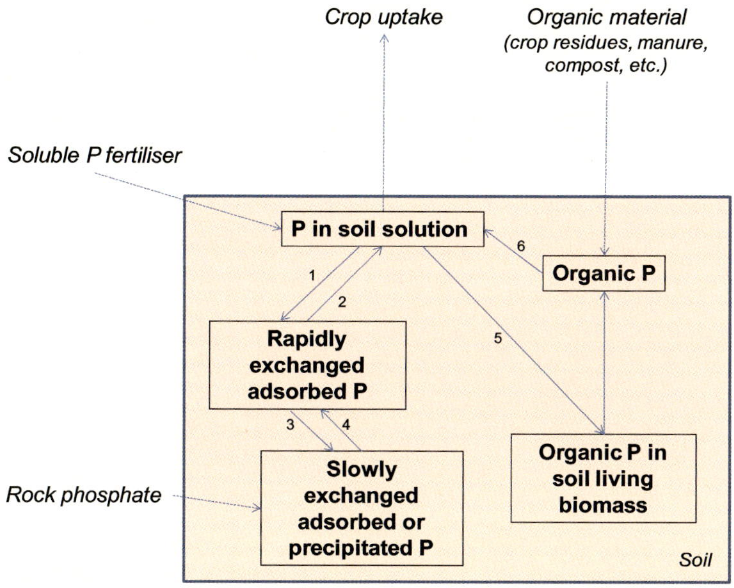

Fig. 2.1 Representation of the different P pools in soils. The different numbers refer to the main process affecting the P pools: 1, adsorption; 2, desorption; 3, precipitation; 4, dissolution; 5, organisation; 6, mineralisation

2.2 Phosphorus Dynamics and Management

2.2.1 The Fate of P in Soils

Phosphorus exists in many different forms in soils that may be classified within five functional groups: P in soil solution; rapidly exchangeable adsorbed inorganic P; slowly exchangeable adsorbed or precipitated inorganic P; organic P; and microbial P (Fig. 2.1). Functionally, soil solution P is of utmost importance since crop roots can take up phosphate ions from this pool alone. However, these pools are interconnected and their respective dynamics are strongly influenced by cropping practices as is shown below.

The sum of the different pools represents the total soil P. Its content varies considerably with soil type and fertiliser history (Richardson et al. 2004; Tiessen 2008). It commonly ranges from 100 to 1000 mg P kg^{-1}, but can be as little as 10–50 mg P kg^{-1} in deeply weathered soils, or reach several thousand mg P kg^{-1} in heavily fertilised soils that can be found in regions of intensive pig farming and pig slurry application in Denmark, the Netherlands, Catalonia in Spain or Brittany in France.

In arable soils, whether farmed organically or conventionally, a major proportion of soil P (up to 80 %) is made up of inorganic P (Pellerin et al. 2003). Inorganic P is bound to a range of P-bearing compounds, namely (i) positively-charged minerals (predominantly, metal oxides and clay minerals) onto which phosphate ions are strongly adsorbed via surface complexation processes and that may be rapidly exchanged with the soil solution (Devau et al. 2011), and (ii) phosphate minerals that slowly release phosphate ions into the soil solution (Frossard et al. 2000; Hinsinger 2001; Kizewski et al. 2011). In neutral to alkaline soils, they are predominantly made up of the least soluble apatite-like calcium phosphates as well as more soluble octocalcium phosphate and dicalcium phosphate (Freeman and Rowell 1981; Lindsay et al. 1989). In acidic soils, iron phosphates (such as strengite) and aluminium phosphates (such as variscite) can occur as well (Hinsinger 2001; Kizewski et al. 2011). Soil pH plays a major role in determining both the equilibrium of dissolution/precipitation and of adsorption/desorption of all these P-bearing minerals and, thus, the availability of inorganic P (Devau et al. 2011; Hinsinger 2001). Soil organic matter can also be involved in surface complexation processes that control the fate of phosphate ions in soils.

Organic matter contains P that makes up the bulk of soil organic P. These organic compounds comprise inositol phosphates (from plants, notably phytate), phospholipids and nucleic acids (DNA, RNA), AMP-ADP-ATP, etc. Their total amount and proportion can vary according to the content of organic matter, fertiliser history and vegetation. Total organic P is greater in forest and grassland soils (amounting to up to 90 % of total P in organic soils) than in arable soils. Turner et al. (2003) showed that organic P represented between 3 and 36 % of total P (in most cases, more than 15 %) within a range of 18 arable soils in semi-arid, conventional agriculture in the US. Organic P is not directly available to plants since it requires hydrolysis by phosphatase-like enzymes, which are produced by plants and, more so, by many soil microorganisms.

Another pool of soil P is the microbial biomass P, i.e., contained in soil microorganisms. It amounts to only 0.4–2.5 % of total P in arable soils, whereas it can reach up to 7.5 % in grassland soils (Bünemann et al. 2011) and 11 % in forest topsoils, excluding the litter layer (Achat et al. 2010). Nevertheless, it can play a major role in soil P availability, especially when its turnover time is short, which is usually the case for a large proportion of microbial P, 80 % of which have a turnover time of 9 days in the study by Achat et al. (2010).

Only orthophosphate ions ($H_2PO_4^-/HPO_4^{2-}$) within the soil solution are taken up by plant roots. However, as explained above, P is strongly bound to the solid fraction of the soil, either as inorganic or organic compounds with low solubility. Thus, their diffusion hardly extends over distances greater than 1 mm over a few days (Hinsinger et al. 2005). As a consequence, the P concentration in the soil solution is much lower than the so-called extractable or labile soil P, and considerably lower than the total soil P content (Hinsinger 2001; Pierzynski et al. 2005). Typical concentrations of phosphate ions in the soil solution range from 0.1 to 10 μM (Hinsinger 2001). This makes the P concentration in soil solution the first key indicator of soil P availability. The phosphate ion concentration in the soil solution is decreased by root uptake but is replenished primarily through desorption of adsorbed ions and diffusion towards roots. The other mechanisms contributing to the replenishment of phosphate ions in the soil solution are the dissolution of phosphate minerals and the mineralisation of organic matter. Thus, the ability of a soil to replenish its P soil solution is referred to as the soil P buffering capacity. It corresponds to the second key indicator of soil P availability. Indeed, the replenishment of the soil solution and the diffusion of phosphate ions are the limiting steps of P acquisition by crop roots, as has been shown for a long time by plant nutrition models (Barber 1995; Tinker and Nye 2000).

2.2.2 Phosphorus Management Principles in Agroecosystems

Soil P status is strongly influenced by cropping practices[1] through plant uptake and removal from the field via crop products, as well as P inputs of both inorganic and organic fertilisers (Fig. 2.1). Regulations on organic farming only allow the input of rock phosphates and P-containing organic materials. However, rock phosphates may be contaminated by cadmium in proportions depending on sedimentary deposits. EU legislation had fixed a maximum of 90 mg cadmium per kg of P_2O_5 in rock phosphates (Commission Regulation (EC) No. 889/2008), but this limit is currently under discussion and might be raised. Additionally, rock phosphates, even finely ground, have poor solubility in all but very acid soils ($pH<5.5$), making them poorly efficient in neutral and alkaline soils that are common in European agricultural regions, as well as under low rainfall conditions such as in a Mediterranean climate, as shown in Australia (Bolland et al. 1997).

Phosphorus-containing organic materials may originate from animal manure, slurry and composts or from organic fertilisers (such as guano, blood, horn, bone and fishbone meals, etc.). These bone and fishbone meals actually largely consist of inorganic P, namely apatite-like calcium phosphate minerals, and therefore exhibit limited solubility in neutral and alkaline soils. Such materials are not necessarily produced under organic certification. However, IFOAM principles and European regulations exclude sewage sludge compost and all organic materials from industrial animal production or livestock fed with genetically-modified crops (Commission Regulation (EC) No. 889/2008).

Thus, P input materials may be forbidden (e.g., chemical fertiliser), expensive and may require approval from an organic certification body if not produced on-farm (e.g., organic fertiliser), poorly efficient (e.g., rock phosphate, bone and fishbone meals) or costly to transport (e.g., manure and slurries). This makes P management in organic systems a critical issue (Guppy and McLaughlin 2009). Indeed, there is a risk of not enough P input at the field scale, resulting in the depletion of soil P on the long-term and, ultimately, in the reduction of agricultural productivity. Alternatively, P may be applied in approved forms but in excess of requirements when P is not available in sufficient quantity (e.g., using rock phosphate in neutral or alkaline soils), resulting in P accumulation in soil, but with low productivity. This is why some authors suggest that regulations should be adapted to the new farming context (e.g., development of stockless farming) or should be more flexible to allow for the use of sewage sludge compost (Cornish and Oberson 2008).

[1] In this chapter, the terms 'crop' and 'cropping' are considered in their general meaning, i.e., relating to arable crops as well as to grasslands pastures and horticultural crops.

Changes in soil total P are related to the soil P budget, calculated as the difference between total P input through supplied materials and output through harvested crop products, as well as environmental losses due to leaching, run-off and erosion (Cobo et al. 2010; Haileslassie et al. 2007) (Fig. 2.1). The budget provides insight into the increase or decrease of soil P reserves, with a negative P budget when soil P reserves decrease. If such a decrease is repeated over long periods of time, it may lead to nutrient scarcity that might negatively affect crop productivity. If the budget is strongly positive over time, it may lead to excess total P in soil, possibly increasing losses through leaching, runoff and erosion.

Available soil P reflects the budget, as well as the dynamics between the various pools of P in the soil. Numerous methods have been developed to assess soil P availability (Harmsen et al. 2005). They are mostly based on soil P extraction by chemical means intended to mimic plants. While none of these methods has proven to be perfect, the Olsen P and ammonium-acetate lactate methods are the most commonly used (Fardeau et al. 1988). However, they are limited because they are not able to precisely predict the actual P bioavailability for a wide variety of plant species and soil types. Indeed, the complex biological, chemical and physical processes that contribute to soil P dynamics are not accounted for when P bioavailability is estimated by chemical soil testing procedures alone. Their adequacy for assessing soil P availability in the context of organic, low input farming is even more questionable, given that such systems rely on the use of P compounds where P is not immediately available (e.g., rock phosphate or animal manure) and that microbial and other biological processes might be enhanced in organic farming (see Sect. 2.3). Therefore, we need to further assess the crop response to supplied P-containing materials in the low available P range to identify the need for a new, more mechanistic soil P test for organic systems.

Crop response curves to increasing doses of fertiliser P were designed several decades ago. They were used to establish threshold values. Basically, extra fertilisation is not recommended when soil P tests are near the threshold, and inputs are recommended just to match outputs (except when P 'fixation', mainly through precipitation, is known to occur. However, in the past decades, much more P was added than required to replace outputs as an insurance against crop loss because P fertilisers were cheap, particularly in industrialised countries.

As a conclusion, there are two general options for maintaining or increasing available P in organically-managed soils, in addition to the minimisation of losses through runoff and erosion: (i) either by using the soil P reserves more efficiently through the management of the equilibrium among soil P pools to draw more P from the slowly available pool (organic and inorganic P). Plant and rhizosphere manipulation may help in this case (see Sect. 2.3); (ii) or by supplying rock phosphate or P-containing materials that are either purchased or come from internal recycling within the farms (see Sect. 2.4). These two options are discussed in the following sections.

2.3 Some Options for Better Soil P Management in Organic Farming Systems

The aim of this section is to review the mechanisms that determine the fate of soil P in organic farming systems, with the objective to show how these processes might be better handled through management practices.

2.3.1 Making Better Use of Plant Functional Diversity to Acquire Soil P

Plants are capable of altering the soil in their rhizosphere where they may either increase or decrease the availability of nutrients through a range of root-induced processes: the release by roots of P-mobilising compounds such as protons/hydroxyls, carboxylates and phosphatases varies with plant species, plant nutritional status and soil properties, and is often triggered by P deficiency (Hinsinger 2001; Raghothama and Karthikeyan 2005; Vance et al. 2003). In addition, the rhizodeposition of carbon-rich compounds stimulates both naturally occurring and inoculated microorganisms in the rhizosphere, which can alter the availability of soil P (Guppy and McLaughlin 2009; Richardson et al. 2009).

There is considerable variation between crop species, as well as for a given species between genotypes, in terms of the capacity to acquire soil P. This means that we need to know more about the traits involved in soil P acquisition in order to make better use of such functional diversity in low input agriculture and organic farming. A difficulty is that those traits that are important for soil P acquisition are below-ground traits (Lynch 2007; Wissuwa et al. 2009) that are not readily measurable, relating either to root architecture and growth, or to root functioning (rhizosphere processes).

Lynch (2007) has stressed the importance of root architecture in relation to the poor mobility of phosphate ions in soils. Hence, plants need to develop a large volume of rhizosphere to access enough P from the soil. This also means that agricultural practices that are prone to maintaining favourable soil physical conditions (low soil compaction) and, hence, root growth, should be implemented (e.g., use of organic amendments, soil tillage). Wissuwa (2005) showed that in rice, an increase of only 20% in the root elongation rate could explain the ability of a P-efficient near-isogenic line of the common cultivar Nipponbare, which is P-inefficient, to take up three times more P under low P conditions. Ge et al. (2000) showed that root architecture had a greater impact when a steep vertical gradient of P availability occurred in the soil profile, which is usually the case in untilled soils. However, Hinsinger et al. (2005) showed that the rhizosphere volume was rather small when considering poorly-mobile nutrients such as P, suggesting the important role of other root traits involved in soil colonisation and access to soil P, such as root hairs and mycorrhiza. However, their quantitative contribution to P acquisition by field-grown plants is difficult to evaluate, and most of our knowledge is derived from controlled growing conditions in pots.

Gahoonia et al. (2001) and Gahoonia and Nielsen (2004a, b) demonstrated the major role of root hairs in extending the volume of the P depletion zone around roots and the potential to explore genotypic variation in traits such as root hair length and density in cereals, e.g., barley. While root hairs can extend up to 1–2 mm away from the root surface, mycorrhizal hyphae can extend from one to about 10 cm (Jakobsen et al. 1992; Thonar et al. 2011) and thus play a greater role in increasing the access to soil P. A few reports suggested that modern genotypes of cereals are less susceptible to mycorrhizal symbiosis than older genotypes, landraces and ancestors (Hetrick et al. 1993; Hetrick et al. 1996; Zhu et al. 2001). Moreover, the contribution of mycorrhizal symbiosis to plant nutrition is likely to be greater in organic than in conventional farming systems due to restricted inputs of P fertilisers and fungicides. In their long term DOK (bio-Dynamic, bio-Organic, and "Konventionell") trial in Switzerland, Mäder et al. (2002) reported that root length colonised by mycorrhizal fungi in organic farming treatments was 40% higher than in the control treatment corresponding to conventional management. However, the actual benefit of such mycorrhizal infections for improving crop productivity under field conditions is still subject to much debate (Smith and Smith 2011).

Besides increasing the size of the rhizosphere volume, there are other potential options for increasing acquisition efficiency in crop species through the manipulation of traits related to plant physiology, including root exudation of P-solubilising compounds, e.g., protons/hydroxyls, carboxylates and phosphatase enzymes (Richardson et al. 2009). Dunbabin et al. (2006) showed that accounting for the exudation of a P-mobilising compound (a surfactant in that case, but the modelling exercise would apply to any) yielded a 14% increase in P uptake in a soil with high P availability, while it amounted to a 50% increase in a soil with low P availability. These rhizosphere processes are not accounted for in plant nutrition models, which adequately predict P uptake in high or moderate P input conditions, whereas they underestimate P uptake under low input conditions (Hinsinger et al. 2011b; Mollier et al. 2008).

These results illustrate that such rhizosphere processes are likely to be of crucial importance in low input and organic farming systems. However, since most crop genotypes have been selected under high input conditions (e.g., high P), their capacity to adapt to low input conditions is therefore questionable (Ismail et al. 2007; Lynch 2007; Rengel and Marschner 2005). Indeed, we have probably counter-selected those genotypes that may perform better under low P input cropping systems. Further work is needed to show if there are likely to be significant benefits from selecting genotypes that perform better in terms of mycorrhizal responsiveness and release of P-solubilising compounds as has been attempted for traits such as rhizosphere acidification (Yan et al. 2004) and carboxylate exudation (Ryan et al. 2001; Vance et al. 2003). If so, breeders may revise breeding schemes in order to select genotypes that are more P-efficient, i.e., that perform best under low soil P availability, as was done in Southern France for organically-grown durum wheat by Desclaux (2005) and Desclaux et al. (2008), and in Europe for other cereals (Wolfe et al. 2008).

In addition to using genotypes that perform better, there are agronomic management options to make use of the functional diversity of plants to access soil P. First, using more diverse species in crop rotations makes sense for more effectively exploiting soil resources, as long as the subsequent crops are functionally diverse in terms of their capacity to explore different soil horizons or P pools. Kamh et al. (1999) showed, for example, in a pot experiment, that white lupine was capable of increasing soil P availability for the benefit of the subsequent maize crop by tapping into pools that would have remained otherwise unavailable to the cereal. Other potential effects for the subsequent crop must be considered, especially when including more legumes in the rotations, which is typically the case in organic farming compared with conventional farming systems. Disentangling the various origins of the observed benefits is far from trivial.

A second option is based on intercropping in which a minimum of two different species are grown simultaneously in the same field (Malezieux et al. 2008). The benefit of such mixed-species systems has been extensively studied in the case of nitrogen efficiency in cereal-legume intercrops (see Bedoussac et al. 2014, Chap. 3), but recent studies have suggested that the yield benefit in such intercropping systems could also result from improved P acquisition (Betencourt et al. 2011; Li et al. 2007). This may be the consequence of either niche complementarity or facilitation (Hinsinger et al. 2011a). Niche complementarity might occur if the two intercropped species make better use of soil P resources by a partitioning of time, space (soil horizons) and P pools (e.g., organic versus inorganic) between the two intercropped species. Betencourt et al. (2011) showed that facilitation occurred in the rhizosphere of durum wheat-chickpea intercrops, and especially under low P input conditions. So far, such processes have been little studied and, to our knowledge, never in the context of organic farming systems. Horst et al. (2001) and McNeill and Penfold (2009) have, however, identified intercropping as one of the agronomic management options for P in low input cropping systems, and Hinsinger et al. (personal communication) are currently testing this option in the context of organic farming in Southern France.

2.3.2 Making Better Use of Soil Organisms Involved in P Dynamics

In addition to the root-mediated rhizosphere processes mentioned above, P availability can also be considerably altered by soil microorganisms and fauna in the rhizosphere as well as in the bulk soil (Guppy and McLaughlin 2009; Richardson et al. 2009). Indeed, to acquire soil P, microorganisms have evolved a whole range of tricks similar to those developed by plants, i.e., releasing P-solubilising compounds such as acids, carboxylates and phosphatase-like enzymes. By producing phosphatases, soil microorganisms play a major role in the fate of organic P in soils and it is noteworthy that Oberson et al. (1996) and Mäder et al. (2002) have reported greater phosphatase activities in organically-managed soils compared with conventionally-managed soils. This is in line with the findings of Mäder et al. (2002) and Oehl et al. (2004) who reported larger microbial biomass in organic farming treatments in their field trials. Oehl et al. (2004) found that microbial biomass C, N and P were consistently larger in organically-managed soils compared with conventionally-managed soils. They also reported an increased basal mineralisation rate of organic

P in organically-managed soils. It should be mentioned, however, that most of these studies considered rather high P input conditions, while low input organic farming conditions have not been very documented in that respect. The roles of microorganisms that are key players in soil P cycling (Bünemann et al. 2011) are therefore likely to be of utmost importance in organic farming, the problem being how to manage such microbial communities in order to improve P use efficiency.

Soil microbial communities can be altered by soil properties (organic matter, pH, availability of nutrients, activity of soil fauna, etc.), climate and farm practices (tillage, fertiliser and pesticide application, etc.), as well as plant cover. Plant species can select their rhizosphere microbial communities, which makes the direct manipulation of the soil microbial community even more complicated (Richardson et al. 2009; Wissuwa et al. 2009). Most attempts to do so for improving P acquisition are based on the use of either mycorrhizal fungi or a whole range of P-solubilising microorganisms (PSM), which belong to many microbial groups. There is an abundant literature on the potential use of PSM as inoculants to improve P acquisition in crops, for both bacteria (Rodriguez and Fraga 1999) and fungi (Wakelin et al. 2007). Yet, most of them showed useful positive effects on crop growth only in pot experiments (Kucey et al. 1989; Richardson et al. 2009; Vessey 2003). In contrast, success stories of PSM inoculants in field-grown plants are rare, as for other plant growth-promoting microorganisms, with the notable exception of rhizobia and other N_2-fixers such as Azospirillum (Richardson et al. 2009). As stressed by Richardson (2001) and Vessey (2003), the inconsistent response of microbial inoculants in various (soil x host plant) combinations is still a major impediment to their widespread application. There are few field studies on the use of mycorrhizal and other microbial inoculants that indicate that this is a direction worth pursuing for its potential application in low input agriculture and organic farming (Mäder et al. 2011).

While our knowledge of the rhizosphere processes involved in P acquisition efficiency of crops has considerably advanced over the recent decade, both at the root and microbial levels, it is still rather difficult to demonstrate and rate their relative contribution under field conditions. Further field assessment of such rhizosphere processes is needed before we can determine the most promising avenues for organic farming under a range of situations, from P-poor to P-rich soils, and P-input options, from strictly organic to inorganic (e.g., phosphate rocks).

2.4 Phosphorus Management on Organic Farms

The aim of this section is to assess organic farmers' practices in terms of P flows, soil P status and resulting crop yields at field and farm levels. At field level, P management results from fertiliser and manure application, crop production, residue management, etc. At farm level, P flows and soil stocks result from interactions between animal and cropping systems, material import and export and spatial distribution of cropping practices (Fig. 2.2).

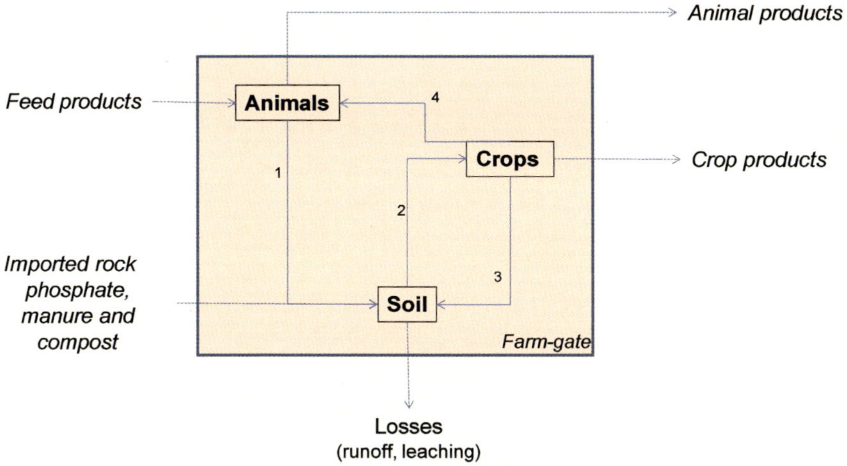

Fig. 2.2 Representation of the P stocks and flows within mixed-farms. The numbers refer to the main internal P flows: 1, animal excretion; 2, crop uptake; 3, crop residues returned to soils; 4, crop products used as feed

2.4.1 Phosphorus Budget at Field Level

At field level, P management is generally assessed by means of a field-gate mass budget. The field is considered as a "black-box" and the P budget is calculated as the difference between total P inflow through imported materials, e.g., fertiliser, manure, compost, animal excretion during grazing, etc., and outflow through exported materials, e.g., crop products, grazed grass, losses due to erosion, etc. (Cobo et al. 2010; Haileslassie et al. 2007). The budget helps to assess the sustainability of a given cropping system in terms of increase or decrease of the soil P reserves. The budget also provides information about the details of the inflows and outflows that contribute the most to the field-gate budget. Therefore, budgets are important and essential criteria to be considered in practical guidelines for P management drawn up for farmers and their advisors. Different kinds of field-gate budgets may be calculated, depending on the limit of the modelled system and the flows under consideration (Watson et al. 2002). However, to integrate the temporal variability in the P budget that might be due to differences in the management of the crops within a rotation, the field-gate P budget is usually performed over the whole duration of the rotation. Indeed, moderate quantities of P fertiliser are applied in some poorly demanding crops such as cereals. However, large quantities of P amendments are used for P-demanding productions such as horticulture, through animal manure or slurry, as well as bone or fishbone meals and ground rock phosphate (Nelson and Janke 2007).

The field-gate budget generally depends on the farm-gate budget: P is usually applied in excess on any field of a given farm when this farm exhibits a largely positive farm-gate P budget, as in the case of intensive dairy or indoor pig production.

However, some variability in field-gate P budget may exist within a given farm due to preferences in allocating animal excreta and organic fertilisers among field plots (Capitaine et al. 2009). For example, in an experimental organic dairy farm in Norway, the farm-gate P budget was -6.3 kg P ha^{-1} year^{-1}, whereas the P budget was positive for pasture and forage rape/Italian rye-grass crops and negative for other crops (Steinshamn et al. 2004). Understanding the relationship between farm- and field-gate budgets would require detailed modelling of farm management, nutrient flows within farms and their corresponding drivers. Except for some original works (Modin-Edman et al. 2007; Vanlauwe et al. 2006), such studies are lacking. They would, however, be useful for assessing the consequences of farm-gate nutrient budgets in terms of hot spots of soil depletion or over build-up.

It is commonly held that an objective of field-gate P budgets is to maintain a balanced P budget. However, the objective should depend on which of the two following situations is best suited. Whenever available soil P is near the optimum threshold for the production system in question (soil, crop type, targeted productivity), then the aim should be to strictly compensate outputs by inputs; a balanced, P budget would then be ideal, although hard to achieve at field level, even though it can be achieved at farm level. In that situation, if a positive P budget is maintained (presumably to maintain crop production), then P would build up in the soil and environmental risks such as losses through erosion would thereafter increase. In contrast, whenever available soil P is below the optimum threshold and some fraction of the P applied will be adsorbed, ending up in the slowly available pool of soil P, a positive P budget is needed for a while. As a consequence, the field-gate P budget is to be considered together with the soil P status.

Phosphorus can be brought to organically-managed soils through crop residues, rock phosphate, organic fertilisers or compost and manure, the latter being either produced on-farm or imported from organic or conventional farms. Of special concern, particularly in the case of disallowance of manure from conventional farms, are the cases of stockless cropping systems that may lead to negative soil P budgets unless inputs of other approved sources of P are increased. Negative P budgets may not affect production for a period of time, depending on the initial soil-P status, but ultimately productivity is likely to decline once available soil P falls below critical values. Indeed, market prices for highly-profitable organic food crops such as cereals, sunflower or soybean have led to specialised cropping systems in stockless farms, with limited supply of composted on-farm manures (David et al. 2005). Such systems were recently assessed in France in two studies.

In the first study (ITAB 2011), 11 typical systems differentiated by the presence of alfalfa in the rotation and the use of irrigation were identified and analysed for the Centre, Ile-de-France, Pays de la Loire, Poitou-Charentes and Rhône-Alpes regions (Table 2.1). Despite variations in length of rotation from three to ten years, the annual P removals varied within a narrow range from 11.4 to 17.5 kg P ha^{-1}year^{-1}. All systems except one received P fertiliser input with frequencies varying from every third year to five years out of six. The average P input at the rotation level varied widely from 0 to 48 kg P ha^{-1}year^{-1}. Alfalfa cutting represented a major removal of P from the field. Thus, all six cropping systems with alfalfa exhibited some P

Table 2.1 Rotational P budget of 11 typical stockless organic cropping systems from five French regions (ITAB 2011). C: Centre region; IDF: Ile-de-France, PC: Poitou-Charentes, PDL: Pays de la Loire; RA: Rhône-Alpes

Cropping system:	C 1	C 2	IDF 1	IDF 2	IDF 3	PC 1	PC 2	PDL 1	PDL 2	RA 1	RA 2
Rotation length (year)	8	8	10	9	6	9	5	3	5	6	3
Presence of alfafa	yes	yes	yes	yes	no	yes	no	No	no	yes	no
Irrigation	no	yes	no	no	no	yes	no	Yes	no	no	yes
Soil productivity level	mean	high	very high	very high	high	mean	mean	High	mean	mean	high
Green manure frequency (year/year)	0/8	1/8	1/10	1/9	1/6	0/9	0/5	0/3	1/5	1/6	1/3
Fertiliser input frequency (year/year)	3/8	4/8	3/10	0/9	5/6	3/9	3/5	2/3	3/5	1/6	1/3
P input (kg P/ha/year)	12.7	13.1	4.4	0	48	8.7	13.1	31.9	13.5	10	12.2
P removals (kg P/ha/year)	16.2	17.5	15.7	16.6	13.5	16.2	11.4	16.6	10.5	15.7	17.5
P balance (kg P/ha/year)	−3.9	−3.9	−11.4	−16.6	34.9	−7.4	1.7	15.7	3.1	−5.7	−5.2

C 1 alfafa (3 years)/winter wheat/triticale/faba bean/winter wheat/winter barley
C 2 alfafa (2 years)/winter wheat/red beet/winter wheat/grain maize/faba bean/winter wheat
IDF 1 alfafa (2 years)/winter wheat/triticale/winter oat/faba bean/winter wheat/spring barley/white clover/winter wheat
IDF 2 alfafa (3 years)/winter wheat/oilseed rape/winter wheat/faba bean/winter wheat/spring barley
IDF 3 faba bean/winter wheat/ grain maize/triticale + peas/ winter wheat
PC 1 alfafa (3 years)/winter wheat/grain maize/faba bean/triticale/sunflower/winter barley
PC 2 faba bean/winter wheat/winter barley sunflower/winter wheat
PDL 1 faba bean/winter wheat/grain maize
PDL 2 faba bean/winter wheat/sunflower/winter barley
RA 1 alfafa (3 years)/winter wheat/winter wheat
RA 2 soybean/winter wheat/grain maize

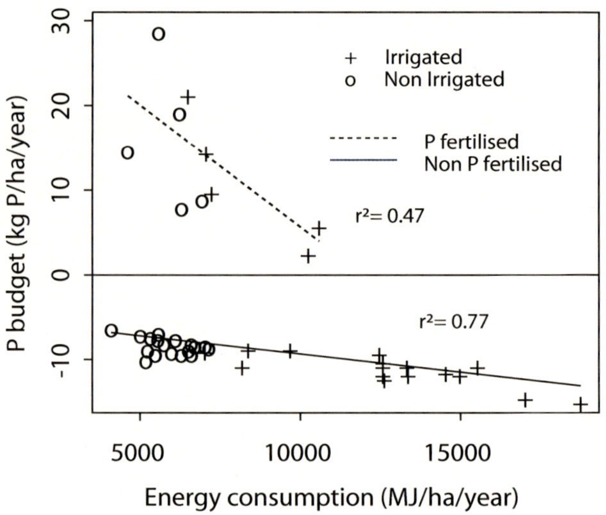

Fig. 2.3 Relationship between the annual P budget and the annual mean energy consumption of 44 organic stockless cropping systems from the French Midi-Pyrenees region. (Colomb et al. 2011)

deficit. The remaining five systems showed a balanced or positive P budget except the irrigated one (RA 2) that was characterised by the highest P removal due to a high productivity level of maize, wheat and soybean.

In the second study (Colomb et al. 2011), 44 stockless cropping systems were analysed in the Midi-Pyrénées region during the 2003–2007 period. The main rotated crops were winter wheat (29 %), soybean (23 %), sunflower (11 %), lentils (9 %) and faba bean (9 %). Half of the systems were irrigated. The average annual P removals amounted to 8.7±3.9 kg P $ha^{-1}year^{-1}$. Only ten cropping systems received at least one P input over the four-year period (5 to 35 kg P $ha^{-1}year^{-1}$), e.g., manure, compost or approved commercial organic fertilisers. The mean annual P budget was +14.3 and −8.6 kg P $ha^{-1}year^{-1}$, respectively, for the P fertilised and unfertilised systems. The P budget decreased for both with increasing intensification as represented by energy consumption (Fig. 2.3). Energy consumption is used as an indicator of the management intensification level of the cropping systems. Increasing energy consumption (via irrigation and mechanical weed control) meant higher yields, which led to higher P removals and lower P budgets in both the P fertilised and the non P fertilised cropping systems.

Both studies showed that field-gate P budgets could vary widely over a rotation. Only a few cropping systems with a balanced or near balanced (within±5 kg P $ha^{-1}year^{-1}$) P budget have been found. These cropping systems belong to arable farms with access to P resources in close proximity. On the contrary, most stockless cropping systems suffered from a negative P budget. Such negative budgets may lead to low soil P status if repeated over long periods. However, their impact on crop yield is often limited in Europe because of the initial high soil P status due to massive use of P fertiliser prior to conversion to organic farming. It is likely to be different in other regions of the world where soil P status is not that favourable.

Soil P status provides useful information on the present soil P availability. When coupled with a determination of the P budget, it offers a trend for this availability. Indeed, the pattern of increase or decrease of soil P status is fairly well explained by the field-gate P budget (Loes and Ogaard 2001; Messiga et al. 2012; Morel 2002). Published studies that assess the soil P status in organic farming are quite rare in Europe, perhaps because of the history of high fertiliser use prior to conversion and, until recently, abundant manure from conventional farming.

In general, these studies show that soil P is moderate or low in many organically-managed soils. In Norwegian dairy farms, the soil P status assessed by the ammonium-acetate lactate method was medium to high but decreased by 1.5 to 2 % per year, particularly in P-rich soils. On the contrary, the P status of the subsoil increased, probably due to some slight P leaching and increased ploughing depth (Loes and Ogaard 2001). In Dordogne (France), the soil P status of 46 organically-managed field plots assessed by the Olsen method ranged from 3.3 to 53 mg P kg^{-1}, but 68 % of the soils sampled exhibited a P status lower than 20 mg P/kg soil, i.e., the threshold below which the yield of low-demanding crops such as wheat is supposed to be reduced (Nesme et al. 2012). In Australia, "paired-farm" comparisons of organic and conventional farms were performed for dairy or extensive crop production systems. They showed that available soil P was consistently lower on organic than on conventional farms. However, a farm-gate P budget was not reported in any of these paired farm studies (Cornish 2009): lower available soil P may result from a negative P budget or from using sources of P that are ineffective in raising available soil P concentrations (e.g., rock phosphate in alkaline soils). Such reports of low P status in organically-managed soils should serve as a warning about potential negative consequences for crop yield. The relationship between crop productivity and soil P status in organic farming systems was recently thoroughly reviewed in Australia (Cornish 2009). It was concluded that many Australian extensive crop farmers experienced a yield reduction with organic farming. However, these yield reductions can generally not be attributed with confidence to the lower soil P status, and may be confused with weed, nitrogen or water stress. Additional research on this topic is definitively needed.

2.4.2 Phosphorus Budgets at Farm Level

Organic principles encourage planned nutrient management across the farming system. Farm-level management aims to benefit from interactions between animal and cropping systems, including the recycling of animal manure to cropland and of crop products fed to animals. Purchased animal feed also introduces P to the farm that is ultimately converted to manure (Fig. 2.2). Conceptually, mixed farming systems are the ideal organic farming systems (Kirchmann et al. 2008), as implied in the EC regulation and by the ecology principle of IFOAM.

However, while this principle encourages the recycling of organic matter and nutrients through livestock to cultivated land in integrated farming systems, the regulations also allow for flexibility in the application of production rules to allow adaptation to local conditions, and this provision has allowed the development of specialised stockless farms in Europe. In Australia, extensive mixed farming systems are common, but many organic farmers appear to have difficulties managing P, even where they approach the ideal integrated crop-animal system, with the result that inputs of rock phosphate significantly exceed outputs in farm production (Cornish 2009). Moreover, some organic farmers, particularly those using biodynamic systems, would like to have, or at least strive to have no farm external inputs i.e., self-sufficiency. In this case, the choice of management of P stocks and flows may become particularly critical and, in the end, soil P status may decrease. This diversity of situations needs to be accounted for to assess the extent to which P management depends on the type of farming system concerned.

At farm level, P management is generally assessed by farm-gate mass budget. Considering the same principle as explained above, the P budget is calculated as the difference between total P inflow through imported materials, e.g., fertiliser, animal feed, manure, straw, etc., and outflow through exported materials, e.g., milk, meat, grain, straw, culled animals, etc. (Cobo et al. 2010; Haileslassie et al. 2007). The budget ultimately provides insight into the increase or decrease of soil P reserves.

Farm-gate budgets have been extensively applied to organic farms both in the scientific literature and by extension services. One objective of these budgets is to assess the capability of organic farming systems to maintain soil P status close to the optimum threshold. Farm-gate P budgets also allow organic systems to be compared to conventional systems in that respect. It is commonly hypothesized that the overall budget is highly dependent on the farming system (Berry et al. 2003; Oehl et al. 2002; Oelofse et al. 2010) and on the livestock density due to the import of feed. For example, in farms surveyed by Kirchmann et al. (2008) in Sweden, those with animals had a slight surplus of $+1$ kg P $ha^{-1}year^{-1}$, whereas stockless farms had negative budgets of -7 kg P $ha^{-1}year^{-1}$, with a risk of soil P depletion. However, extensive surveys of organic farms demonstrated that farm-gate P budgets could be positive or negative on both stock and stockless farms, and that the budget really reflected individual management rather than the type of farming system *per se* (Watson et al. 2002).

The comparison of 13 organic vs. 25 conventional dairy farms in Denmark showed that organic farms imported P through animal feed concentrate and manure for crops. However, for a given livestock density, organic systems exhibited lower P surplus (8 ± 3.7 vs. 14 ± 2.9 kg P.ha^{-1}.yr^{-1}) due to smaller feed import. Moreover, even if crop and animal product exports were smaller than on conventional farms, their overall ratio of P in exported products to P in imported products was higher (68 ± 26% in organic vs. 46 ± 20% on conventional farms). However, large variability in P budgets was observed among each farm type (Nielsen and Kristensen 2005).

One of the largest studies of P budgets on organic farms comprised three different counties (Skåne, Halland and Västra Götaland) and three different farm types in Sweden (Wivstad et al. 2009) (Table 2.2). This illustrates some interesting local variations in farming practices. However, overall, the organic farms studied showed a small surplus of P in crop, dairy and meat production systems. The positive results for organic crop farms reveal that 60% of these farms brought in manure or

Table 2.2 Phosphorus annual budget of organic and conventional crop, dairy and meat farms in three counties in Sweden based on data for 2001–2006

	Arable farms		Dairy farms		Meat farms	
	Number	*P kg ha^{-1}*	*Number*	*P kg ha^{-1}*	*Number*	*P kg ha^{-1}*
All farms						
Organic	76	6.1	107	2.3	93	2.8
Conventional	1535	−0.8	1517	4.0	267	4.1
p-value[1]		*<0.0001*		*0.0112*		*ns*
Skåne						
Organic	32	4.2	18	−0.3	31	0.6
Conventional	1017	−2.5	661	2.8	113	3.9
p-value		*0.0022*		*ns*		*0.0478*
Halland						
Organic	10	8.7	14	4.1	15	3.5
Conventional	66	3.9	157	6.5	26	7.0
p-value		*ns*		*ns*		*ns*
Västra Götaland						
Organic	15	5.4	35	2.6	23	1.1
Conventional	189	2.9	335	5.7	48	3.7
p-value		*ns*		*0.0016*		*0.0439*

[1] p-value indicates significance level of difference; *a* p-value of 0.05 indicates a significance level of 5 %; *p*-value > 0.05 is considered not significant (ns)

specialist organic fertilisers. In contrast, negative P budgets reflect export of P with very few or even no compensation by P inflow. Such results, yielding positive P budgets in organic, stockless crop farms associated with high rates of manure or organic fertiliser import, have already been reported by several authors (Nesme et al. 2012; Pellerin et al. 2003). For example, in Dordogne (south-western France), stockless organic farms had an average positive P budget of 17 kg P ha^{-1}year^{-1} due to massive import of manure, compost or organic fertiliser from neighbouring farms or urban sources, whereas stock farms had an average P budget of only 4 kg P ha^{-1}year^{-1} (Nesme et al. 2012). These results confirmed that farm-gate P budget depends more on individual management that determines farm inflows through import of materials (feedstuffs, straw, manure, compost and organic fertiliser) than on the type of farming system (stock vs. stockless, organic vs. conventional).

Farm inflows of organic materials depend on the availability of such materials in the agricultural geographic context. For example, in Dordogne, farm inflows were made possible by the characteristics of the region where materials could be easily exchanged among stock and stockless farms. More generally, such inflows are more common in Europe due to a higher concentration of livestock farming than in other regions oriented toward broad acre agriculture or extensive grazing systems such as Australia. Material exchanges among farms may also involve conventional farms (e.g., through import of manure or bedding materials), thus contributing to the import of P from conventional systems and, ultimately, from conventional P fertiliser. This point has already been stressed by various authors (Kirchmann et al. 2008;

Nesme et al. 2012; Oelofse et al. 2010). Studies that assess the flow from conventional to organic farming systems, with conclusions about the real self-sufficiency of organic farming systems for P, are lacking. However, recent changes in the EU regulation on organic farming that came into effect in January 2009 (834/2007) meant that 100% of the feed for organically-produced ruminant livestock must come from organic farms.

The gradual shift over time from allowing the import of some non-organic feed to 100% organic feed, as well as changes in the regulation associated with manure import, mean that nutrient budgets calculated on organic farms in the past may not be relevant today. For example, Fowler et al. (1993) described an organic farm in the mid-1980s, which relied on the import of non-organic poultry manure. This would no longer be allowed since the EC Regulation 889/2008 bans the use of manure from 'industrial' livestock systems. Such a change in the regulation also has consequences beyond the farm-gate budgets in terms of stressing the importance of ensuring that manures produced on organic farms are used on organic land, and that organic livestock feed is fed to organic livestock. This is at risk of yielding an overall depletion of P from organic land as a whole since the ability to bring in P from outside (i.e., from conventional agricultural products) is becoming more limited. It also means that organic crop farmers will need to start using rock phosphate and other approved inorganic inputs. However, references to their efficiency in cropping systems are clearly lacking under European conditions where access to cheap manure-P and already high-P soils has meant little dependency on the direct application of rock phosphate.

2.5 Conclusion

As shown above, different options exist for managing P and they depend on site-specific conditions. They are summarised in Table 2.3. Phosphorus management and the resulting farm-gate P budgets depend on the type of farming system (stock vs. stockless, organic vs. conventional) but farmers may counter-balance this relationship through their management practices. However, organic farming systems generally exhibit moderate to low farm- and field-gate positive P budgets since many of them try to move toward nutrient self-sufficiency. Where these budgets are negative, the result will be a decrease in soil P availability, possibly limiting crop yield if repeated over long periods. Tightening regulations on the use of manure from non-organic sources may further lead to negative P budgets. In Europe, soil P levels are generally high as a legacy of high inputs in the past, but in organic systems with negative P budgets, P-deficiency will ultimately occur unless P is accessed from less-available soil resources and/or approved inputs are used.

A range of mechanisms to help access the slowly available P were reviewed. These include root characteristics, root/mycorrhizal fungi interactions, proton efflux or enzyme excretion by roots, and enhanced microbial rhizosphere activity that influences P availability in the soil close to roots. All these processes may represent a promising way for the more effective use of soil P reserves and for designing P-efficient crop genotypes, although the benefits apparent under controlled conditions

Table 2.3 Options for P management at field level

		Positive field P budget	Negative field P budget
	Typical situation	High rate of organic fertiliser application (e.g., for horticultural production) High rate of manure application produced on-farm (e.g., resulting from feed import and high livestock density)	No or small import of P-containing materials while significant crop product exports
High soil P status	Massive use of soluble P fertiliser prior to conversion to organic farming	Consequences: High P build-up in soils and environmental risks (runoff).	Consequences: decrease in available soil P reserve.
	Long-term farming with positive P budget	Strategy: change in feeding regime or livestock density	Strategy: maintain negative P budget for a while (how long?) and make use of crop and soil microbial functional diversity for mining soil P reserves
Low soil P status	Long-term farming with negative P budget or with use of unavailable P forms (e.g., rock phosphate in alkaline soils)	Consequences: some P will be adsorbed and end up in the slowly available soil P pool; enhanced role of microbial functional biodiversity.	Consequences: decrease in available soil P reserves and risk for crop productivity; enhanced role of microbial functional biodiversity
		Strategy: maintain positive P budget for a while	Strategy: change fertilisation strategy and/or crop and animal interaction at farm level

are not always reproduced under field conditions. Moreover, these mechanisms allow plants to access the slowly available P, which itself is a finite resource. These rhizosphere mechanisms enable crops to operate at lower P levels so that less P is tied up in soil, but the crop P requirement will be the same if production of the same products is to be maintained at the same level. This points to another major avenue for improving P efficiency that has been much less explored in comparison with P acquisition efficiency: internal use efficiency. Identifying crops and genotypes capable of producing high yields at lower internal P concentrations is another route towards decreasing P outputs in farming systems.

Several studies reported that rhizosphere processes such as phosphatase activities and mycorrhizal colonisation of roots would be triggered by low soil P availability and, therefore, would possibly be enhanced under organic farming conditions. However, studies that would quantify the contribution of such mechanisms to the crop P uptake are still missing. Indeed, the extension of these processes to the whole crop and to field conditions is still a big challenge. Such studies would undoubtedly be useful for organic farming systems as well as for many other farming systems that may be inspired by organic production principles in the context of future fertiliser P scarcity.

References

Achat DL, Morel C, Bakker MR, Augusto L, Pellerin S, Gallet-Budynek A, Gonzalez M (2010) Assessing turnover of microbial biomass phosphorus: Combination of an isotopic dilution method with a mass balance model. Soil Biol Biochem 42:2231–2240

Barber SA (1995) Soil nutrient bioavailability: a mechanistic approach. 2nd Ed. John Wiley, New York

Berry PM, Stockdale EA, Sylvester-Bradley R, Philipps L, Smith KA, Lord EI, Watson CA, Fortune S (2003) N, P and K budgets for crop rotations on nine organic farms in the UK. Soil Use Manag 19:112–118

Betencourt E, Duputel M, Colomb B, Desclaux D, Hinsinger P (2011) Intercropping promotes the ability of durum wheat and chickpea to increase rhizosphere phosphorus availability in a low P soil. Soil Biol. Biochem. 46:181–190

Bolland MDA, Lewis DC, Gilkes RJ, Hamilton LJ (1997) Review of Australian phosphate rock research. Aust J Exp Agric 37:845–859

Bünemann EK, Prusisz B, Ehlers K (2011) Characterization of Phosphorus forms in soil microorganisms. In: Bünemann EK, Oberson A, Frossard E (eds.) Phosphorus in Action. Springer-Verlag, Berlin

Capitaine M, Boisdon I, Alvarez G, Vassal N (2009) L'utilisation des engrais de ferme sur prairie de montagne en agriculture bioologique et conventionnelle: des similitudes. Fourrages 197:75–84

Cobo JG, Dercon G, Cadisch G (2010) Nutrient balances in African land use systems across different spatial scales: a review of approaches, challenges and progress. Agricu, Ecosys and Environ 136:1–15

Colomb B, Aveline A, Carof M (2011) Une évaluation multicritère qualitative de la durabilité des systèmes de grandes cultures biologiques: quels enseignements? Restitution des programmes RotAB et CITODAB. Document d'analyse PSDR3 Midi-Pyrénées-Projet CITODAB et CASDAR RotAB

Cordell D, Drangert J-O, White S (2009) The story of phosphorus: global food security and food for thought. Global Environ Chan 19(2):292–305

Cornish PS (2009) Phosphorus management on extensive organic and low-input farms. Crop Past Sci 60:105–115

Cornish PS, Oberson A (2008) New approaches to Phosphorus regulation and management. In: D. Neuhoff et al. (eds), 16th IFOAM Organic World Congress, Modena, Italy

David C, Jeuffroy M-H, Henning J, Meynard JM (2005) Yield variation in organic winter wheat: a diagnostic study in the Southeast of France. Agron Sustain Develop 25:213–223

Desclaux D (2005) Participatory plant breeding methods for organic cereals. In: E.T. Lammerts Van Bueren and H. Ostergard (eds.), Proceedings of the COST SUSVAR/ECO-PB workshop on organic plant breeding strategies and the use of molecular markers, Driebergen (NK), pp. 17–23

Desclaux D, Nolot JM, Chiffoleau Y, Gozé G, Leclerc C (2008) Changes in the concept of genotype x environment interactions to fit agriculture diversification and participatory plant breeding: pluridisciplinary point of view. Euphytica 163:533–546

Devau N, Le Cadre E, Hinsinger P, Colomb B, Gérard F (2011) Fertilization and pH effects on processes and mechanisms controlling dissolved inorganic phosphorus in soils. Geochim Cosmochim Acta 75:2980–2996

Dunbabin VM, McDermott S, Bengough AG (2006) Upscaling from rhizosphere to whole root system: Modelling the effects of phospholipid surfactants on water and nutrient uptake. Plant Soil 283:57–72

Fardeau J-C, Morel C, Boniface R (1988) Pourquoi choisir la méthode Olsen pour estimer le phosphore "assimilable" des sols? Agronomie 8:577–584

Fowler S, Waston CA, Wilman D (1993) N, P and K on organic farms - Herbage and cereal production, purchases and sales. J Agric Sci 120:353–360

Freeman JS, Rowell DL (1981) The adsorption and precipitation of phosphate onto calcite. J Soil Sci 32:75–84

Frossard E, Condron LM, Oberson A, Sinaj S, Fardeau JC (2000) Processes governing phosphorus availability in temperate soils. J Environ Qual 29:12–53
Gahoonia TS, Nielsen NE (2004a) Barley genotypes with long root hairs sustain high grain yields in low-P field. Plant Soil 262:55–62
Gahoonia TS, Nielsen NE (2004b) Root traits as tools for creating phosphorus efficient crop varieties. Plant Soil 260:47–57
Gahoonia TS, Nielsen NE, Joshi PA, Jahoor A (2001) A root hairless barley mutant for elucidating genetic of root hairs and phosphorus uptake. Plant Soil 235:211–219
Ge Z, Rubio G, Lynch JP (2000) The importance of root gravitropism for inter-root competition and phosphorus acquisition efficiency: results from a geometric simulation model. Plant Soil 218:159–171
Guppy CN, McLaughlin MJ (2009) Options for increasing the biological cycling of phosphorus in low-input and organic agricultural systems. Crops Pasture Sci 60:116–123
Haileslassie A, Priess JA, Veldkamp E, Lesschen JP (2007) Nutrient flows and balances at the field and farm scale: exploring effects of land-use strategies and access to resources. Agric Sys 94:459–470
Harmsen J, Rulkens W, Eijsackers H (2005) Bioavailability: concept for understanding or tool for predicting? Land Contamin Reclam 13:161–171
Hetrick BAD, Wilson GWT, Cox TS (1993) Mycorrhizal dependence of modern wheat cultivars and ancestors—a synthesis. Can J Bot 71:512–518
Hetrick BAD, Wilson GWT, Cox TS (1996) Mycorrhizal response in wheat cultivars: Relationship to phosphorus. Can J Bot 74:19–25
Hinsinger P (2001) Bioavailability of soil inorganic P in the rhizosphere as affected by root-induced chemical changes: a review. Plant Soil 237:173–195
Hinsinger P, Betencourt E, Bernard L, Brauman A, Plassard C, Shen JB, Tang YY, Zhang FS (2011a) P for two, sharing a scarce resource—Soil phosphorus acquisition in the rhizosphere of intercropped species. Plant Physiol 156:1078–1086
Hinsinger P, Brauman A, Devau N, Gerard F, Jourdan C, Laclau JP, Le Cadre E., Jaillard B, Plassard C (2011b) Acquisition of phosphorus and other poorly mobile nutrients by roots. Where do plant nutrition models fail? Plant Soil 348:29–61
Hinsinger P, Gobran GR, Gregory PJ, Wenzel WW (2005) Rhizosphere geometry and heterogeneity arising from root-mediated physical and chemical processes. New Phytol 168:293–303
Horst WJ, Kamh M, Jibrin JM, Chude VO (2001) Agronomic measures for increasing P availability to crops. Plant Soil 237:211–223
Ismail AM, Heuer S, Thomson JT, Wissuwa M (2007) Genetic and genomic approaches to develop rice germplasm for problem soils. Plant Mol Biol 65:547–570
ITAB (2011) Rotations en grandes cultures biologiques sans élevage: 8 fermes types, 11 rotations; repères agronomiques, économiques, techniques et environnementaux., CAS-DAR 7055 RotAB, Paris
Jakobsen I, Abbott LK, Robson AD (1992) External hyphae of vesicular-arbuscular mycorrhizal fungi associated with Trifolium subterraneum l.1. Spread of hyphae and phosphorus inflow into roots. New Phytol 120:371–380
Jasinski SM (2011) Phosphate rock, USGS
Kamh M, Horst WJ, Amer F, Mostafa H, Maier P (1999) Mobilization of soil and fertilizer phosphate by cover crops. Plant Soil 211:19–27
Kirchmann H, Kätterer T, Bergström L (2008) Nutrient supply in organic agriculture: plant availability, sources and recycling. In: Kirchmann H, Bergström L (eds.) Organic crop production: ambitions and limitations. Springer, Dordrecht, The Netherlands, pp 89–116
Kizewski F, Liu YT, Morris A, Hesterberg D (2011) Spectroscopic Approaches for Phosphorus Speciation in Soils and Other Environmental Systems. J Environ Qual 40:751–776
Kucey RMN, Janzen, H.H. and Leggett mE (1989) Microbially mediated increases in plant-available phosphorus. Adv Agron 42:199–228
Li L, Li SM, Sun JH, Zhou LL, Bao XG, Zhang HG, Zhang FS (2007) Diversity enhances agricultural productivity via rhizosphere phosphorus facilitation on phosphorus-deficient soils. Proc Natl Acad Sci USA 104:11192–11196

Lindsay WL, Vlek PLG, Chien SH (1989) Phosphate minerals. In: J.B. Dixon and S.B. Weed (Editors), Minerals in soil environment, 2nd edn. Soil Science Society of America, Madison, pp. 1089–1130

Loes A-K, Ogaard AF (2001) Long-term changes in extractable soil phosphorus (P) in organic dairy farming systems. Plant Soil 237:321–332

Lynch JP (2007) Roots of the second green revolution. Aust J Bot 55:493–512

Mäder P, Fliessbach A, Dubois D, Gunst L, Fried P, Niggli U (2002) Soil fertility and biodiversity in organic farming. Science 296:1694–1697

Mäder P, Kaiser F, Adholeya A, Singh R, Uppal HS, Sharma AK, Srivastava R, Sahai V, Aragno M, Wiemken A, Johri BN, Fried PM (2011) Inoculation of root microorganisms for sustainable wheat-rice and wheat-black gram rotations in India. Soil Biol Biochem 43:609–619

Malezieux E, CrozatY, Dupraz C, Laurans M, Makowski D, Ozier-Lafontaine H, Rapidel B, de Tourdonnet S, Valantin-Morison M (2009) Mixing plant species in cropping systems: concepts, tools and models. A review. Agronomy Sustain Develop 29:43–62

McNeill AM, Penfold CM (2009) Agronomic management options for phosphorus in Australian dryland organic and low-input. Crops Pasture Sci 60:163–182

Messiga AJ, Ziadi N, Bélanger G, Morel C (2012) Process-based mass-balance modelling of soil phosphorus availability in a grassland fertilized with N and P. Nutr Cycl Agroecosyst 92:273–287

Modin-Edman A-K, Öborn I, Sverdrup H (2007) FARMFLOW—A dynamic model for phosphorus mass flow, simulating conventional and organic management of a Swedish dairy farm. Agric Syst 94:431–444

Mollier A, de Willigen P, Heinen M, Morel C, Schneider A, Pellerin S (2008) A two dimensional simulation model of phosphorus uptake including crop growth and P-response. Ecological Modelling 210:453–464

Morel C (2002) Caractérisation de la phytodisponibilité du P du sol par la modélisation du transfert des ions phosphate entre le sol et la solution. HDR Thesis, INPL-ENSAIA, Nancy, France, 80p

Nelson NO, Janke RR (2007) Phosphorus sources and management in organic production systems. Hort Technology 17(4):442–454

Nesme T, Toublant M, Mollier A, Morel C, Pellerin S (2012) Assessing phosphorus management among organic farming systems: a farm input, output and budget analysis in Southwestern France. Nutr Cycl Agroecosyst 92:225–236

Nielsen AH, Kristensen IS (2005) Nitrogen and phosphorus surpluses on Danish dairy and pig farms in relation to farm characteristics. Livestock Prod Sci 96:97–107

Oberson A, Besson JM, Maire N, Sticher H (1996) Microbiological processes in soil organic phosphorus transformations in conventional and biological cropping systems. Biol Fertil Soils 21:138–148

Oehl F, Frossard E, Fliessbach A, Dubois D, Oberson A (2004) Basal organic phosphorus mineralization in soils under different farming systems. Soil Biol Biochem 36:667–675

Oehl F, Oberson A, Tagmann HU, Besson JM, Dubois D, Mader P, Roth HR, Frossard E (2002) Phosphorus budget and phosphorus availability in soils under organic and conventional farming. Nutr Cycl Agroecosyst 62(1):25–35

Oelofse M, Hogh-Jensen H, Abreu LS, Almeida GF, El-Araby A, Hui QY, de Neergaard A (2010) A comparative study of farm nutrient budgets and nutrient flows of certified organic and non-organic farms in China, Brazil and Egypt. Nutr Cycl Agroecosyst 87:455–470

Pellerin S, Le Clech B, Morel C, Linères M (2003) Gestion de la fertilité phospho-potassique en agriculture biologique: questions posées et premiers résultats. Compte Rendu de l'Académie d'Agriculture de France, 89(1):30–34

Pierzynski GM, McDowell RW, Sims JT (2005) Chemistry, cycling, and potential movement of inorganic phosphorus in soils. In: J.T. Sims and A.N. Sharpley (Editors), Phosphorus: agriculture and the environment. American Society of Agronomy, Madison, USA, pp. 53–86

Raghothama KG, Karthikeyan AS (2005) Phosphate acquisition. Plant Soil 274:37–49

Rengel Z, Marschner P (2005) Nutrient availability and management in the rhizosphere: exploiting genotypic differences. New Phytol 168:305–312

Richardson AE (2001) Prospects for using soil microorganisms to improve the acquisition of phosphorus by plants. Aust J Plant Physiol 28:897–906

Richardson AE, Barea JM, McNeil AM, Prigent-Combaret C (2009) Acquisition of phosphorus and nitrogen in the rhizosphere and plant growth promotion by microorganisms. Plant Soil 321:305–339

Richardson SJ, Peltzer DA, Allen RB, McGlone MS, Parfitt RL (2004) Rapid development of P limitation in temperate rainforest along the Franz Joseph soil chronosequence. Oecologia 139:267–276

Rodriguez H, Fraga R (1999) Phosphate solubilizing bacteria and their role in plant growth promotion. Biotechnol Adv 17:319–339

Ryan PR, Delhaize E, Jones DL (2001) Function and mechanism of organic anion exudation from plant roots. Ann. Rev. Plant Physiol. Plant Mol Biol 52:527–560

Smith FA, Smith SE (2011) What is the significance of the arbuscular mycorrhizal colonisation of many economically important crop plants? Plant Soil 348:63–79

Steinshamn H et al (2004) Utilization of nitrogen (N) and phosphorus (P) in an organic dairy farming system in Norway. Agriculture, Ecosystems and. Environment 104:509–522

Thonar C, Schnepf A, Frossard E, Roose T, Jansa J (2011) Traits related to differences in function among three arbuscular mycorrhizal fungi. Plant Soil 339:231–245

Tiessen H (2008) Phosphorus in the global environment. In: P.J. White and J.P. Hammond (eds.), The ecophysiology of plant-Phosphorus interactions. Springer, pp. 1–7

Tinker BT, Nye PH (2000) Solute movement in the rhizosphere. Oxford University Press, Oxford

Turner BL, Cade-Menun BJ, Westermann DT (2003) Organic phosphorus composition and potential bioavailability in semi-arid arable soils of the Western United States. Soil Sci Soc Am J 67:1168–1179

Van Vuuren DP, Bouwman AF, Beusen AHW (2010) Phosphorus demande for the 1970-2100 period: a scenario analysis of resource depletion. Global Environn Chan 20:428–439

Vance CP, Uhde-Stone C, Allan DL (2003) Phosphorus acquisition and use: critical adaptations by plants for securing a nonrenewable resource. New Phytol 157:423–447

Vanlauwe B, Tittonel P, Mukalama J (2006) Within-farm soil fertility gradients affect response of maize to fertiliser application in western Kenya. Nutr Cycl Agroecosyst 76:171–182

Vessey JK (2003) Plant growth promoting rhizobacteria as biofertilizers. Plant Soil 255:571–586

Wakelin SA, Gupta VVSR, Harvey PR, Ryder MH (2007) The effect of Penicillium fungi on plant growth and phosphorus mobilization in neutral to alkaline soils from southern Australia. Can J Microbiol 53:106–115

Watson CA, Bengtsson H, Ebbesvik M, Loes AK, Myrbeck A, Salomon E, Schroder J, Stockdale EA (2002) A review of farm-scale nutrient budgets for organic farms as a tool for management of soil fertility. Soil Use Manag 18:264–273

Wissuwa M (2005) Combining a modelling with a genetic approach in establishing associations between genetic and physiological effects in relation to phosphorus uptake. Plant Soil 269:57–68

Wissuwa M, Mazzola M, Picard C (2009) Novel approaches in plant breeding for rhizosphere-related traits. Plant Soil 321:409–430

Wivstad M, Salomon E, Spångberg J, Jönsson H (2009) Organic Production - Possibilities to Reduce Eutrophication, Swedish University of Agriculture, Centre for Sustainable Agriculture, Uppsala

Wolfe MS, Baresel JP, Desclaux D, Goldringer I, Hoad S, Kovacs G, Loschenberger F, Miedaner T, Ostergard H, van Bueren ETL (2008) Developments in breeding cereals for organic agriculture. Euphytica 163:323–346

Yan X, Liao H, Beebe SE, Blair MW, Lynch JP (2004) QTL mapping of root hair and acid exudation traits and their relationship to phosphorus uptake in common bean. Plant Soil 265:17–29

Zhu YG, Smith SE, Barritt AR, Smith FA (2001) Phosphorus (P) efficiencies and mycorrhizal responsiveness of old and modern wheat cultivars. Plant Soil 237:249–255

Chapter 3
Eco-Functional Intensification by Cereal-Grain Legume Intercropping in Organic Farming Systems for Increased Yields, Reduced Weeds and Improved Grain Protein Concentration

Laurent Bedoussac, Étienne-Pascal Journet, Henrik Hauggaard-Nielsen, Christophe Naudin, Guénaëlle Corre-Hellou, Loïc Prieur, Erik Steen Jensen and Eric Justes

Abstract Intercropping, i.e., simultaneously growing two (or more) species in the same field for a significant period of time but without necessarily concomitant sowing or harvest, is a practice aimed at eco-functional intensification.

This chapter integrates a comprehensive amount of original data from field experiments conducted since 2001 on spring and winter cereal-grain legume intercrops in experimental and farm contexts in France and Denmark, in an attempt to generalise the findings and draw up common guidelines. We have shown that intercrops appear to be a useful agronomic solution for organic arable cropping,

L. Bedoussac (✉)
ENFA-INRA, UMR1248 AGIR,
2 route de Narbonne, CS 22687, 31326 Castanet-Tolosan, Cedex, France
e-mail: laurent.bedoussac@educagri.fr

É.-P. Journet
CNRS, UMR2594 LIPM, INRA, UMR1248 AGIR,
Auzeville, CS 52627, 31326 Castanet-Tolosan, Cedex, France

H. Hauggaard-Nielsen
Department of Chemical and Biochemical Engineering, Technical University of Denmark,
2800 Kgs. Lyngby, Denmark

C. Naudin · G. Corre-Hellou
LUNAM Université, Groupe ESA, UR LEVA, SFR 4207 QUASAV,
55 rue Rabelais, 49007 Angers Cedex 01, France

L. Prieur
CREAB Midi-Pyrénées, LEGTA Auch-Beaulieu, 32020 Auch, France

E. S. Jensen
Department of Biosystems and Technology, Swedish University of Agricultural Sciences,
PO Box 103, SE-230 53 Alnarp, Sweden

E. Justes
INRA-Université de Toulouse-INPT, UMR1248 AGIR,
Auzeville, CS 52627, 31326 Castanet-Tolosan, Cedex, France

S. Bellon, S. Penvern (eds.), *Organic Farming, Prototype for Sustainable Agricultures*,
DOI 10.1007/978-94-007-7927-3_3,

particularly in low-N input systems, to enhance: (i) yields because of a general improvement of environmental resource use; (ii) cereal grain protein concentration due to a non-proportional competition for soil mineral N and other plant growth factors; and (iii) weed control compared to legume sole crops.

Therefore, intercropping can be a way to successfully produce organic grain legumes and cereals. However, it is difficult to propose generic crop technical protocols because of the multitude of production objectives and, hence, of combinations of species, varieties, densities, structure and manuring strategies.

Consequently, it should be emphasized that: (i) the species and varietal traits suited to intercropping and organic farming will make it necessary to reconsider the varietal selection criteria; (ii) further mechanistic understanding of the behaviour of intercropping systems is required to be integrated into crop models; and (iii) the development of intercrops cannot take place without the participation of all of the actors in the value chain because of lock-in mechanisms.

Keywords Environmental resource use · Management system · Nitrogen · Eco-functional intensification · Cereal-grain legume intercrop · Protein concentration · Weed · Yield

3.1 Introduction

Organic farming is based on a higher cropping system diversity than its conventional counterpart and is regarded as a prototype capable of enhancing the sustainability of agriculture and cereal-rich cropping systems. Nevertheless, organic arable crop rotations in temperate regions consist mainly of sole crops (SC; pure stands), with the exception of diverse pastures in farming systems with livestock (Hauggaard-Nielsen et al. 2001b).

In organic farming, nitrogen (N) availability can be limiting, especially in the absence of livestock (David et al. 2005a, b), and leads to decreases in cereal yield and lower protein concentration. For these reasons, integrating legumes with symbiotic fixation of atmospheric N_2 is essential for balancing nitrogen exports from the system. New agronomic solutions should be developed that address multifunctionality, including: (i) higher yields; (ii) improved quality; (iii) supply of ecosystem services; and (iv) the adaptation of production systems to climate change (IAASTD 2009). Intercropping (IC) cereals and legumes, i.e., simultaneously growing two (or more) species in the same field for a significant period of time but without necessarily sowing or harvesting them at the same time (Willey 1979; Vandermeer et al. 1998; Malézieux et al. 2008), is a practice for eco-functional intensification, which is considered as a means to enhance yields in organic farming (Niggli et al. 2008). However, due to the intensification of agriculture over the last 50 years (Crews and Peoples 2004), annual intercropping is now rare in European countries (except for animal feeds) and elsewhere in intensive farming systems (Anil et al. 1998; Malézieux et al. 2008). Nevertheless, because of the numerous ecosystem services provided by introducing cereal-legume intercropping (Hauggaard-Nielsen and Jensen 2005), there seems to be a renewed interest in cereal/legume intercrops in

Europe, notably in organic farming (Anil et al. 1998; Malézieux et al. 2008) for the purpose of eco-functional intensification.

Intercropping has been shown to increase and stabilise yields (Hauggaard-Nielsen et al. 2009b; Lithourgidis et al. 2006) and to increase cereal grain protein concentration and baking quality compared to sole crops (Gooding et al. 2007), particularly in low-N input systems and organic farming where N can be a limiting resource (Corre-Hellou et al. 2006; Bedoussac and Justes 2010a, b; Naudin et al. 2010). Intercropping has also been shown to: (i) improve soil conservation (Anil et al. 1998); (ii) favour weed control (Vasilakoglou et al. 2005; Banik et al. 2006; Corre-Hellou et al. 2011); (iii) reduce pests and diseases (Trenbath 1993; Altieri 1999); and (iv) provide better lodging resistance (Anil et al. 1998). In contrast, grain legumes such as peas (*Pisum sativum* L.), grown as sole crops, are known to be weak competitors towards weeds (Wall et al. 1991; Townley-Smith and Wright 1994; Mcdonald 2003), and weed infestations have been shown to severely limit the N nutrition and grain yield of organically-grown grain legumes (Hauggaard-Nielsen et al. 2001b; Corre-Hellou and Crozat 2005). Moreover, grain legume sole crops are sensitive to lodging and affected by numerous pests and diseases, which can cause serious yield losses in organic farming where pesticide use is forbidden. Thus, from these perspectives, intercropping can be a way to successfully produce organic grain legumes (Hauggaard-Nielsen et al. 2007).

The main objective of this study was to analyse and describe the potential advantages of cereal-grain legume intercrops for grain yield, grain protein concentration and weed control in organic cropping systems. This chapter integrates a comprehensive amount of original data from field experiments conducted since 2001 in France (southern and western, with contrasting soil and climatic conditions), and in Denmark, in experimental and farm contexts, on spring and winter cereal-grain legume intercrops (Table 3.1), in an attempt to generalise the findings in order to draw up more common guidelines.

The intercrops evaluated were as follows: (i) spring barley (*Hordeum vulgare*)-spring pea (*Pisum sativum*); (ii) spring barley-spring faba bean (*Vicia faba*); (iii) soft wheat (*Triticum aestivum*)-winter pea; (iv) soft wheat-spring faba bean; (v) durum wheat (*Triticum turgidum*)-winter pea; and (vi) durum wheat-winter faba bean. The experiments cover a wide range of management practices to evaluate their effects on competition, such as: (i) with or without N fertilisation (up to 100 kg mineral N ha^{-1}); (ii) sowing in separate rows or mixing within the same row; and (iii) different cereal/legume sowing proportions. Intercrops were always compared with the corresponding sole crops sown on the same date, receiving the same N fertilisation and harvested at crop maturity (that of the later crop in intercrops).

3.2 Yield Advantages and Cereal Quality Improvement

Fulfilling the cereal N demand is crucial for obtaining profitable yield and grain quality (Garrido-Lestache et al. 2004). Consequently, cereals are generally fertilised with high levels of N using considerable amounts of organic inputs like animal and

Table 3.1 List of field experiments from which data are derived. Six intercrops (*HW* Durum wheat, *SW* Soft wheat, *B* Barley, *F* Faba bean, *P* Pea) were evaluated in France (southern and western) and Denmark at 13 different sites representing 58 treatments. For each trial, we indicate the cereal and legume densities in intercrops (as a percentage of the sole crop densities), N treatment (*N0* no N-fertilisation, *N* organic N-fertilisation) and cultivars. More information about experiments can be found in Jensen et al. (2006), Hauggaard-Nielsen et al. (2007), Hauggaard-Nielsen et al. (2009a, b), Knudsen et al. (2004) and Naudin et al. (2009)

Crops	Species	Year	Location	Sites (per location)	N treatment	IC densities (% of sole crop)	Cultivar: Cereal/ Legurne	Number of treatments
Winter crops	HW-P	2009	France (Toulouse-south area)	1	N	58–93	Dakter/Enduro	1
				1	NO	58–93	Acalou/Livia	1
				1			Dakter/Enduro	1
		2010	France (Toulouse-south area)	1	N	58–72	Dakter/Cartouche	1
				1	NO	58–72	Dakter/Enduro	1
	HW-F	2009	France (Toulouse-south area)	1	N	58–49	Duetto/Irena	1
				2	NO	58–49	L1823/Irena	2
				1			Duettu/Irena	1
		2010	France (Toulouse-south area)	2	NO	66–50	Dakter and L1823/ Castel	2
	SW-P	2003 and 2005	France (Angers-west area)	1	NO	50–100 and 50–50	Apache/Lucy	4
		2006	France (Toulouse-south area)	1	N and NO	50–50	Caphorn/Arthur	2
		2009	France (Toulouse-south area)	1	NO	30–70 and 50–50	PR22R58/Livia	2
		2010	France (Toulouse-south area)	2	N	58–72	Aérobic/Enduro	2
				1	NO	30–70 and 50–50	PR22R58/Enduro	2
				2	NO	58–72	Aérobic/Enduro	2
Spring crops	SW-F	2003 and 2004	Denmark (Taastrup)	1	NO and N	100–100 and 50–50		8
	B-P	2001; 2002; 2003	Denmark (Taastrup)	1	NO	50–50	Otira/Agadir and Bohatyr	6
		2003. 2004: 2005	France (Angers-west area) and Denmark	1	NO	50–100 and 50–50	Scarlett/Baccara	12
		2009; 2010	France (Toulouse-south area)	1	NO	30–70 and 50–50	Nevada/Livia	4
	B-F	2001; 2002; 2003	Denmark (Taastrup)	1	NO	50–50	Otira/Columho	3

green manuring. However, in lower N input systems, an eventual limiting N level makes it difficult to reach a sufficient grain yield and protein concentration as required by the agro-food industries both for soft wheat to make bread and for durum wheat to make semolina and pasta. Cereal-legume intercrops might be a way to increase total grain yield per area and grain quality, in particular, protein concentration (e.g., Gooding et al. 2007; Bedoussac and Justes 2010a; Naudin et al. 2010), which are the most obvious advantages emphasized when trying to convince farmers to adopt intercropping strategies in organic farming systems.

3.2.1 Intercropping Increases Total Grain Production

Over a wide range of intercropping studies, the total grain yield of the intercrop (cereal plus legume) is on average 3.3 ± 1.0 Mg ha^{-1}, which is: (i) nearly always (in 91 % of our trials) more than the mean yield of the respective sole crops (2.7 ± 0.9 Mg ha^{-1}; Fig. 3.1a); (ii) greater (in 64 % of our trials) than the sole cropped cereal yield (2.9 ± 0.9 Mg ha^{-1}; Fig. 3.1b); and (iii) greater (in 83 % of our trials) than the sole cropped legume yield (2.4 ± 1.4 Mg ha^{-1}; Fig. 3.1c). Independent of cropping strategy, the cereal is most often more productive than the legume. Furthermore, the proportion of cereal in the intercrop is greater than that calculated on the basis of the sole crops, which indicates that the cereal is more competitive (Vandermeer et al. 1998). The relative advantage of the intercrops seems to be greater when the yield of the respective sole crops is quite low and when the quantity of soil mineral N is limited.

3.2.2 Intercropping Improves the Protein Concentration of the Cereal Grain

Our results confirm that the protein concentration of the intercropped cereal is almost always greater than that of the respective cereal sole crop (11.1 ± 1.7 % and 9.8 ± 1.7 %, respectively; Fig. 3.2a). The complementarity between the cereal and legume is observed when the cereal sole crop protein concentration is at the low end. In the case of legumes, there is no difference between the intercrop and the sole crop condition in grain protein concentration (24.8 ± 3.9 % and 24.9 ± 4.3 %, respectively; Fig. 3.2b).

Our results confirmed those obtained both in conventional agriculture and organic farming, showing a general improvement of environmental resource use when intercropping (e.g., Jensen 1996a; Bedoussac and Justes 2010a; Hauggaard-Nielsen et al. 2009b). Moreover, present results confirm that the relative advantage of the intercrops seems to be greater when the yield of at least one of the sole crops is limited in one way or another, which can quite often happen in organic farming and low-N systems (Hauggaard-Nielsen et al. 2003; Corre-Hellou et al. 2006; Bedoussac and Justes 2010b).

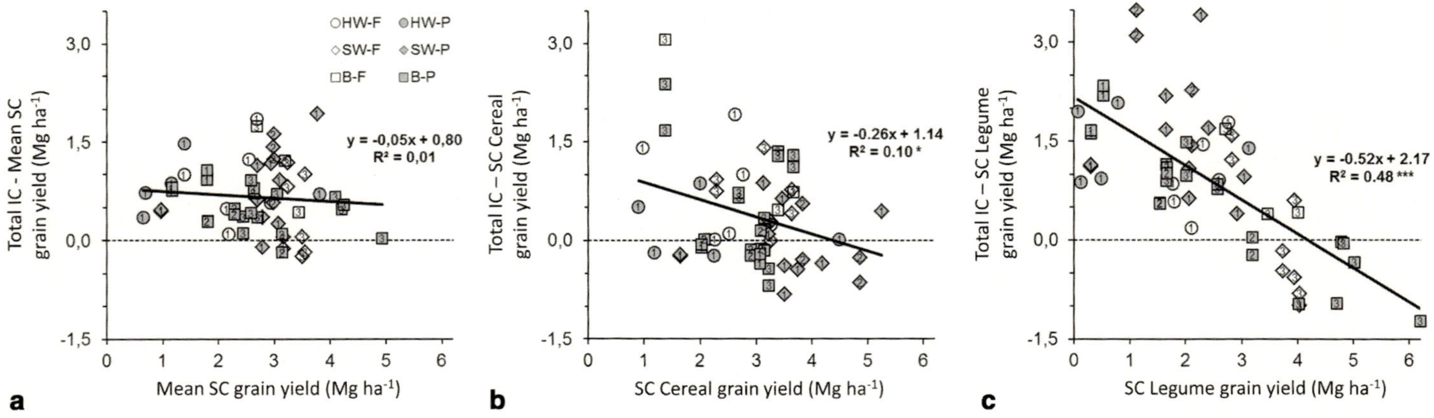

Fig. 3.1. Intercropping increases total grain production. Relationship between total grain yield of the intercrop (*IC* cereal + legume) and **a** mean sole crop (SC), **b** cereal SC and **c** legume SC. Numbers inside the symbols indicate the experimental site (*1* Southern France; *2* Western France; *3* Denmark). *HW* Durum wheat, *SW* Soft wheat, *B* Barley, *F* Faba bean, *P* Pea. Single asterisks (*) and triple asterisks (***) indicate that linear regression is significant at $P=0.05$ and $P=0.001$, respectively ($N=58$)

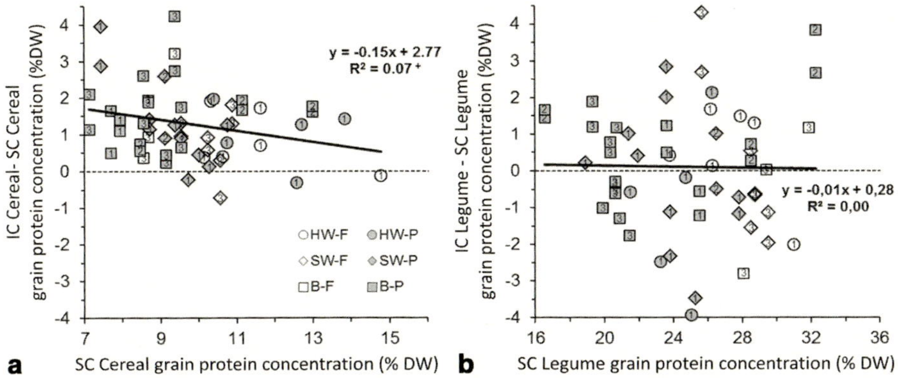

Fig. 3.2 Intercropping improves the protein concentration of the cereal grain. Relationship between grain protein concentration in intercrops and **a** the sole cropped (SC) cereal and **b** the SC legume. The grain protein concentration was calculated by multiplying the nitrogen concentration by 6.25 for the legume and the barley (animal consumption) and by 5.7 for soft and durum wheat (human consumption). Numbers inside the symbols indicate the experimental site (*1* Southern France; *2* Western France; *3* Denmark). *HW* Durum wheat, *SW* Soft Wheat, *B* Barley, *F* Faba bean, *P* Pea. A single plus (+) indicates that linear regression is significant at $P=0.10$ ($N=58$)

3.3 Complementarity and Competition Between Associated Species for Use of Resources

3.3.1 Improved Light Interception

In the absence of limiting or reducing abiotic and biotic factors such as water or nutrient availability, pests, diseases and weeds, the crop dry matter yield depends mainly on the radiation absorbed (Loomis and Williams 1963), and this applies both to the sole crop (Shibles and Weber 1966; Monteith 1977; Kiniry et al. 1989) and intercrop growing conditions (Natarajan and Willey 1980; Sivakumar and Virmani 1984; Bedoussac and Justes 2010b).

Intercrops are known to be more efficient compared to sole crops for light interception and use (Jahansooz et al. 2007) because of species complementarity in space—when crops differ in their shoot architecture—and time—when crop life cycles differ (Trenbath 1986; Tsubo et al. 2001; Tsubo and Walker 2002). These differences and interspecific complementarities allow a better dynamic occupation of the space and, hence, an increase in light interception throughout the growth of the intercrop and, finally, higher global biomass and grain yield.

3.3.2 Non-Proportional Competition for Soil Mineral N and Other Plant Growth Factors in the Intercrop Results in Higher Soil Nitrogen Availability Per Cereal Grain

To increase yield and improve grain protein concentration, it is necessary to obtain more remobilised N in the grain during the final part of the crop cycle. Using a simplified theoretical scheme representing crop yield and mineral N available for the intercrops and cereal sole crops (Fig. 3.3), it can be demonstrated that a greater quantity of N per kg of grain is available for the intercropped cereal than for the pure cereal, i.e.:

$$\left(\frac{\text{Nmin}_{\text{IC-Cereal}}}{\text{Y}_{\text{IC-Cereal}}} > \frac{\text{Nmin}_{\text{SC-Cereal}}}{\text{Y}_{\text{SC-Cereal}}}\right), \text{ only if } \frac{\text{Y}_{\text{SC-Cereal}} - \text{Y}_{\text{IC-Cereal}}}{\text{Y}_{\text{SC-Cereal}}} > \frac{\text{Ndfsoil}_{\text{IC-Legume}}}{\text{Nmin}_{\text{SC-Cereal}}}$$

where $\text{Nmin}_{\text{SC-Cereal}}$ and $\text{Nmin}_{\text{IC-Cereal}}$ are the quantity of available soil mineral N for the SC cereal and IC cereal, respectively, $\text{Ndfsoil}_{\text{IC-Legume}}$ is the mineral N absorbed by the IC legume, and $\text{Y}_{\text{SC-Cereal}}$ and $\text{Y}_{\text{IC-Cereal}}$ are the grain yield of the SC cereal and IC cereal, respectively.

For a partial data set ($N=19$[1]), we found that on average, these conditions were verified because: (i) $(\text{Y}_{\text{SC-Cereal}}-\text{Y}_{\text{IC-Cereal}})/\text{Y}_{\text{SC-Cereal}}=0.33$[2]; and (ii) soil mineral N accumulated in the shoots of the intercropped legume ($\text{Ndfsoil}_{\text{IC-Legume}}$) represented barely 17 ± 14 % (on average 21 ± 24 kg N ha^{-1})[3] of the total available soil mineral N[4] to a first approximation.

The greater efficiency generally observed in intercrops can be explained by the fact that the two intercropped species use N sources (mineral soil N and atmospheric N_2) in a complementary way (Jensen 1996a; Bedoussac and Justes 2010a; Corre-Hellou et al. 2006). Indeed, the legume is forced to rely on N_2 fixation because the cereal is more competitive for soil mineral N (Hauggaard-Nielsen et al. 2001a; Bellostas et al. 2003), which leads to a rapid decrease in the quantity of available mineral N in the surface soil layer (the zone of symbiotic fixation), causing an increase in the N_2-fixing activity of the legume compared with sole crops (Jensen 1996a; Corre-Hellou et al. 2006; Hauggaard-Nielsen et al. 2009b; Naudin et al. 2010). When combining the different experiments and growing conditions in organic farming, our results confirmed a higher percentage of N derived from

---

[1] We considered the data subset for which all the variables needed for the calculation were available.

[2] $\text{Y}_{\text{SC-Cereal}}=2.9 \pm 0.6$ Mg ha^{-1} and $\text{Y}_{\text{IC-Cereal}}=2.0 \pm 0.7$ Mg ha^{-1}on average.

[3] The nitrogen accumulated in the shoots of the intercropped legume was on average 54 ± 36 kg N ha^{-1}, of which only 21 ± 24 kg N ha^{-1} came from the soil (the percentage of plant N derived from N_2 fixation was determined using the ^{15}N natural abundance method for unfertilised treatments, according to Amarger et al. (1979), Unkovich et al. (2008) and Bedoussac and Justes (2010a).

[4] Total available nitrogen (112 ± 38 kg N ha^{-1}) was estimated as the sum of the N accumulated by the SC cereal (62 ± 21 kg N ha^{-1}) and the soil N residue at harvest of the SC cereal (50 ± 28 kg N ha^{-1}).

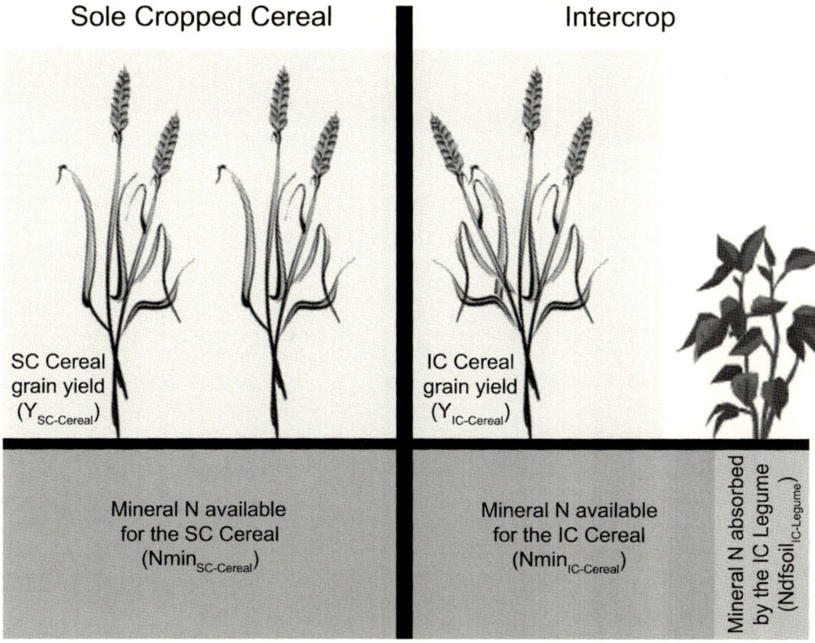

Fig. 3.3 Theoretical scheme linking grain production and the availability of mineral nitrogen for an intercropped and sole cropped cereal

air (%Ndfa) in intercropped legumes than in sole crops (on average, 75±18% and 62±16%, respectively).

A second hypothesis (which does not exclude the first one) that explains the improvement in the protein concentration of the intercropped cereal is based upon a better fit of the N availability to the cereal requirements, depending on the developmental stage and the yield level. This supports the previous explanation that the effect of intercropping is small or absent when large quantities of soil mineral N are available.

These hypotheses might only be part of the explanation because several authors have shown the effects of the legume on facilitating the absorption of soil mineral N by the cereal (Stern 1993; Xiao et al. 2004) and the transfer of N from the legume to the cereal (Jensen 1996b). However, in view of the total quantity of N available in agricultural systems, these processes of N transfer from the legume to the cereal are regarded as small, even if they can contribute up to 15% of the N absorbed by barley in intercrops with peas (Jensen 1996b).

3.3.3 Less Light and Nitrogen Available to Weeds

Intercrops can potentially reduce weeds (Vasilakoglou et al. 2005; Banik et al. 2006; Corre-Hellou et al. 2011), often regarded as key factors influencing crop production (Liebman 1988; Liebman and Davis 2000; Hauggaard-Nielsen et al. 2001b).

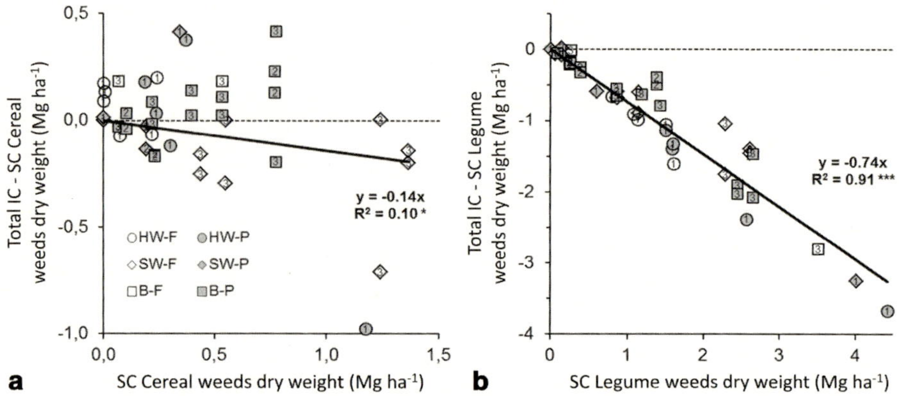

Fig. 3.4 Intercropping improves the weed control of the legume. Relationship between weed dry weight below intercrops (IC) and **a** the sole cropped (SC) cereal and **b** the SC legume. Weed dry weight below intercrops (IC) as a function of **a** Cereal sole crops (SC) and **b** Legume SC. Numbers inside the symbols indicate the experimental site (*1* Southern France; *2* Western France; *3* Denmark). *HW* Durum wheat, *SW* Soft wheat, *B* Barley, *F* Faba bean, *P* Pea. Single asterisks (*) and triple asterisks (***) indicate that linear regression is significant at $P=0.05$ and $P=0.001$, respectively ($N=43$)

In particular, intercrops can help to suppress weeds in not very competitive crops such as peas and other grain legumes, even with a low percentage of cereal in the total biomass, as observed in pea-barley intercrops (Corre-Hellou et al. 2011). In our experiments, the weed biomass within the intercrops or the cereal sole crops are comparable (0.40 Mg ha^{-1}; Fig. 3.4a) and significantly lower than within the legume sole crops (1.38 Mg ha^{-1}; Fig. 3.4b).

This weed reduction can be explained by improved resource use leaving less resources available for the weeds. Nitrogen and light are two main growth parameters involved in such weed suppression because of the intercropped species complementarity such as: (i) use of N (soil mineral N and atmospheric N_2); (ii) capture of light energy (e.g., Bedoussac and Justes 2010b); and (iii) soil cover (Anil et al. 1998).

3.4 Designing Appropriate Intercrop Management Systems

Designing crop management systems—the logical and sequentially arranged techniques applied on a farm field to achieve a given production objective (Sebillotte 1974)—is much the same for intercrops and sole crops, except that the choices have to be made for several crops instead of just one.

In multi-species mixtures (two or more), the interactions between species can be represented as the effect of one species on the environment and the response of the other (one or more) species to this change (Vandermeer 1989; Goldberg 1990). The interactions are complex, occur dynamically over time and space (Connolly et al. 1990) and depend, *inter alia*, on the availability of nutrients, soil-climatic conditions and the companion species and cultivars.

As discussed by Naudin et al. (2010), adoption of intercropping strategies might be guided by several production objectives such as: (i) *improving the quality of the cereal* by maximising the availability of soil mineral N and by increasing the symbiotic fixation rate of the legume; or (ii) *producing legumes using intercrops* by reducing weed pressure and spread of diseases and pests because of a cereal physical barrier effect, and by providing mechanical support to avoid pea lodging.

The choices of species, varieties, plant densities, patterns and N fertilisation levels are regarded as the determining factors of the functioning and performance of intercrops. Interactions between these various technical choices in relation to the production objective make generalisations rather difficult. However, two general rules can be defined: (i) improve the use of light energy; and (ii) improve the use of N sources.

With respect to light, the dominant species should have a shoot architecture and biomass production that allows a reasonable amount of light to reach the understorey (Berntsen et al. 2004; Jahansooz et al. 2007). In the case of durum wheat/winter pea intercrops (Bedoussac 2009), a short-strawed durum wheat variety would be favoured to intercrop with winter peas, and a long-strawed one for IC with faba beans. Moreover, with the objective of improving the protein concentration for the cereal, a cereal variety with good sole crop characteristics (grain protein concentration, vitreousness, bread-making quality, etc.) would be preferable, but at the same time, should have sufficient sensitivity to leguminous interspecific competition to secure complementary interactions.

In intercrops, the optimal total plant density can be greater than that of each of the sole crops because of the complementarity between species (e.g., maize/bean mixtures) (Willey and Osiru 1972). The increase in plant density increases the competition between the components of the mixture, which, as Willey (1979) noted, tends to favour the dominant species. Consequently, an increase in the density of the dominated species would be favoured (more than 50 % of that in sole cropping) and/or a reduction of that of the dominant species (less than 50 % of that in sole cropping) to manage competitive effects.

Apart from species, varieties and densities, variations in spatial structure of intercrops (such as mixtures within the row or alternate rows or strips of varying width) and row orientation will modify the distribution of radiation, water and nutrients. Such effects were reported on maize-pigeon pea mixtures (Dalal 1974), maize/soya and sorghum/soya mixtures (Mohta and De 1980) or barley/pea intercrops (Chen et al. 2003). Consequently, densities should be chosen according to the spatial arrangement of the species, their competitivity and the production objectives.

Nitrogen availability as a result of organic N fertilisation strongly affects species complementarity. Increased availability of soil mineral N in early growth stages will result in: (i) reduced amounts of fixed N; (ii) reduced legume yield; and (iii) a correspondingly increased cereal yield. Conversely, late availability of soil N will have little or no effect on the overall symbiotic fixation and yield of the legume but will improve the protein concentration of the cereal. Unlike mineral N, which is immediately available, organic manures undergo soil microbial mineralisation. Consequently, only early applications of organic N from animal manure, green manuring, etc., can have an effect on the behaviour of the intercrop and, in particular, on the proportions of the two species at harvest. Early competitive advantages are often found to form the basis for a competitive dominance throughout the growing season (Andersen et al. 2007).

Mechanical weeding using a tine harrow (an effective tool widely used in organic farming) can be very efficient provided that the operation is correctly timed. However, the optimal growth stages for its use on each of the two species can differ enough so that the time window for using the tine harrow in an intercrop is shorter. Hence, this technique must be applied with care and certainly requires more technical skill when applied to intercrops.

Evaluation of intercrops should not only be considered in terms of crop management practices but should also include the cropping system. Integration of intercrops within traditional rotations and their subsequent crop effects and minimum time of return between two intercrops, among other issues, needs to be clarified in future studies. For example, if the intercrops significantly reduce the pest and disease pressure, it may be possible to reduce the return times compared with sole crops. It is also reasonable to imagine the successive cropping of different cereal/grain legume intercrops whose possible combinations are numerous and, for the more southerly climates of Europe, to consider summer crops (e.g., sunflower/soya).

3.5 What is the Economic Benefit of Intercropping?

Crop rotation, soil fertility, commodity price and the availability of a market, etc., are some factors that influence crop preference by farmers and the adoption of intercrops. The potential economic advantage of intercrops depends on the selling prices of the crops and, in particular, on the differential between cereals and legumes, which is a difficult figure to obtain when prices are volatile. In general, we observe that the sale price of organic grain legumes is higher than that of standard quality wheat and comparable to that of high quality wheat.

From the micro-economic point of view, there is an economic advantage of intercropping in organic farming due to the increase in total grain yields in intercrops compared to the respective sole crops, especially the grain legume sole crops and, particularly, for years when one of the respective sole crops produces low yields. In some years, intercropping might lead to an intermediate net income for the farmer, but it is regarded as a better safeguard for the farmer's earnings compared to sole grain legume cropping. Indeed, grain legumes have a reputation for low yield and low yield stability in organic crop rotations, which is linked to several factors such as water stress intolerance, harvest difficulties due to lodging or late maturity, diseases (e.g., *Ascochyta* spp., *Botrytis* spp., *Erypsiphe* spp.) or because they are weak competitors for weeds.

The economic value of intercrops will increase through quality improvements such as increased wheat grain protein concentration and reduced hard wheat vitreousness, giving access to the market for direct human consumption with higher selling prices. Focusing on wheat-faba bean intercrops over five regions across Europe and three seasons, Gooding et al. (2007) showed an economic benefit of intercrops, despite a 25–30 % reduction in wheat yield. This resulted from the added value of a higher crude protein concentration of intercropped wheat, combined with the effective marketing of the legume crop.

However, intercrops can be sold for the human consumption market only if crops can be correctly sorted. For that reason, the main obstacle to the development of intercrops for the companies collecting and storing the seeds is the capacity for sorting large volumes efficiently, quickly and cheaply. On the basis of a preliminary survey of French companies that collect and store the seeds, it seems possible to correctly separate the grains of the two species, provided that they sufficiently differ in size and/or shape and that the mixture does not contain too many broken grains. To reach the latter objective, it has to be ensured that: (i) the species and varieties reach maturity at similar dates; and (ii) the combined harvester adjustments are made to suit the more fragile species (at the risk of losing some of the grain of the other species). Another option is that the companies collecting and storing the seeds adjust already available equipment to deal with seed mixtures, obviously at some cost for the farmer.

This practical question thus raises various issues in terms of the choice of machinery and its adjustment, as well as from the logistic point of view for the companies collecting and storing the seeds. Indeed, their organisational structure can play the role of a self-reinforcement mechanism that reduces the incentives to adopt new practices (Fares et al. 2012). Conversely, the adoption of intercropping to produce animal feed on farms seems less problematic as it is possible to either crudely sort the grain or else to adjust the diet by adding either one of the two species to the harvested mixture.

3.6 Conclusions and Perspectives

We have shown that intercrops present numerous advantages and appear to be a useful agronomic solution for organic arable cropping. However, it is difficult to propose scientifically proven and generic crop technical protocols because of the multitude of possible production objectives and, hence, of combinations of species, varieties, densities, structure and organic manuring strategies. Therefore, it is necessary to emphasize:

- That the identification of the species and varietal traits suited to intercropping and, more generally, to low-input systems and organic farming is therefore an important issue that will make it necessary to reconsider the varietal selection criteria. Indeed, those used for sole crops are probably not ideal for intercrops, and especially for organic farming systems, as illustrated by Carr et al. (1998), who showed that the yields of barley-peas or oats-pea forage intercrops were higher when the varieties used had been selected in multi-species stands.

- The limitations of experiments and the value of modelling multi-species cropping systems (Brisson et al. 2004; Corre-Hellou et al. 2009; Launay et al. 2009). In fact, for a given production objective, this would allow: (i) the performance and behaviour of intercrops to be evaluated under a wide range of conditions; (ii) to help with the determination of varietal characteristics suited to intercropping; (iii) to optimise the crop technical protocols according to multiple criteria; and (iv) to devise a decision-aid model. However, this requires a better mechanistic understanding of the behaviour of multi-species cropping systems and the integration of this knowledge into current crop models or the development of new models that correctly represent the inter- and intraspecific competition (Launay et al. 2009).
- That the development of intercrops cannot take place without the assent and participation of all the actors in the value chain because the low degree of integration of the supply chain can be viewed as a lock-in mechanism (Fares et al. 2012) with, in particular: (i) farmers who need technical support since the new generation of farmers may not possess the know-how; (ii) companies that collect and store the seeds that will have to adapt their collecting, sorting and storage equipment in order to satisfy the processors' quality demands; (iii) breeders who are expected to select varieties suited to intercropping; (iv) technical institutions that must acquire technical and cognitive knowledge; (v) national and European authorities who must consider relevant policies and subsidies to help reintroduce these cropping strategies; and (vi) research institutions.

References

Altieri M (1999) The ecological role of biodiversity in agroecosystems. Agric Ecosyst Environ 74:19–31

Amarger N, Mariotti A, Mariotti F, Durr J, Bourguignon C, Lagacherie B (1979) Estimate of symbiotically fixed nitrogen in field grown soybeans using variations in ^{15}N Natural abundance. Plant Soil 52:269–280

Andersen MK, Hauggaard-Nielsen H, Weiner J, Jensen ES (2007) Evaluating competitive dynamics in two and three component intercrops of Pisum sativum, Hordeum vulgare and Brassica napus. J Appl Ecol 44:545–551

Anil L, Park J, Phipps RH, Miller FA (1998) Temperate intercropping of cereals for forage: a review of the potential for growth and utilization with particular reference to the UK. Grass Forage Sci 53:301–317

Banik P, Midya A, Sarkar BK, Ghose SS (2006) Wheat and chickpea intercropping systems in an additive experiment. Advantages and weed smothering. Europ. J. Agron 24:325–332

Bedoussac L (2009). Analyse du fonctionnement des performances des associations blé dur-pois d'hiver et blé dur-féverole d'hiver pour la conception d'itinéraires techniques adaptés à différents objectifs de production en systèmes bas-intrants. Mémoire de thèse, INP-Toulouse, 220 pages + annexes

Bedoussac L, Justes E (2010a) The efficiency of a durum wheat-winter pea intercrop to improve yield and wheat grain protein concentration depends on N availability during early growth. Plant Soil 330:19–35

Bedoussac L, Justes E (2010b) Dynamic analysis of competition and complementarity for light and N use to understand the yield and the protein content of a durum wheat-winter pea intercrop. Plant Soil 330:37–54

Bellostas N, Hauggaard-Nielsen H, Andersen M, Jensen ES (2003) Early interference dynamics in intercrops of pea, barley and oilseed rape. Biol Agric Hort 21:337–348

Berntsen J, Hauggaard-Nielsen H, Olesen JE, Petersen BM, Jensen ES, Thomsen A (2004) Modelling dry matter production and resource use in intercrops of pea and barley. Field Crops Res 88:69–83

Brisson N, Bussière F, Ozier-Lafontaine H, Tournebize R, Sinoquet H (2004) Adaptation of the crop model STICS to intercropping. Theoretical basis and parameterisation. Agronomie 24:409–421

Carr PM, Martin GB, Caton JS, Poland W (1998) Forage and nitrogen yield of barley-pea and oat-pea intercrops. Agronomy J 90:79–84

Chen Y, Yu S, Yu Z (2003) Relationship between amount or distribution of PAR interception and grain output of wheat communities. Acta Agronomica Sinica 29:730–734

Connolly J, Wayne P, Murray R (1990) Time course of plant-plant interactions in experimental mixtures of annuals—Density, frequency, and nutrient effects. Oecologia 82:513–526

Corre-Hellou G, Crozat Y (2005) N2 fixation and N supply in organic pea (Pisum sativum L.) cropping systems as affected by weeds and peaweevil (Sitona lineatus L.). Eur J Agron 22:449–458

Corre-Hellou G, Fustec J, Crozat Y (2006) Interspecific competition for soil N and its interaction with N_2 fixation, leaf expansion and crop growth in pea-barley intercrops. Plant Soil 282:195–208

Corre-Hellou G, Faure M, Launay M, Brisson N, Crozat Y (2009) Adaptation of the STICS intercrop model to simulate crop growth and N accumulation in pea-barley intercrops. Field Crops Res 113:72–81

Corre-Hellou G, Dibet A, Hauggaard-Nielsen H, Crozat Y, Gooding M, Ambus P, Dahlmann C, von Fragstein P, Pristeri A, Monti M, Jensen ES (2011) Competitive ability of pea-barley intercrops against weeds and interactions with crop productivity and soil N availability. Field Crops Res 122:264–272

Crews TE, Peoples MB (2004) Legume versus fertilizer sources of nitrogen: ecological tradeoffs and human needs. Agric Ecosyst Environ 102:279–297

Dalal RC (1974) Effects of intercropping maize with pigeon peas on grain yield and nutrient uptake. Exp Agric 10:219–224

David C, Jeuffroy MH, Laurent F, Mangin M, Meynard JM (2005a) The assessment of a decision making tool for managing the nitrogen fertilization of organic winter wheat. Eur J Agron 23:225–242

David C, Jeuffroy MH, Henning J, Meynard JM (2005b) Yield variation in organic winter wheat: a diagnostic study in the Southeast of France. Agron Sust Dev 25:213–223

Fares M, Magrini MB, Triboulet P (2012) Transition agroécologique, innovation et effets de verrouillage: le rôle de la structure organisationnelle des filières. Le cas de la filière blé dur française. Cah Agric 21:34–45

Garrido-Lestache E, López-bellido RJ, López-bellido L (2004) Effect of N rate, timing and splitting and N type on bread-making quality in hard red spring wheat under rainfed Mediterranean conditions. Field Crops Res 85:213–236

Goldberg D (1990) Components of resource competition in plant communities. Perspectives on plant competition. Academic Press, San Diego, pp 27–50

Gooding MJ, Kasynova E, Ruske R, Hauggaard-Nielsen H, Jensen ES, Dahlmann C, von Fragstein P, Dibet A, Corre Hellou G, Crozat Y, Pristeri A, Romeo M, Monti M, Launay M (2007) Intercropping with pulses to concentrate nitrogen and sulphur in wheat. J Agric Sci 145:469–479

Hauggaard-Nielsen H, Ambus P, Jensen ES (2001a) Temporal and spatial distribution of roots and competition for nitrogen in pea-barley intercrops—a field study employing P-32 technique. Plant Soil 236:63–74

Hauggaard-Nielsen H, Ambus P, Jensen ES (2001b) Interspecific competition, N use and interference with weeds in pea-barley intercropping. Field Crops Res 70:101–109

Hauggaard-Nielsen H, Ambus P, Jensen ES (2003) The comparison of nitrogen use and leaching in sole cropped versus intercropped pea and barley. Nutr Cycl Agroecosyst 65:289–300

Hauggaard-Nielsen H, Jørnsgaard B, Kinane J, Jensen ES (2007) Grain legume-cereal intercropping: the practical application of diversity, competition and facilitation in arable and organic cropping systems. Ren Agric Food Syst 23:3–12

Hauggaard-Nielsen H, Gooding M, Ambus P, Corre-Hellou G, Crozat Y, Dahlmann C, Dibet A, von Fragstein P, Pristeri A, Monti M, Jensen ES (2009a) Pea-barley intercropping and short-term subsequent crop effects across European organic cropping conditions. Nutr Cycl Agroecosyst 85:141–155

Hauggaard-Nielsen H, Gooding M, Ambus P, Corre-Hellou G, Crozat Y, Dahlmann C, Dibet A, von Fragstein P, Pristeri A, Monti M, Jensen ES (2009b) Pea-barley intercropping for efficient symbiotic N2-fixation, soil N acquisition and use of other nutrients in European organic cropping systems. Field Crops Res 113:64–71

Hauggaard-Nielsen H, Jensen ES (2005) Facilitative root interactions in intercrops. Plant Soil 274:237–250

IAASTD (2009) International Assessment of Agricultural knowledge, Science and Technology for Development. Executive summary of the synthesis report. Intergovernmental plenary, 7–11 April 2008, Johannesburg, South Africa

Jahansooz MR, Yunusa I, A M, Coventry DR, Palmer AR, Eamus D (2007) Radiation- and water-use associated with growth and yields of wheat and chickpea in sole and mixed crops. Eur J Agron 26:275–282

Jensen ES (1996a) Grain yield, symbiotic N2 fixation and interspecific competition for inorganic N in pea-barley intercrops. Plant Soil 182:25–38

Jensen ES (1996b) Barley uptake of N deposited in the rhizosphere of associated field pea. Soil Biol Biochem 28:159–168

Jensen ES, Ambus P, Bellostas N, Boisen S, Brisson N, Corre-Hellou G, Crozat Y, Dahlmann C, Dibet A, von Fragstein P, Gooding M, Hauggaard-Nielsen H, Kasyanova E, Launay M, Monti M, Pristeri A (2006). Intercropping of cereals and grain legumes for increased production, weed control, improved product quality and prevention of N-losses in European organic farming systems. European Joint Organic Congress, 30–31 May, Odense, Denmark, 180–181

Kiniry J, Jones C, O'Toole J, Blanchet R, Cabelguenne M, Spanel D (1989) Radiation-use efficiency in biomass accumulation prior to grain-filling for five grain-crop species. Field Crops Res 20:51–64

Knudsen T, Hauggaard-Nielsen H, Jørnsgard B, Jensen ES (2004) Comparison of interspecific competition and N use in pea-barley, faba bean-barley and lupin-barley intercrops grown at two temperate locations. J Agric Sci 142:617–627

Launay M, Brisson N, Satger S, Hauggaard-Nielsen H, Corre-Hellou G, Kasynova E, Ruske R, Jensen ES, Gooding MJ (2009) Exploring options for managing strategies for pea-barley intercropping using a modeling approach. Eur J Agron 31:85–98

Liebman M (1988) Ecological suppression of weeds in intercropping systems: a review. In: Altieri M, Liebman M (eds) Weed management in agroecosystems: ecological approaches. CRC Press, pp. 197–212

Liebman M, Davis A (2000) Integration of soil, crop and weed management in low-external-input farming systems. Weed Res 40:27–47

Lithourgidis AS, Vasilakoglou IB, Dhima KV, Dordas CA, Yiakoulaki MD (2006) Forage yield and quality of common vetch mixtures with oat and triticale in two seeding ratios. Field Crops Res 99:106–113

Loomis RS, Williams WA (1963) Maximum crop productivity: an estimate. Crop Sci 3:67–72

Malézieux E, Crozat Y, Dupraz C, Laurans M, Makowski D, Ozier-Lafontaine H, Rapidel B, de Tourdonnet S, Valantin-Morison M (2008) Mixing plant species in cropping systems: concepts, tools and models. A review. Agron Sustain Dev 29:43–62

Mcdonald GK (2003) Competitiveness against grass weeds in field pea genotypes. Weed Res 43:48–58

Mohta NK, De R (1980) Intercropping maize and sorghum with soya beans. J Agric Sci 95:117–122

Monteith J (1977) Climate and the efficiency of crop production in Britain. Phil Trans R Soc Lond B 281:277–294

Natarajan M, Willey R (1980) Sorghum-pigeonpea intercropping and the effects of plant population density. 2. Resource use. J Agric Sci 95:59–65
Naudin C, Aveline A, Corre-Hellou G, Dibet A, Jeuffroy M-H, Crozat Y (2009) Agronomic analysis of the performance of spring and winter cereal-legume intercrops in organic farming. J Agric Sci Technol 3:17–28
Naudin C, Corre-Hellou G, Pineau S, Crozat Y, Jeuffroy MH (2010) The effect of various dynamics of N availability on winter pea-wheat intercrops: crop growth, N partitioning and symbiotic N_2 fixation. Field Crops Res 119:2–11
Niggli U, Fliessbach A, Hepperly P (2008) Low greenhouse gas agriculture: mitigation and adaptation potential of sustainable farming systems. FAO, Rome
Sebillotte M (1974) Agronomie et agriculture. Essai d'analyse des tâches de l'agronome. Cahiers ORSTOM série Biol 24:3–25
Shibles RM, Weber CR (1966) Interception of solar radiation and dry matter production by various soybean planting patterns. Crop Sci 6:55–59
Sivakumar MV, Virmani S (1984) Crop productivity in relation to interception of photosynthetically active radiation. Agric Forest Meteorol 31:131–141
Stern W (1993) Nitrogen fixation and transfer in intercrop systems. Field Crops Res 34:335–356
Townley-Smith L, Wright AT (1994) Field pea cultivar and weed response to crop seed rate in western Canada. Can J Plant Sci 76:907–914
Trenbath BR (1986) Resource use by intercrops. In: Francis CA (ed) Multiple cropping systems. MacMillan, New York, pp 57–81
Trenbath BR (1993) Intercropping for the management of pests and diseases. Field Crops Res 34:381–405
Tsubo M, Walker S (2002) A model of radiation interception and use by a maize-bean intercrop canopy. Agric Forest Meteorol 110:203–215
Tsubo M, Walker S, Mukhala E (2001) Comparisons of radiation use efficiency of mono-/intercropping systems with different row orientations. Field Crops Res 71:17–29
Unkovich M, Herridge D, Peoples M, Cadisch G, Boddey R, Giller K, Alves B, Chalk P (2008) Measuring plant-associated nitrogen fixation in agricultural systems. Clarus design, Canberra
Vandermeer J (1989) The ecology of intercropping. Cambridge University Press, Cambridge
Vandermeer J, van Noordwijk M, Anderson J, Ong C, Perfecto I (1998) Global change and multispecies agroecosystems: concepts and issues. Agric Ecosyst Environ 67:1–22
Vasilakoglou IB, Lithourgidis AS, Dhima KV (2005) Assessing common vetch-cereal intercrops for suppression of wild oat, Proceedings of 13th International Symposium, Session S5 European Weed Research Society, Bari, Italy
Wall D, Friesen GH, Bhati TK (1991) Wild mustard interference in traditional and semi-leafless field peas. Can J Plant Sci 71:473–480
Willey R (1979) Intercropping—its importance and research needs. 1. Competition and yield advantages. Field Crop Abstr 32:1–10
Willey R, Osiru D (1972) Studies on mixtures of maize and beans (Phaseolus vulgaris) with particular reference to plant population. J Agric Sci 79:517–529
Xiao Y, Li L, Zhang F (2004) Effect of root contact on interspecific competition and N transfer between wheat and fababean using direct and indirect 15N techniques. Plant Soil 262:45–54

Chapter 4
Regulatory Framework for Plant Protection in Organic Farming

Bernhard Speiser, Lucius Tamm and Franco P. Weibel

Abstract Plant protection in organic farming has to simultaneously comply with two sets of regulations: regulations on organic production and pesticide legislation. This chapter describes the organic approach to plant protection, including the role of systems management versus direct interventions, the range of authorised substances and the procedures for authorising new substances and the withdrawal of old substances.

External factors not related to organic farming also influence the availability of plant protection products. Scientific, regulatory and economic aspects may limit the registration of substances in a given country. On the other hand, there is an alternative route for the registration of fertilisers and plant strengtheners in some countries. As a result, the range of plant protection products available to organic farmers varies from one country to another. The history of the authorisation of sodium bicarbonate, spinosad, copper fungicides, clay minerals and granulosis viruses illustrates how the two sets of regulations can interact in very different ways, creating different patterns of availability.

The practice of plant protection is illustrated for the prevention and control of apple scab, fire blight and codling moth in organic apple orchards. At the end of the chapter, research perspectives for a 'self-regulating' apple orchard where plant protection fully relies on systems management are presented. The level of environmental friendliness already achieved by organic plant protection is discussed, and approaches with the potential for improvement are identified.

Keywords Apple orchard · Authorisation criteria · Organic farming regulation · Pesticide registration · Plant protection

4.1 Introduction

Organic farming has developed a set of comprehensive rules that define the crop protection measures allowed and, in particular, the substances that are authorised for use in organic farming. These rules are drawn up into private standards and laws.

B. Speiser (✉) · L. Tamm · F. P. Weibel
Research Institute of Organic Agriculture (FiBL), Ackerstrasse 113,
5070 Frick, Switzerland
e-mail: bernhard.speiser@fibl.org

S. Bellon, S. Penvern (eds.), *Organic Farming, Prototype for Sustainable Agricultures*,
DOI 10.1007/978-94-007-7927-3_4,

Nevertheless, in practice, organic plant protection is not entirely under the control of these organic regulations and standards because other regulations, particularly those related to pesticides, are also applicable. In addition, economic considerations influence the availability or non-availability of plant protection products in a given country. Organic plant protection is thus the result of an interaction between the *organic views* of plant protection and *'external' factors* (i.e., factors not related to organic farming).

A central concern of organic farming is that all management practices should be as environmentally-friendly as possible. At the same time, they must be economically feasible. Carrying out plant protection according to this dictum is a challenge. For economic reasons, crops should be effectively protected, which requires that the measures taken have some effect on the target organisms. For environmental reasons, negative side effects on the environment and on non-target organisms should be avoided. Several plant protection methods, mainly biocontrol agents, can fulfill both requirements reasonably well. However, such an ideal solution is not available for all pests and diseases. In many cases, plant protection in organic farming is a compromise between what is environmentally desirable and what is economically necessary.

The major regulations and processes are presented in this chapter, and the outcomes of their interaction are briefly described for a few selected substances. The resulting plant protection practices are discussed in relation to apple orchards, together with promising research perspectives. Finally, we discuss the situation regarding environmentally-friendly organic plant protection today, and what can be done to make improvements in the future.

4.1.1 Principles of Organic Plant Protection

Plant health should be primarily maintained with preventive measures such as the choice of adapted species and varieties, crop rotation, cultivation techniques, thermal processes and the protection and/or release of natural enemies. Direct plant protection methods should only be considered if these methods are insufficient. However, only a very limited range of substances are authorised for use. Thus, the difference between organic and conventional plant protection is not primarily *which* pesticides are used ('pesticide substitution'), but also *whether* pesticides are used. A more detailed description of the hierarchy of plant protection measures is given in Speiser et al. (2006) and Deguine and Penvern (2014, Chap. 6 of this book).

4.1.2 The Limits of Regulation

Inspection and certification ensure that organic farmers rigorously follow the standards and regulations for organic production, including plant protection. The *choice* of pesticides (i.e., only authorised products) is easy to verify. In contrast, it is difficult to verify whether the application was truly *necessary*, and whether it was done at the *optimal moment* with the best *equipment* and with the necessary *care*. Similarly, preventive plant protection with the *systems approach* is a process of individual optimisation. Such practices are at the limits of regulation. Regulation can postulate that they must be carried out, but *how well* they are carried out cannot be controlled and must be left to the farmer's initiative. Because this chapter describes the *regulation* of organic plant protection, its focus is on the range of pesticides authorised. Other aspects are equally important from an environmental point of view, but are not within the scope of this chapter.

4.1.3 The General Regulatory Framework

Plant protection, organic or not, is subject to the general regulation concerning pesticide registration and use. Thus, plant protection in organic farming has to comply with two sets of regulations at the same time: (i) regulations on organic production methods, and (ii) pesticide legislation. Depending on the situation, each of these two sets of regulations can be limiting for the availability of plant protection products (see below).

4.2 How Organic Farming Regulates Plant Protection

The discussion on regulation is mainly based on the European situation. Because other regulatory schemes are similar, they are only briefly mentioned at the end of this chapter.

In the EU, a legal definition of organic farming practices was first given in 1991 (EC 1991). This regulation has subsequently been amended many times. After a thorough revision, a new set of regulations has been in force since January 2009. The new 'organic regulation' consists of a 'framework regulation' (EC 2007), complemented by 'implementation rules'.

4.2.1 Systems Management and Direct Interventions

The framework regulation states that plant protection in organic farming shall primarily rely on the protection by natural enemies, the choice of species and varieties, crop rotation, cultivation techniques and thermal processes, while plant protection products may only be used in the case of an established threat to a crop (EC 2007, Art. 12). The same hierarchy of plant protection measures is established in all other standards discussed in this chapter.

4.2.2 Currently Authorised Substances for Organic Plant Protection

Only a small proportion of all the pesticides authorised for general agriculture are permitted for use in organic farming (listed in Annex II of the implementation rule 889/2008 (EC 2008b). These include substances of plant or animal origin, microorganisms, one substance produced by microorganisms (spinosad), substances to be used in traps and/or dispensers or to be surface-spread between cultivated plants, and a few other substances traditionally used in organic farming. For some substances, the authorisation is limited to certain conditional requirements or conditions for use. Borderline cases classified as fertilisers or as plant strengtheners are discussed below.

4.2.3 How the Range of Authorised Substances Can Be Changed

The range of authorised substances reflects the state of the art in organic plant protection at any given time. It is crucial for organic farming that the range of authorised substances be adapted to new developments in science, regulation, consumer needs or public opinion.

Procedures for Authorisation of Substances

The same procedures apply to allow the use of new substances for organic farming in the EU, to change their conditions of use, or to withdraw them from the list of authorised substances. Requests for 'organic authorisation' can only be made by the EU member states, but not by manufacturers. In the past, requests for authorisation of new active substances were usually made by the member states where the manufacturer was located, and/or by member states where organic farmers experienced a great need for the substance.

In the evaluation of these requests, the Commission is assisted by the 'expert group for technical advice on organic production' (EGTOP) (EC 2009a, 2010). EGTOP provides reports[1] with a discussion of all relevant aspects of a request. EGTOP reports form the basis for discussions between the Commission and the member state delegates (Standing Committee on Organic Farming). If a change in current practices is required, the implementation rules have to be changed with a Commission regulation.

EGTOP was only recently constituted. At the time of this writing, it had produced one report on pesticides, dealing with requests for the inclusion of laminarin, kaolin (aluminium silicate), sheep fat, sodium hypochlorite (for seed disinfection) and with the use of UV light. Changes in the implementation rules are under way.

[1] All EGTOP reports are published on the website: http://ec.europa.eu/agriculture/organic/eu-policy/expert-recommendations/expert-group_en.

Evaluation Criteria

For the authorisation of new substances, the framework regulation (EC 2007, Art. 16) specifies the following criteria:

- The authorisation is subject to the *objectives and principles of organic farming* (EC 2007, Art. 3–5). These include the health of soil, water, plants and animals, high levels of biodiversity, responsible use of energy and natural resources, animal welfare and exclusion of GMOs.
- The products are *necessary* for sustained production and *essential* for their intended use (i.e., biological, physical or breeding alternatives or other effective management practices are not available).
- The substances shall be of *plant, animal, microbial or mineral origin* (exceptions are possible under specific circumstances if the conditions for use of the product preclude any direct contact with the edible parts of the crop).

There is no explicit requirement concerning the *efficacy* of products, but there is a requirement for necessity. A product can be considered necessary (i) if it controls pests or diseases at least partially; (ii) if there are no authorised alternative methods or products, or at least none which are practical and economic; and (iii) if the pest or disease it controls is important for organic farming. There is also no explicit requirement concerning the *sustainability* of products, but the evaluation criteria cover environmental, economic and social aspects, along with other aspects such as principles of organic farming. The above criteria are evaluated *as a whole*. For example, limited negative side effects may be tolerated if a product is highly necessary (e.g., side effects of spinosad on certain non-target organisms).

Generically Authorised Substances—A Special Case

Three groups of substances are listed in the implementation rules in a generic way: (i) 'plant oils (e.g., mint oil, pine oil, caraway oil)'; (ii) 'microorganisms (bacteria, viruses and fungi)'; and (iii) 'pheromones'. Thus, all plant oils that are registered as pesticides are automatically authorised for organic farming, and the same is true for microorganisms and for pheromones (including chemically-synthesized pheromones). In recent years, a considerable number of microorganisms and pheromones have been developed and registered as pesticides. Because of the generic listing, no separate authorisation for organic farming was necessary, so that they could be immediately adopted in organic farming systems.

Beneficial Arthropods—Another Special Case

The framework regulation requires that the prevention of damage shall rely primarily on the protection by natural enemies (EC 2007, Art. 12), but does not explicitly mention mass-bred, commercially available beneficial insects and mites. Because beneficial arthropods are not subject to EU pesticide legislation, there is no requirement for them to be listed in the implementation rules. They are thus implicitly authorised as a generic category. In some EU member states, beneficial arthropods are subject to national pesticide legislation and may require the same registration processes as pesticides. Nevertheless, this does not affect their status as a generically authorised category.

4.2.4 Major Standards for Organic Farming Outside the EU

In the United States, organic farming is regulated by the 'National Organic Program' (hereafter referred to as NOP). Contrary to the EU system, there is no closed list of allowed substances in the NOP; instead, the legislation states that 'synthetic substances' are generally prohibited, while 'non-synthetic substances' are generally allowed (with a few exceptions in both cases). Despite this formal difference, the range of authorised substances is fairly similar in the USA and in Europe (for a brief overview, see Baker 2004). A major difference is that the NOP allows the use of several substances such as herbicides, while no substance with such a use is currently authorised in the EU.

The Codex Alimentarius is a joint food standards programme of FAO/WHO (United Nations Food and Agriculture Organisation and World Health Organisation). Its 'guidelines for the production, processing, marketing and labelling of organically produced foods' (hereafter referred to as 'Codex guidelines') were first published in 1999 (Codex Alimentarius Commission 1999), and have since been revised several times. The Codex guidelines represent a broad international consensus about the nature of organic production. They are not legally binding but have a strong influence on national and international regulations. For small, developing countries, it is particularly useful to follow the Codex guidelines because it facilitates international recognition of the organic status of the products. Plant protection is regulated in a similar way as in Europe, but due to the global nature of this standard, a number of substances are authorised that are not in use in Europe.

The International Federation of Organic Agriculture Movements (IFOAM) is the worldwide umbrella organisation of the organic sector. The 'IFOAM Basic Standards for Production' were first published in 1980. The most recent edition is part of the 'IFOAM norms for organic production and processing' (IFOAM 2006). The IFOAM basic standards are a private initiative and have no legal standing, but their political and practical impact has been huge (Blake 2004). Because they are the oldest standards and are well-rooted within the organic sector, they have directly or indirectly served as blueprints in the development of all other standards and regulations worldwide. The IFOAM basic standards are not directly implemented at the farm level, but indirectly through regional standards of IFOAM-accredited organisations. The IFOAM basic standards allow a range of substances similar to that of the EU regulation.

In conclusion, it can be noted that these standards have achieved a remarkable degree of harmonisation in the area of plant protection.

4.3 Influence of External Factors

The organic regulation clearly specifies which substances can be used in organic farming, but organic farming is also subject to other, non-organic legislation (also referred to as 'transversal legislation' in the European organic sector) that has an impact on the availability of plant protection products. As a result, farmers in different European countries have access to a very unequal line-up of pesticides (Speiser and Schmid 2004). The major factors responsible for this are described below.

4.3.1 Influence of Pesticide Registration on Product Availability

Plant protection was originally under the authority of the European member states and has been gradually harmonised and centralised. At present, active substances must be approved at the EU level, while plant protection products containing approved substances are registered at the member state level. Plant protection in Europe is governed by Council Regulation No. 1107/2009 (EC 2009d), with details specified by Commission Regulation No. 540/2011 (EC 2011).

Scientific and Regulatory Aspects

A large proportion of the products authorised for organic plant protection are plant extracts, pheromones or microorganisms. Each of these groups has specific characteristics that greatly differ from most synthetic pesticides. These products therefore are not consistent with the requirements of pesticide legislation, which often causes misunderstandings and delays during the registration process. A few years ago, the EU-funded 'REBECA' project analysed the situation in Europe and developed proposals for improvement (see below).

Plant extracts are usually complex mixtures of substances. It is thus difficult to completely describe their chemical composition, which may vary over time, regions or races. Furthermore, several substances in a plant extract may contribute to the pesticide effect. For example, three esters of chrysanthemic acid and three esters of pyrethric acid contribute to the insecticidal activity of pyrethrum (Isman 2006). The REBECA project proposed to facilitate registration by (i) drawing up a comprehensive guidance document for 'botanicals' (i.e., pesticides based on plant extracts); (ii) using representative lead substances (markers) if an active substance cannot be identified; (iii) requiring identification and analytical methods only for constituents of toxicological concern, instead of all constituents; (iv) adequately taking the history of safe use into account (e.g., use of lecithin in the food industry) (Tamm et al. 2011b).

Pheromones are used to control populations of insect pests through mating disruption, 'attract and kill', mass trapping or monitoring. Their mode of action and their application are completely different from those of insecticides. Pheromones are often used at doses similar to those naturally emitted by the pest insects. It was recognised long ago that pheromones require adapted registration requirements (OECD 2001). The REBECA project proposed: (i) simplified procedures for 'straight-chained lepidopteran pheromones' (SCLPs), a group of well-known pheromones with a long history of safe use; and (ii) flexible trial protocols for efficacy tests (Speiser et al. 2011). For example, mating disruption is only effective when used over relatively large areas. It is therefore impossible to install a trial with the usual number of replicates. Adapted data requirements were recently published (EPPO 2008; Chemicals Regulation Directorate 2010).

Among the microorganisms, baculoviruses have been used for plant protection for more than two decades (Andermatt 2008), without any indications of negative side effects. The OECD already proposed facilitated registration requirements in 2002 (OECD 2002). These proposals were integrated into those of the REBECA project (Hauschild 2011) and into a recent guidance document of the European Commission (SANCO 2008).

Bacterial and fungal biocontrol agents are a highly diverse group, comprising safe biocontrol agents as well as potential human or animal pathogens. The main efforts during registration are thus directed to exclude potential risks for human health. The REBECA project proposed adapted methods for assessing infectivity, toxicity and sensitisation (Strauch et al. 2011).

In general, plant extracts, pheromones and microorganisms are less effective than their synthetic counterparts. Registration authorities are often reluctant to register such products because their effectiveness is lower than that of synthetic pesticides and thus insufficient according to the standards of conventional farming. For organic farmers, however, such a comparison is inadequate, and these products are useful in the context of organic plant protection strategies. It is crucial for organic farming that registration also be possible for products with lower effectiveness than synthetic pesticides. The Swiss system might serve as a model: in such cases, products can be registered, but there is a mention that they have only 'partial effectiveness'.

Economic Aspects

Organic farmers are a minority in all countries: on average, 4.7 % of the European farmland is organically managed (Willer 2011). Thus, organic plant protection is a relatively small potential market. This creates only limited economic incentives to undergo the financial risks of the pesticide registration process (costs for dossier preparation and registration fees). In addition, pesticide registration may take several years, leading to a significant delay between investments in registration costs and potential payback from product sales. Finally, uncertainties as to whether registration will be granted at all create a significant financial risk. If registration costs and financial risks outweigh the potential gains, manufacturers will not register substances. Such economic mechanisms have greatly limited the availability of plant protection products for organic farmers in many European countries (Ehlers 2011).

Recent Developments in EU Pesticide Legislation

The new EU pesticide regulation No. 1107/2009 (EC 2009d) contains special clauses concerning registration of 'basic substances' (substances primarily used for purposes other than plant protection, e.g., lecithin) and for 'low-risk substances' (details yet to be defined). In addition, it governs mutual recognition of registrations within the same climatic zone, which will result in more homogeneous registrations across Europe. It can be expected that these measures should improve the access of organic farmers to products that are authorised by the organic regulation.

The sustainable use directive No. 2009/128/EC (EC 2009c) aims to achieve a sustainable use of pesticides. Among other measures, non-chemical alternatives to pesticides are promoted. These principles have to be implemented by the EU member states at the national level. For example, the French action plan 'Ecophyto 2018' (Ministère de l'Agriculture et de la Pêche 2008) aims to reduce the use of chemical pesticides by 50 % from 2008 to 2018, while promoting the use of alternative solutions. To achieve this ambitious goal, combined efforts in research, education and legislation have been considered. For example, it is noteworthy that 'biopesticides' are subject to lower taxes than synthetic pesticides.

At the time of this writing, practical experience with these new clauses is still very limited (for the new pesticide regulation and for the sustainable use directive and the corresponding national action plans). It is thus unclear whether these measures are sufficient to overcome the economic hurdles to registration.

4.3.2 Availability of Products for Plant Health

Plant health is the result of complex interactions between plants and their environment. The use of pesticides for direct control of pests and diseases has been discussed in the previous sections of this chapter. In addition, numerous substances used for agronomic purposes have the potential to indirectly influence plant health (Tamm et al. 2011a). Thus, there is a grey zone where it is not always possible to determine whether a given substance acts as a fertiliser, a plant strengthener or a plant protection product. European countries have taken different regulatory approaches to the use of such products (Speiser and Schmid 2004).

Fertilisers

Farmers and home gardeners claim that a number of fertilisers improve crop health. For products registered as fertilisers, manufacturers are not allowed to make 'plant protection claims' (i.e., to mention control of a specific pest or disease on the product label). The implementation of this rule may vary from case to case. In France, an intervention against the marketing of 'nettle broth' (*purin d'ortie*) in 2006 upset a wide circle of home gardeners and caused the authorities to review their policies towards these kinds of products within the framework of the national action plan, 'Ecophyto 2018'. Products registered as fertilisers fall within the scope of the organic regulation, with a list of authorised substances and clear specifications. Registration of fertilisers is relatively simple and fast, and entails very limited costs. It is therefore not a limiting factor for the availability of products.

Plant Strengtheners

Only several European countries have a separate legislation for plant strengtheners. The German system, which is the oldest and best-known, is discussed below. German registrations of plant strengtheners are valid not only in Germany but also in some other countries that recognise German registrations, e.g., Austria.

The German list of plant strengtheners comprises several hundred commercial products that contain substances such as plant extracts, hydrolised proteins, stone meal, kieselgur, chitosan, etheric oils, microorganisms, homeopathic preparations, humic acids, sugars, waxes, plant oils, kaolin, potassium and sodium bicarbonate[2]. Some of these substances have a long tradition of use in organic farming. Because plant strengtheners are neither plant protection products nor fertilisers from a legal point of view, their use is not limited by the organic regulation and they are therefore generally allowed in organic farming. Registration of plant strengtheners is relatively simple and fast, and entails very limited costs, but this type of registration is possible only in a few countries.

4.3.3 A Brief History of Authorisation for Selected Substances

When the European 'Organic Regulation' was first published in 1991 (EC 1991), 19 products or groups of products were authorised for plant protection. Since then, a number of products were added to the list, while others were withdrawn and/or the conditions for their use were changed. Approximately one-half of the products that were originally authorised are still authorised today (pyrethrins, quassia, sulphur, Bordeaux mixture, soft soap, pheromones, *Bacillus thuringiensis*, granulosis viruses, plant oils, paraffin oil). The history of authorisation is discussed below for a few selected substances.

Sodium bicarbonate was originally authorised for organic farming and later withdrawn because of a lack of registered plant protection products. However, plant strengtheners based on sodium and/or potassium bicarbonate continued to be used as plant strengtheners in Germany. After potassium bicarbonate was registered as a pesticide, its authorisation for organic farming became possible, and was given in 2008 (EC 2008a). There were hardly any controversies in the organic sector regarding the authorisation of potassium bicarbonate. The major arguments in favour of

[2] The category of plant strengtheners has recently been revised, and does not contain all of the mentioned substances any more

its authorisation were the ubiquitous occurrence of potassium and bicarbonate in nature, its very low toxicity to humans and the environment, as well as the previous authorisation of sodium bicarbonate ('traditional use').

Spinosad is an insecticidal substance produced by the soil bacterium *Saccharopolyspora spinosa*. It was registered as a pesticide in Europe in 2007. Spinosad was evaluated together with potassium bicarbonate, and authorised for EU organic farming at the same time (EC 2008a; Forster et al. 2008). Unlike potassium bicarbonate, however, some private label organisations decided not to authorise spinosad at all, or to set stricter restrictions (e.g., longer pre-harvest intervals). The reasons for this are: (i) spinosad is a very effective insecticide that is also frequently used in conventional farming and is therefore sometimes considered as a 'conventional pesticide'; (ii) spinosad is toxic for some non-target arthropods (e.g., honey bees) and may present a risk for them if applied incorrectly; and (iii) spinosad is included in multi-residue pesticide screenings, and some label organisations are afraid that spinosad residues could be detected in organic foods sold with their labels.

Copper fungicides were traditionally used in organic farming. Since copper can accumulate in soils, its reduction, replacement or withdrawal has been intensively discussed in the organic sector for a long time. Because it became evident that there was still a need for copper fungicides at the time, its use was re-confirmed in 2002 (EC 2002). However, it was progressively restricted, down to a maximum 6 kg copper per hectare per year. National regulations and private standards often set lower limits (e.g., 1.5 kg in apples according to Bio Suisse standards). The registration of copper compounds as pesticides was reviewed and scrutinised before it was prolonged in 2009 (EC 2009b). Thus, organic farmers in Southern and Central Europe can use copper fungicides. In Scandinavian countries, however, copper fungicides are not registered and therefore not available to farmers. Nevertheless, the search for alternatives and the replacement of copper are still a priority in organic farming research[3].

Clay minerals are currently the most important substances for copper replacement in German-speaking countries. In Switzerland, they are registered as fungicides and explicitly authorised by the organic farming ordinance. They are neither registered as pesticides nor authorised for organic farming at the EU level. In Germany, they were registered as plant strengtheners and thus available for organic farmers until recently.

The *Cydia pomonella* granulosis virus (CpGV) was the first granulosis virus worldwide to be registered for food production (Andermatt 2008). It has been registered in Switzerland since 1987 but was not authorised in some European countries for a long time. It is frequently used in German and Swiss organic apple orchards. Around 2005, some growers in Southern Germany and Switzerland experienced reduced effectiveness of CpGV due to evolving resistance (Asser-Kaiser et al. 2007). A new strain of CpGV was developed and registered within a short time (Zingg and Kessler 2008). Since spinosad is allowed, organic growers can alternate CpGV with spinosad to prevent the build-up of resistance.

[3] The European Commission has recently funded a research project for 'innovative strategies for copper-free low input and organic farming systems'; see http://ec.europa.eu/research/bioeconomy/agriculture/projects/co-free_en.htm. Accessed on 2012/09/14.

The authorisation history of these substances illustrates how the two sets of regulations can interact in very different ways, creating different patterns of availability.

4.4 Organic Plant Protection in Practice: Case Study on Organic Apple Orchards

Organic apple orchards were chosen for this case study because of the complexity of pest and disease problems, and because of the multitude of control approaches practiced. In Central Europe, apple is attacked by a range of pathogens and pests (insects and small rodents). Overviews of the major diseases and pests of organic apples in Central Europe are given by Tamm et al. (2004), Häseli et al. (2005) and Holb et al. (in press). Because of limited space, only scab, fire blight and codling moth are discussed below. The aim of this section is not to give practical instructions to the reader, but to illustrate the combined use of preventive measures and plant protection products in organic plant protection. The last section describes research activities that aim to improve the systems approach in organic apple production.

4.4.1 Apple Scab

Apple scab is the apple disease of the greatest economic importance in humid climates. It is caused by the fungal pathogen, *V. inaequalis*, which attacks leaves and fruits (Holb et al., in press). Today's market tolerates only very little scab on apples.

Cultivation of resistant varieties is the most straightforward method of prevention. Some one hundred scab-resistant apple varieties exist at this time (Holb et al., in press), but only a few of these satisfy the quality requirements of the market. The bottleneck to grow more resistant cultivars is the market's reluctance to accept new varieties that are unknown to consumers. In Switzerland, a marketing system was established that categorises apples primarily by their organoleptic properties and less prominently by varieties (Weibel and Leder 2007; Weibel et al. 2007). This 'Flavour Group Concept' allows a relatively easy market introduction of new varieties because they are primarily marketed by their flavour group and not by their (still unknown) variety name. This made it possible to increase the proportion of scab-resistant organic apples to 30 % by the year 2008 (Silvestri et al. 2008). Where no specific efforts for market introduction of resistant varieties are made, their percentage is much lower. For example, it is estimated at 3–4 % for Italy (Gessler et al. 2006). Another method for scab prevention is the removal of fallen leaves, which reduces primary inoculum sources in the next spring (Holb et al., in press). Organic apple growers in the region of Lake Constance (Germany) use mechanical leaf harvesters for this purpose today.

A typical fungicide spraying scheme in a Swiss organic orchard with *scab-susceptible* apple varieties includes applications of copper early in the season, and of sulphur later in the season. Acidified clays and potassium bicarbonate are also used, often in tank-mix with sulphur. In an orchard with *scab-resistant* varieties, copper can be omitted and sulphur greatly reduced. The remaining applications are either for resistance management or for the control of secondary diseases, i.e., mildew and sooty blotch (Speiser et al. 2012).

4.4.2 Fire Blight

Fire blight is the most devastating disease of apples and pears. It is caused by the bacterial pathogen *E. amylovora*. Fire blight originated in America and has been spreading in Europe since the middle of the 20th century. Because *E. amylovora* is classified as a quarantine organism, the measures for its control are governed by quarantine measures as a third set of legislation. The pathogen can infect flowers, leaves or shoots. Severe infections are lethal for trees, and severe outbreaks of fire blight can threaten the existence of apple growers.

Only a very limited number of apple varieties with reduced susceptibility against fire blight are currently available, but breeding is under way. As a preventive measure, alternative hosts of fire blight (native species from the rose family) are eliminated in the vicinity of commercial orchards. Meadow trees are also often potential sources of inoculum (Landwirtschaftliches Zentrum SG 2006). In the case of moderate infection, sanitation pruning may control the disease; in the case of severe infection, trees have to be cleared (Schärer 2000).

There are a number of substances that have at least a partial effect against fire blight. In countries where they are registered as pesticides, the microbial biocontrol agents, *Bacillus subtilis* and *Aureobasidium pullulans*, may be used by organic farmers. Acidified clays may only be used in a few countries (see section on the history of regulation above). Laminarin might also be used; a request for its authorisation is pending at the time of this writing.

4.4.3 Codling Moth

The codling moth (Lepidoptera: *Cydia pomonella*) is one of the most severe pests of apples. In Swiss organic farming, pheromones for mating disruption, *Cydia pomonella* granulosis virus (CpGV) and spinosad, are authorised for its control. In general, mating disruption is used over larger areas and in cases of moderate pest pressure, while CpGV is used over small areas and in cases of high pest pressure (Speiser et al. 2006). Spinosad is not frequently used but is important to manage resistance against CpGV. A typical control strategy against the codling moth may also combine mating disruption and CpGV (Speiser et al. 2012).

Systems management aims to increase the number of beneficial insects in organic apple orchards. These measures are directed against various arthropod pests, but not specifically against the codling moth (for details, see next section).

4.4.4 Research on Improvements of the Systems Approach

The Research Institute of Organic Agriculture (FiBL) in Switzerland is currently experimenting with a 'self-regulating', pesticide-free orchard, to explore the horizons of sustainable plant protection beyond the current state of the art (Weibel et al. 2010). The orchard covers 1 ha and is managed similarly to a commercial, organic orchard. All measures of system design documented in the literature that can increase the self-regulation of pests and diseases are implemented in this orchard (Simon et al. 2014; Chap. 5 of this book). Unlike a typical, organic orchard, no pesticides authorised for organic fruit production are applied in this orchard. On one-half of the orchard, however, commercial biocontrol agents such as granulosis virus and *Bacillus thuringiensis* are applied. The aim of the self-regulating orchard is to test the feasibility of pesticide-free apple production under near-to practical conditions (Colour plate 04).

The 'self-regulating orchard' incorporates a large number of management practices that enhance self-regulation, most of which are not normally applied in commercial orchards. It combines hedges with selected plants (plants for alternative hosts, flowering shrubs), flowering plants in the tree rows, companion plants in the alleyways, pest- and disease-tolerant apple cultivars, soil quality enhanced by the addition of compost, and a number of artificial nesting sites for predators (birds, bats, parasitoids) to study the feasibility of pesticide-free apple production (Weibel et al. 2010). Although each of the measures alone is already known and, in most cases, scientifically studied, their combined effects in a near-to practical orchard system have not been studied to date. The measures are not static; they are continuously improved during the orchard's lifetime.

At the time of this writing, the self-regulating orchard has not yet reached full yield, and conclusive results are not available. Preliminary assessments suggest that most pests are sufficiently controlled in the majority of years. However, diseases of secondary importance such as sooty blotch are currently building up and causing losses. Exceptional climatic conditions may lead to pest populations that exceed the system's capacity for self-regulation. For example, cold and humid weather in the spring of 2010 slowed down the reproduction and activity of beneficials, which in turn favoured the proliferation of the rosy apple aphid (*Dysaphis plantaginea*) (Weibel, unpublished).

4.5 How to Move Towards Environmentally-Friendly Organic Plant Protection

Organic plant protection today is often a compromise between protection of the environment and economic feasibility. The case study of apple orchards shows that organic plant protection has already achieved a good level of environmental friendliness. Nevertheless, it is in the interest of organic farming to improve environmental friendliness of plant protection beyond the state already achieved, as long as the system remains economically feasible. In this last section, we briefly discuss how improvements could be made. The discussion is based on the case study of organic apple orchards, but many of the findings are also true for other crops.

In direct plant protection, improvements may be achieved by research on new plant protection products, improved application technologies or forecasting systems. In addition, improvements of regulatory processes may lead to the better availability of authorised plant protection products to organic farmers so that they can more effectively select the most appropriate and environmentally-friendly alternatives. Among the products mentioned in the case study above, acidified clays, potassium bicarbonate, *B. subtilis*, *A. pullulans* and CpGV have very few negative side effects on the environment, while copper fungicides may accumulate in the soil, and sulphur and spinosad are hazardous to some non-target arthropods (e.g., Biondi et al. 2012). The greatest improvements can thus be achieved by reducing applications of copper, sulphur and spinosad.

The use of preventive methods for plant protection has almost no negative side effects on the environment. There is general agreement that the systems approach can be useful. At the moment, however, the systems approach alone is often not sufficient to solve problems with pests or diseases.

All of the organic farming standards described in this chapter assume that the systems approach must have first priority in organic plant protection. However, the systems approach is difficult to impose by way of *regulation* and should preferably be initiated by *motivation*. Moreover, it should be acknowledged that self-regulating systems must be adapted to local agronomic conditions (e.g., soil, climate, pest and disease pressure) and to the farm's socio-economic situation (e.g., availability of labour and machinery, marketing channels, financial pressure). Last but not least, if no plant protection products are available, farmers have a greater motivation to choose systems management.

Systems such as the self-regulating orchard described above have the potential for the greatest improvements of the environmental friendliness of organic plant protection. However, the success of the systems approach depends not only on innovative research, but requires combined efforts of the *entire organic sector* as well. Researchers should develop such systems, agricultural schools and agri-environmental schemes should promote them, farmers should install and manage them, retailers should accept selling new varieties with better resistance or tolerance traits, and consumers should be prepared to pay the price for more laborious production and/or accept minor quality defaults (e.g., slight scab infection of apples).

Acknowledgments We warmly thank Christian Schader for his expert advice on sustainability, and the anonymous referees for their helpful suggestions.

References

Andermatt M (2008) 20 years for MADEX—20 years for Andermatt Biocontrol AG. AG J 2008:2–3

Asser-Kaiser S, Fritsch E, Undorf-Spahn K, Kienzle J, Eberle KE, Gund NA, Reineke A, Zebitz CPW, Heckel DG, Huber J, Jehle JA (2007) Rapid emergence of baculovirus resistance in codling moth due to dominant, sex-linked inheritance. Science 317:1916–1918

Baker B (2004) Plant protection products in organic farming in the USA. In: Speiser B, Schmid, O (eds) Current evaluation procedures for plant protection products used in organic agriculture. Proceedings of a workshop held on 25–26 September, 2003 in Frick, Switzerland. FiBL, Frick, Switzerland, pp 66–72

Biondi A, Desneux N, Siscaro G, Zappala L (2012) Using organic-certified rather than synthetic pesticides may not be safer for biological control agents: selectivity and side-effects of 14 pesticides on the predator *Orius laevigatus*. Chemosphere 87:803–812

Blake F (2004) IFOAM policies concerning inputs evaluation. In: Speiser B, Schmid O (eds) Current evaluation procedures for plant protection products used in organic agriculture. Proceedings of o workshop held on 25–26 September, 2003 in Frick, Switzerland. FiBL, Frick, Switzerland, pp 74–79

Chemicals Regulation Directorate (2010) Data requirements and trials design for mating disruption pheromone products. v2 04/11/2010

Codex Alimentarius Commission (1999) Guidelines for the production, processing, labelling and marketing of organically produced foods. GL 32-1999. Rev 4-2007. Rome

Deguine JP, Penvern S (2014) Agroecological crop protection in organic farming: relevance and limits. In: Bellon S, Penvern S (eds) Organic farming, prototype for sustainable agricultures, Editions Springer, Chap. 6, pp 107–130

EC (1991) Council regulation No 2092/91 of June 24 1991 on organic production of agricultural products and indications referring thereto on agricultural products and foodstuffs. Official Journal of the European Communities L 198:1–15

EC (2002) Commission Regulation (EC) No 473/2002 of 15 March 2002, amending Annexes I, II and VI to Council Regulation (EEC) No 2092/91 on organic production of agricultural products and indications referring thereto on agricultural products and foodstuffs, and laying down the detailed rules as regards the transmission of information on the use of copper compounds. Official Journal of the European Communities L 75:21–24

EC (2007) Council Regulation (EC) No 834/2007 of 28 June 2007 on organic production and labelling of organic products and repealing Regulation (EEC) No 2092/91. Official Journal of the European Communities L 189:1–23

EC (2008a) Commission Regulation (EC) No 404/2008 of 6 May 2008 amending Annex II to Council Regulation (EEC) No 2092/91 on organic production of agricultural products as concerns the authorization of spinosad, potassium bicarbonate and copper octanoate and the use of ethylene. Official Journal of the European Communities L 120:8–10

EC (2008b) Commission Regulation (EC) No 889/2008 of 5 September 2008, laying down detailed rules for the implementation of Council Regulation (EC) No 834/2007 on organic production and labelling of organic products with regard to organic production, labelling and control. Official Journal of the European Communities L 250:1–84

EC (2009a) Commission decision of 3 June 2009 establishing the expert group for technical advice on organic production. Official Journal of the European Union L 139:29–31

EC (2009b) Commission directive 2009/37/EC of 23 April 2009 amending council directive 91/414/EEC to include chlormequat, copper compounds, propaquizafop, quizalofop-P, teflubenzuron and zeta-cypermethrin as active substances. Official Journal of the European Communities L 104:23–32

EC (2009c) Directive 2009/128/EC of the European parliament and of the council of 21 October 2009 establishing a framework for community action to achieve the sustainable use of pesticides. Official Journal of the European Union L 309:71–86

EC (2009d) Regulation (EC) No 1107/2009 of the European parliament and of the council of 21 October 2009 concerning the placing on the market of plant protection products and repealing council directives 79/117/EEC and 91/414/EEC. Official Journal of the European Communities L 309:1–50

EC (2010) Commission decision of 28 September 2010 appointing the members of the expert group for technical advice on organic production and drawing up the pool list. Official Journal of the European Union C 262:3–4

EC (2011) Commission implementing regulation (EU) No 540/2011 of 25 May 2011 implementing regulation (EC) No 1107/2009 of the European parliament and of the council as regards the list of approved active substances. Official Journal of the European Union L 153:1–186

Ehlers RU (2011) Cost-benefit, risk and trade-off analysis of regulation. In: Ehlers RU (ed) Regulation of biological control agents. Springer, pp 139–153

EPPO (2008) EPPO Guideline 'Mating disruption pheromones' (PP 1/264). Bulletin OEPP/EPPO Bulletin 38:322–325

Forster R, Micheloni C, Regouin E, Speiser B, Viñuela E (2008) Report of the ad-hoc expert group on pesticides in organic food production. Meeting on 22–23 January 2008. European Commission, Directorate F.5. Organic farming, Brussels. http://ec.europa.eu/agriculture/organic/eu-policy/expert-recommendations_en. Accessed 12 Jan 2011

Gessler C, Patocchi A, Sansavini S, Tartarini S, Gianfranceschi L (2006) *Venturia inaequalis* resistance in apple. Cr Rev Plant Sci 25:473–503

Häseli A, Weibel FP, Daniel C, Schmid A, Tamm L, Wyss E (2005) Pflanzenschutz im Biokernobstanbau. FiBL, Frick

Hauschild R (2011) Facilitations in the regulation of plant protection products containing baculoviruses. In: Ehlers RU (ed.) Regulation of biological control agents. Springer, pp 259–266

Holb I, Heijne B, Tamm L (in press) Organic apple disease management. In: Finckh MR, van Bruggen AHC, Tamm L (eds) Plant diseases management in organic agriculture. American Phytopathological Society, St. Paul

IFOAM (2006) IFOAM norms for organic production and processing. IFOAM, 136p

Isman MB (2006) Botanical insecticides, deterrents and repellents in modern agriculture and an increasingly regulated world. Annu Rev Entomol 51:45–66

Landwirtschaftliches Zentrum SG (2006) Neupflanzung von Hochstammobstbäumen im Rahmen von Vernetzungsprojekten und Pflanzaktionen Ministère de l'Agriculture et de la Pêche (2008) The Ecophyto 2018 plan

Ministère de l'agriculture et de la pèche (2008) The écophyto 2018 plan. http://agriculture.gouv.fr/IMG/pdf/PLAN_ECOPHYTO_2018_eng.pdf. Accessed 21 Nov 2013

OECD (2001) Guidance for registration requirements for pheromones and other semiochemicals used for arthropod pest control. Environment directorate. Organisation for Economic Co-operation and Development, Paris

OECD (2002) Consensus document on information used in the assessment of environmental applications involving baculoviruses. ENV/JM/MONO (2002) 1. OECD

SANCO (2008) Guidance Document on the assessment of new isolates of baculovirus species already included in Annex I of council directive 91/414/EEC. European Commission, Health and Consumer Protection Directorate-General. SANCO/0253/2008 rev. 2

Schärer HJ (2000) Feuerbrand, ein Dauerbrenner. Agrarforschung 7:404–409

Silvestri G, Egger S, Székely T, Weibel F (2008) Schorfresistente Sorten überzeugen Konsumentinnen und Konsumenten. Schweizerische Zeitung für Obst- und Weinbau 11/08: 10–13

Simon S, Rusch A, Wyss E, Sarthou JP (2014) Organic farming as an inspirational model for developing conservation biocontrol. In: Bellon S, Penvern S (eds) Organic farming, prototype for sustainable agricultures. Springer, Chap. 5, pp 83–105

Speiser B, Schmid O (2004) Current evaluation procedures for plant protection products used in organic agriculture. Proceedings of a workshop held 25–26 September 2003 in Frick, Switzerland. FiBL, Frick

Speiser B, Wyss E, Maurer V (2006) Biological control in organic production: first choice or last option? In: Eilenberg J, Hokkanen H (eds) An ecological and societal approach to biological control. Springer, Dordrecht, pp 27–46

Speiser B, Tamm L, Mattock S (2011) Proposals for regulation of semiochemicals. In: Ehlers RU (ed) Regulation of biological control agents. Springer, pp 305–321

Speiser B, Stolze M, Oehen B, Gessler C, Weibel FP, Bravin E, Kilchenmann A, Widmer A, Charles R, Lang A, Stamm C, Triloff P, Tamm L (2012) Sustainability assessment of GM crops in a Swiss agricultural context. Agron Sus Dev 33(1):21–61

Strauch O, Strasser H, Hauschild R, Ehlers RU (2011) Proposals for bacterial and fungal biocontrol agents. In: Ehlers RU (ed) Regulation of biological control agents. Springer, pp 267–288

Tamm L, Häseli A, Fuchs JG, Weibel FP, Wyss E (2004) Organic fruit production in humid climates of Europe: bottlenecks and new approaches in disease and pest control. Acta Horticulturae 638:333–339

Tamm L, Thürig B, Fliessbach A, Goltlieb AE, Karavani S, Cohen Y (2011a) Elicitors and soil management to induce resistance against fungal plant diseases. NJAS—Wagen J Life Sc 58:131–137

Tamm L, Speiser B, Mercier T (2011b) Proposals for regulation of botanicals. In: Ehlers RU (ed) Regulation of biological control agents. Springer, pp 289–304

Weibel FP, Daniel C, Hammelehle A, Pfiffner L, Wyss E (2010) Potential and limits of pesticide free apple growing by a self-regulating orchard set-up: project presentation and first experience. Proceedings of the 14th international conference on organic fruit growing, Fördergemeinschaft Oekologischer Obstbau e.V. (FOEKO), Weinsberg, 22–24 Feb, pp 292–296

Weibel FP, Leder A (2007) Experiences with the Swiss (organic) method how to introduce new apple varieties into retail market: flavour group concept and variety team. Compact Fruit Tree 40:1–5

Weibel FP, Tamm L, Wyss E, Daniel C, Häseli A, Suter F (2007) Organic fruit production in Europe: successes in production and marketing in the last decade, perspectives and challenges for the future development. Proc. 1st Int. seminar on organic apple and pear. Acta Horticulturae 737:163–172

Willer (2011) Organic agriculture worldwide—the results of the FiBL/IFOAM survey. In: Willer H, Kilcher L (eds) The world of organic agriculture—statistics and emerging trends 2011. IFOAM/ FiBL, Bonn, pp 34–60

Zingg D, Kessler P (2008) Madex Plus und Madex I12 break virus resistance of codling moth. AG J 2008:10

Chapter 5
Conservation Biocontrol: Principles and Implementation in Organic Farming

Sylvaine Simon, Adrien Rusch, Eric Wyss and Jean-Pierre Sarthou

Abstract Conventional agricultural systems have become more intensive and pesticide-dependent over the last few decades. The contamination of the environment by pesticides, the use of mineral fertilisers, and habitat loss in many agroecosystems have led to a drastic decrease in plant and animal biodiversity. Ecosystem services provided by functional biodiversity (e.g., pollination, biological pest control) have also been negatively impacted. Conservation biocontrol aims to preserve and promote natural enemies to enhance pest control, avoid pest outbreaks and reduce pesticide reliance. However, despite a consensus on the main underlying principles, intentional practical applications are still rare. It is assumed that the diversity of habitats and resources in agroecosystems enhances the diversity and/or effectiveness of the natural enemies of pests. In this article, we argue that organic farming (OF) provides a promising framework for increasing conservation biocontrol at field and farm scales in agricultural landscapes. We looked at most of the commonly used OF practices at different spatio-temporal scales and discussed their effects on pest populations, natural enemy communities and biocontrol in agroecosystems. Several OF management practices such as crop diversification, use of organic fertilisers, diversification of resource plants at the field or landscape scales

S. Simon (✉)
INRA, UE695 UERI Unité Expérimentale Recherches Intégrées,
Domaine de Gotheron, 26320 Saint-Marcel-lès-Valence, France
e-mail: sylvaine.simon@avignon.inra.fr

A. Rusch
Institut des Sciences de la Vigne et du Vin,
33883 Villenave d'Ornon, France

INRA-Université de Bordeaux, UMR1065 Santé et Agroécologie du Vignoble,
33883 Villenave d'Ornon, France

E. Wyss
Research Institute of Organic Agriculture (FiBL), Ackerstrasse, 5070 Frick, Switzerland

GLOBE Switzerland, c/o Federal Office for the Environment, 3003 Berne, Switzerland

J.-P. Sarthou
Université de Toulouse-INP-ENSAT-INRA, UMR1248 AGIR,
31326 Castanet Tolosan, France

S. Bellon, S. Penvern (eds.), *Organic Farming, Prototype for Sustainable Agricultures*,
DOI 10.1007/978-94-007-7927-3_5,

and land-use management are examined in our review. We particularly focused on possible strategies to enhance pest control measures in two case studies (i.e., orchard and annual crops) and discussed how and at which scales such strategies should be implemented. In the end, we identified knowledge gaps and bottlenecks that, if resolved, would help to enhance conservation biocontrol and applications in OF systems that aim to maximise both bottom-up (through plants) and top-down (through natural enemies) processes.

Keywords Arthropod · Conservation biocontrol · Crop protection · Functional biodiversity · Landscape · Natural enemy · Organic farming · Pest · Plant diversity · Semi-natural habitat

Definition and/or common usage of some terms used in agroecology

Biodiversity: *"Variability among living organisms from all sources... and the ecological complexes of which they are part: this includes diversity within species, between species and of ecosystems"* (UN 1992). *"Composition, structure, and function"* of biodiversity are considered here at different scales and levels of organisation (Noss 1990). **Agro-biodiversity** is related to the diversity of cultural practices and to the use within time and space of various species managed to supply agricultural products (Le Roux et al. 2008). **Functional biodiversity** is the part of the total biodiversity that provides *ecosystem services* such as soil structure stability, plant nutrition, pollination or pest suppression, and *ecosystem disservices* such as insect pests and disease damage, or biological invasions (Zhang et al. 2007).

Biocontrol (or **biological control**): *"The use of living organisms to suppress the population density or impact of a specific pest organism, making it less abundant or less damaging than it would otherwise be"* (Eilenberg et al. 2001). **Classical biocontrol**: *"The intentional introduction of an exotic, usually co-evolved, biocontrol agent for permanent establishment and long-term pest control"* (Eilenberg et al. 2001). **Conservation biocontrol**: see below.

Community: An assemblage of populations of different species that occupy the same area at the same time and interact (Loreau 2010).

Conservation biocontrol (or **conservation biological control**): *"Conservation biological control involves manipulation of the environment to enhance the survival, fecundity, longevity, and behaviour of natural enemies to increase their effectiveness."* (Landis et al. 2000). *"Modification of the environment or existing practices to protect and enhance specific natural enemies or other organisms to reduce the effect of pests"* (Eilenberg et al. 2001). Specialist (feeding upon one or a few prey/hosts) and/or generalist (wide range of prey/hosts) natural enemies are considered for conservation biocontrol.

Diversity indices: **Richness** is the number of taxa within a community. **Diversity** (e.g., Shannon index) measures both the richness and the distribution of individuals in the different taxa of a community. **Evenness** (or equitability) is the ratio of the real to the maximum theoretical diversity in a community.

Ecosystem service: "*Ecosystem services are natural functions that can be secondarily used for human benefit. These services involve biological, chemical and geological processes and include nutrient recycling, water and gas regulation, biological control, genetic resources and pollination, as well as the scenic beauty explored in ecotourism.*" (De Marco and Coelho 2004 and references therein).

Guild: Assemblage of species or taxa that prey on the same class of environmental resources in a similar way (Root 1967).

Landscape: "*A level of organisation of ecological systems, superior to the ecosystem level, mainly characterised by its heterogeneity and a dynamics partly determined by human activities*" (Burel and Baudry 1999). **Landscape complexity** is defined as the proportion and the spatial configuration of non-crop habitats in the landscape (Concepcion et al. 2008). **The functional landscape connectivity** measures the possibilities for an individual to move from a patch of vegetation to another and is a function of both landscape composition and the dispersal abilities of a species (Tischendorf and Fahrig 2000). "**Landscape ecology** *considers the development and dynamics of spatial heterogeneity, spatial and temporal interactions and exchanges across heterogeneous landscapes, influences of spatial heterogeneity on biotic and abiotic processes, and management of spatial heterogeneity.*" (Risser et al. 1984).

Population: A set of individuals of the same taxon sharing a habitat in time and/or space.

Resilience: In ecology, the capacity of an ecosystem to recover its initial state after a perturbation. There is no adaptation of species but a restoration of the state that prevailed before the perturbation.

Semi-natural habitats (also called ecological infrastructures): Components of the landscape such as hedgerows, ditches, field edges, etc., which are favourable to ecological processes such as trophic interactions or dispersal.

Top-down process: A process mediated by natural enemies that affects lower trophic levels, mainly pests, e.g., conservation biocontrol. Conversely, a **bottom-up process** is a process mediated by the plant and affecting upper trophic levels, mainly pests.

5.1 Introduction

Conventional agriculture mainly relies on the use of chemical inputs, namely fertilisers and pesticides. The contamination of the environment by pesticides coupled with habitat loss in many agroecosystems has led to a drastic decrease in plant and animal biodiversity in cultivated landscapes (Krebs et al. 1999; Benton et al. 2003). Biodiversity is essential for mankind for both patrimonial and functional roles. In agroecosystems, biodiversity sustains ecosystem services such as plant nutrition, pollination or pest suppression, which are important services for crop production (Zhang et al. 2007). The preservation and enhancement of biodiversity to maximise the related ecosystem services are thus a challenge for sustainable agriculture. Organic Farming (OF) and Integrated Pest Management (IPM) principles are based on such approaches, even though the methods to achieve such aims are not always documented. The complex relationships between biodiversity, associated ecosystem services and agriculture have been widely developed in Le Roux et al. (2008). The present chapter focuses on pest control afforded by arthropod natural enemies in conservation biocontrol approaches ('top-down' processes), and discusses to what extent OF systems, considered to be the most favourable to biodiversity (Hole et al. 2005), can both promote and benefit from the ecosystem service of pest suppression. After general consideration of the principles, scales and underlying ecological processes at stake in conservation biocontrol, the perspectives and limits of conservation biocontrol are discussed in view of case studies of some perennial and annual crops.

5.2 Diversity as the Cornerstone of Conservation Biocontrol

5.2.1 Diversity and Biocontrol Processes

Higher *richness* (Letourneau et al. 2009) and *evenness* (Crowder et al. 2010) of natural enemies have been shown to promote pest control in most cases. Several complex processes may interfere (Lawton 1994):

1. Pest suppression resulting from the combined activities of several natural enemy species can be equal (*additivity*) or greater (*synergy*) than the summed mortality caused by each species (Letourneau et al. 2009). This outcome is based on the *niche complementary hypothesis,* which includes resource partitioning (Loreau and Hector 2001) such as predation at different life stages, at different periods or spatial locations, or foraging in ways that facilitate predation by other species.
2. Higher diversity in natural enemy communities may also increase the stability of various ecological functions, thus contributing to the *resilience* of the system (Bengtsson et al. 2003). The *insurance hypothesis* (Yachi and Loreau 1999)

postulates that ecosystem properties are more stable in rich assemblages since the higher species diversity will be able to compensate to a greater degree for any perturbations in fluctuating environments.
3. When one or a few species cause most pest suppression (Tscharntke et al. 2005), the *sampling effect hypothesis* suggests that the probability that a given arthropod assemblage includes such species increases with the natural enemy species richness (Loreau and Hector 2001).
4. However, under the same hypothesis, the probability of including inefficient or even disruptive natural enemies may also increase with species richness (Straub et al. 2008). An increase in natural enemy richness may indeed enhance negative interactions among natural enemy species, including *intraguild predation* (i.e., predation among predators sharing the same prey), *hyperparasitism* (i.e., the development of a secondary parasitoid in a primary parasitoid parasitising a host insect) and *behavioural interference.*

Although favourable to diversity *per se*, an increase in plant and associated animal biodiversity relies on complex interacting processes that may alter the expected results: (i) pest arthropods may also be favoured by the presence of host plants in the vicinity of the crop (Solomon 1981; Rusch et al. 2010); and (ii) negative interactions among natural enemies may be unfavourable to pest biocontrol (see above) even though most of these negative interactions are attenuated in structurally complex habitats (Finke and Denno 2006). The diversity of functional traits in a community rather than the species richness is to be considered to understand biocontrol processes. Diversifying resources in the crop environment to manage functional diversity and enhance conservation biocontrol is thus a challenge.

5.2.2 Provision of Supplementary and/or Complementary Resources to Favour Natural Enemies

The increase in natural enemy abundance and diversity relies on the presence and permanency of local trophic resources and adequate living conditions provided by cultural practices and the management of resources and/or habitats. Both plant- and detritus-mediated resources are the basis of food webs that comprise predatory trophic levels. Supplementary and/or complementary food (nectar, pollen, honeydew, hosts and prey) and suitable habitats for reproduction, diapause, overwintering, refuge, etc. (Landis et al. 2000), maintain and favour a pool of beneficial species, whereas pests, which are generally specialised on one or a few crops, rarely benefit from such resources. Favouring natural enemies then contributes to the pest suppression function, especially when host and prey density is low in the field (Denys and Tscharntke 2002). For instance, floral and extra-floral nectars have been found to increase longevity, fecundity, foraging abilities and parasitism rates of many parasitoid species and to limit pest populations (Wäckers et al. 2007). In agroecosystems, the most common types of plant resource diversification that favour natural enemies are hedgerows, ground cover plants, wild or sown flower strips, other companion plants, etc. (Gurr et al. 2003). The increase in natural enemies permitted by organic matter supply is less well documented.

From the Field to the Landscape Scale

The functional scale of these underlying processes may be far beyond the cultivated field, depending on the biology and life traits of natural enemy species (Tscharntke et al. 2005). A landscape-based perspective is needed to understand population dynamics and biocontrol mechanisms. Indeed, both insect pest and natural enemy species usually exploit several resources from cultivated and more or less distant uncultivated habitats during their life cycle to feed, reproduce, lay eggs, diapause, overwinter, etc. (Keller and Häni 2000). *Semi-natural habitats* are more stable and less disturbed than crop fields to provide natural enemies with key resources. They are particularly important to enable the overwintering of natural enemy populations likely to disperse towards the crop in the following spring (Landis et al. 2000; Rand et al. 2006). Moreover, semi-natural habitats are generally assumed to harbour larger proportions of neutral and beneficial arthropods than detrimental ones (Denys and Tscharntke 2002).

The movements of natural enemies across the crop/semi-natural habitat interface produce *spillover effects* (Rand and Tscharntke 2007) that affect population dynamics (Tscharntke et al. 2005; Rand et al. 2006). Semi-natural habitats can be seen as the starting point of field colonisation by natural enemies that then spill over into adjacent crops, leading to a greater aggregation of natural enemies and a more efficient biocontrol at field edges (Thies and Tscharntke 1999; Tylianakis et al. 2004; Tscharntke et al. 2007). However, reciprocally, crop fields can also provide key resources to natural enemies, mainly hosts and prey for their offspring. The direction and magnitude of spillover effects are thus constrained by differences in primary productivity, temporal variation and complementarity in resources between semi-natural habitats and adjacent crops (Rand et al. 2006).

Landscape ecology brings new insights to understand processes affecting population dynamics at the landscape scale (Dunning et al. 1992) and trophic interactions in agroecosystems. Studies on *metapopulation dynamics* explain the regional persistence of a population by a stochastic balance between the extinction of local populations and the colonisation of previously empty habitat patches (Hanski 1999). Habitat fragmentation constrains species abundance and richness (Rand and Tscharntke 2007), and *landscape connectivity*, in conjunction with the dispersal abilities of species, is therefore a key element to understand population dynamics (Hanski 1999; Tischendorf and Fahrig 2000). Agricultural landscapes with higher levels of heterogeneity generally support more diverse and abundant natural enemy communities (Tscharntke et al. 2005; Chaplin-Kramer et al. 2011), exhibit more efficient natural pest control (Chaplin-Kramer et al. 2011) and a better ability to recover after disturbances (resilience) than simple landscapes (Tscharntke et al. 2005, 2007).

It is thus necessary to base conservation biocontrol on the life traits of both pests and natural enemies (such as dispersal abilities and specialisation), and on intra- and inter-species interactions at different scales to privilege: (i) the effectiveness of each natural enemy species (e.g., avoiding intra-guild predation); (ii) additive and synergistic interactions between species; and (iii) interface effects between complementary habitats.

5.3 Applications of Conservation Biocontrol

5.3.1 Conservation Biocontrol at Field and Farm Scales

Farmers manage agricultural and semi-natural habitats at field and farm scales. The field is the place where agro-biodiversity within space (e.g., crop species, cultivars and within-field plant assemblages) and time (e.g., crop rotation), and field margins are managed. Moreover, *cultural practices* such as compost and mulch applications have significant benefits for the management of insect pests (Brown and Tworkoski 2004). *Intercropping* (including *agroforestry*) and *trap cropping* are approaches that use plant diversity to limit pest damage through olfactory and/or visual stimuli (Altieri and Nicholls 2004). The focus of *crop rotation* is mainly a weed- and disease-oriented management, but crop rotation also contributes to insect pest control, either through bottom-up effects that hamper insect pest reproduction or nutrition (Büchs et al. 1997), or via top-down effects related to natural enemy dispersal between successive or adjacent crops (e.g., parasitoids, Rusch et al. 2011a).

Field margins are man-made, semi-natural and generally herbaceous habitats that have an important function as refugia for biodiversity (Landis et al. 2000; Marshall and Moonen 2002) and beneficial effects on crop pollinators, pest predators and parasitoids within the fields (Pfiffner and Wyss 2004; Winkler et al. 2009). They serve as overwintering sites (Pfiffner and Luka 2000) and provide many natural pest enemies with essential food sources (Jervis et al. 2004). Plant species have been tested to select those that more adequately fulfil the food requirements of pest enemies when sown in field margins or as companion plants within the field (Pfiffner and Wyss 2004; Wäckers et al. 2007; Winkler et al. 2009). Field margins are also functional corridors linking different habitats (Holzschuh et al. 2009).

The land-use system at the farm scale is important for designing and interconnecting habitats. Farmscaping (Smukler et al. 2010) refers to a management system that aims to increase ecosystem services. Farmers' decision-making on the short- and long-term also affects ecosystem services. In the last decades, changes in crop protection management at the farm level in a shift towards IPM rules have positively influenced the natural enemies of pests as a result of either the lower toxicity of the chemicals used (Thomson and Hoffmann 2006) or reduced chemical use, as observed by Häni et al. (1998). Comparisons of the effects of organic vs. conventional farming systems on arthropod and bird communities are widely documented,

Table 5.1 Main organic farming strategies and practices likely to promote and preserve biodiversity in agroecosystems

Management	Strategies or practices commonly used on organic farms	Prevailing scale	Direct and indirect contribution to biodiversity[b]	Presumed effects with regard to pest suppression
Cropping system [a]	Long crop rotation (3 to > 7 years)	Farm	Higher plant agro-biodiversity within space and time	Cycle of specialised insect pests is broken
	Diversified cultivars (e.g., risk insurance, diversification in direct sale)	Farm		Lower crop availability and location by pests; lower plant attractivity to polyphagous insects
	Leguminous (or grass/leguminous mixture) as associated or previous crops	Field		Higher abundance, richness and/or effectiveness of NE
	Local/hardy cultivars	Field		
	Mixed cropping/intercropping	Field		
Fertilisation	Organic matter supply	Field	More resources for scavengers and polyphagous NE	Contribution to the permanency of food webs
Direct pest and disease control	Less pesticide use, use of pesticides of natural origin, low environmental contamination with some exceptions (e.g., copper)	Field	Preservation of NE	Higher abundance and/or richness of NE
Weed control	Less drastic weed control	Field	Higher plant diversity (e.g., segetal plants)	Higher abundance, richness and/or effectiveness of NE
Non-crop surface areas	Field margins	Field	Higher plant diversity	Higher abundance, richness and/or effectiveness of NE
	Habitat management and farmscaping	Farm	Higher plant and habitat diversity	

NE natural enemies of pests

[a] Cropping system management also contributes to the management of both fertilisation and pest (*lato sensu*) control, but all types of crop management were grouped within this line on the table for clearness

[b] An increase in both plant and habitat diversity provides NE with a higher diversity of resources

mainly to the benefit of OF practices (Mäder et al. 2002; Gabriel and Tscharntke 2007; Holzschuh et al. 2007; Rösler 2007). Hence, real benefits exist for farmers when they adapt their crop management, farming systems and farm landscape structure in a way that both minimises pest development and maximises the potential for pest control by natural enemies.

5.3.2 Organic Farms and Conservation Biocontrol

Alternative agricultural systems such as OF have been presented as being more favourable to biodiversity (Bengtsson et al. 2005; Hole et al. 2005; Letourneau and Bothwell 2008), as adopting more environmentally-friendly practices (Bengtsson et al. 2005) than conventional ones, and as providing numerous ecosystem services (Sandhu et al. 2010). Since the OF methods used for the direct control of pests, diseases and weeds are rare and/or less efficient than synthetic pesticides, farmers have to adopt various strategies to reduce crop damage or weed competition in re-designed cropping systems (Zehnder et al. 2007; Table 5.1): long and complex crop rotations, hardy cultivars and/or population varieties, development and environmentally-friendly management of field margins and semi-natural habitats. Most OF practices are also favourable for preserving and/or increasing plant and animal biodiversity:

- Because mechanical weeding is less efficient than herbicides, diversified plants associated with crops (e.g., segetal plants) are present and more abundant in OF fields (Gibson et al. 2007).
- Organic fertilising inputs and compost supply the soil with organic matter, thus favouring scavengers, ecosystem engineers and, therefore, the permanency of food webs and biodiversity (Mäder et al. 2002; Birkhofer et al. 2008). Below- and aboveground food webs are also interconnected through the plant (Blossey and Hunt-Joshi 2003) and/or generalist predators that feed on both scavengers and herbivores (Scheu 2001). Tillage practices also contribute to soil aeration and are favourable to many arthropods that live belowground (Birkhofer et al. 2008).
- Pesticide use is one of the most disruptive practices in agriculture (Geiger et al. 2010). Some organic pesticides such as copper and spinosad also have detrimental effects on the environment and/or animal communities (Extoxnet 1996; Cisneros et al. 2002), e.g., the neuro-toxic spinosad has a lethal effect on and reduces natural enemy populations when applied in fields to control pests. However, the rapid binding to soil aggregates (Extoxnet 1996) and UV or biological degradation of most organic compounds (Isman 2006), the tolerance of pests in crops and less direct measures against pests generally cause less disruptive effects in organic compared to conventional systems (Birkhofer et al. 2008; Crowder et al. 2010). In the case study of fruit tree production, which is one of the most extensively treated crops to prevent pest and disease damage to fruit,

OF orchards host a more abundant arthropod fauna (Rösler 2007), which is (Penvern et al. 2010) or not (Simon et al. 2007) more diversified than in conventional orchards at the family taxonomic level. However, this higher abundance did not (Simon et al. 2007) or barely (Penvern et al. 2010) lead to a higher level of pest control. We suppose that negative interactions (i.e., intraguild predation and cannibalism) may occur among natural enemies when abundance and/or richness increase. It is also likely that the pest multiplication rate is so highly favoured in high-density monoclonal orchards that natural pest control cannot occur. This would thus attest to the limits of such agricultural designs.

Thus, in OF, both within-time (crop rotations, higher agro-biodiversity) and within-space (higher agro-biodiversity, semi-natural habitats) plant designs and management (leading to less direct mortality and sub-lethal effects) contribute to the increase in biodiversity through higher plant richness and associated fauna than in conventional systems (Hole et al. 2005). Ecological compensation areas in Switzerland are also larger by two-thirds in OF compared to IPM farms (Schader et al. 2008). Moreover, OF systems were reported to maintain a higher level of structural complexity at both local and landscape scales compared to conventional ones (Norton et al. 2009). Nevertheless, various authors have reported that organic or low-intensity farming systems have little or no effect on populations in complex landscapes, whereas they have a substantial influence in simple landscapes (Holzschuh et al. 2007; Bengtsson et al. 2005; Tscharntke et al. 2005). Under the *moderated conservation effectiveness hypothesis*, Tscharntke et al. (2005) suggest more effective local management to enhance biological control in simple landscapes (i.e., 1–20 % proportion of semi-natural habitats) than in cleared (< 1 %) or complex (> 20 %) landscapes. This hypothesis has been recently confirmed in croplands where agri-environmental management was more effective in enhancing species richness in simple than in complex landscapes (Batáry et al. 2011). Cleared landscapes with few semi-natural habitats do not support a sufficient number of species to obtain a significant enhancement of natural enemies in response to local management. Conversely, complex landscapes rich in semi-natural habitats already support abundant and diverse populations of natural enemies (Thies et al. 2003; Tscharntke et al. 2005). Information is also still lacking on the level of pest suppression associated with a generalised regional development of OF systems. Lastly, taking the diversity of practices within OF systems into account can be of major importance when assessing the effects of farming systems on associated ecosystem services.

Of course, as outlined by Hole et al. (2005), biodiversity is not an exclusivity of OF, and conventional farming systems may potentially harbour high levels of biodiversity at the farm scale, provided that resource-rich habitats are present and that IPM practices are adopted. This reveals that landscape structure together with farm management are powerful but interdependent levers to drive biodiversity and ecosystem services, provided that sufficient and relevant information is available on the life traits of the species present, and that adequate organisation levels, both in terms of landscape structure and farm management, are understood (Le Roux et al. 2008; Rusch et al. 2010; Médiène et al. 2011).

5.4 Case Studies of Conservation Biocontrol

Theories and cases studies (see case boxes) have made it possible to identify key issues related to conservation biocontrol approaches, which also address the significance of landscape effects:

1. Why do results vary among studies?
2. What are the expected benefits for growers?

Orchards

The specificity of perennial crops like fruit orchards is related to longevity (persistence of the host plant), multi-strata designs and heterogeneity due to branching structures and within-field distribution of tree rows and alleys (Simon et al. 2010). The application of conservation biocontrol measures as indirect plant protection methods is therefore complex and should be carefully planned within a long-term perspective. Various types of plant resource diversification have been studied within orchards: *ground-cover crops* (Bugg and Waddington 1994); *herbaceous flowering companion plants* to either promote a complex of predators or enhance the parasitism rate due to one or several species in apple orchards (Wyss 1995; Wyss et al. 1995; Stephens et al. 1998; Bostanian et al. 2004; Irvin et al. 2006); *interplanting extrafloral nectar-bearing peach trees* (Brown and Mathews 2007); *alternative prey/hosts on a non-crop plant* (Pfannenstiel and Unruh 2004; Bribosia et al. 2005) to enhance in-field production of parasitoids by providing substitute hosts on non-crop plant or tree species. The effects of *hedgerows* (Solomon 1981; Pfannenstiel et al. 2010) and the *surrounding vegetation* (Debras et al. 2006) were also investigated. The design of tree species assemblages providing natural enemies with successive resources throughout the year is also documented (Simon et al. 2009).

Speiser et al. present in this book (Chap. 4) an experimental prototype orchard designed to maximise conservation biocontrol in operation since 2006 at the Swiss Research Institute of Organic Agriculture. It combines hedges with selected plants (plants for alternative hosts, flowering shrubs), flowering plants in the tree rows, companion plants in the alleyways, pest- and disease-tolerant apple cultivars, soil quality enhanced by the addition of compost, and a number of artificial nesting sites for predators (birds, bats, parasitoids) to study the feasibility of pesticide-free apple production. The first unpublished results show that compared to orchards under usual OF and IPM management, the number of pest and natural enemy species is the highest in the prototype orchard. However, the diverse pests observed in the prototype orchard are usually controlled by the abundant natural enemies, attesting to the effectiveness of regulation processes promoted by friendly practices and plant and habitat management. Whether or not tree and fruit damage that occurs, not only as a result of pests but of diseases such as sooty blotch as well, makes pesticide-free apple production economically feasible, is still under study (Speiser et al. Chap. 4).

Arable Crops

Conservation biocontrol of arthropod pests is of particular importance for annual (i.e., arable and vegetable) crops that represent a discontinuous habitat for arthropods and occupy large surface areas in Europe in the case study of arable crops. Although different types of plant resource diversification devoted to conservation biocontrol have been studied for over 20 years now in arable annual crops (Andow 1988; Nentwig 1988), there are very few field applications, partly because typical landscapes in annual crop regions are lacking in semi-natural habitats and because the insights provided by landscape ecology have only recently become available to help us to understand large-scale ecological processes (Thies and Tscharntke 1999; Tscharntke et al. 2007).

Specific practices of arable crops contribute to conservation biocontrol. At the field scale, *cultural practices* such as reduced tillage or no tillage at all enhance the overwintering survival of parasitoids and also benefit polyphagous predators (ground beetles, spiders, rove beetles) overwintering within the fields (Hokkanen 2008), thus decreasing the reproduction rates of pest species in organic oilseed rape (Büchs and Katzur 2004). Due to increased predator pressure, *intercropping* has been shown to reduce damage to oilseed rape taproots due to the maggot *Delia radicum* (L.) when the proportion of wheat increases in the intercrops (Hokkanen 2008; Hummel et al. 2009). In most cases, *mixed crops* decrease cereal aphid damage (Andow 1991) and some minor weeds that provide pollen and nectar can be left in the cereal field without any real damage to the wheat (Norris and Kogan 2005). *Decaying organic debris* on the soil from a previous crop or intercrop provides both shelter and alternative food via scavengers (mainly Collembola) to generalist ground-dwelling predators, and enhances aphid biological control in wheat (Schmidt et al. 2004). Moreover, polyphagous ground-dwelling insects benefit at the synergistic level from the activity of flying natural enemies that make some aphids fall to the ground when eating or laying eggs (Losey and Denno 1998). Management of the surrounding field environment to provide suitable overwintering habitats such as the establishment of *beetle banks* (raised grass strips sown with tussocky species, MacLeod et al. 2004) promotes winter survival and favours field re-colonisation the following spring (Hokkanen 2008). Old *field margin* strips, old fallow habitats (Thies and Tscharntke 1999) and adjacent wild flower strips (Büchi 2002) increased parasitism rates of pollen beetle larvae through enhanced fecundity and longevity.

A recent study (Colour plate 5) designed to explore the relative importance of oilseed rape crop management and landscape context on pollen beetle density, damage and parasitism rates in France revealed a strong positive effect of landscape complexity, particularly of grasslands, on the parasitism rate of pollen beetles (Rusch et al. 2011a). However, this study also revealed a positive effect of landscape complexity, particularly of woodland, on adult pest populations. Indeed, this study identified woodland as the main overwintering habitat for pollen beetles. It therefore illustrates the so-called ambivalent effect of landscape complexity and the need for a clear description of the semi-natural habitats and their functions for natural enemy and pest populations. Moreover, this study also found that there is a positive effect on the parasitism rate of pollen beetles in terms of the proportion of the previous year's oilseed rape crop in the surrounding landscape (i.e., the overwintering habitat of parasitoids) with reduced soil tillage. The importance of the nitrogen status of the crop on its ability to recover from pest damage has also been demonstrated. This work revealed the importance of taking both farming practices and semi-natural habitats scattered within the landscape into account to understand biological control in agroecosystems and thus opens new perspectives in terms of IPM strategies that maximise biological control at the landscape scale (Rusch et al. 2011a, 2011b).

5.4.1 Promising but Variable Results

Considering the case study of perennial crops, the plant assemblage sown in companion plant strips within or at the border of organic Swiss orchards successfully increased predation rates of the rosy apple aphid in the spring and fall (Wyss 1995; Wyss et al. 1995). Flowering plants increased the parasitoid fauna several-fold in Canadian apple orchards during the study, which resulted in 90.8% undamaged fruits at harvest in the fifth year of the study vs. 67.5% in the control where one insecticide per year was allowed to control plum curculio (Bostanian et al. 2004). Irvin et al. (2006) established a higher tortricid parasitism rate and less damage in New Zealand apple orchards in the presence of floral understories compared to the control. In North American apple orchards, parasitoids migrating from hedges sheltering alternative overwintering hosts to the adjacent orchards helped control leafroller pests in spring (Pfannenstiel and Unruh 2004; Pfannenstiel et al. 2010). However, ground cover crops provide both benefits and disadvantages (Bugg and Waddington 1994). In some cases, predators were favoured, whereas in other cases, the number of phytophagous arthropods increased. Woody species in hedgerows favour natural enemies (Sarthou 1995) but can also harbour pests (Solomon 1981; Jeanneret 2000).

A review of orchards (Simon et al. 2010) showed that the effect of plant diversification in orchards on pest control was mostly positive (16 cases) or null (9), but also negative in some cases (5). Studies on spontaneous cover crops, alfalfa and grass in the alleys of untreated or organic orchards (Brown and Glenn 1999) showed no significant effects either on the presence of natural enemies or on apple pests.

In arable crops, cultural practices such as reduced tillage can enhance both pest (e.g., insect pests overwintering in the soil) and natural enemy populations, depending on their life cycle (Rusch et al. 2010). Moreover, increasing within-field plant diversity in annual crops has been shown to have a strong positive effect on natural enemies, pest control and crop damage suppression, even if there are also some cases where no particular effects were reported (Rusch et al. 2010; Letourneau et al. 2011). Different studies on oilseed rape have demonstrated the positive effect of flower strips (wild and sown strips) on the parasitism rates of the pollen beetle (Büchi 2002; Scheid et al. 2011). Moreover, the density and species richness of aphidophagous syrphids was significantly higher, not only in narrow and broad sown flower strips compared to grassy strips and wheat–wheat boundaries, but also within wheat fields adjacent to the broad sown flower strips than in the others. This suggests that these fields benefit from a potential biocontrol of cereal aphids from distant plant management (Haenke et al. 2009). Enhanced conservation biocontrol due to landscape complexity is also observed, e.g., higher levels of biological control of the pollen beetle due to parasitoids (Thies et al. 2003) and higher parasitism rates of aphids on wheat (Thies et al. 2005) and of Lepidoptera on maize (Marino and Landis 1996) in heterogeneous compared to homogeneous landscapes. However, cases of plant diversification or mosaic landscapes were also favourable to some crop pests and/or diseases (Thies et al. 2005; Rusch et al. 2010). Indeed, the effects of landscape complexity on natural enemies and pest populations may reveal ambivalent effects since insect pests may also benefit from an increased amount of non-crop habitats in the landscape (Thies et al. 2005; Zaller et al. 2008).

More detailed knowledge is therefore needed on the arthropod community of plant species used in conservation biocontrol to provide clear advice to farmers on how to enhance natural enemies without promoting pests. To consider the whole pest complex of crops, the biological requirements of both pests and natural enemies of a given crop (Irvin et al. 2006) and the diversity of functional traits that will be present in a community are thus a prerequisite to understand the processes at work. Lastly, the sometimes ambivalent effect of landscape complexity highlights the importance of implementing adapted farming practices at the local scale to enhance natural enemy populations only.

5.4.2 Expected Benefits from Conservation Biocontrol

Each measure of resource diversification mentioned above contributes, to some extent, to the control of pests, but is generally not as effective in preventing damage as pesticides since higher predation or parasitism rates do not always make it possible to avoid direct control measures. Brown and Mathews (2007) showed that interplanting extrafloral nectar-bearing peach trees within an apple orchard increased the presence of aphid predators in the very early spring when stem mothers of the rosy apple aphid are hatching. However, despite some significant reduction of aphids by ladybird beetles, this type of conservation biocontrol measure was not sufficient to reliably control this pest. Conservation biocontrol measures often require several years to build up the beneficial arthropod fauna to constitute an effective biocontrol force (Bostanian et al. 2004), but cultural practices (including pesticide use) may durably alter this process (Geiger et al. 2010). Application of conservation biocontrol measures is more effective for pests with a high treatment threshold (e.g., mites or pear psyllids) (Simon et al. 2010). To control pests in perennial crops, either conservation biocontrol measures are accompanied by additional biocontrol techniques (inundative or inoculative biocontrol) and/or cultural and alternative measures to decrease pesticide use reliance are combined, as is the case in the Swiss experimental prototype orchard (see Text Box). This brings us back to the problem of (i) the limits of conservation biocontrol to control pests whose populations cannot be tolerated in crops, even at low levels, e.g., most tortricids in orchards, and (ii) the current and unfavourable farming and food systems that are devoted to production and standardisation of agricultural products with poor consideration of ecological processes and tri-trophic relationships.

Lastly, null or small added value for ecological achievements (ecologically-driven subsidies), cosmetic standards in the case of the fruit market, a high number of pests and diseases, constantly increasing production costs and/or low prices for products are factors that limit the farmer's motivation to develop conservation biocontrol. An increasing societal demand for products that are free of pesticide residues could be favourable to stimulate the development of indirect control measures such as conservation biocontrol (Pearson et al. 2011).

5.5 Perspectives on Conservation Biocontrol

OF is the agricultural system that has been largely under focus at both the scientific and technical levels to develop global approaches that maximise 'plant-mediated' bottom-up processes and 'natural enemies-mediated' top-down processes that are both related to the preservation and promotion of biodiversity (Letourneau and Bothwell 2008). This provides perspectives on the importance of the development of innovative agricultural systems and their within-landscape distribution to both contribute to biodiversity in agroecosystems and to optimise ecosystem services.

5.5.1 Need for Basic Research

Whereas the basic principles of conservation biocontrol are recognised at this time, field applications are still rare (Letourneau and Bothwell 2008). There is a challenge to develop a thorough understanding of the functional life traits of pests and natural enemies (such as dispersal abilities and specialisation), and interactions between species at different space and time scales: suitable alternative food, periods of food shortage, fitness of natural enemies according to food, prey/host preferences and conditioning, dispersal ability and range of dispersal, and intra- and inter-species interactions at different embedded scales. For example, it is well accepted that conservation biocontrol actions have to be based on the simplistic 'resource diversity and availability' principle, regardless of the natural enemy, rather than on precise resources (specific nectar or alternative prey/host), undoubtedly much more difficult to identify but more accurately fitted to the requirements of natural enemies. Empirical but promising applications have been reported: while conservation biocontrol implementation was largely empirical before 2000, Gurr et al. (2000) calculated that more than 78 % of the published conservation biocontrol management operations from 1990 onwards had successful effects on natural enemies, and more than 63 % succeeded in decreasing pest densities. The challenge is therefore to identify the ecological processes involved in these successful management systems. Better knowledge about functional biodiversity life traits and multi-scale relationships between and within species, and between abiotic environmental parameters and species could significantly improve the effectiveness of conservation management. In addition to diversity *per se*, some studies have also pointed out the importance of investigating other aspects of biodiversity (e.g., natural enemy evenness) to explain pest control (Crowder et al. 2010).

5.5.2 Developing Practical Applications of Conservation Biocontrol

The reliability of field applications is a prerequisite for farmers to adopt conservation biocontrol. Moreover, enhanced predation or parasitism due to conservation biocontrol measures may be significant from an ecological (and statistical) point of view, but the effect alone can be insufficient from an agricultural point of view, e.g., when the pest infestation level remains above the treatment threshold. To help farmers to design plant assemblages for conservation biocontrol, experiments could be proposed to assess: (i) candidate selective food plants; (ii) the effectiveness of natural enemies to control pests using exclusion experimental designs or paired-fields, combined with gut-content analysis of natural enemies and/or modelling; and (iii) which measures could be of the greatest benefit when farmscaping. Although very few applications have been reported, the case study of protected vegetable crops could help us to understand processes and to develop natural enemy reservoir systems: banker plant systems maintain and multiply native as well as released natural enemies, which emerge in a permanent release process within the greenhouse (Frank 2010).

One major problem encountered today is that information about the spontaneous level of pest suppression and the benefit of the introduction of plant diversity within the agroecosystem is lacking at this time. This highlights three main points. First, there is the need to develop taxonomic skills and long-term experiments (Scherber et al. 2010) to acquire a large number of references on arthropod populations hosted by various plants and plant assemblages. Second, there is also the need to develop accurate and generic indicators at species, community and/or landscape levels to satisfy the requirements of both field and *ex-ante* assessments. Last, interactions between local conservation biocontrol measures and landscape context need to be taken into account (see below).

5.5.3 Towards the Redesign of Production Systems to Enhance Conservation Biocontrol

A combination of levers has to be used to achieve natural pest regulation. Indeed, pest control can seldom rely on natural enemies alone. Top-down processes related to conservation biocontrol need to be completed by direct measures and/or bottom-up processes to control pests. However, this more generally leads to questions about the design and management of current production systems—albeit organic—and associated plant diversity. To increase conservation biocontrol, guidelines could be proposed for developing or enhancing: (i) multi-strata designs in perennial crops; (ii) within-field structural and genetic heterogeneity (this leads to new challenges and stakes both in farm organisation and the food processing industry, to deal with multi-cultivar and multi-species crops); and (iii) habitat management at the farm scale adapted to the landscape context. Indeed, new insights provided by landscape ecology should be included in conservation biocontrol experiments even though the relative effects of local plant resource diversification or cultural practices are very seldom disassociated from landscape effects (Debras et al. 2006; Rusch et al. 2011a). Because landscape management is hardly possible on the short-term, the challenge will be more focused on designing appropriate conservation biocontrol measures (e.g., hedgerow and flower strip planting, semi-natural habitat preservation or restoration) adapted to the landscape context (Batáry et al. 2011; Kleijn et al. 2011) so that they will be the most effective in terms of pest suppression.

This overview of the interactions between biodiversity and conservation biocontrol shows that OF systems are adequate models to enhance as well as to provide ecosystem services related to biodiversity in the agroecosystem. Organic farmers already make use of and benefit from biodiversity. However, conservation biocontrol must be further developed to improve effectiveness and reliability for farmers. A participatory approach to combine results from basic research and promising field applications would provide a solid foundation to develop and promote conservation biocontrol.

Acknowledgments The authors are grateful to Dr. M.W. Brown for his helpful comments on a previous draft of this manuscript, and to A. Lefèvre (INRA Alenya) for sharing her experience on vegetable crops.

References

Altieri MA, Nicholls CI (2004) Biodiversity and pest management in agroecosystems, Issue 2. Haworth Press Inc. Binghamton, NY, USA

Andow DA (1988) Management of weeds for insect manipulation in agro-ecosystems. In: Weed management in agro-ecosystems: ecological approaches. CRC Press, Inc., Boca Raton, Florida, USA, pp. 265–301

Andow DA (1991) Vegetational diversity and arthropod population response. Ann Rev Entomol 36:561–586

Batáry P, Báldi A, Kleijn D, Tscharntke T (2011) Landscape-moderated biodiversity effects of agri-environmental management: a meta-analysis. Proc R Soc Biol Sci 278:1894–1902

Bengtsson J, Angelstam P, Elmqvist T, Emanuelsson U, Folke C, Ihse M, Moberg F, Nystrom M (2003) Reserves, resilience and dynamic landscapes. AMBIO 32(6):389–396

Bengtsson J, Ahnström J, Weibull AC (2005) The effects of organic agriculture on biodiversity and abundance: a meta-analysis. J Appl Ecol 42:261–269

Benton TG, Vickery JA, Wilson JD (2003) Farmland biodiversity: is habitat heterogeneity the key? Trends Ecol Evol 18(4):182–188

Birkhofer K, Bezemer TM, Bloem J, Bonkowski M, Christensen S, Dubois D, Ekelund F, Fließbach A, Gunst L, Hedlund K, Mäder P, Mikola J, Robin C, Setälä H, Tatin-Froux F, Van der Putten WH, Scheu S (2008) Long-term organic farming fosters below and aboveground biota: Implications for soil quality, biological control and productivity. Soil Biol Biochem 40:2297–2308

Blossey B, Hunt-Joshi TR (2003) Belowground herbivory by insects: influence on plants and aboveground herbivores. Annu Rev Entomol 48:521–547

Bostanian NJ, Goulet H, O'Hara J, Masner L, Racette G (2004) Towards insecticide free apple orchards: flowering plants to attract beneficial arthropods. Biocontrol Sci. Technol 14:25–37

Bribosia E, Bylemans D, Migon M, van Impe B (2005) In-field production of parasitoids of *Dysaphis plantaginea* by using the rowan aphid *Dysaphis sorbi* as substitute host. BioControl 50:601–610

Brown MW, Glenn DM (1999) Use of ground cover plants as a pest management tool in apple orchards. J Econ Entomol 92:899–905

Brown MW, Tworkoski T (2004) Pest management benefits of compost mulch in apple orchards. Agr Ecosyst Environ 103:465–472

Brown MW, Mathews CR (2007) Conservation biocontrol of rosy apple aphid, *Dysaphis plantaginea* (Passerini), in Eastern North America. Environ Entomol 36:1131–1139

Büchi R (2002) Mortality of pollen beetle (*Meligethes* spp.) larvae due to predators and parasitoids in rape fields and the effect of conservation strips. Agric Ecosyst Environ 90:255–263

Büchs W, Katzur K (2004) Cultivation techniques as means to control pests in organic oilseed rape production. IOBC/WPRS Bull. 27(10): 225–236

Büchs W, Harenberg A, Zimmermann J (1997) The invertebrate ecology of farmland as a mirror of the intensity of the impact of man?—An approach to interpreting results of field experiments carried out in different crop management intensities of a sugar beet and an oil seed rape rotation including set-aside. Biol Agric Hort 15(1-4):83–107

Bugg RL, Waddington C (1994) Using cover crops to manage arthropod pests of orchards: a review. Agric Ecosyst Environ 50:11–28

Burel F, Baudry J (1999) Ecologie du paysage: concepts, méthodes et applications. Lavoisier, Paris

Chaplin-Kramer R, O'Rourke ME, Blitzer EJ, Kremen C (2011) A meta-analysis of crop pest and natural enemy response to landscape complexity. Ecol Lett 14:922–932

Cisneros J, Goulson D, Derwent LC, Penagos DI, Hernandez O, Williams T (2002) Toxic effects of spinosad on predatory insects. Biol Control 23:156–163

Concepcion ED, Diaz M, Baquero RA (2008) Effects of landscape complexity on the ecological effectiveness of agri-environment schemes. Landscape Ecol 23:135–148

Crowder DW, Northfield TD, Strand MR, Snyder WE (2010) Organic agriculture promotes evenness and natural pest control. Nature 466:109–112

Debras J-F, Torre F, Rieux R, Kreiter S, Garcin M-S, Van Helden M, Buisson E, Dutoit T (2006) Discrimination between agricultural management and the hedge effect in pear orchards (southeastern France). Ann Appl Biol 149:347–355

De Marco P, Coelho FM (2004) Services performed by the ecosystem: forest remnants influence agricultural cultures' pollination and production. Biodiv Conserv 13(7):1245–1255

Denys C, Tscharntke T (2002) Plant-insect communities and predator-prey ratios in field margin strips, adjacent crop fields, and fallows. Oecologia 130(2):315–324

Dunning JB, Danielson BJ, Pulliam HR (1992) Ecological processes that affect populations in complex landscapes. Oikos 65(1):169–175

Eilenberg J, Hajek A, Lomer C (2001) Suggestions for unifying the terminology in biological control. BioControl 46:387–400

Extoxnet (1996) Copper sulfate [Online]. Pesticide Information Profiles. Extension Toxicology Network. Available at: http://extoxnet.orst.edu/pips/coppersu.htm (accessed 18 March 2010)

Finke DL, Denno RF (2006) Spatial refuge from intraguild predation: implications for prey suppression and trophic cascades. Oecologia 149(2):265–275

Frank SD (2010) Biological control of arthropod pests using banker plant systems: Past progress and future directions. Biol Control 52:8–16

Gabriel D, Tscharntke T (2007) Insect pollinated plants benefit from organic farming. Agric Ecosyst Environ 118:43–48

Geiger F, Bengtsson J, Berendse F, Weisser WW, Emmerson M, Morales MB, Ceryngier P, Liira J, Tscharntke T, Winqvist C, Eggers S, Bommarco R, Pärt T, Bretagnolle V, Plantegenest M, Clement LW, Dennis C, Palmer C, Oñate JJ, Guerrero I, Hawro V, Aavik T, Thies C, Flohre A, Hänke S, Fischer C, Goedhart PW, Inchausti P (2010) Persistent negative effects of pesticides on biodiversity and biological control potential on European farmland. Basic Appl Ecol 11(2):97–105

Gibson RH, Pearce S, Morris RJ, Symondson WOC, Mermmott J (2007) Plant diversity and land use under organic and conventional agriculture: a whole-farm approach. J Appl Ecol 44:792–803

Gurr GM, Wratten SD, Barbosa P (2000) Success in conservation biological control of arthropods. In: Gurr GM, Wratten SD (Eds) Biological control: measures of success. Kluwer Academic Publishers, Dordrecht, pp. 105–132

Gurr GM, Wratten SD, Luna JM (2003) Multi-function agricultural biodiversity: pest management and other benefits. Basic Appl Ecol 4:107–116

Haenke S, Scheid B, Schaefer M, Tscharntke T, Thies C (2009) Increasing syrphid fly diversity and density in sown flower strips within simple vs. complex landscapes. J Appl Ecol 46(5):1106–1114

Häni FJ, Boller EF, Keller S (1998) Natural regulation at the farm level. In: Pickett CH, Bugg RL (Eds) Enhancing biological control—habitat management to promote natural enemies of agricultural pests. University of California Press, Berkeley, pp. 161–210

Hanski I (1999) Habitat connectivity, habitat continuity, and metapopulations in dynamic landscapes. Oikos 87(2):209–219

Hokkanen HMT (2008) Biological control methods of pest insects in oilseed rape. EPPO Bull 38(1):104–109

Hole DG, Perkins AJ, Wilson JD, Alexander IH, Grice PV, Evans AD (2005) Does organic farming benefit biodiversity? Biol Conserv 122:113–130

Holzschuh A, Steffan-Dewenter I, Kleijn D, Tscharntke T (2007) Diversity of flower-visiting bees in cereal fields: effects of farming system, landscape composition and regional context. J Appl Ecol 44:41–49

Holzschuh A, Steffan-Dewenter I, Tscharntke T (2009) Grass strip corridors in agricultural landscapes enhance nest-site colonization by solitary wasps. Ecol Appl 19(1):123–132

Hummel JD, Dosdall LM, Clayton GW, Harker KN, O'Donovan JT (2009) Effects of canola-wheat intercrops on *Delia* spp. (Diptera: Anthomyiidae) oviposition, larval feeding damage, and adult abundance. J Econ Entomol 102(1):219–228

Irvin NA, Scarratt SL, Wratten SD, Frampton CM, Chapman RB, Tylianakis JM (2006) The effects of floral understoreys on parasitism of leafroller (Lepidoptera: Tortricidae) on apples in New Zealand. Agric. Forest Entomol 8:25–34

Isman MB (2006) Botanical insecticides, deterrents, and repellents in modern agriculture and an increasingly regulated world. Annu Rev Entomol 51:45–66

Jeanneret P (2000) Interchanges of a common pest guild between orchards and surrounding ecosystems: a multivariate analysis of landscape influence. In: Ekbom B, Irwin ME, Robert Y (eds) Interchanges of insects between agricultural and surrounding landscapes. Kluwer Academic, Dordrecht, pp. 85–107

Jervis MA, Lee JC, Heimpel GE (2004) Use of behavioural and life-history studies to understand the effects of habitat manipulation. In: Gurr GM, Wratten SD, Altieri MA (eds) Ecological engineering for pest management, advances in habitat manipulation for arthropods. CSIRO Publishing, pp. 65–100

Keller S, Häni F (2000) Ansprüche von Nützlingen und Schädlingen an den Lebensraum. In: Nentwig W (ed) Streifenförmige ökologische Ausgleichsflächen in der Kulturlandschaft: Ackerkrautstreifen, Buntbrache, Feldränder. Verlag Agrarökologie, Bern, Hannover, pp. 199–217

Kleijn D, Rundlöf M, Scheper J, Smith HG, Tscharntke T (2011) Does conservation on farmland contribute to halting the biodiversity decline? Trends Ecol Evol 26(9):474–481

Krebs JR, Wilson JD, Bradbury RB, Siriwardena GM (1999) The second silent spring? Nature 400:611–612

Landis DA, Wratten SD, Gurr GM (2000) Habitat management to conserve natural enemies of arthropod pests in agriculture. Annu Rev Entomol 45:175–201

Lawton JH (1994) What do species do in ecosystems. Oikos 71(3):367–374

Le Roux X, Barbault R, Baudry J, Burel F, Doussan I, Garnier E, Herzog F, Lavorel S, Lifran R, Roger-Estrade J, Sarthou JP, Trommetter M (2008) Agriculture and biodiversity: promoting synergies; Agriculture et biodiversité: Valoriser les synergies. Expertise scientifique collective, synthèse du rapport. INRA, Paris, France

Letourneau DK, Bothwell SG (2008) Comparison of organic and conventional farms: challenging ecologists to make biodiversity functional. Front Ecol Environ 6(8):430–438

Letourneau DK, Jedlicka JA, Bothwell SG, Moreno CR (2009) Effects of natural enemy biodiversity on the suppression of arthropod herbivores in terrestrial ecosystems. Ann Rev Ecol Evol Syst 40:573–592

Letourneau DK, Armbrecht I, Salguero Rivera B, Montoya Lerma J, Jiménez Carmona E, Martha CD, Selene Escobar S, Víctor Galindo V, Gutiérrez C, Duque LS, López MJ, Acosta Rangel AM, Herrera Rangel J, Rivera L, Saavedra CA, Torres AM, Reyes Trujillo A (2011) Does plant diversity benefit agroecosystems? A synthetic review. Ecol Appl 21:9–21

Loreau M (2010) Linking biodiversity and ecosystems: towards a unifying ecological theory. Philos Transact R Soc B-Biol Sci 365(1537):49–60

Loreau M, Hector A (2001) Partitioning selection and complementarity in biodiversity experiments. Nature 412:72–76

Losey JE, Denno RF (1998) Positive predator-predator interactions: enhanced predation rates and synergistic suppression of aphid populations. Ecology 79(6):2143–2152

MacLeod A, Wratten SD, Sotherton NW, Thomas MB (2004) 'Beetle banks' as refuges for beneficial arthropods in farmland: long-term changes in predator communities and habitat. Agr Forest Entomol 6(2):147–154

Mäder P, Fliessbach A, Dubois D, Gunst L, Fried P, Niggli U (2002) Soil fertility and biodiversity in organic farming. Science 296:1694–1697

Marino PC, Landis DA (1996) Effect of landscape structure on parasitoid diversity and parasitism in agroecosystems. Ecol Appl 6(1):276–284

Marshall EJP, Moonen AC (2002) Field margins in northern Europe: their functions and interactions with agriculture. Agr Ecosyst Environ 89:5–21

Médiène S, Valantin-Morison M, Sarthou J-P, de Tourdonnet S, Gosme M, Bertrand M, Roger-Estrade J, Aubertot J-N, Rusch A, Motisi N, Pelosi C, Doré T (2011) Agroecosystem management and biotic interactions: a review. Agro Sust Develop 31:491–514

Nentwig W (1988) Augmentation of beneficial arthropods by strip-management. 1. Succession of predacious arthropods and long-term change in the ratio of phytophagous and predacious arthropods in a meadow. Oecologia 76(4):597–606

Norris RF, Kogan M (2005) Ecology of interactions between weeds and arthropods. Ann Rev Entomol 50:479–503

Norton L, Johnson P, Joys A, Stuart R, Chamberlain D, Feber R, Firbank L, Manley W, Wolfe M, Hart B, Mathews F, MacDonald D, Fuller RJ (2009) Consequences of organic and non-organic farming practices for field, farm and landscape complexity. Agr Ecosyst Environ 129:221–227

Noss RF (1990) Indicators for monitoring biodiversity: a hierarchical approach. Conserv Biol 4:355–364

Pearson D, Henryks J, Jones H (2011) Organic food: what we know (and do not know) about consumers. Renew Agr Food Syst 26(2):171–177

Penvern S, Bellon S, Fauriel J, Sauphanor B (2010) Peach orchard protection strategies and aphid communities: Towards an integrated agroecosystem approach. Crop Prot 29:1148–1156

Pfannenstiel RS, Unruh TR (2004) Conservation of leafroller parasitoids through provision of alternate hosts in near-orchard habitats. In: Van Driesche RG (Ed) Proceedings of the International Symposium on Biological Control of Arthropods, FEHTET-2003-05, Honolulu, Hawaii, 14-18 Jan. 2002. USDA, Forest Service: 256–262

Pfannenstiel RS, Unruh TR, Brunner JF (2010) Overwintering hosts for the exotic leafroller parasitoid, *Colpoclypeus florus*: Implications for habitat manipulation to augment biological control of leafrollers in pome fruits. J Insect Sci 10:75

Pfiffner L, Luka H (2000) Overwintering of arthropods in soils of arable fields and adjacent semi-natural habitats. Agric Ecosyst Environ 78:215–222

Pfiffner L, Wyss E (2004) Use of sown wildflower strips to enhance natural enemies of agricultural pests. In: Gurr GM, Wratten SD, Altieri MA (eds) Ecological engineering for pest management, advances in habitat manipulation for arthropods. CSIRO Publishing, pp. 167–188

Rand TA, Tylianakis JM, Tscharntke T (2006) Spillover edge effects: the dispersal of agriculturally subsidized insect natural enemies into adjacent natural habitats. Ecol Lett 9(5):603–614

Rand TA, Tscharntke T (2007) Contrasting effects of natural habitat loss on generalist and specialist aphid natural enemies. Oikos 116(8):1353–1362

Risser PG, Karr JR, Formann RTT (1984) Landscape ecology: Directions and Approaches, Special Publication 2. Illinois Natural History Survey, Champaign, Illinois

Root RB (1967) The niche exploitation pattern of the bluegrey gnatcatcher. Ecol Monog 37:317–350

Rösler S (2007) Natur- und Sozialverträglichkeit des Integrierten Obstbaus. Ein Vergleich des integrierten und des ökologischen Niederstammobstbaus sowie des Streuobstbaus im Bodenseekreis unter besonderer Berücksichtigung ihrer historischen Entwicklung sowie von Fauna und Flora. Dissertation Universität Kassel, 2. Auflage

Rusch A, Valantin-Morison M, Sarthou J-P, Roger-Estrade J (2010) Biological control of insect pests in agroecosystems: effects of crop management, farming systems and semi-natural habitats at the landscape scale. A review. Adv Agro 109:219–259

Rusch A, Valantin-Morison M, Sarthou J-P, Roger-Estrade J (2011a) Multi-scale effects of landscape complexity and crop management on pollen beetle parasitism rate. Landscape Ecol 26(4):473–486

Rusch A, Valantin-Morison M, Sarthou J-P, Roger-Estrade J (2011b) Effect of crop management and landscape context on insect pest populations and crop damage. Agr Ecosyst Environ. doi:10.1016/j.agee.2011.05.004

Sandhu HS, Wratten SD, Cullen R (2010) Organic agriculture and ecosystem services. Environ. Sci. Policy 13:1–7

Sarthou J-P (1995) Haies composites et protection biologique: l'entomofaune associée aux essences ligneuses. In: Corroyer N, Garapon D, Arnal A (Eds) Mise en place et développement de haies composites en arboriculture biologique. Groupe de Recherche en Agriculture Biologique, Avignon, France: Annexe 2

Schader C, Pfiffner L, Schlatter C, Stolze M (2008) Umsetzung von Ökomassnahmen auf Bio- und ÖLN-Betrieben. Agrarforschung 15:506–511

Scheid BE, Thies C, Tscharntke T (2011) Enhancing rape pollen beetle parasitism within sown flower fields along a landscape complexity gradient? Agr Forest Entomol 13(2):173–179

Scherber C, Eisenhauer N, Weisser WW, Schmid B, Voigt W, Fischer M, Schulze ED, Roscher C, Weigelt A, Allan E, Beßler H, Bonkowski M, Buchmann N, Buscot F, Clement LW, Ebeling A, Engels C, Halle S, Kertscher I, Klein AM, Koller R, König S, Kowalski E, Kummer V, Kuu A, Lange M, Lauterbach D, Middelhoff C, Migunova VD, Milcu A, Müller R, Partsch S, Petermann JS, Renker C, Rottstock T, Sabais A, Scheu S, Schumacher J, Temperton VM, Tscharntke T (2010) Bottom-up effects of plant diversity on multitrophic interactions in a biodiversity experiment. Nature 468:553–556

Scheu S (2001) Plants and generalist predators as links between the below-ground and above-ground system. Basic Appl Ecol 2(1):3–13

Schmidt MH, Thewes U, Thies C, Tscharntke T (2004) Aphid suppression by natural enemies in mulched cereals. Entomol Exp Appl 113(2):87–93

Simon S, Defrance H, Sauphanor B (2007) Effect of codling moth management on orchard arthropods. Agric Ecosyst Environ 122:340–348

Simon S, Sauphanor B, Defrance H, Lauri PE (2009) Manipulations des habitats du verger biologique et de son environnement pour le contrôle des bio-agresseurs - Des éléments pour la modulation des relations arbre-ravageurs-auxiliaires. Innov Agron 4:125–134

Simon S, Bouvier JC, Debras JF, Sauphanor B (2010) Biodiversity and pest management in orchard systems. A review. Agro Sustain Develop 30:139–152

Smukler SM, Sánchez-Moreno S, Fonte SJ, Ferris H, Klonsky K, O'Geen AT, Scow KM, Steenwerth KL, Jackson LE (2010) Biodiversity and multiple ecosystem functions in an organic farmscape. Agr Ecosyst Environ 139:80–97

Solomon G (1981) Windbreaks as a source of orchard pests and predators. In: Thresh JM (ed) Pests, pathogens and vegetation: the role of weeds and wild plants in the ecology of crop pests and diseases. Pitman Books Ltd., London: pp. 273–283

Stephens MJ, France CM, Wratten SD, Frampton C (1998) Enhancing biological control of leafrollers (Lepidoptera: Tortricidae) by sowing buckwheat (*Fagopyrum esculentum*) in an orchard. Biocontrol Sci. Technol 8:547–558

Straub CS, Finke DL, Snyder WE (2008) Are the conservation of natural enemy biodiversity and biological control compatible goals? Biol Control 45(2):225–237

Thies C, Tscharntke T (1999) Landscape structure and biological control in agroecosystems. Science 285:893–895

Thies C, Steffan-Dewenter I, Tscharntke T (2003) Effects of landscape context on herbivory and parasitism at different spatial scales. Oïkos 101(1):18–25

Thies C, Roschewitz I, Tscharntke T (2005) The landscape context of cereal aphid-parasitoid interactions. Proc Royal Soc B-Biol Sci 272(1559):203–210

Tischendorf L, Farhig L (2000) How should we measure landscape connectivity? Landscape Ecol 15(7):633–641

Thomson LJ, Hoffmann AA (2006) Field validation of laboratory-derived IOBC toxicity ratings for natural enemies in commercial vineyards. Biol Control 39(3):507–515

Tscharntke T, Klein AM, Kruess A, Steffan-Dewenter I, Thies C (2005) Landscape perspectives on agricultural intensification and biodiversity—ecosystem service management. Ecol Lett 8(8):857–874

Tscharntke T, Bommarco R, Clough Y, Crist TO, Kleijn D, Rand TA, Tylianakis JM, Van Nouhuys S, Vidal S (2007) Conservation biological control and enemy diversity on a landscape scale. Biol Control 43:294–309

Tylianakis JM, Didham RK, Wratten SD (2004) Improved fitness of aphid parasitoids receiving resource subsidies. Ecology 85(3):658–666

UN United Nations (1992) Convention on Biological Diversity. Rio de Janeiro, 5 juin 1992

Wäckers F, Romeis J, van Rijn P (2007) Nectar and pollen feeding by insect herbivores and implications for multitrophic interactions. Ann Rev Entomol 52:301–323
Winkler K, Wäckers F, Pinto DM (2009) Nectar-providing plants enhance the energetic state of herbivores as well as their parasitoids under field conditions. Ecol Entomol 34:221–227
Wyss E (1995) The effects of weed strips on aphids and aphidophagous predators in an apple orchard. Entomol Exp Appl 75:43–49
Wyss E, Niggli U, Nentwig W (1995) The impact of spiders on aphid populations in a strip-managed apple orchard. J Appl Entomol 119:473–478
Yachi S, Loreau M (1999) Biodiversity and ecosystem productivity in a fluctuating environment: The insurance hypothesis. Proc Natl Acad Sci USA 96(4):1463–1468
Zaller JG, Moser D, Drapela T, Schmoger C, Frank T (2008) Insect pests in winter oilseed rape affected by field and landscape characteristics. Basic Appl Ecol 9:682–690
Zehnder G, Gurr GM, Kühne S, Wade MR, Wratten SD, Wyss E (2007) Arthropod pest management in organic crops. Ann Rev Entomol 52:57–80
Zhang W, Ricketts TH, Kremen C, Carney K, Swinton SM (2007) Ecosystem services and disservices to agriculture. Ecol Econ 64(2):253–260

Chapter 6
Agroecological Crop Protection in Organic Farming: Relevance and Limits

Jean-Philippe Deguine and Servane Penvern

Abstract Plant protection is one of the major issues in organic farming. Organic crop protection (OCP) strategies often rely on a limited number of methods that provide only partial control of pests and that induce lower yields and economic performances. As a result, farmers hesitate to adopt these strategies and doubts are cast on the ability of organic agriculture to feed the world. This chapter questions how agroecological concepts may contribute to OCP, while taking the different alternative schemes already developed to manage, integrate and design crop protection strategies into account. As demonstrated by a bibliographic analysis, Integrated pest management (IPM) remains the leading paradigm in crop protection. It also provides its foundational basis, giving priority to bioecological processes and alternative techniques to reduce pesticide use. Beyond IPM, agroecology is characterised by a holistic approach and the importance given to the design of a "healthy" agroecosystem. In practice, all these concepts are subject to various interpretations, and organic farming includes a variety of practices, ranging from intensive input-substitution to a comprehensive integrated approach. This paper provides key elements for crop protection in OF on the basis of the adaptation of the agroecological crop protection approach. Based on a successful case study of fruit fly management in OF in Reunion Island (France), we highlight three major pillars to design pest management strategies: sanitation, habitat manipulation and conservation biological control. Finally, in the field of crop protection, this paper shows that organic farming can be both a prototype for designing innovations and a source of practices to be extended to other types of agroecosystems.

Keywords Organic farming · Agroecology · Integrated pest management · Agroecological crop protection

---

J.-P. Deguine (✉)
CIRAD, UMR PVBMT, 3P, 7 chemin de l'IRAT 97410
Saint-Pierre, La Réunion, France
e-mail: jean-philippe.deguine@cirad.fr

S. Penvern
INRA, UR0767 Écodéveloppement, 228 route de l'Aérodrome,
Domaine Saint Paul, Site Agroparc, CS 40509, 84914 Avignon Cedex 9, France

S. Bellon, S. Penvern (eds.), *Organic Farming, Prototype for Sustainable Agricultures,*
DOI 10.1007/978-94-007-7927-3_6,

6.1 Introduction

Plant protection has been constantly evolving over the past century, wavering between the steady improvements in the use of modern pesticides (high concentration and ultra-low volume, spraying conditions and application timing) and the search for more rational and ecologically compatible methods of pest control. Because pesticides have been available and effective until now, agricultural systems were designed with the focus on productivity and/or income rather than on adaptation to the local pest and disease pressure. This resulted in input-dependent systems now regarded as non-sustainable. They indeed favour pest resistance, resurgences and the emergence of secondary pests, as well as environmental and human health problems due to chemical residues (Deguine et al. 2009; Krebs et al. 1999). Therefore, and despite a real socio-historical "lock-in" that has impeded the widespread diffusion of alternative techniques and strategies (Vanloqueren and Baret 2009), a switch towards improved knowledge and management of ecological (and biological) processes that occur in agroecosystems is now taking place.

Organic farming (OF) may be considered as a prototype for alternative crop production. In general, organic farming rejects synthetic compounds assumed to disrupt the complex dynamics of agroecosystems, and instead promotes natural regulation processes and relies on a restrictive list of active substances registered under organic agriculture guidelines (in Europe, EC 2092/91; Annex II B). The active ingredients must be of natural origin (derived from plants, animals or minerals, or of microbial origin), unless the extraction is unacceptable, in which case it may exceptionally be synthetically produced as long as it is chemically identical to the natural compound (e.g., synthesized pheromones) (Speiser et al. 2014, Chap. 4). The list is prone to controversies due to the lack of a scientific rationale for including or excluding certain substances, and the lack of knowledge about the properties and impacts of certain compounds (e.g., Biondi et al. 2012). Whereas some new synthetic compounds may be considered sustainable because of low environmental concern, energy savings and efficacy, some organically registered compounds are often of limited efficacy, must often be applied preventively and, in some cases, in relatively large quantities, leading to input-intensive practices with environmental hazards (Dayan et al. 2009; Sauphanor et al. 2009; Simon et al. 2014, Chap. 5). The most common example is sulphur that has deleterious side effects on soil microflora and fauna, requires more frequent spraying, and is sometimes unsuccessful at maintaining fungal diseases under control. Such a gap between organic principles and practices casts doubts on the possible contribution of organic farming to sustainable development (Darnhofer et al. 2010; Trewavas 2001).

This chapter questions how different alternative schemes that have already been developed to manage or design plant protection strategy concepts could contribute to organic pest management. Agroecological crop protection (ACP) and integrated pest management (IPM) are other alternative schemes that also promote ecologically-based pest management to reduce reliance on external inputs and increase sustainability. Both share several principles with organic crop protection (OCP):

preventive measures, a combination of techniques, biodiversity conservation, etc. Care must however be taken when comparing different concepts, given the variety of interpretations and the multiple forms they may take. In the first part of this paper, we present a bibliographic analysis of the relationships between each concept in terms of the way they are used in the literature concerning the crop protection sciences. In the second part, we propose an adaptation of the ACP in OF, based on three key elements: sanitation, habitat manipulation and conservation biological control. A 5-step strategy for crop protection in organic farming is then proposed, using the example of a successful case study: the agroecological management of fruit flies in organic farming in Reunion Island (France).

6.2 Relationships Between Integrated, Agroecological and Organic Pest Management: A Review of the Literature

This analysis is based on scientific views published in peer-reviewed journals and addresses the relationships between the three different concepts: OCP, ACP and IPM, based on the literature concerning the crop protection sciences. Assuming that concepts that co-occur more frequently tend to be related (Whittaker et al. 1989), a global bibliographic analysis based on co-word analysis was performed. We compared co-occurrences between Boolean topic search queries mentioning each or multiples of the "concepts" from the Web of Science database over a period of 36 years (1975–2012). The topic search queries, indicated in the caption below, were built to identify publications that referred to a minimum of one of the most commonly used terms of the concepts in their title, abstract or keywords. This method allows (i) a dynamic quantitative study of the co-occurrence frequency of terms, and (ii) a qualitative analysis of the way each concept is used.

IPM largely dominates the corpus and is also the oldest (first publication in 1975 compared to AE and OF that began in 1992), proving its role as the paradigm leader in the crop protection sciences (Fig. 6.1), whereas the number of papers referring to ACP in plant protection is still relatively low (63).

6.2.1 Relationships Between Integrated Pest Management and Organic Crop Protection

IPM and Organic Crop protection (OCP) Share a Common Characteristic: Integration

According to the standards and theories, organic crop protection (OCP) and IPM share common principles. Both aim to avoid the use of pesticides and give priority to self-regulating processes and preventive measures to regulate pests and patho-

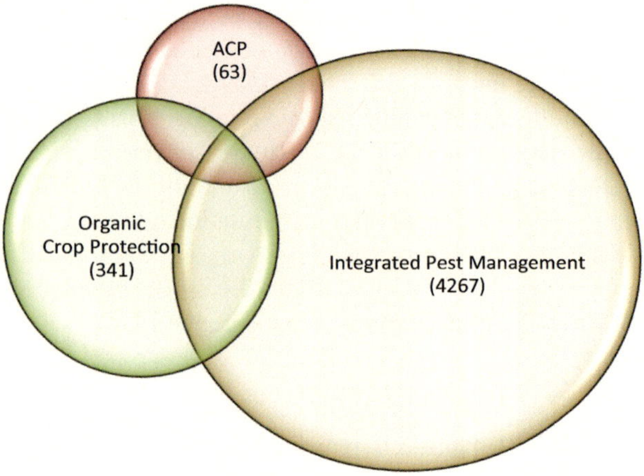

Fig. 6.1 Venn diagram for the occurrence of the three defined concepts in the literature concerning pest management. Overlapping circles display the number of publications that refer to two or three of the concepts. *ACP: agroecological crop protection. Topic search queries in the Web of Science database of: #1: AE (& PM) Topic = (("agroecolog*" OR "agro-ecolog*") AND ("crop protection" OR "pest management" OR "plant protection" OR "pest control")); #2: IPM (& PM) Topic = (("ntegrated pest management" OR "IPM" OR "integrated crop protection"OR "integrated plant protection"OR "integrated pest control") AND ("crop protection" OR "pest management" OR "plant protection" OR "pest control")); #3: OF (& PM) Topic = (("organic* agr*" OR "organic* farm*" OR "organic* and conventional*" OR "organic system" OR "organically-grown" OR "conventional* and organic*" OR "organic, conventional" OR "Organic Fruit Production" OR "organic prod*") AND ("crop protection" OR "pest management" OR "plant protection" OR "pest control")); #1 AND #2: AE & IPM (& PM)*

gens. IPM has been the fundamental paradigm in plant protection since the late 1960s, with the following key ideas: do not spray poisons unless it is necessary, and manage the ecosystem in such a way that it does not become necessary (Vandermeer 1995). Likewise, organic farming, as defined by the international food standards of the Codex Alimentarius, "*emphasizes the use of management practices in preference to the use of off-farm inputs [...] accomplished by using agronomic, biological, and mechanical methods, as opposed to using synthetic materials, to fulfill any specific function within the agroecosystems*" (FAO/WHO Codex Alimentarius Commission 1999). This principle is applied in plant protection and has been translated into protection rules in the current European organic production standards that mention the following specific principles (Article 5 (f) and Article 12 (g); RCE 834/2007): "*the prevention of damage caused by pests, diseases and weeds shall rely primarily on the protection by natural enemies, the choice of species and varieties, crop rotation, cultivation techniques and thermal processes*", and advising the use of pesticides only "*in the case of an established threat to a crop*".

Nevertheless, in the abundant literature on the adoption or failure of IPM, some authors report how IPM practices are not always consistent with these principles and why the effective reduction of pesticides is often limited. "Integration" remains the real conceptual foundation of IPM, with the understanding that IPM integrates

(i) a set of available control tactics, (ii) the estimated impacts of the various pest categories (arthropod and vertebrate pests, plant pathogens and weeds); and (iii) the scales of agricultural units targeted for pest management. However, IPM is a flexible concept and exists as a continuum where integration ranges from rather low to very high levels (Kogan 1998). Many authors describe IPM implementation as integrated *pesticide* management (Ehler 2006), mainly using *a single control method* to overcome *one limiting factor* in *a single pest approach* (Altieri et al. 2000; Kogan and Hilton 2009). The authorised access to cheap and efficient pesticides and easy ways to get rid of pests partly explain farmers' preferences for the use of pest control tools such as genetically-modified plants (GMP), insecticidal seed treatments or the follow-up management of pest populations with the use of *"therapeutic" materials*(Ferron and Deguine 2005; Lewis et al. 1997).

On the contrary, synthetic pesticides are strictly limited in organic farming. This restriction should lead practitioners to adopt alternatives and foster ecosystem services. Moreover, these alternatives are most often less efficient and must be combined to achieve successful control of pests, ensuring an integrated approach. However, pest management practices in OCP also display a continuum where "integration" ranges from rather high to low levels. The available pesticides are often of limited efficacy, must often be applied preventively and, in some cases, in relatively large quantities, leading to input-intensive practices (Dayan et al. 2009; Sauphanor et al. 2009). In orchard production, Penvern et al. (2010) identified two distinct pest management strategies among organic peach growers in southeastern France. Half of the growers used a preventive strategy based on input-substitution (substituting synthetic for natural products), resulting in more treatments than in conventional farming due to the lack of efficient alternatives. In contrast, the other half adopted an "integrated" protection strategy that primarily relied on cultural, biological and mechanical methods, resulting in fewer treatments and a higher diversity and abundance of natural enemies than conventional ones.

IPM and OCP share common principles and both present in practice different degrees of integration up to a limited number of methods and targets. In these cases, practices are not very consistent with their principles, and the differences with "conventional farming" are not very clear either, even in the case of input-substitution. Other factors must thus be involved in the adoption of an integrated approach

Results from the Co-Occurrence Analysis Show Different Interpretations and Distinct Relationships

A total of 28.15% of the publications on OCP also refer to IPM, whereas only 2.25% of the publications on IPM refer to OCP (Fig. 6.1). If OCP and IPM share common principles, three types of relations can be distinguished from the literature: inclusion, comparison and transition.

Inclusion: Organic farming constitutes an appropriate model (or "framework") to implement IPM strategies or practices, including crop rotations, choice of cultivars, prophylactic biocontrol, insect-trapping methods, field and pest scouting, use of less toxic pesticides and information-based decision tools. Those techniques are recognised as conditions for successful OCP (Delate and Friedrich 2004). OCP is also considered in many cases as a prototype that may catalyse IPM principles given its constraints on chemical-synthesized pesticides that force organic practitioners to adopt prophylactic measures. For example, Jacobson et al. (2003) demonstrated in their study that organic farmers performed more pest scouting than conventional ones. Many techniques developed in IPM are thus particularly appropriate and implemented in OF (e.g., Boisclair and Estevez 2006), making it a relevant research object to investigate predator-prey interactions (Furlong and Zalucki 2010).

Comparison: In a majority of publications, OF and IPM are used interchangeably in comparative studies, since they are both considered as alternative farming systems to conventional (e.g. Mzoughi 2011) or agricultural intensification (Peterson et al. 2000). Techniques are applicable in both systems (e.g., application of kaolin in Daniel et al. (2005); mating disruption in Vanbuskirk et al. (2008); biodiversity in Xu et al. (2011)). Such comparisons display various results, either in favour of OCP or in favour of IPM. Following are some examples in favour of OCP: on pesticide residues (e.g. Mladenova and Shterevad 2009), of fruit nutritional properties (Fernandes et al. 2012) or of the lower incidence of protection methods on natural enemies (Rajapakse 2000). Botanical insecticides used in OCP are thus considered as an alternative to IPM (Mumford 1992). In contrast, recent studies emphasize the toxicity of organic pesticides (e.g. Biondi et al. 2012) and the lack of efficiency of OF performances (Elliot and Mumford 2002), supporting the hypothesis that overall benefits should be achieved by the use of legislation to restrict more harmful technologies.

Transition: This relationship concerns a small number of publications (5.2%). At the individual scale, IPM is considered as an intermediate phase that may facilitate organic conversion (Lamine 2011; Zinati 2002). An abrupt transition from conventional to organic may be risky if the number of pests is high and if alternative practices are not yet in place, inducing crop losses during the transitional period. McSorley (2002) refers to a hybrid system where conventional tactics are decreased in favour of organic ones.

Conclusion: IPM and Of Enrich One Another

The development of IPM has contributed to OCP in three different ways. As emphasized above, it first contributed in terms of research and development for new pest management tools and technologies. Its integration into national development and research programmes fostered the development of techniques to (i) increase pesticide use efficiency, particularly through decision tools and spraying materials, (ii) reduce the drawbacks of protection methods and environmental hazards, and (iii) alternatives for pesticide substitution. Second, conventional agriculture is no longer the reference in system comparison and performance assessment. Since both systems share common principles, criteria and indicators are more consistent with

OCP principles and practices. Third, it contributed in terms of transition pathways, where IPM can be considered a step towards organic conversion.

Conversely, organic farming represents a good prototype for the implementation and testing of IPM principles. It constitutes a framework with specific constraints that may catalyse technological innovations for more sustainable pest management.

6.2.2 Contributions of Agroecology to Organic and Integrated Crop Protection

To Design Healthy Agroecosystems

In the 1970s, agroecology initially dealt with crop production and protection aspects, in particular, using the ecological sciences to design and manage "healthy" agro-ecosystems, i.e., more resilient and less susceptible to pests (Harper 1974; Gliessman 2007; Wezel et al. 2009). Agroecology therefore promotes an understanding of agroecosystems for a "positive plant-mediated" approach (as opposed to "negative pest-mediated") to manage the system for beneficial processes and cycles, and keep pest populations in check (Altieri and Nicholls 2000; Gliessman 2007)[1]. In order to do so, and beyond the classical techniques of integrated crop protection, ACP relies on the *design* of diversified agroecosystems (crops, trees and animals in spatial and temporal arrangements) and on the use of low-input technologies to favour the establishment of a functional biodiversity that performs key ecological services such as biological control, nutrient recycling and pollination. ACP thus operates as of the conception of the system and at larger scales in time and space than IPM, from a single crop cycle to several years, and from a single field to an agroecosystem or a landscape. This transition towards system design/redesign corresponds to a major change in the pest management paradigm (Hill et al. 1999; Hill, Chap. 22).

Agroecology is recognised by many authors as an umbrella for alternative farming approaches, of which OF is often mentioned as a relevant model (Francis 2009b; Lotter 2003). In fact, there are similarities with the first vision of organic farming. Sir Albert Howard, one of its founders, defined pests as nature's censors: "*Insects and fungi are not the real cause of plant diseases but only attack unsuitable varieties or crops imperfectly grown. Their true role is that of censors for pointing out the crops that are improperly nourished and so keeping our agriculture up to the mark.*" (Howard 1943). "Health" and "ecology" are still two of the four basic principles of OF defined by the IFOAM[2] in 2005. In practice, many authors relate that organically-managed agroecoystems are also designed to promote beneficial biotic and abiotic processes (Lampkin 1990; Letourneau and Goldstein, 2001; Garratt

---

[1] see Table 16.1, p. 219.

[2] The IFOAM is the International Federation of Organic Agriculture Movements. It defines four basic principles of organic agriculture: health, ecology, fairness and care; http://www.ifoam.org/en/organic-landmarks/principles-organic-agriculture.

et al. 2011). Crop rotations, soil structuring practices, plant nutrition and host plant resistance are examples of practices frequently implemented in OF.

However, differences between ACP and OCP exist and rely firstly on the lack of codification and regulations that provide the framework for OCP, whereas synthetic pesticides and GMP are, by definition, not excluded in AE. It should, however, be emphasized that the most commonly used GMP for plant protection (one or several genes coding for a resistance to a specific insect, a gene coding for tolerance to a specific herbicide) do not fit the general agroecological principle that aims at addressing crop protection in its entirety (Altieri and Rosset 1999; Altieri 2005). Moreover, some researchers criticize the trend observed in OF towards the replacement of agroecological practices such as rotations, vegetation design, and knowledge-intensive techniques by a set of energy and capital intensive organic "technology packages", and denounce the deterioration of organic standards and core values (Rosset and Altieri 1997; Trewavas 2001; Guthman 2004; Darnhofer et al. 2010; Hill, Chap. 22). There are large-scale commercial organic farms that do not considerably differ from their conventional counterparts, using power-dependent machinery, pest-susceptible crop varieties and adopting monocultures (Rosset and Altieri 1997). These simplified systems lack natural regulatory mechanisms and therefore remain dependent on external inputs to substitute functions of pest control (Altieri 1999).

Results from the Co-Occurrence Analysis Stresses Its Role of Umbrella for Alternative Pest Management and Research Programmes

A total of 38 % and 25.4 % of the 63 publications on ACP refer to IPM and OCP, respectively. Both are interpreted as effective prototypes for the implementation of agroecological principles. IPM is generally referred to as a strategy (or "programme") based on technologies that are compatible with agroecological principles. OF offers an appropriate framework or model of production where agroecological principles may be implemented.

Two major interpretations of ACP can be distinguished according to the way publications refer to the term "agroecology".

First, ACP provides *principles* and protection methods that promote agroecoystem management and a design that fosters functional biodiversity (Tscharntke et al. 2012) as well as natural regulatory processes. The main pillars are focused on diversified cropping agroecosystems (including the landscape) (Ratnadass et al. 2012) and preventive methods (Deguine and Ferron 2006). Farmer knowledge and information transfer are another important aspect, especially for publications dealing with the adoption or progression within the different stages of IPM (Ponce 2007). Nicholls and Altieri (1997) suggest dissemination, broadening education and outreach for successful ecological pest management in IPM. To manage agroecosystem functioning assumes an understanding of the complexity of interactions and local specificities to adequately adapt practices (Tscharntke et al. 2012).

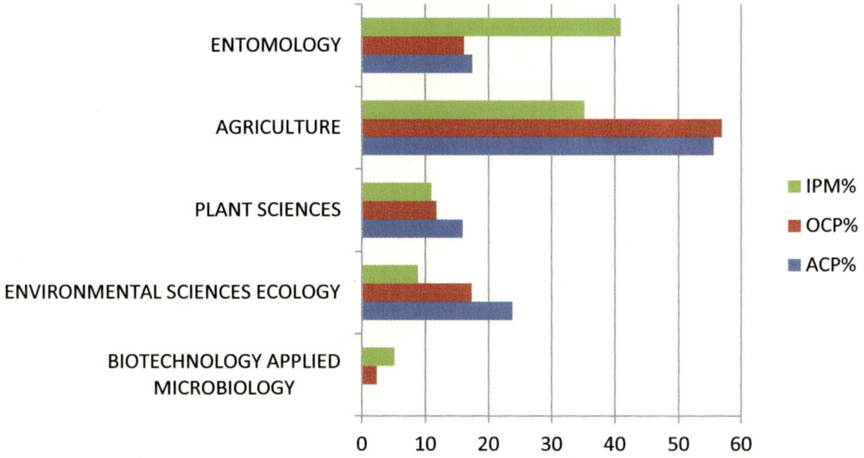

Fig. 6.2 Distribution of the publications (in % of the total number for each corpus) on the four principal research areas for each concept corpus, using the WOS classification

Second and more importantly, authors from both corpuses refer to ACP as an *approach.* The first publication by M. Altieri dates back to 1992 and defines agroecology as "the scientific bases of alternative ecologically based agriculture" (Altieri 1992). ACP promotes the integration of the multiple components of agroecosystems and long-term comparison (Delate 2002), multi-disciplinarity (Valenzuela and Defrank 1995), particularly biological and ecological disciplines, and a wider systemic approach, including uncultivated areas and landscape approaches (Deguine and Ferron 2006; Deguine et al. 2008). The environment, in the general sense of the milieu where bioecological processes take place, is recurrent. An analysis of the research area using Web of Science tools (Fig. 6.2) shows the importance of environmental sciences and ecology. A total of 46 % of the 26 publications on ACP and IPM refer to agroecology as an entity or "zone" where functional biotic and abiotic processes occur (e.g., Rodriguez et al. 2012). "Agroecological conditions" are then necessary to ensure adoption (Holland et al. 2009) and efficiency of site-specific practices (Kremen and Miles 2012).

Third, and derived from the two former interpretations, AE may be used as a *referential* for pest management evaluation, especially among OCP publications (e.g., Peck et al. 2010; Mena et al. 2012).

Conclusion: A Holistic Approach for Local Specific Practices

One major contribution of agroecology to organic and integrated pest management relies on the reinforcement of bioecological processes through *system design* to achieve resilient agroecosystems and to thus minimise pest occurrence. OCP should intrinsically be based on a holistic approach (e.g., healthy soil to feed the plant and contribute to plant health, preservation of the ecosystem service of pest suppression, etc.) and should not be reduced to a ban on synthetic pesticides. This transition towards system design/redesign corresponds to a major change in the pest management paradigm (Hill et al. 1999) that also occurs in the ecological intensification of agriculture (Malezieux et al. 2012; Tscharntke et al. 2012). Ecological knowledge, constant monitoring and adaptation capacities are key prerequisites to manage the complexity and dynamics of agroecosystems (Bird et al. 2009). Research approaches and agendas must thus be reviewed to respond to these challenges.

6.2.3 At the Crossroads of the Three Concepts

The number of publications at the intersection between the three concepts is too low to allow for interpretation but reflects the aforementioned analyses. OF is used as a framework for IPM technologies and strategy implementations (Sharma et al. 2011; Lotter 2003), while AE provides the approach and principles for sustainable pest management. Ponce (2007) refers to agroecological knowledge as one condition for progression through the four stages of IPM.

IPM and AE approaches appear to be complementary for designing organic crop protection strategies. Whereas IPM recommends a parsimonious use of inputs and ecological engineering to foster regulatory processes through biological control, AE focuses on the design of resilient and healthy agroecosystems that assume a holistic understanding of the interacting processes that occur inside and outside the agroecosystem. Prevention, biodiversity preservation and the knowledge of shareholders appear as common key elements for such ecologically sound pest management.

6.3 Application of Agroecological Crop Protection to Organic Farming

6.3.1 Adopting an Agroecological Strategy for Organic Crop Protection

The implementation of agroecological principles for OCP was proposed by Deguine et al. (2009), by prioritising preventive measures and by giving priority to the promotion of biodiversity and soil health. The application of this strategy to OF is presented below. The choice of farm location as well as the location of fields to be cultivated within the farm are of utmost importance and constitute the first stages of a pest management strategy (Wyss et al. 2005). They, in fact, determine the framework and profile of the pest management strategy. Furthermore, the strategy corresponds to the development of a management plan for the farm at different time and spatial scales, taking the agroecological characteristics of the overall agroecosystem into account. Another prerequisite is to comply with crop protection regulations applied to a given crop at the international, national and regional scales. These three preliminary conditions are not integrated into the following stages because they do not entail a choice on the farmer's part in most of the situations.

Improved Plant Health Through the Implementation of Measures to Ensure Healthy Soil

- Agronomic measures aimed at optimising soil quality and functioning (structural stability of the soil, availability of water for primary production, soil fertility, regulation of the microclimate). These measures depend on farmers' strategies and pedo-climatic conditions and may include rational irrigation as well as organic amendments.

Implementation of Management Practices for Plant Communities (Cultivated or Not) to Reduce Populations and the Impact of Bioagressors and/or to Increase Populations and the Impact of Beneficial Fauna

- Choice of species and varieties cultivated for increased plant tolerance or resistance, or reduced susceptibility;
- more extensive use of rotation crops, associated crops or intercropping;
- practices that encourage conservation biological control (e.g., non-toxic pesticides applied outside of the activity time of natural enemies);
- plant biodiversity management practices from the field to the landscape scales/levels: systems under permanent plant cover, minimum tillage, weed management, insertion of trap plants in the field or around the field, establishment of refuge areas for natural enemies (addition of grass or flower strips, field borders, and restoration of corridors).

Implementation of Prevention Measures for Preventing Infestation

- Use of healthy registered plant varieties;
- regular collection and systematic elimination of crop residues or other sources of disease or pest contamination;
- implementation of concerted practices at the local level in relation to both time and space;
- biodiversity management: avoid host plants of major bioagressors and favour trap plants in the vicinity of the field.

Assessment of Socio-Economic, Environmental and Health Risks and Decision-Making for Curative Measures If Needed

- Assessment of production losses for existing crops, even at scales other than the cultivated field, e.g., on other crops in the area and on future crops, and taking negative collateral impacts into consideration;
- taking environmental and health indicators into account;
- use of sampling techniques adapted to a field, to a group of fields, to a farm and to overall agroecosystems scales, with the possible assistance of regional agricultural extension services;
- decision-making using decision support and consensus-building tools, taking local and evolutive multicriteria intervention thresholds (economic, social, environmental) into account, as well as the risk of the occurrence of resistance phenomena.

Only in the Case of Absolute Necessity: Use of Curative Intervention Measures

- Compliance with OA specifications, particularly concerning practices at the farm and agroecosystem scales;
- at this stage, priority is given to cultivation techniques (e.g., defoliation, topping), inundative biological control, physical control, biotechnical control, mating disruption, etc.
- as a last resort, use of biopesticides allowed in OA (biological or mineral) with the lowest ecological impact, chosen to avoid the occurrence of resistance phenomena or secondary effects on non-target organisms.

6.3.2 A Case Study: Fruit Fly Management in Organic Farming in Reunion Island

The application of agroecological crop protection to a case study requires consideration of both the pests and the context, including the three main components of agroecological crop protection (prevention, plant biodiversity manipulation and conservation biological control) and the different steps of the proposed strategy.

Fruit flies (Diptera: Tephritidae) are among the most destructive and widespread pests of horticultural systems in the tropical and subtropical areas of the world (White and Elson-Harris 1992). Although they have been the subject of many studies because of their economic impact, their control remains problematic in most cases and requires large amounts of pesticides. This situation is exacerbated under insular and tropical conditions, as is the case of Reunion Island and, of course, in OF. The melon fly *Bactrocera cucurbitae* (Coquillett), the Indian Ocean cucurbit fly *Dacus demmerezi* (Bezzi) and the Ethiopian cucurbit fly *Dacus ciliatus* (Loew) are the major pests of horticultural crops in Reunion Island. They are able to attack several species of cultivated cucurbits such as zucchini (*Cucurbita pepo*), pumpkin (*Cucurbita maxima*), chayote (*Sechium edule*) and cucumber (*Cucumis sativus*) that are the most commonly cultivated crops. Among the three species, the melon fly causes the most economic damage, and crop losses can reach 100%. During the last years, research has allowed us to gather knowledge about the bioecology of fruit flies (Ryckewaert et al. 2010; Veyssières 1999), making it possible to apply the principles of agroecological crop protection not only in "conventional" agriculture, but in OA as well (Atiama-Nurbel and Deguine 2010).

Agroecological Practices that Improve Soil and Plant Health

Permanent covering of the soil with local weeds or chosen plant species and good management of field surroundings are promoted for their positive impact in terms of soil health (fertility, structure, control of erosion and evapotranspiration). These practices complement the positive impact of the techniques described below in terms of crop protection.

Habitat Manipulation Through Trap Crops

Several plants are known as roosting sites for cucurbit flies, and can be used to concentrate the populations of adult flies (McQuate and Vargas 2007). Some plants have been tested in Reunion Island, including corn and cane grass that can thus become the central location of fruit fly management. Three types of trap plant systems have been designed using corn: borders around the field, patches within the field and strips within the field. In some situations, natural tree borders can also be used as trap plants (Colour plate 06). The population of adults roosting on the corn border of a zucchini field was approximately a thousand times higher than the populations counted on the crop itself (Fig. 6.3) a result confirmed at different sites over several years (Deguine et al. 2012a).

After concentrating adult flies on corn, it is necessary to avoid gravid females from moving to host plants (the crop) and laying eggs on the fruits. A curative method is therefore proposed (see "Assisted Push Pull" below).

Preventive Measures: Sanitation

Each fruit may host many fruit fly eggs in the vegetation bordering crops. One cucurbit fruit, infested and fallen to the ground, can allow several hundred adults to emerge, (i.e., more than 200/kg of cucumber, more than 340/kg of pumpkin, and more than 500/kg of zucchini; (Deguine, unpublished data). Our sanitation approach mainly relies on risk prevention using an augmentorium (Deguine et al.

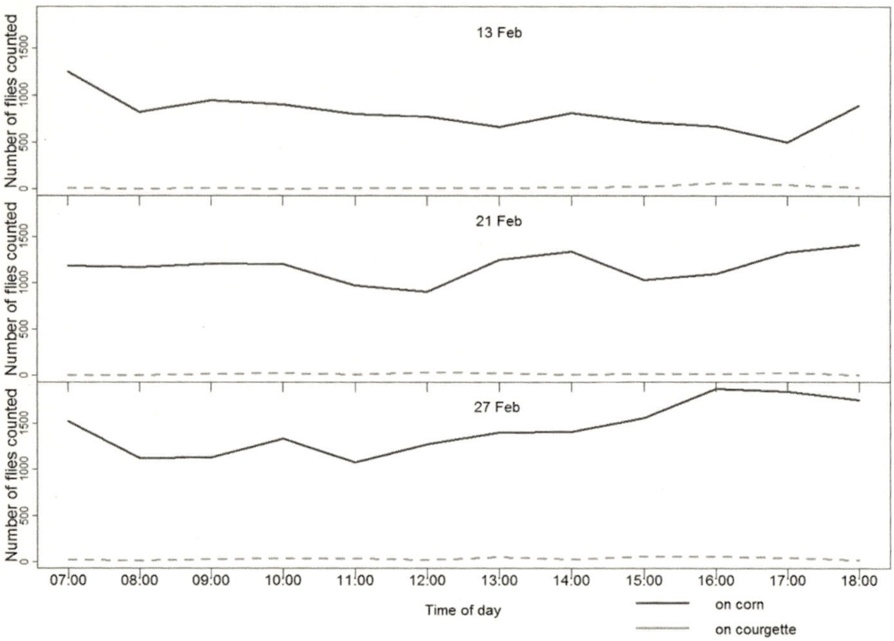

Fig. 6.3 Concentration per hour of adult fly populations (*Bactrocera cucurbitae, Dacus ciliatus and Dacus demmerezi*) during the day (7:00–18:00) on a corn border around a zucchini crop (Piton Bloc 2008; Atiama-Nurbel, unpublished data)

2011) (Colour plate 06). This tent-like structure confines rotting fruits collected on the ground of the field, preventing the agroecosystem re-infestation by the next fly generation. The first role of the augmentorium is sanitation (retaining flies that emerge from the infested fruits collected). Furthermore, it has an opening fitted with a mesh netting on its top, which prevents adult flies from escaping, while allowing the escape of beneficial parasitoids. It can thus be considered as a tool for conservation biological control. Organic farmers readily accept the technique and some of them also use the augmentorium to produce compost, mixing cucurbit fruits with organic matter and sugar cane stems. Farmers consider that it is simple, effective, environmentally safe and not time-consuming. In the future, the use of augmentoriums can be considered in fields, rural villages and towns, linking agroecological crop protection within OF to urban ecology.

Monitoring and Curative Techniques

Male Annihilation Technique. Populations of *B. cucurbitae* and *D. demmerezi* were monitored during the year using male cue-lure bait traps without insecticide, in order to evaluate the critical period for management. In addition, a design of "experimental reference fields" was proposed, including fields of cucurbits untreated

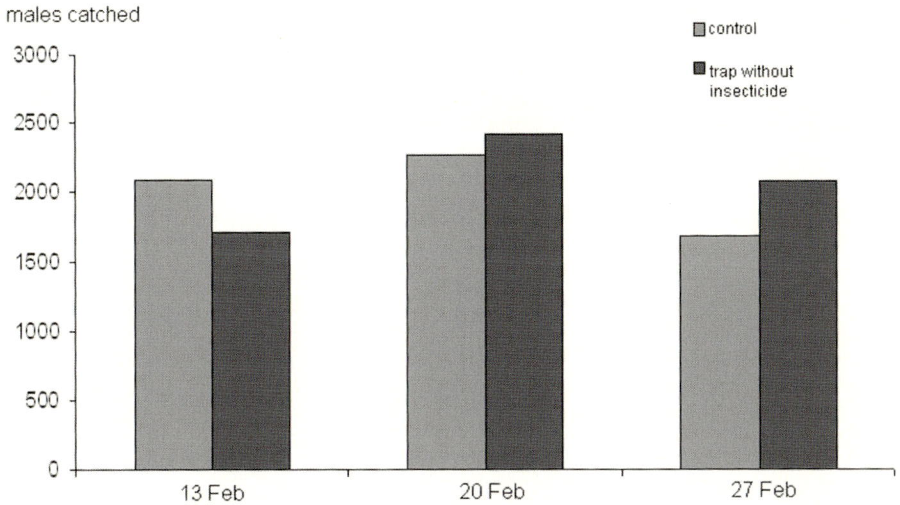

Fig. 6.4 Number of males of *B. cucurbitae* caught at three dates by the control trap (cue-lure and diclorvos) in *grey*, and by the implemented trap for OA (cue-lure without insecticide) in *black* (Les Avirons 2008). ANOVA (six replications per date) did not show significant differences between these two types of traps

throughout the year where observations of adult flies are performed once a week. This experimental design made it possible to evaluate the seasonal abundance of the populations, the relative abundance and the sex ratio of the three species within the communities (Deguine et al. 2012a). A similar male cue-lure trap free of insecticide is used in a Male Annihilation Technique (M.A.T.). Under the M.A.T., male cue-lure bait traps are used in fields at high densities to catch the maximum number of males and to disrupt the equilibrium and the sex ratio of the population. In "conventional farming", the traps used until now associated male cue-lure to attract males and an insecticide to kill them. In OCP, these traps must be free of insecticide, so a prototype of lure-trap has been implemented using a recycled plastic bottles (Colour plate 06). This trap is effective and easy to implement, and it is also now used even on "conventional" farms (Fig. 6.4). These results have been extended to *Bactrocera zonata* (using methyl-eugenol) and *Dacus demmerezi* (using cue-lure) in different situations, both for monitoring and for M.A.T.

Assisted Push-Pull. It was previously observed that adult flies can be trapped on corn borders around the cultivated field. To avoid females going from corn borders to the host crop and laying eggs on the host fruits, the technique of "attract and kill" is used. "Syneïs-appât"® (Dow AgroSciences, U.S.A.), approved for OF, is a combination of an attractant and feeding stimulant (99.98 %) and of Spinosad (a biological insecticide) (0.02 %). This product can be applied in very low quantities of active ingredients/ha on trap plants (instead of on cultivated plants) as a spot spray, to attract adult flies and kill them. Syneïs-appât is known to be attractive to flies

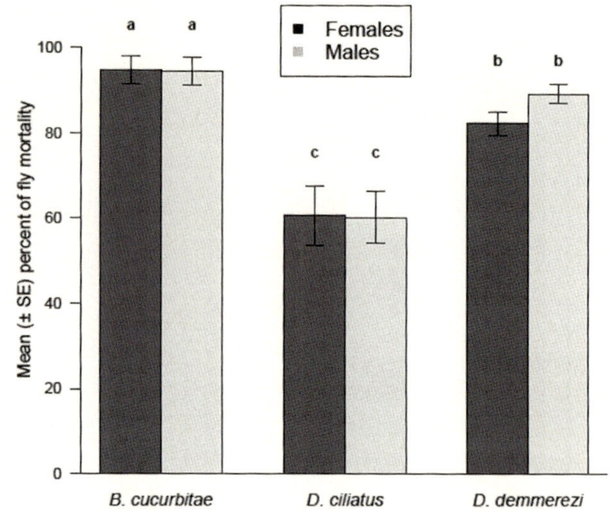

Fig. 6.5 Mortality of adults of the three species 7h following application of Synéis-appât® (Saint-Pierre 2009) (Deguine, unpublished data). *500 males and 500 females of each species were analysed, five replications. Bars with different letters are significantly different (α = 0.05. The ANOVA results of the mortality rate of protein-deprived flies (7-days-old) of the three species (males + females) showed that there was a significant species effect ($F = 47.20$; $df = 2$; $P < 10^{-8}$)*

within a 5–10-meter radius, depending on the wind conditions. Despite a significant species effect (Deguine et al. 2012b), Synéis-appât® represents an appropriate Attract & Kill tool to manage the populations of flies concentrated on trap plants in an overall Assisted Push-Pull System (Fig. 6.5). However, the risk of pesticide resistance build-up in fruit flies could result in the recommendation, for the sake of sustainability, to alternate with non-biological pesticides in combination with the food attractant (in which case it would not comply with the scope statement of organic farming).

Conservation Biological Control: Promising Perspectives?

Overall, conservation biological control was favoured by promoting cultural practices that improve the development of populations of natural enemies, particularly terrestrial predators (ants, spiders, staphylinids, etc.). For example, their populations are significantly higher when a permanent plant cover is present in the field. In addition, the parasitoid *Psyttalia fletcheri* (Hymenoptera, Braconidae, Opininae), was introduced from Hawaii at the beginning of the 2000s and is now well established in Reunion Island but does not contribute to the control of melon fly, probably because of the chemical pressure in intensive cropping areas. However, its impact could be significant in organic farming. Likewise, beneficial entomofauna for conservation biological control is expected to rise with the strong decrease in pesticide spraying/application, coupled with the combination of augmentoriums and border plants. This is therefore an outcome of the previous practices that do not represent an additional workload for the farmer (Colour plate 06). We are currently evaluating entomological diversity in some areas to assess this impact.

The Case of Chayote

In Reunion Island, chayote is traditionally and extensively cultivated for its leaves and is also cultivated under trellis for fruit production. Farmers usually ascribe yield losses to the three species of cucurbit flies (Tephritidae). The agroecological approach implemented in such seasonal-perennial agroecosystems stresses the relevance of maintaining permanent vegetable cover on the ground and enhancing conservation biological control by offering favourable habitats to natural enemies, especially generalist predators. Our studies, carried out over three years on organic farms and on untreated chayote crops, showed the following results: (i) there is no incidence of blemishes caused by cucurbit flies on the growth of the fruit; (ii) only 6 % of fruits support the emergence of adult flies (on a sample of 587 chayote fruits collected); (iii) the fall of the fruit from the arbour to the soil is not due to cucurbit flies (93 % of 197 fallen fruit were not blemished by the flies); (iv) *D. ciliatus* is the only species that can emerge from the fruit in significant numbers. It is now accepted that chayote cultivation does not require any insecticide protection aimed at cucurbit flies. As a result, most of the chayote producers are moving, via the "agroecological" approach, from the "conventional" to the "organic system".

Efficacy of This Agroecological Protection and Socio-Economic Considerations

This ACP has been implemented since 2009 on conventional farms and on four experimental organic farms in Reunion Island, in a programme called Gamour (http://gamour.cirad.fr). Two years after starting field operations, the results are considered highly encouraging (Augusseau et al. 2011). Monitoring data show that populations of *D. demmerezi* and *B. cucurbitae* have generally been maintained at low levels from the beginning of the programme onwards. However, we still observe limited punctual outbreaks that seem to be correlated with climatic parameters. The impact of the flies is now very low and no insecticide is used. Organic farmers are technically satisfied with the agroecological package. Yield losses are very low compared to the previous situation. The pilot areas totaled approximately 50 ha of vegetable crops, of which 10 ha were devoted to chayote and a variable part to other cucurbits (mainly zucchini, pumpkin and cucumber). The socio-economical results were compared to data obtained from "conventional farms" where conventional (=chemical) protection was applied (Deguine et al. 2011). The yields tend to be slightly higher under agroecological protection (average of 19.3 t/ha for the agroecological package vs. 13.1 t/ha for conventional-chemical protection), and losses due to fly infestations appear to be lower than in conventional protection (average of 13 % vs. 34 %, respectively). Table 6.1 presents the results concerning the protection (quantity of active ingredients applied, cost, time needed per week). These results confirm that ACP, compatible with OF requirements, is cheaper and more efficient that the conventional protection based on chemical insecticide use.

Table 6.1 Comparison of the protection modalities between conventional and agroecological protection from 2009–2011 (on 24 zucchini crop cycles) (in Deguine et al. 2011)

Modalities of protection	Conventional	Agroecological
N of applications/ week	1–2	2
Commercial product	Cyperfor-Danadim	Syneïs-appât
Active ingredients	Cypermethrin-dimethoate	Spinosad
Volume applied/week	1–2 l/ha	0.4 l/ha
Quantity of active ingredient applied/ week	100–800 g/ha	0.008 g/ha
Place of application	All crop plants	Spots on the trap plants
Time to spray one hectare/week	3–6 h	1 h
Cost of protection/ week	€ 44–88	€ 21–37

In conclusion, the case study of ACP against cucurbit flies in Reunion Island confirms that it is compatible with the OF approach and requirements. However, it is not always the case, e.g., with conservation agriculture that may rely on both external inputs such as synthetic herbicides and bioecological processes that make it "agroecological", but not "organic". The technical package is now going to be extended to all cucurbit farmers in Reunion Island. In this respect, OF can be considered as a prototype for carrying out studies for both organic and conventional farming. Furthermore, new initiatives based on the agroecological approach are being proposed, including the cultivation of mangoes without insecticide, representing a major step towards organic mango production.

6.4 Conclusion and Perspectives: OF as a Research Laboratory and a Prototype for Designing Innovations in Crop Protection

As a result of the design of diversified agroecosystems and the use of "low input" technologies, agroecology aims at establishing stable yields, biological soil fertility and natural plant regulation within a balanced bioecological environment. Agroecological crop protection therefore appears to be relevant for pest control in OF. The pest control strategies and the application presented in this chapter confirm this hypothesis. OCP relies by definition on input substitution applied according to IPM recommendations for a better efficiency of input use. A major contribution of agroecology to agriculture has also been to base all phases of the production system on sound ecological principles, with the aim of designing economically and ecologically sustainable agroecosystems (Altieri et al. 2000; Gliessman 2007; Hill et al. 1999; Vandermeer 1995; Zehnder et al. 2007). The case study of fruit fly agroecological management in Reunion Island shows that ACP can be applied in OF conditions, even under tropical and insular conditions, where insect populations and the number of biological cycles are greater than in temperate conditions.

Plant protection research has followed the same pathways as plant protection strategies in practice. Most research programmes have heretofore focused to single pests, single techniques, and at the plant or field level. The theory guiding conventional agricultural management is based on a "command-and-control" approach that emphasizes the simplification of natural systems, leading to the development of practices and inputs aimed at reducing variation and uncertainty. This reductionist approach has deconstructed cropping systems into specific components that are most often studied separately for scientific purposes (Drinkwater 2009). Yet, the application of adaptive, ecosystem-based management can only be supported by research that is grounded within an agroecological conceptual framework. Only the combination of a holistic and functional approaches makes it possible to organise and understand the biocomplexity inherent to all natural systems and agroecosystems.

In addition, the rules of OF make it necessary to find technical solutions without synthetic insecticides (e.g., traps for MAT). In this respect, organic farming systems represent laboratories for specific studies and research on bioecological processes (Furlong and Zalucki 2010). The absence of treatments with synthetic pesticides in organic farming does not generally lead[3] to strong artificial disturbances or to adverse effects on the functional biodiversity, particularly on natural enemies of pests (predators, parasitoids), "soil engineers" (earthworms, ants, termites, moles), microbial soil fauna or pollinators. These conditions make it possible to carry out cognitive research on the bioecological processes that determine agroecosystem functioning. For example, the services provided by functional biodiversity can be measured at high levels in these "laboratories", whereas they may be low in other types of farming systems. The results obtained are generally well adapted to other types of agriculture (e.g., pheromone traps for fruit flies in Reunion Island). The research results obtained in OF can therefore be applied to both organic and conventional farming.

Because only a limited range of suppressive pest control tactics are available for organic growers, knowledge-intensive cultural practices form the basis of an agroecological organic pest management programme (Nicholls and Altieri 1997). Farmers are therefore more prone to revise their protection strategies towards predominantly preventive processes rather than curative ones. Research is also on the rise today in OF systems, and many technical innovations can be expected as a result of the evolution of protection practices. The challenge for organic farmers and researchers has been to identify sets of context-specific practices that, in combination, are effective in preventing economic pest damage (Francis 2009a; Zehnder

[3] Some mineral insecticides, even if they are allowed in organic farming, can have collateral effects on non-target insects. For example, although Spinosad has low toxicity for most beneficial insects, initial acute laboratory tests indicated that Spinosad is intrinsically toxic to pollinators (Mayes et al. 2003) and biological control agents (Biondi et al. 2012).

et al. 2007). Current national policies to decrease pesticide use and new market opportunities to promote sustainable agriculture might further encourage empirical and experimental work in applied ecology. The growing restrictions in the area of pest control where the use of pesticides is increasingly limited or unauthorised catalyse the research and development of innovative biopesticides, pesticide application technology and more extensive training of farmers and advisers.

Acknowledgments We would like to thank all of the readers who helped us to improve the quality of this paper and who contributed to the definition of each of the concepts herein. We are also grateful to Guillaume Ollivier for sharing his knowledge about bibliometry with us and his invaluable help in these discussions.

The case study relies on a vast effort provided by many people who unfortunately cannot all be mentioned here, although we acknowledge each and every one of them. We particularly thank Marie-Ludders Moutoussamy, Cédric Ajaguin Soleyen, Toulassi Atiama-Nurbel and Pascal Rousse for technical assistance in conducting experiments. We also acknowledge Dow AgroSciences (France) for permission to conduct trials on Synéïs-appât. Gamour was operated by a fruitful collaboration between ASP, DAAF, the Chamber of Agriculture of Reunion Island, CIRAD, Farre Reunion, FDGDON, GAB, Université de La Réunion, Takamaka Industries, SCA Terres Bourbon and Vivéa Réunion. Funding for this research was provided by Odeadom (Office de Développement de l'Economie Agricole des Départements d'Outre-mer), the European Commission, the Regional Council of Reunion Island, the General Council of Reunion Island, CIRAD and the French Ministry of Food, Agriculture and Fishing through a CAS-DAR grant.

References

Altieri MA (1992) Sustainable agricultural-development in latin-america—exploring the possibilities. Agric Ecosyst Environ 39(1–2):1–21

Altieri MA, Nicholls CI (2000) Applying agroecological concepts to development of ecologically based pest management strategies. National Research Council, Washington DC

Altieri MA, Rosset P (1999) Ten reasons why biotechnology will not ensure food security, protect the environment and reduce poverty in the developing world. In: Sherlock R, Moorey JD (eds) Ethical issues in biotechnology. Rowman & Littlefield

Altieri MA (2005) The myth of coexistence: why transgenic crops are not compatible with agroecologically based systems of production. B Sci Technol Soc 25(4):361

Atiama-Nurbel T, Deguine JP (2010) Agroecological crop protection in organic agriculture: the case of Tephritid fruit flies in Reunion Island, international conference on organic agriculture in scope of environmental problems. Famagusta, Cyprus, pp 107–109

Augusseau X, Deguine JP, Douraguia E, Duffourc V, Gourlay J, Insa G, Lasne A, Le Roux K, Poulabssia E, Rousse P, Roux E, Suzanne W, Tilma P, Trules E (2011) Gamour, l'agroécologie en action à la Réunion. Phytoma, pp 33–37

Biondi A, Desneux N, Siscaro G, Zappala L (2012) Using organic-certified rather than synthetic pesticides may not be safer for biological control agents: Selectivity and side effects of 14 pesticides on the predator Orius laevigatus. Chemosphere 87(7):803–812

Bird GW, Grieshop M, Hepperly P, Moyer J (2009) Climbing Mt. organic: an ecological approach. In: Francis C (ed) Organic farming: the ecological system. American society of agronomy, crop science society of America, Soil science society of america, pp 197–214

Boisclair J, Estevez B (2006) Insect pest management in organic agriculture: acting in harmony with complexity. Phytoprotection 87(2):83–90

Daniel C, Pfammatter W, Kehrli P, WYSS E (2005) Processed kaolin as an alternative insecticide against the European pear sucker, Cacopsylla pyri (L.). J Appl Entomol 129(7):363–367

Darnhofer I, Lindenthal T, Bartel-Kratochvil R, Zollitsch W (2010) Conventionalisation of organic farming practices: from structural criteria towards an assessment based on organic principles. A review. Agron Sustain Dev 30:67–81. DOI: http://dx.doi.org/10.1051/agro/2009011
Dayan FE, Cantrell CL, Duke SO (2009) Natural products in crop protection. Bioorg Med Chem 17:4022–4034. DOI: 10.1016/j.bmc.2009.01.046
Deguine JP, Ferron P (2006) Crop protection, preservation of biodiversity, respect of the environment. Cah Agric 15(3):307–311
Deguine JP, Ferron P, Russell D (2008) Sustainable pest management for cotton production. A review. Agron Sustain Dev 28(1):113–137
Deguine JP, Ferron P, Russell D (eds) (2009) Crop protection: from agrochemistry to agroecology. Science, Enfield, p 378.
Deguine JP, Atiama-Nurbel T, Quilici S (2011) Net choice is key to the augmentorium technique of fruit fly sequestration and parasitoid release. Crop Prot 30:198–202. DOI: 10.1016/j.cropro.2010.10.007
Deguine J-P, Atiama-Nurbel T, Douraguia E, Chiroleu F, Quilici S (2012a) Species diversity within a community of the Cucurbit fruit flies Bactrocera cucurbitae, Dacus ciliatus and Dacus demmerezi roosting in corn borders near cucurbit production areas of Reunion Island. J Insect Sci 12(32). insectscience.org/12.32
Deguine J-P, Douraguia E, Atiama-Nurbel T, Chiroleu F, Quilici S (2012b) Cage study of spinosad-based bait efficacy on Bactrocera cucurbitae, Dacus ciliatus and Dacus demmerezi in Reunion Island. J Econ Entomol 105:1358–1365
Deguine J-P, Rousse P, Le Roux K, Augusseau X (2011) Agroecological crop protection in Reunion: first results in commercial farm conditions. Commun Agric Appl Biol Sci 76:107–118
Delate K (2002) Using an agroecological approach to farming systems research. Horttechnology 12(3):345–354
Delate K, Friedrich H (2004) Organic apple and grape performance in the Midwestern US. In: Bertschinger L, Anderson JD (coord.) (eds) Sustainability of horticultural systems in the 21st century. Leuven 1: International Society Horticultural Science, pp 309–320
Drinkwater LE (2009) Ecological knowledge:foundation for sustainable organic agriculture. In: Francis C (ed) Organic farming: the ecological system. American society of agronomy, crop science society of America, soil science society of America, pp 19–47
Ehler LE (2006) Integrated pest management (IPM) definition, historical development and implementation, and the other IPM. Pest Manag Sci 62:787–789. DOI: 10.1002/ps.1247
Elliot SL, Mumford JD (2002) Organic, integrated and conventional apple production: why not consider the middle ground? Crop Prot 21(5):427–429.
FAO/WHO Codex Alimentarius Commission (1999) Organic agriculture. Committee on Agriculture, Rome
Fernandes VC, Domingues VF, De Freitas V, Delerue-Matos C, Mateus N (2012) Strawberries from integrated pest management and organic farming: phenolic composition and antioxidant properties. Food Chem 134(4):1926–1931
Ferron P, Deguine JP (2005) Crop protection, biological control, habitat management and integrated farming. A review. Agron Sustain Dev 25:17–24. DOI: 10.1051/agro:2004050
Francis C (2009a) Education in organic farming and food systems. In: Francis C (ed) Organic farming: the ecological system, American society of agronomy, crop science society of America. Soil Science Society of America, pp 283–299
Francis C (Eds) (2009b) Organic farming: the ecological system. American Society of Agronomy, Inc, Madison
Furlong MJ, Zalucki MP (2010) Exploiting predators for pest management: the need for sound ecological assessment. Entomol Exp Appl 135(3):225–236.
Garratt MPD, Wright DJ, Leather SR (2011) The effects of farming system and fertilisers on pests and natural enemies: A synthesis of current research. Agricu Ecosyst Environ 141(3–4):261–270
Gliessman SR (Eds) (2007) Agroecology: the ecology of sustainable food systems, CRC Press, pp 384

Guthman J (2004) The trouble with 'organic lite' in California: A rejoinder to the 'conventionalisation' debate. Sociol Ruralis 44:301–316

Harper JL (1974) Agricultural ecosystems. Agro-Ecosyst 1:1–6. DOI: 10.1016/0304–3746(74)90002-x

Hill SB, Vincent C, Chouinard G (1999) Evolving ecosystems approaches to fruit insect pest management. Agric Ecosyst Environ 73:107–110

Hill SB (2014) Considerations for enabling the ecological redesign of organic and conventional agriculture: a sociology and psychosocial perspective. In: Bellon S, Penvern S (eds) Organic agriculture, prototype for systainable agricultures. Springer, Chap. 22, pp 401–422

Holland JM, Birkett T, Southway S (2009) Contrasting the farm-scale spatio-temporal dynamics of boundary and field overwintering predatory beetles in arable crops. BioControl 54(1):19–33

Howard A (1943) An agricultural testament. Oxford University Press, p 228

Jacobson SK, Sieving KE, Jones GA, Van Doorn A (2003) Assessment of farmer attitudes and behavioral intentions toward bird conservation on organic and conventional Florida farms. Conserv Biolo 17(2):595–606

Kogan M (1998) Integrated pest management: historical perspectives and contemporary developments. Ann Rev Entomol 43:243–270. DOI: doi:10.1146/annurev.ento.43.1.243

Kogan M, Hilton RJ (2009) Conceptual framework for integrated pest management (IPM) of tree-fruit pests. In: Aluja M et al (eds) Biorational tree-fruit pest management. CABI, pp 1–31

Krebs JR, Wilson JD, Bradbury RB, Siriwardena GM (1999) The second silent spring? Nature 400:611–612

Kremen C, Miles A (2012) Ecosystem services in biologically diversified versus conventional farming systems: benefits, externalities, and trade-offs. Ecol Soc 17(4):40

Lamine C (2011) Transition pathways towards a robust ecologization of agriculture and the need for system redesign. Cases from organic farming and IPM. J Rural Stud 27:209–219. DOI: 10.1016/j.jrurstud.2011.02.001

Lampkin N (1990) Organic farming. Farming Press, Ipswich

Letourneau DK, Goldstein B (2001) Pest damage and arthropod community structure in organic vs. conventional tomato production in California. J Appl Ecol 38:557–570. DOI: 10.1046/j.1365–2664.2001.00611.x

Lewis WJ, vanLenteren JC, Phatak SC, Tumlinson JH (1997) A total system approach to sustainable pest management. Proceedings of the national academy of sciences of the United States of America 94:12243–12248

Lotter DW (2003) Organic agriculture. J Sustain Agricu 21:59–128

Malezieux E (2012) Designing cropping systems from nature. Agron Sustain Dev 32(1):15–29

Mayes MA, Thompson G.D., Husband B, Miles MM (2003) Spinosad toxicity to pollinators and associated risk, reviews of environmental contamination and toxicology, vol 179. Springer, New York, pp 37–71

McQuate GT, Vargas RI (2007) Assessment of attractiveness of plants as roosting sites for the melon fly, Bactrocera cucurbitae, and oriental fruit fly, Bactrocera dorsalis. J Insect Sci 7:57

Mcsorley R (2002) Nematode and insect management in transitional agricultural systems. Horttechnology 12(4):597–600

Mena Y, Nahed J, Ruiz FA, Sanchez-Munoz JB, Ruiz-Rojas JL, Castel JM (2012) Evaluating mountain goat dairy systems for conversion to the organic model, using a multicriteria method. Animal 6(4):693–703.

Mumford J D (1992) Economics of integrated pest-control in protected crops. Pestic Sci 36(4):379–383

Mladenova R, Shtereva D (2009) Pesticide residues in apples grown under a conventional and integrated pest management system. Food Additives and Contaminants Part A-Chemistry Analysis Control Exposure & Risk Assessment 26(6):854–858

Mzoughi N (2011) Farmers adoption of integrated crop protection and organic farming: Do moral and social concerns matter? Ecol Econ 70(8):1536–1545

Nicholls CI, Altieri MA (1997) Conventional agricultural development models and the persistence of the pesticide treadmill in Latin America. Int J Sustain Dev World Ecol 4(2):93–111

Peck GM, Merwin IA, Brown MG, Agnello AM (2010) Integrated and organic fruit production systems for 'liberty' apple in the northeast United States: a systems-based evaluation. Hortscience 45(7):1038–1048

Penvern S, Bellon S, Fauriel J, Sauphanor B (2010) Peach orchard protection strategies and aphid communities: Towards an integrated agroecosystem approach. Crop Prot 29:1148–1156. DOI: 10.1016/j.cropro.2010.06.010

Peterson G, Cunningham S, Deutsch L, Erickson J, Quinlan A, Raez-Luna E, Tinch R, Troell M, Woodbury P, Zens S (2000) The risks and benefits of genetically modified crops: A multidisciplinary perspective. Conserv Ecol 4(1):13

Ponce NLC (2007) Evaluation of knowledge agroecologiques in organic and conventional horticulturists of the north zone of Cartago, Costa Rica. Cuadernos De Desarrollo Rural 3(58):37–48

Rajapakse R, BCPC (2000) The management of major insect pests Bactocera cucurbitaceae and Aulacaphora spp. in cucurbits under 3 intensive systems: Integrated chemical and organic agriculture in Southern Sri Lanka. Bcpc Conference: Pests & Diseases 2000, vols 1–3, Proceedings. British Crop Protection Council, Farnham, p 981–985

Ratnadass A, Fernandes P, Avelino J, Habib R (2012) Plant species diversity for sustainable management of crop pests and diseases in agroecosystems: a review. Agron Sustain Dev 32(1):273–303

Rodriguez E, Gonzalez B, Campos M (2012) Natural enemies associated with cereal cover crops in olive groves. Bulletin of Insectology 65(1):43–49

Rosset PM, Altieri MA (1997) Agroecology versus input substitution: a fundamental contradiction of sustainable agriculture. Soc Nat Resour 10:283–295

Ryckewaert P, Deguine JP, Brevault T, Vayssieres JF (2010) Fruit flies (Diptera: Tephritidae) on vegetable crops in Reunion Island (Indian Ocean): state of knowledge, control methods and prospects for management. Fruits 65:113–130. DOI: 10.1051/fruits/20010006

Sauphanor B, Simon S, Boisneau C, Capowiez Y, Rieux R, Bouvier JC, Defrance H, Picard C, Toubon JF (2009) Protection phytosanitaire et biodiversité en agriculture biologique. Le cas des vergers de pommiers. Innovations Agronomiques 4:217–228

Sharma OP, Bhosle BB, Kamble KR, Bhede BV, Seeras NR (2011) Management of pigeonpea pod borers with special reference to pod fly (Melanagromyza obtusa). Indian J Agric Sci 81(6):539–543

Simon S, Rusch A, Wyss E, Sarthou JP (2014) Organic farming as an inspirational model for developing conservation biocontrol. In: Bellon S, Penvern S (eds) Organic farming, prototype for sustainable agricultures. Springer, Chap. 5, pp 83–105

Speiser B, Tamm L and Weibel FP (2014) Regulatory framework for plant protection in organic farming. In: Bellon S, Penvern S (eds) Organic agriculture, prototype for sustainable agricultures. Springer, Chap. 4, pp 65–82

Tscharntke T, Clough Y, Wanger TC, Jackson L, Motzke I, Perfecto I, Vandermeer J, Whitebread A (2012) Global food security, biodiversity conservation and the future of agricultural intensification. Biol Cons 151(1):53–59

Trewavas A (2001) Urban myths of organic farming. Nature 410:409–410

Valenzuela HR, Defrank J (1995) Agroecology of tropical underground crops for small-scale agriculture. Cr Rev Plant Sci 14(3):213–238

Vanbuskirk P.D., Hilton R.J., Benbow J., Basile S (2008) Mating disruption and Cydia pomonella Granulosis virus for control of codling moth in pear. In: Webster AD, Oliveira CM (coord.) (eds) Proceedings of the Xth international pear symposium, vols 1–2. Leuven 1: International Society Horticultural Science, p 947–953

Vandermeer J (1995) The ecological basis of alternative agriculture. Ann Rev Ecol Syst 26:201–224

Vanloqueren G, Baret PV (2009) How agricultural research systems shape a technological regime that develops genetic engineering but locks out agroecological innovations. Res Policy 38:971–983. DOI: 10.1016/j.respol.2009.02.008

Veyssières JF (1999) Les relations plantes-insectes chez les Dacini (Diptera-Tephritidae) ravageurs des Cucurbitaceae à La Réunion. University Paris XII, Paris

Wezel A, Bellon S, Dore T, Francis C, Vallod D, David C (2009) Agroecology as a science, a movement and a practice. A review. Agron Sustain Dev 29:503–515. DOI: 10.1051/agro/2009004
White IM, Elson-Harris MM (1992) Fruit flies of economic significance: their identification and bionomic. CAB International, Wallingford
Whittaker J, Courtial JP, Law J (1989) Creativity and conformity in science—titles, keywords and co-word analysis. Soc Stud Sci 19:473–496
Wyss E, Luka H, Pfiffner L, Schlatter C, Gabriela U, Daniel C (2005) Approaches to pest management in organic agriculture: a case study in European apple orchards "IPM in Organic Systems", XXII International Congress of Entomology, Brisbane, Australia
Xu QC, Fujiyama S, Xu HL (2011) Biological pest control by enhancing populations of natural enemies in organic farming systems. J Food Agric Environ 9(2):455–463
Zehnder G, Gurr GM, Kuhne S, Wade MR, Wratten SD, Wyss E (2007) Arthropod pest management in organic crops. Ann Rev Entomol 52:57–80
Zinati GM (2002) Transition from conventional to organic farming systems: I. Challenges, recommendations, and guidelines for pest management. Horttechnology 12(4):606–610

Chapter 7
Adapting Apple Ideotypes to Low-Input Fruit Production Agro-Ecosystems

Luciana Parisi, Laurent Jamar, Marc Lateur, François Laurens and Pierre-Eric Lauri

Abstract Current commercial apple growing is highly dependent on off-farm inputs and it is urgent to develop new strategies to remedy this situation. The challenge for the future is to achieve lower-input apple orchards, whether under Integrated Fruit Production (IFP) or Organic Fruit Production (OFP) systems. This paper analyses the different agronomic factors that play key roles both in the current and future 'More Sustainable Orchard', with particular attention on plant protection. Firstly, the concept of 'ideotype' is developed, emphasizing the most important characteristics of optimal ideotypes for apple. Secondly, current knowledge on the relationships between genotype, cultural practices and the environment is presented and discussed. This paper deals with properties that need to be combined at plant material and orchard levels to optimise the IFP and OFP low-input systems. The focus is on: (a) the main characteristics of apple ideotypes; (b) breeding strategies; and (c) adapted cultural practices and control measures in the orchards.

Keywords Apple breeding · Cultural practices · Pest and disease control · Integrated fruit production (IFP) · Low-input systems · *Malus x domestica* · Organic fruit production (OFP) · Sustainable orchard

---

L. Parisi (✉)
INRA, UR407 Pathologie Végétale, CS 60094, 84143 Avignon Cedex 9, France
e-mail: luciana.parisi@avignon.inra.fr

L. Jamar · M. Lateur
Department of Life Sciences, Unit Breeding & Biodiversity, CRA-W, rue de Liroux, 4–5030, Gembloux, Belgium

F. Laurens
INRA, UMR IRHS, 42 rue Georges Morel, BP60057, 49071, Beaucouzé Cedex, France

P.-E. Lauri
INRA, UMR AGAP, 34398, Montpellier Cedex 05, Montpellier, France
e-mail: lauri@supagro.inra.fr

S. Bellon, S. Penvern (eds.), *Organic Farming, Prototype for Sustainable Agricultures*,
DOI 10.1007/978-94-007-7927-3_7,

7.1 Introduction

Defining and creating apple ideotypes adapted to alternative systems is a necessary step to improve the sustainability of fruit growing in Europe. For both organic and conventional agriculture, the choice of cultivar is one of the most important elements for the design of low chemical, fertilisation and water input systems that ensure regular and high quality production. Most cultivars today are adapted to intensive agriculture: they are productive, but to ensure high yield, they require high inputs of pesticides, fertilisers, water and, to a lesser extent, growth regulators (Brun et al. 2008). This situation is a consequence of the reigning paradigm of the fruit industry. For decades, most breeders have neglected traits such as durable pest and disease resistance and tree adaptability to the environment, concentrating their efforts primarily on high yields, fruit aesthetics and shelf life. The fruits must have an attractive and homogeneous appearance and good eating quality. Most of these cultivars are susceptible to major diseases and require a high level of skills for tree management. However, these disadvantages have not yet limited their success. The dominant mainstream cultivars today are generally adapted to major areas of production around the world where their high input needs can be met. There was a decrease in the quantity of active substances of PPP (Plant Protection Products) applied per hectare on fruits and vegetables in Europe over the period 1994–2003. However, these crops are still strong users of chemicals (EuroStat 2007; Simon et al. 2011). Apple is the most intensively sprayed fruit crop, with an average Frequency Treatment Index (FTI) of 40 in Europe in the past years. Almost half of these treatments were against the major disease of apple, scab (*Venturia inaequalis*) (Demeyere and de Turck 2002; Spruijt-Verkerke et al. 2004). The ecological footprint of intensive commercial fruit growing is therefore particularly high and unsustainable. Even in organic growing systems, very intensive spray schemes are required to comply with the phytosanitary and cosmetic demands of the market (Brauwer and Balkhoven 2000; Jamar et al. 2008).

Apple growing and orchard management reflect and respond to diverse factors including natural (climate, soil characteristics, pest and disease) and human-driven (orchard layout, tree support system, hail net, training and pruning practices) factors. At both levels, the economic balance of an orchard is based on choices made by each grower at spatial and temporal scales. The first key decision starts with orchard establishment (e.g., site, row orientation, rootstock/cultivar combination, support system) and is then repeated annually throughout the orchard life span and within each growing season (e.g., irrigation and nutrient application scheduling, phytosanitary treatments). Like all managers of an enterprise, fruit growers usually grow the cultivars that fetch the highest prices, even if such cultivars are dependent on high inputs of plant protection products, fertilisers and tree management. Technical operations are generally motivated by market fruit traits, without much regard for tree characteristics. Integrated apple production has developed in recent years as the main challenge. It strives for greater sustainability based upon linking desirable fruit characteristics with knowledge of the physiology of fruit trees, especially in relation to regular bearing from one year to the next, within local abiotic and biotic environments (Lauri et al. 2009). Indeed, it has been shown that tree manipulation can be a relevant although partial lever to decrease infection and infestation (Simon et al. 2012).

In this chapter, after having defined the Integrated Fruit Production (IFP) and Organic Fruit Production (OFP) low-input systems (Box 7.1), the ideotype concept in tree crops will be introduced and developed for the apple. The key characteristics of apple ideotypes adapted to these systems will be presented. The breeding objectives and strategies, and the cultural practices and pest and disease control measures adapted to this plant material will then be described. Economic, environmental and social aspects will not be dealt with here although they are part of the definition of an ideotype. Instead, we will focus on agronomic aspects with particular attention to plant protection.

Box 7.1: Integrated and Organic Fruit Production vs. Low Input Fruit Production

Integrated Production: "Integrated Production is a farming system that produces high quality food and other products by using natural resources and regulation mechanisms to replace polluting inputs and to secure sustainable farming" (IOBC 2004).

Organic Agriculture: "Organic agriculture is a holistic production management system which promotes and enhances agro-ecosystem health, including biodiversity, biological cycles and soil biological activity. It emphasizes the use of management practices in preference to the use of off-farm inputs, taking into account that regional conditions require locally adapted systems. This is accomplished by using, where possible, agronomic, biological, and mechanical methods, as opposed to using synthetic materials, to fulfil any specific function within the system" (FAO 1998).

On the other hand **Low Input Farming systems** "seek to optimize the management and use of internal production inputs (i.e., on-farm sources) and to minimize the use of production inputs (i.e., off-farm inputs), such as purchased fertilizers and pesticides, wherever and whenever feasible and practicable, to lower production costs, to avoid pollution of surface and groundwater, to reduce pesticide residues in food, to reduce a farmer's overall risk, and to increase both short- and long-term farm profitability" (Parr et al. 1990).

For fruit growing, the first two systems will hereafter be referred to as IFP (Integrated Fruit Production) and OFP (Organic Fruit Production), respectively. It is important to mention that this chapter is focused on low-input IFP and low-input OFP. However, there is independence between the IFP and OFP farming systems, on the one hand, and the input level, high versus low in each of those farming systems, on the other. Indeed, an IFP or OFP system may have either high inputs or low inputs in terms of off-farm resources.

7.2 The Ideotype Concept

The need to integrate, within the same plant, different characteristics related not only to the quality of the harvested product but also to the characteristics of the whole plant itself in relation to its cultural system was first conceptualised by Donald (1968). This cereal breeder defined the crop ideotype as "a plant model which is expected to yield a greater quantity or quality of grain, oil or other useful products when developed as a cultivar". This concept has been used extensively and adapted to various contexts (e.g., a crop ideotype to reduce invasiveness; Anderson et al. 2006). Dickman et al. (1994) broadened this definition to include tree crops and defined an ideotype as "a model tree that will produce an economic yield that approaches the maximum in a particular environment (or on a certain site), using a prescribed cultural system and assuming a well-defined end use for the harvested products". The advantages of such a broad definition are twofold. Firstly, it clearly states that a plant ideotype cannot be proposed without comprehensive knowledge of the socio-economic, ecological, including abiotic and biotic components, and agronomic context in which the harvested product(s) will be promoted. Secondly, it fosters a large field of collaborative fundamental and applied biological as well as socio-economic research. As a consequence, if the proposal of an ideotype is initially consumer-driven, the biological elements that should be integrated into the plant to reach the economic objectives have to be carefully assessed and compatible. As stated by Dickman et al. (1994), "an ideotype will be no better than the scientific foundation upon which it rests". It follows from this set of definitions that the concept of ideotype is complex by nature, integrating several elements from socio-economy into plant, pest and disease biology. This implies, firstly, that objectives that have to be solved by the ideotype have to be restricted through prioritisation and, secondly, that choices have to be made as to whether these objectives should be reached through genetics and/or cultural practices.

The relationship between plant architecture and fruit production represents one such research area for optimising yield improvement (Lauri and Laurens 2005). The apple provides a good example for contrasting the relationships between vegetative growth patterns (shoot length, branching density) and fruiting (Lauri and Costes 2005; Costes et al. 2006). A first ideotype was proposed by Lespinasse and Delort (1986), based on the observations that regular fruiting cultivars generally have a high proportion of long fruiting shoots, namely between 15 and 25 cm, and produce fruit in the terminal position on these long shoots. Another apple ideotype was proposed by Dickman et al. (1994) that was adapted to a "high-density orchard under Michigan (USA) or similar environmental conditions". They associate the ability to have "high fruit productivity" with the "spurred habit", i.e., with a high proportion of shoots less than 10 cm long and that produce fruit in the terminal position on these short shoots. However, studies conducted on apple (Lauri and Trottier 2004) and on other tree species (e.g., mango: Normand et al. 2009, etc.) strongly support the idea that there is a biological discrepancy between the "spurred habit" promoted by Dickman et al. and regular fruiting. These findings suggest that as far as regular fruiting is looked for in progenies, apple tree architecture should combine, among other traits, fruiting in the terminal position on long shoots (Lauri et al. 2011). These objectives may be reached through breeding and/or cultural practices (see below).

7.3 Characteristics of Apple Ideotypes Adapted to IFP and OFP Low-Input Systems

The characteristics of cultivars that are well adapted to low input and even more to organic orchards are different from those of mainstream cultivars currently cultivated with high inputs. Firstly, as stated above, it is not possible to define one or even a few ideotypes adapted to diverse local situations. Each ideotype will be adapted to particular climatic conditions, cultural practices and economic situations. To meet this new challenge, a new alternative strategy consists of broadening the genetic biodiversity and promoting a greater diversity of cultivars, not just those such as Golden Delicious, Red Delicious, Gala and Fuji that are widely grown throughout the world. The second important point is that these cultivars will have to provide good cropping with regular high quality production without major tree manipulations to reduce labour costs, treatments or fertilisers. To reach this aim, the ideotypes adapted to low-input IFP and OFP systems will require certain characteristics:

- The ability to ensure a **regular income** for the grower, i.e., high regularity of fruit production with a good cropping of marketable fruit size. Tree architecture that requires less tree training is also an important characteristic.
- **Tolerance and durable resistance** to the most important pests and diseases. Tolerance to disease may be defined as the capacity of a given cultivar to minimise yield or quality loss due to disease or pathogen development compared with other cultivars (Schafer 1971). This characteristic, often encountered in old cultivars, may become a new key factor for breeding and screening cultivars that are better adapted to low-input IFP and OFP systems. Indeed, selection pressures occurred in the past when fungicides were unknown, leading to a large diversity of apple cultivars. Many of these were formerly grown in extensive, highly-branched standard tree orchards, and exhibited quantitative traits such as high tolerance to disease (for example, in Northern France and Belgium: Reinette Hernaut, Reinette des Capucins, Cabarette, Belle-Fleur Large Mouche). Surveys have pointed out that many landraces are already safeguarded in repository orchards and need to be properly evaluated with the aim of actively using some of them either as parents in breeding programmes or occasionally as cultivars (Populer et al. 1998; Lateur 2003). Durability of resistance is especially important in perennial crops like apple, and this objective is not achieved in most of the new cultivars (Box 7.2).
- **High and homogeneous fruit quality** including visual and eating quality, fruit size, and prolonged storage potential without physiological disorders. Visual and gustative quality is likely to vary across agricultural systems and different mar-

kets. For example, some irregularities in appearance can be tolerated in OFP sold directly on the farm.

- **Hardiness** is the ability to produce under limiting conditions. It encompasses resilience to abiotic, i.e., soil and climate constraints, and increasing fertiliser use efficiency by the plant. This important point is not yet currently taken into account in fruit breeding programmes. Since climate is expected to become less predictable with more extreme events, it is relevant to adapt fruit growing to future challenges such as drought and nutrient stress.

Box 7.2: Resistance to Scab and Durability

The importance of resistance durability can be illustrated by the example of resistance to scab, the most important fungal disease of apple. Until now, the most important breeding programmes throughout the world have worked with a single resistance source, the *Vf* gene from *Malus floribunda* 821. This gene was overcome by the pathogen in the 1990s (Parisi et al. 1993, 2004), and did not provide durable resistance. Strategies aimed to diversify the genes employed and to pyramid these genes in a single cultivar are under development (Gessler et al. 2006), but the question of durability of such approaches has not been solved. Breeding programmes have so far provided a few cultivars with partial resistance to scab, despite the studies undertaken on the expression and genetic determinants of this resistance by several teams (Visser et al. 1974; Durel et al. 2003; Calenge et al. 2004; Lefrancq et al. 2004). This lack of modern cultivars adapted to low-input IFP and OFP systems is a real problem. On the other hand, partial or quantitative resistance to scab was present in some old cultivars and has apparently not failed (Lateur and Populer 1994; Lateur et al. 1999; Didelot et al. 2007; Brun et al. 2008). Unfortunately, many of these old cultivars are not always adapted to modern fruit quality standards and/or do not reach sufficient yield (Lateur 2000; Jamar et al. 2010). Resistance to scab is particularly important in regions with a wet climate. However, in the search for an ideotype adapted to a biotic environment with high disease pressure, resistance or tolerance to scab must be associated with tolerance and/or durable resistance to the other pests and diseases mainly present in the area of cultivation: powdery mildew (*Podosphaera leucotricha*), European canker (*Nectria galligena*), prevalent post-harvest diseases (*Botrytis* sp., *Gloeosporium* sp., *Penicilium* sp.), fire blight (*Erwinia amylovora*), rosy apple aphids (*Dysaphis plantaginea*) and codling moth (*Cydia pomonella*).

Globally, it is clear that each of these characteristics needs to be ranked differently for each specific situation, soil and climatic conditions, type of growing system, choice of low-input system (IFP vs. OFP), and intended markets. **Therefore, an ideotype could be defined as a hardy and reliable productive genotype**

that is well adapted to its specific cultural, socio-economic and environmental conditions. At present, only a few small apple breeding programmes aim to achieve these goals (Warlop et al. 2010).

7.4 Breeding for Apple Ideotypes Adapted to IFP and OFP Low-Input Systems

Breeding for low input apple cultivars following the ideotype concept will require changing the selection process in mainstream breeding programmes. While the main objectives of selection are still focused on high fruit quality, disease and pest resistance (mainly scab), and high and regular cropping, novel strategies will have to be implemented to take durable/sustainable resistance/tolerance to many pests and diseases into account, as well as the adaptation to specific geographical environments, cultural practices and socio-economic systems. This will demand changes in the selection process itself. For example, it will be necessary to enlarge the genetic diversity of the breeding gene pool. The evaluation of resistance to various pests and diseases in germplasm collections showed a great potential in old and local cultivars (Lateur and Populer 1994; Laurens et al. 2004). However, including local cultivars in breeding programmes means that in addition to the favourable and original traits they bring, they may also carry some deleterious characteristics (biennial bearing, less adaptability to dwarfing rootstocks, higher acidity, old-fashioned fruit textures and aromas, irregular shapes and sizes, shorter period of optimal fruit quality, etc.) against which the breeder will have to select. It will also be necessary to avoid preliminary selection that provides a rigorous screening for major genes of resistance (for example, the scab that is currently challenging seedlings in a greenhouse), but also discards potentially interesting individuals with partial resistance to a pathogen. To face these new challenges, breeders will have to implement new methodologies and new screening methods. The use of molecular markers is promising, but some preliminary studies are needed to decipher the genetic control of the main agronomic traits. So far, most studies have been performed on fruit quality traits and scab resistance (Arus et al. 2007; Gardiner et al. 2007; Bus et al. 2009). Molecular markers are actually rarely used in current breeding programmes to select for quantitative traits because of two main bottlenecks, the low density of genetic maps and the lack of information on allelic diversity. The publication of the entire apple genome sequence (Velasco et al. 2010), and new ongoing projects (Laurens et al. 2010) may allow a significant and efficient use of molecular markers in subsequent future breeding programmes. Another important point is that the evaluation of new apple ideotypes needs to be done under conditions that make it possible to test their adaptability to low-input IFP and OFP systems in different regions. This will require in-depth studies to better understand cultivar adaptability to various conditions and to decipher the complex interactions between genotype, environment, cultural practices and socio-economic context. Experimental networks that take these various factors into account will have to be built. New cultivar testing trials will need to integrate these features into their strategies in accordance with the basic objectives of low-input IFP and OFP systems. Similarly, special attention will have to be paid to pest and disease tolerance traits, better water- and fertiliser-use efficiency, and post-harvest disease resistance in future research programmes.

7.5 Cultural Practices and Control Measures Adapted to IFP and OFP Low-Input Systems

Current high-input intensive fruit growing practices have promoted the cultivation of several modern and disease-susceptible cultivars, which resulted in losses of genetic diversity of cultivated apple trees. This situation has exacerbated the development of pests and diseases, some of which have the potential to genetically recombine each year and to improve their parasitic fitness over time (Parisi et al. 2004). Consequently, even the best apple ideotype cultivar could also become unsustainable under a long-term intensive mono-cultivar context. Management practices used for apple pest and disease control have the potential to disrupt or halt the pests' or pathogens' reproductive and dispersive strategies. Some of these practices adapted to low-input systems are illustrated below.

7.5.1 Tree Management, Training and Pruning

Successful management of apple trees in all planting systems depends on maintaining a balance between vegetative growth and fruiting. If vegetative vigour is too low, an imbalance occurs, leading to excessive fruit load, fruit size decline and biennial bearing increase. Moreover, the tree fails to fill its allotted space within the row, resulting in a decrease of orchard profitability. If vigour is excessive, then flowering and fruiting are reduced and containment of the tree within its allotted space becomes problematic. Pruning and tree-training strategies are the primary management methods, along with fertilisation strategies, that are used to achieve this balance between vegetative growth and cropping throughout the orchard's life. Unsuitable combinations of soil, rootstock and cultivar can easily result in overly vigorous growth, a biennial bearing habit or too much of a dwarfing effect, each resulting in unproductive trees. Improved nutrient uptake abilities, tolerance to weed competition and mechanical control, and sufficient anchorage without staking or trellising, should be important tree traits from the point of view of low-input IFP and OFP fruit growers (Weibel and Häseli 2003).

Training and pruning strategies are usually conceived as uniform recipes applied in the same way to all cultivars, and even for various fruit tree species (Jackson 1999). As concerns the apple, the recognition of two main fruiting patterns, where fruits are preferentially borne in the terminal position of either short or long shoots, has emphasized the interest in adapting cultural practices to these architectural

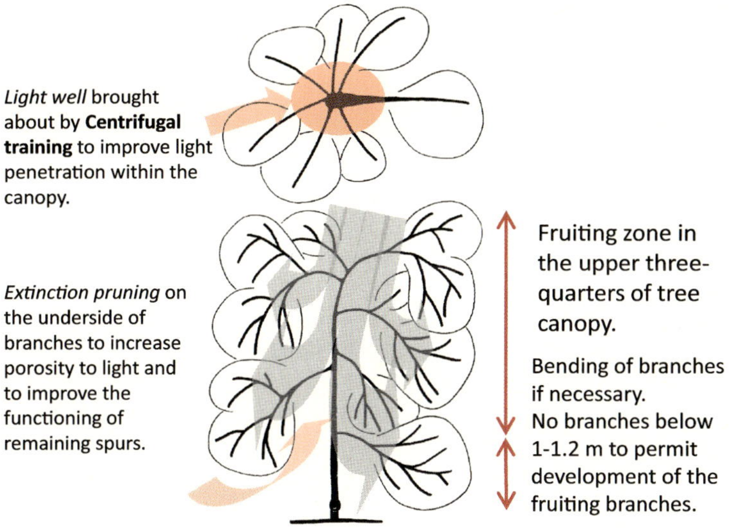

Fig. 7.1 Centrifugal Training System—Basic principles. (From Lauri 2002)

characteristics (Forshey et al. 1992). However, a large spectrum of architectural diversity exists among apple cultivars, suggesting various underlying physiological mechanisms (Lespinasse and Delort 1986; Lespinasse 1992). The concept of architectural types has emerged to better take advantage of the variability of combinations between vegetative and reproductive traits, some of which are especially relevant in the search for regular bearing, for example (Lauri and Laurens 2005). A vision of the tree as an integrated living organism has been developed from these seminal works, with the novel objective of developing training and pruning systems adapted to the intrinsic characteristics of each cultivar, thus reducing deleterious reactions to manipulations, e.g., strong vegetative reactions to excessive pruning that delay fruiting (Lauri et al. 2011; Box 7.3, Fig. 7.1)

Box 7.3: A Recent Example of an Apple Tree Cultural Ideotype, the "Centrifugal Training System"

The Centrifugal Training System is inspired from the natural architecture of existing cultivars with regular fruiting. This concept has shown its potential to homogenise fruit size and to improve fruit colour, as well as to improve the regularity of fruiting for some cultivars (Lauri et al. 2004; Lauri et al. 2009). The centrifugal training ideotype is an evolving concept based on current expertise and scientific knowledge. Adapting the concept to specific cultivar-rootstock combinations and environments, and especially to optimising growth during the vegetative season in relation to flower induction, is a priority for which fundamental and applied research is necessary (Lauri et al. 2011).

The *light well* results from the removal of all water shoots and spurs along the trunk and at the bottom of branches. *Extinction pruning* is the removal of vegetative and fruiting spurs on branches. Depending on the context, bending of the trunk at the desired height may be replaced by heading to obtain a better distribution of branches at the top of the tree.

For a given cultivar with its own characteristics of partial resistance or tolerance to pests and diseases, the manipulation of tree architecture through pruning and fertilisation and irrigation management can affect the spread of diseases and pests (Simon et al. 2007a). A four-year experiment on organic Golden Delicious trees shows a significant decrease in some pest infestation and pathogen infection in centrifugal-trained trees compared to a reference training system characterised by higher fruiting spur density (Solaxe-trained trees; Simon et al. 2006). These studies reinforce the idea that factors other than genetic parameters may influence pest infestation and disease infection. Tree architecture manipulation such as pruning and bending may also have significant effects in determining resource availability (which directly affects organ attractiveness) and topological relationships (e.g., distance) between organs, resulting in significantly different patterns of pest and disease dissemination within the tree crown. These studies also revealed interactions between pests (e.g., *Aphis pomi* vs. *Dysaphis plantaginea*), which may lead to variability in infestation patterns depending on the year (Simon et al. 2006). This highlights how difficult it may be to satisfy all criteria. Certainly, there is no single management practice that could durably negatively affect all pests and pathogens. As an example, in the above-mentioned experiment, the decreased branching density favours within-tree oviposition and infestation of the codling moth (*Cydia pomonella*), especially when there are high population densities of this pest (Simon et al. 2007b). However, although there is no data quantifying the benefits of training alone because it is used in combination with other means such as predators or sexual mating, tree architecture manipulation is likely to be an efficient although partial means to reduce phytosanitary inputs.

7.5.2 Sanitation

With regard to apple diseases, in addition to selecting cultivars with disease resistance characteristics, preventive strategies include chemical, biological and physical sanitation methods aimed at reducing first pathogen dispersions. In the case of apple scab, infected leaf litter contains the pseudothecia of this fungus and is the origin of ascospore ejections in spring (MacHardy 1996). Sanitation practices including either the destruction of the fungus in leaf litter or the reduction of the leaf litter present in the orchard after leaf fall in autumn could play an important role for the reduction of scab inoculum in the orchard. Examples include: (i) physically removing fallen leaves from the orchard using a lawn sweeper; (ii) raking and then ploughing the leaf litter into the soil using a cover-crop machine (Gomez et al. 2007); (iii) shredding the leaf litter using a flail mower; and (iv) treating the leaf litter with substances or microorganisms that hasten decomposition. It is therefore recommended to think and adapt the conception of the orchard designs in order to improve the effectiveness of these sanitation practices.

For apple powdery mildew and canker, it is also possible to suppress infected shoots during pruning in winter. Concerning powdery mildew, between green tip and bloom, the removal of primary infected buds could be achieved while performing centrifugal-tree training through the artificial extinction procedure (Brun et al. 2010). This practice reduces primary inoculum, but there has been no quantitative evaluation of the efficacy of this method.

7.5.3 Fertilisation

Several studies have shown that plant resistance to pest and disease is linked to the optimal physical, chemical, and, perhaps most importantly, biological properties of soil (Altieri and Nicholls 2003). Fertilisation strategies that are used to achieve the balance between vegetative growth and cropping throughout the orchard's life can influence tree pest and disease susceptibilities. For example, enhanced vegetative growth of apple trees by nitrogen fertilisation often leads to increased susceptibility to aphids, fire blight (Neilsen and Neilsen 2003) or apple scab (Leser and Treutter 2005). Several researchers have reported lower numbers of pests and diseases on crops grown with organic compared to synthetic sources of fertiliser. Subsequent experiments supported the mineral balance hypothesis, suggesting that the organic matter and microbial activity associated with organically-managed soils provide a buffering capability to maintain optimal nutrient and mineral balance in plants, which in turn affects the performance of phytophagous pests (Zehnder et al. 2007). Further investigations are needed to determine the optimal balance for micro- and macronutrient status in environmentally-friendly apple orchards and to find the rootstock-cultivar combinations with high physiological efficiency adapted to low-input fertilisation levels. The cultivar selection methods need to be in line with these new objectives, e.g., by application of low-inputs and organic growing approaches inside the experimental plots.

7.5.4 Orchard Layout

In temperate regions, the choice of site is important for low-input OFP and IFP apple production because of the effects of frost, hail, wind, sunshine and soil on disease and productivity. Locations on south-facing slopes with good ventilation greatly facilitate control of fungal diseases. Thorough knowledge of the microclimate history is necessary for an optimum choice of site. The relationships between tree density, rootstock vigour, canopy ventilation, sunshine duration and pest and disease development have been widely recognised (Holb 2005; Stoeckli et al. 2008). In areas with a high incidence of hail, for instance, it is necessary to use protective netting that has the side effects of raising humidity and excluding birds that would otherwise control rodent populations. The role of the rootstock in scab development is evident in the fact that the more dwarfing the rootstock is, the more readily the ascospores from the dead leaves on the ground (primary inoculum) will come into contact with the tree canopy (Aylor 1998; MacHardy et al. 2001).

Various experiments with cultivar mixtures have proved to be effective in reducing disease epidemics. It has been shown that increasing host genetic diversity in orchards, especially through cultivar selection, combining resistant cultivars with susceptible cultivars, limits the spread of scab on susceptible cultivars (Didelot et al. 2007). Overall, mixing within the row produces better results than mixing by alternate rows (Bousset et al. 1997; Didelot et al. 2007), although the former mixture is difficult to manage in terms of phytosanitary protection and harvest. However, since increasing host genetic diversity in orchards could improve management of scab and probably other diseases as well (Parisi et al. 2013), it will be useful to develop further research to find the best way to display this host diversity in innovative orchard designs (Speiser et al. 2014, Chap. 4; Simon et al. 2012).

The ground cover canopy is another aspect that may influence pest and disease development. Chemical weed control or mechanical tillage beneath fruit trees as well as regularly mowing and cutting the grass in the inter-rows, widely used in fruit production, could give newly discharged ascospores a clear path towards the tree canopy (Aylor 1998) and could disturb the useful micro- and macrofauna normally present in the herbaceous ground-cover strata. Field experiments and modelling efforts are currently under way to determine the magnitude of these effects.

7.5.5 Chemical Disease Control

Defensive strategies for disease control include fungicide treatments from bud break, in spring, during summer and sometimes until harvest (MacHardy et al. 2001). To make the best possible use of partial resistances of hardy ideotypes and to reduce the number and impact of these treatments, control methods must be adapted to the characteristics of each cultivar. The most difficult aspect of this effort is to establish thresholds that permit this adaptation without excessive risk for the quality of the crop. Since no effective curative products are available in the organic production system, a new successful strategy was experimented for seven years in an OFP orchard in Belgium, involving spraying during the fungal infection or germination process, before fungal penetration into the leaf, using only contact fungicides allowed by organic guidelines (Jamar 2011). The evidence presented in this work suggests a model for an efficient timing of treatments in apple orchards for primary scab control. This strategy is in full phase with a low-input approach since it avoids unnecessary preventive treatments and includes low frequency of treatments, never exceeding 12 treatments per season. The method is even very effective on high scab-susceptible cultivars under high disease pressure. Moreover, the experiment showed that the amount of product used for each treatment could be successfully reduced on medium and low scab-susceptible cultivars. Conversely, on monogenic scab-resistant cultivars, the effective scab control decreased with the reduction of the amount of product used for each treatment, mainly because the major *Vf* gene protection had been broken down by the appearance of new virulent races in the experimental orchard (Jamar et al. 2010). As a whole, this suggests that the tolerance thresholds have to be established depending on the cultivar as well as on the farm location (i.e., disease pressure) and on the farmer's objectives.

7.5.6 Interactions Between Cultivar Properties, Cultural Practices and Pesticide Applications to Control Pests and Diseases

The amount of phytosanitary products needed to reach optimal disease protection is related to the resistance traits of each cultivar. It has been shown that partial resistance of the cultivar is the first step for the control of apple scab involving less fungicide either under controlled conditions (Jamar 2011) or in experimental orchards (Didelot et al. 2010; Jamar 2011).

These results were confirmed in a system approach study, conducted from 2005 onwards at the INRA experimental station, "UERI de Gotheron" (Drôme, France). The study reports the effects of the level of pesticide use on the agri-environmental performances (yield, fruit damage) of three protection systems in experimental apple orchards: (1) conventional; (2) low-input IFP; and (3) organic farming (Simon et al. 2011). Moreover, to assess the significance of the effect of the cultivar in decreasing pesticide use, these protection systems were combined with three cultivars that differed in scab susceptibility: 'Ariane' (*Vf*-resistant), 'Melrose' (medium susceptibility) and 'Golden Delicious' (high susceptibility). The level of pesticide use is the highest in 'Golden Delicious' plots, regardless of the protection system. A 43–56 % decrease in pesticide use is observed on 'Melrose' and 'Ariane' in both low-input IFP and organic farming protection systems compared to conventional 'Golden Delicious', used as a reference. Only low-input 'Melrose' and low-input 'Ariane' systems achieved a level of yield and fruit damage similar to the same cultivars in the conventional system, permitting reduced environmental impacts (Alaphilippe et al. 2013). This example illustrates how the cultivar choice is a key option to reduce pesticide use in all agricultural systems and why the range of commercial apple cultivars should be renewed to offer more hardy cultivars with an acceptable level of disease susceptibility.

7.6 Conclusion and Prospects

In this chapter, we focused on the characteristics of ideotypes for the apple, an important fruit species in temperate latitudes, adapted to low-input IFP and OFP systems, and on the management strategies of these cultivars to obtain the best performances in these systems. It is obvious that the choice of plant material, both the scion and the rootstock, is a first and very important step to promote durable and competitive orchard systems. These ideotypes must be able to fruit under conditions with biotic and abiotic constraints, and to ensure a regular and good-quality fruit production. They should also have a durable tolerance or resistance to the complex of major pests and diseases in each region. The main concept we support is that a given cultivar must be adapted to the growing conditions in which it will have to grow, i.e., the local soil and climatic environment. This also holds true for cultural practices and strategies to control major pests and diseases that need to be adapted to specific cultivars. Therefore, there are not one but several ideotypes, each one being chosen for the specific environmental and socio-economic conditions in which they must thrive. This concept can be achieved only if there are strong links between all stakeholders along the fruit supply chain through participatory breeding programmes that involve researchers, breeders, technicians and growers with IFP or OFP expertise at the national or even worldwide scale (Warlop et al. 2010; Chable et al. 2013 in Chapter 21).

Beside the need for better adapted cultivars, better rootstock/cultivar adaptation and better training and pruning management, this study emphasizes the necessity of further exploiting the strategies of new orchard eco-design that place more emphasis on practices that enhance soil fertility and functional biodiversity (Penvern et al. 2012). Concepts of balanced eco-systems rely on the knowledge and integration of many complementary key factors, either positively or even negatively related, that will have a differential importance in accordance with specific environmental and socio-economic situations (Desclaux and Nolot 2013 in Chapter 20). Eventually, as a key element of new pest and disease control strategies, all possible prophylactic sanitation techniques must be fully utilised and taken advantage of. To dispose of overly reductive indicators of agri-environmental and societal performance, different methods of monitoring agro-ecosystems already exist (Bockstaller et al. 2008), but they need to be better implemented and adapted to future apple growing practices.

References

Alaphilippe A, Simon S, Brun L, Hayer F, Gaillard G (2013) Life cycle analysis reveals higher agroecological benefits of organic and low-input apple production. Agron Sustain Dev 33:581–592

Altieri MA, Nicholls C (2003) Soil fertility and insect pests: harmonizing soil and plant health in agroecosystems. Soil Tillage Res 72:203–211

Anderson NO, Gomez NA, Galatowitsch SM (2006) A non-invasive crop ideotype to reduce invasive potential. Euphytica 148:185–202

Arús P, Gardiner S (2007) Genomics for improvement of Rosaceae temperate tree fruit. In: Varshney RK, Tuberosa R (eds) Genomics-Assited Crop Improvement, vol 2. Genomics applications in Crops. Springer, Dordrecht, pp 357–397

Aylor DE (1998) The aerobiology of apple scab. Plant Dis 82:838–849

Bockstaller C, Guichard L, Makowski D, Aveline A, Girardin P, Plantureux S (2008) Agri-environmental indicators to assess cropping and farming systems. A review. Agron Sustain Dev 28(1):139–49

Bousset L, Blaise P, Kellerhals M, Gessler C (1997) Mixtures of apple cultivars in orchards: effect on the scab epidemics. IOBC/WPRS Bull 20:42–48

Brauwer G, Balkhoven H (2000) Biologische telere gaan de uitdaging aan. Fruitteelt 29:12–13

Brun L, Didelot F, Parisi L (2008). Effects of apple cultivar susceptibility to *Venturia inaequalis* on scab epidemics in apple orchards. Crop Prot 27(6):1009–1019

Brun L, Guinaudeau J, Gros C, Parisi L, Simon S (2010) Assessment of fungicide protection strategies in experimental apple orchards. IOBC/WPRS Bull 54:103–107

Bus VGM, Esmenjaud D, Buck E, Laurens F (2009) Application of genetic markers in rosaceous crop. In: Folta KM, Gardiner SE (eds) Genetics and Genomics of Rosaceae, Plant Genetics and Genomics: Crops and Models 6:563–599

Calenge F, Faure A, Goerre M, Gebhardt C, Van De Weg WE, Parisi L, Durel CE (2004) Quantitative Trait Loci (QTL) analysis reveals both broad-spectrum and isolate-specific QTL for scab resistance in an apple progeny challenged with eight isolates of *Venturia inaequalis*. Phytopathol 94:370–379

Chable V, Dawson J, Bocci R, Goldringer I (2014) Seed for Organic Agriculture: the Development of the Participatory Plant Breeding and the Farmers' Networks in France. In: Penvern S, Bellon S (eds) Organic farming, prototype for sustainable agricultures. Springer, Chap. 21, pp 383–400

Costes E, Lauri PÉ, Regnard JL (2006) Analysing fruit tree architecture. Implication for tree management and fruit production. Horticultural Reviews, 32:1–61

Demeyere A, de Turck R (2002) Use of crop protection products in Belgium's main crops over the period 1991–2000. Ministère des Classes Moyennes et de l'Agriculture, Bruxelles, pp. 39

Desclaux D, Nolot JM (2014) Seed sector and breeding methods: challenges for crop organic varieties. In: Penvern S, Bellon S (eds) Organic farming, prototype for sustainable agricultures. Springer, Chap. 20, pp 367–382

Dickman DI, Gold MA, Flore JA (1994) The ideotype concept and the genetic improvement of tree crops. Plant Breed Rev 12:163–193

Didelot F, Brun L, Parisi L (2007) Effects of cultivar mixtures on scab control in apple orchards. Plant Pathol 56:1014–1022

Didelot F, Caffier V, Baudin M, Orain G, Lemarquand, Parisi L (2010) Integrating scab control methods with partial effects in apple orchards: the association of cultivar resistance, sanitation and reduced fungicide schedules. IOBC/WPRS Bull 54:525–528

Donald CM (1968) The breeding of crop ideotypes. Euphytica 17:385–403

Durel CE, Parisi L, Laurens F, Van De Weg WE, Liebhard R, Jourjon MF (2003) Genetic dissection of partial resistance to race 6 of *Venturia inaequalis* in apple. Genome 46:224–23

Eurostat (2007) Eurostat Statistical Book: the use of plant protection products in the European Union, Data 1992–2003. http://epp.eurostat.ec.europa.eu/cache/ITY_OFFPUB/KS-76–06-669/EN/KS-76–06-669-EN.PDF

FAO (1998) Evaluating the potential contribution of organic agriculture to sustainability goals. FAO's technical contribution to IFOAM's Scientific Conference Mar del Plata, Argentina, 16–19 November 1998. ftp://ftp.fao.org/docrep/fao/003/ac116e/ac116e00.pdf

Forshey CG, Elfving DC, Stebbins RL (1992) Training and pruning apple and pear trees. Am Soc Hort Sci, Alexandria, 166p

Gardiner SE, Bus VGM, Rusholme RL, Chagné D, Rikkerink EHA (2007) Apple. In: Kole C (ed) Genome mapping and molecular breeding in plants. Fruits and Nuts, vol 4. Springer, Berlin, pp 1–62

Gessler C, Patocchi A, Sansavini S, Tartarini S, Gianfranceschi L (2006) *Venturia inaequalis* resistance in apple. Crit Rev Plant Sci 25:473–503

Gomez C, Brun L, Chauffour D, De Le Vallée D (2007) Effect of leaf litter management on scab development in an organic apple orchard. Agric Ecosys Environ 118:249–255

Holb IJ (2005) Effect of pruning on apple scab in organic apple production. Plant Dis 89:611–618

IOBC (2004) Integrated Production: principles and Technical Guidelines. IOBC/wprs Bull 27:2

Jamar L, Lefrancq B, Fassotte C, Lateur M (2008). A 'during-infection' spray strategy using sulphur compounds, copper, silicon and a new formulation of potassium bicarbonate for primary scab control in organic apple production. Eur J Pl Pathol 122:481–492

Jamar L, Cavelier M, Lateur M (2010) Primary scab control using a 'during-infection' spray timing and the effect on fruit quality and yield in organic apple production. Biotechnol Agron Soc Environ 14:423–439

Jamar L (2011) Innovative strategies for the control of apple scab (*Venturia inaequalis* [Cke.] Wint.) in organic apple production (PhD thesis). 188 p. http://orbi.ulg.ac.be. University of Liege—Gembloux Agro-Bio Tech, Belgium

Jackson DJ (1999) Grapevines. In: Jackson DJ, Looney NE (eds) *Temperate and subtropical fruit production*, 2nd edn. CABI Publishing, London, pp 70–77

Lateur M, Populer C (1994) Screening fruit tree genetic resources in Belgium for disease resistance and other desirable characters. Euphytica 77:147–153

Lateur M, Wagemens C, Populer C (1999) Evaluation of fruit tree genetic resources: use of the better performing cultivars as sources of polygenic scab resistance in an apple breeding programme. Acta Hortic 484:35–42

Lateur M (2000) Fruit tree genetic resources and integrated fruit production. Acta Hortic 525:317–323

Lateur M (2003) The integration of different sectors is a key factor for the conservation, the evaluation and the utilisation of our Belgian fruit tree biodiversity. Bulletin de l'Institut Royal des Sciences Naturelles de Belgique—Biologie, 73-SUPPL:85–95

Laurens F, Chevalier M, Dolega E, Gennari F, Goerre M, Fischer C, Kellerhals M, Lateur M, Lefrancq B, Parisi L, Shouten HJ, Tartarini S (2004) Local European cultivars as sources of durable scab resistance in apple. Acta Hortic 663:115–122

Laurens F, Durel CE, Patocchi A, Peil A, Salvi S, Tartarini S, Velasco R, Van de Weg E (2010) Review on apple genetics and breeding programmes and presentation of a new European initiative to increase fruit breeding efficiency. J Fruit Sci 27:102–107

Lauri PÉ (2002) From tree architecture to tree training—an overview of recent concepts developed in apple in France. J Korean Soc Hort Sci 43:782–788

Lauri PÉ, Trottier C (2004) Patterns of size and fate relationships of contiguous organs in the apple (*Malus domestica* Borkh.) crown. New Phytol 163:533–546

Lauri PÉ, Willaume M, Larrive G, Lespinasse JM (2004) The concept of centrifugal training in apple aimed at optimizing the relationship between growth and fruiting. Acta Hortic 636:35–42

Lauri PÉ, Costes E (2005) Progress in Whole-Tree Architectural Studies for Apple Cultivar Characterization at INRA, France – Contribution to the Ideotype approach. Acta Hortic 663:357–362

Lauri PÉ, Laurens F (2005) Architectural types in apple (*Malus X domestica* Borkh.). In: Dris R (ed) Crops: growth, quality and biotechnology. World Food Limited, Helsinki, pp 1300–1314

Lauri PÉ, Costes E, Regnard JL, Brun L, Simon S, Monney, Sinoquet H (2009) Does knowledge on fruit tree architecture and its implications for orchard management improve horticultural sustainability? An overview of recent advances in the apple. Acta Hortic 817:243–249

Lauri PÉ, Hucbourg B, Ramonguilhem M, Méry D (2011) An Architectural-based tree training and pruning—identification of key features in the apple. Acta Hortic 903:589–596

Lefrancq B, Lateur M, Rondia A (2004) Screening method for polygenic scab resistance within an apple breeding programme: relationship between early greenhouse screening test on young seedlings and their scab susceptibility in natural field conditions. Acta Hortic 663:793–798

Leser C, Treutter D (2005) Effects of nitrogen supply on growth, contents of phenolic compounds and pathogen (scab) resistance of apple trees. Physiol Plant 123:49–56

Lespinasse JM, Delort F (1986) Apple tree management in vertical axis: appraisal after ten years of experiments. Acta Hortic 160:120–125

Lespinasse Y (1992) Breeding apple tree: aims and methods. Proc. Joint Conf. of the EAPR Breeding and varietal assessment section and the Eucarpia Potato section:103–11

MacHardy WE (1996) Apple Scab Biology, Epidemiology, and Management. APS Press, St Paul, Minnesota

MacHardy WE, Gadoury DM, Gessler C (2001) Parasitic and Biological Fitness of *Venturia inaequalis*: Relationship to Disease Management Strategies. Plant Dis 85:1036–1051

Neilsen GH, Neilsen D (2003) Nutritional requierements of apple. In: Ferree DC, Warrington IJ (eds) Apples, botany, production and uses. CABI publishing

Normand F, Pambo Bello AK, Trottier C, Lauri PÉ (2009) Is axis position within tree architecture a determinant of axis morphology, branching, flowering and fruiting? An essay in mango. Ann Bot 103:1325–1336

Parisi L, Lespinasse Y, Guillaumes J, Kruger J (1993) A new race of *Venturia inaequalis* virulent to apples with resistance due to the *Vf* gene. Phytopathol 83(5):533–537

Parisi L, Fouillet V, Schouten HJ, Groenwold R, Laurens F, Didelot F, Evans K, Fischer C, Gennari F, Kemp H, Lateur M, Patocchi A, Thissen J, Tsipouridis C (2004) Variability of the pathogenicity of *Venturia inaequalis* in Europe. Acta Hortic 107–113

Parisi L, Gros C, Combe F, Parveaud CE, Gomez C, Brun L (2013) Impact of cultivar mixture on scab, powdery mildew and rosy aphid in an organic apple orchard. Crop Prot 43:207–212

Parr JF, Papendick RI, Youngberg IG, Meyer RE (1990) "Sustainable Agriculture in the United States" in Sustainable Agricultural Systems. In: Clive A. Edwards et al. (Ankeny, Iowa: Soil and Water Conservation Society, 1990), p. 52

Penvern S, Jamar L, Dapena E, Lateur M, Simon S, Bellon S (2012) Sustainable orcharding through eco-design and co-design. Symposium of the European Society for Horticultural Science, Angers, July 2012

Populer C, Lateur M, Wagemans C (1998) Ressources génétiques et résistance aux maladies des arbres fruitiers. Biotechnol Agron Soc Environ 2 (1):46–58

Schafer JF (1971) Tolerance to plant disease. Annu Rev Phytopathol 9:235–52

Simon S, Lauri PÉ, Brun L, Defrance H, Sauphanor B (2006) Does fruit-tree architecture manipulation affect the development of pests and pathogens? – a case study in apple orchard. J Hortic Sci Biotechnol 81(4):765–773

Simon S, Sauphanor B, Lauri PÉ (2007a) Control of fruit tree pests through manipulation of tree architecture. Pest Technol 1(1):33–37

Simon S, Miranda C, Brun L, Defrance H, Lauri PÉ, Sauphanor B (2007b) Effect of centrifugal tree training on pests and pathogens in apple orchards. IOBC/wprs Bull 30(4):237–245

Simon S, Brun L, Guinaudeau J, Sauphanor B (2011). Pesticide use in current and innovative apple orchard systems. Agron Sust Dev 31:541–555

Simon S, Morel K, Durand E, Brevalle G, Girard T, Lauri PÉ (2012) Aphids at crossroads: when branch architecture alters aphid infestation patterns in the apple tree. Trees Struct Funct 26:273–282

Speiser B, Tamm L, Weibel FP (2014) Regulatory framework for plant protection in organic farming. In: Penvern S, Bellon S (Eds) Organic farming, prototype for sustainable agricultures. Springer, Chap. 4, pp 65–82

Spruijt-Verkerke J, Schoorlemmer H, van Woerden S, Peppelman G, de Visser M, Vermeij I (2004) Duurzaamheid van de biologische landbouw. Wageningen, Praktijkonderzoek Plant & Omgeving, pp.118

Stoeckli S, Mody K, Dorn S (2008) Influence of canopy aspect and height on codling moth (Lepidoptera: Tortricidae) larval infestation in apple, and relationship between infestation and fruit size. J Econ Entomol 101:81–89

Velasco R, Zharkikh A, Affourtit J, et al (2010) The apple genome: emergence of the species and its domestication [J]. Nat Genet 42:833–839

Visser T, Verhaegh JJ, De Vries DP (1974) Resistance to scab (*Venturia inaequalis*) and mildew (*Podosphaera leucotricha*) and fruiting properties of the offspring of the apple cultivar Antonovka. Euphytica 23:353–364

Warlop F, Dapena E, Lateur M, Bastiaanse H, Blázquez MD, Fillatre JY, Gomez C, Jamar L, Leterme E., Libourel G, Miñarro M, Parveaud CE, Pissard A, Rondia A, Stievenard R (2010) Urgent need for new apple breeding methods better adapted to low-input agro ecosystems. *In*:

Breeding for Resilience: a strategy for Organic and low-input farming systems? Eucarpia 2nd Conference of the "Organic & Low-input Agriculture" Section. Paris, p 106–110

Weibel F, Häseli A (2003) Organic apple production—with emphasis on European Experiences. In: Ferree DC, Warrington IJ (eds) Apples, botany, production and uses. CABI publishing

Zehnder G, Gurr GM, Kuhne S, Wade MR, Wratten S, Wyss E (2007) Arthropod pest management in organics crops. Annu revi Entomol 52:57–80

Chapter 8
Alternatives to Synthetic Chemical Antiparasitic Drugs in Organic Livestock Farming in Europe

Hervé Hoste, Smaragda Sotiraki, Helena Mejer, Felix Heckendorn, Veronika Maurer and Stig Thamsborg

Abstract Gastrointestinal parasitism with helminths and protozoa remains a major threat in cattle and small ruminant production and has been "rediscovered" in pig and poultry production systems operating under organic farming rules because of the requirement to grant outdoor access to the animals. The control of these parasitic infections is a key issue for the economic viability of farms and for animal welfare. Control solutions aim at: (i) providing the agronomical bases for parasite control through rational management of pastures; (ii) stimulating the host immune response through different means; (iii) exploring and evaluating the efficacy of new drugs (i.e., phytotherapy, homeopathy and nutraceuticals); and (iv) developing new concepts of application of chemical antiparasitic drugs (e.g., targeted selective treatments). These four control principles will be illustrated with examples taken from the different livestock production systems. The perspectives of current research are to provide both organic and conventional farmers with a "basket of options" to be adapted to the various situations, in order to achieve more sustainable, integrated approaches to parasite control.

Keywords Parasite control · Helminths · Protozoa · Livestock · Hygiene · Alternative treatment · Immune prophylaxis

---

H. Hoste (✉)
INRA-Université de Toulouse-ENVT, UMR1225 IHAP Interactions Hôte-Agents Pathogènes
23 Chemin des Capelles, 31076 Toulouse Cedex, France
e-mail: h.hoste@envt.fr

S. Sotiraki
Veterinary Research Institute (VRI),
Hellenic Agricultural Organisation "Demeter" (former NAGREF)
NAGREF Campus, 57001 Thermi, Thessaloniki, Greece

H. Mejer · S. Thamsborg
Department of Veterinary Disease Biology, Faculty of Health and Medical Sciences,
University of Copenhagen, 1870 Frederiksberg, Denmark

F. Heckendorn · V. Maurer
Department of Livestock Science, Research Institute of Organic Agriculture (FiBL),
Ackerstrasse, Postfach, CH 5070 Frick, Switzerland

S. Bellon, S. Penvern (eds.), *Organic Farming, Prototype for Sustainable Agricultures*,
DOI 10.1007/978-94-007-7927-3_8,

8.1 Comparing Parasitic Risks and Their Control in OF in Ruminants, Pigs and Poultry

There is no doubt that organic farming (OF) rules that promote access to pastures or runs for livestock may enhance some parasitic risks in farming systems. On the other hand, other basic OF rules (e.g., reduced stocking rates of animals kept outdoors compared to indoor stocking rates or to conventional systems) are also known to reduce parasitic risks and to favour animal welfare. Therefore, when focusing on gastrointestinal (GI) helminths or protozoan parasites, different hypotheses concerning the consequences of introducing OF rules can be made between:

1. The transmission and level of infections of parasites whose infectious life stages (either as free-living eggs or larvae, or in intermediate hosts/vectors) are found in the outdoor environment (e.g., a whole range of GI helminths from the three main taxonomical groups: nematodes (strongyles), cestodes (taenia) and trematodes (flukes) (Taylor et al. 2007). In this case, OF rules are likely to increase the parasitic risks. The case of transmission of potential zoonoses (e.g., trichinellosis, toxoplasmosis) whose prophylaxis was previously controlled by keeping livestock indoors has to be examined in terms of food safety and human health.
2. The transmission of some parasites (e.g., certain intestinal protozoa such as eimeriosis in poultry and isosporosis in pigs) whose biological traits, rates of infection and associated pathological consequences are strongly linked to the concentration of animals. In this case, the application of OF rules is instead expected to reduce the parasitic risk.

To some extent, these concepts are overstated. For example, it is well known that some helminth transmission occurs indoors (strongyloidosis). On the other hand, severe cases of coccidiosis have been observed outdoors in calves and chickens (Nielsen et al. 2003). Also, the differences in parasitic risks between OF vs. conventional systems depend on the initial status. Major differences in parasitism have been found: (1) when conversion to OF concerns conventional production systems for which, on a large scale, animals were usually kept indoors (pigs and poultry); or (2) when compared to those where grazing was still the common rule (ruminants).

In the first cases and in most places, the required outdoor access for all or a part of the animals' lives has resulted in deep changes in management when compared to conventional, intensive production. The associated OF changes in nutrition, biosecurity and practices are reflected in changes in disease patterns and food safety risks in OF. In many situations, this has been linked to the "re-emergence" of parasite infections whose life cycles cannot take place indoors. For example, although the availability of data on diseases in organic pig production is still limited, several studies have targeted GI parasites as a significant disease problem in the EU (Thamsborg and Roepstorff 2003; Roepstorff et al. 2011). In contrast, the OF requirements for outdoor access do not fundamentally change the overall husbandry and associated parasitic risks in ruminants for which grazing, at least for part of the animal's life, is the general situation. This was also reported in a few cases of traditional pig farming systems with outdoor farms in the UK and Spain (Hovi et al. 2003; Trujillo and Mata 2000). However, some other general rules and recommendations in OF (e.g., to replace the usual allopathic antiparasitic treatments by phytotherapy and homeopathy when they have demonstrated their efficacy) can contribute to modulating the intensity or diversity of parasitic infections (Cabaret et al. 2002).

In both ruminant and monogastric hosts, the presence of GI parasites affects animal health and welfare and represents a major economic issue because host-parasite interactions are usually chronic infections that provoke long-term, insidious production losses. Therefore, the control of these parasitic infections is a key issue for the economic viability of farms and for animal welfare. However, in contrast to conventional systems that, until now, almost exclusively relied on chemotherapy to control parasites, the general goal promoted in OF is to achieve a more sustainable, integrated approach by combining solutions related to three main objectives linked to three key targets in the parasites' life cycles.

1. *the environment (hygiene):* The aim is to reduce contact between the hosts and the infective parasitic stages through rational pasture management systems (outdoors, for helminths and protozoa) and by rigorous application of hygienic principles (indoors, mainly for protozoa).
2. *the farm animals (host protective reactions):* The aim is to improve the host response against parasitic stages (either resistance or resilience; see Box 8.1) by means of either the use of potential vaccines (immunoprophylaxis), alternated with genetic selection, or well-balanced nutrition.
3. *the parasites (treatment):* The aim is to eliminate or to modulate the biology of various parasitic stages in the host, either by evaluating the efficacy of novel, recommended approaches (phytotherapy, nutraceuticals and homeopathy) or by proposing more sustainable ways to apply synthetic antiparasitic drugs.

Box 8.1: Host Resistance and Host Resilience

Both concepts correspond to two components of the host response when infected with parasites. These two concepts have been particularly illustrated in the case of gastrointestinal infections with parasitic nematodes.

- The host resistance is usually described as the ability of a host to affect the nematode biology either by decreasing the establishment of the infective L3, by delaying the worm growth, by reducing the female worm fertility and egg excretion and/ or by expelling existing adult worm populations. It is suspected that the host resistance is mainly dependent on the immune mechanisms acquired after repeated contact with the worms. It is usually estimated through measurements characterising the parasite populations
- The host resilience is a second component of the host response to parasitism. It is described as the ability of an host to withstand the negative pathological effects due to the presence of worms inhabiting the digestive tract. This ability is usually measured by pathophysiological or production parameters. The underlying mechanisms for the host resilience remain obscure, although the role of some physiological feedback or adaptive

mechanisms aiming at maintaining the host tissular and blood homeostasis is suspected.

- Usually, a better resistance of the host is associated with a better resilience because processes affecting the worm biology and/or number also reduce the lesions and functional disturbances imposed to the host. However, a few examples have been described indicating that a higher host resistance could sometimes be associated with a lower host resilience, confirming the idea that the immune response as "a cost" (Colditz 2002)

It is noteworthy that a trend towards such integrated approaches is increasingly promoted today in conventional husbandry systems, mainly because of the rapid development and widespread diffusion of resistances to chemical drugs in parasite populations (Kaplan 2004; Chapman et al. 2010). In this respect, solutions developed in OF production systems can also largely benefit conventional systems and vice versa. For example, many of the innovative technical options explored in pigs and poultry have benefited from solutions previously explored in ruminants.

Therefore, this review aims at summarising and at illustrating:

1. How these various approaches that refer to the three main principles of disease control can represent the cornerstones of parasite control under OF rules: (i) in different livestock species (ruminants, pigs and poultry, focusing on laying hens); and (ii) within a wide range of epidemiological European conditions,
2. How these solutions need to be adapted depending on the mode of production and/or of some parasitic biological traits.

The framework of this review will be restricted to the main GI helminths and protozoan (coccidian) infections. For pig and poultry production, specific information on "re-emerging" parasites will be provided. For ruminants, because of the continued use of grasslands, the reader can find the list of the main GI helminth and protozoan parasites in the literature (Taylor et al. 2007).

8.2 The Importance of Parasites in Organic Livestock Farming

8.2.1 In Ruminants

As previously stated, in ruminants, the changes introduced by the conversion to OF rules do not basically modify the patterns of parasitic infection but are instead associated with changes either in the overall intensity of infection or in higher species diversity (Cabaret et al. 2002). The same dominant helminths and protozoan species are encountered in OF and conventional systems. Most studies have been dedicated to gastrointestinal nematodes (GINs), although trematodes, and to a lesser degree, cestodes, can also severely impair animal health and production by decreasing the quantity or downgrading the quality of products.

8.2.2 In Pigs

The introduction of outdoor access, whether it be pastures or small, concrete covered runs, is a major risk factor for a range of GI helminths and protozoa with direct life cycles, (e.g., helminth genera *Ascaris, Trichuris*, *Strongyloides*, *Hyostrongylus* and *Oesophagostomum* spp). Furthermore, outdoor access is a necessity for the completion of a helminth's indirect life cycle, e.g., *Metastrongylus* spp. or spiruroids with earthworms and dung beetles, respectively, as intermediate hosts (Taylor et al. 2007).

In contrast, in many intensive indoor herds, *Isospora suis* is the most pathogenic coccidian, causing severe diarrhea in piglets within the first weeks of life. In contrast to helminth infections, the prevalence of *I. suis* is lower in OF herds and associated clinical problems are uncommon (Roepstorff et al. 1992).

Studies in Northern Europe on a limited number of farms have identified *Ascaris suum*, *Oesophagostomum* spp. and *Trichuris suis* as the most common and important parasites in OF production (Carstensen et al. 2002). Overall, compared with conventional farms, the prevalence and levels of helminth infection were higher in OF systems. However, some differences have been observed between the studies, depending on the worm species and/or host factors (e.g., piglets vs. adult pigs). These results illustrate how differences between OF and conventional systems with regards to parasite infections depend (i) on the parasitic species, (ii) on the local production practices (e.g., whether growing pigs or farrowing sows are pastured) and management, and (iii) on the epidemiological conditions (Lindgren et al. 2008; Carstensen et al. 2002).

When examined in detail, there is no indication of the reintroduction of parasites on OF farms that were suspected because of the potential occurrence of wild boar reservoirs (Lindgren et al. 2008). In Swedish studies, the stomach worm, *Hyostrongylus rubidus* and the lungworm *Metastrongylus* spp were absent due to long history of intensification, and indoor production was not found (Lindgren et al. 2008). spp. A currently on-going survey in eight EU countries is expected to complete these data (Roepstorff et al. 2011).

Besides the recommendation to use outdoors runs, other factors may limit the options for control of parasitism in pigs. They include limitations on preventive use of antiparasitic drugs, use of deep litter bedding and, in general, lower levels of hygiene. Although not proven, the potential shortage of certain amino acids may also affect susceptibility to infections. There are indications that the use of permanent pastures as compared to pastures within the crop rotation may result in more parasites (Carstensen et al. 2002; Lindgren et al. 2008). Other risk factors for helminth infections, based on knowledge acquired from conventional farms, include: solid floors, straw bedding, infrequent dung removal, high indoor humidity, and high intake of insoluble fibre (roughage/forage). On the other hand, the latter may limit problems with certain bacterial infections.

The clinical implications and potential impact of helminths on production remain largely unknown since no intervention studies have been performed on organic farms. Clinical trichuriosis has been reported on an organic farm (Carstensen et al. 2002), but even on conventional farms, a pathological impact as measured in experimental studies has been difficult to demonstrate. However, indirect effects such as a reduced protective response after vaccination against unrelated agents (e.g., *Mycoplasma hyopneumoniae*) should not be neglected (Steenhard et al. 2009).

8.2.3 In Poultry

The risk of helminth infestations is higher in hens in free-range systems compared to systems without outdoor runs (Permin et al. 1999) (Fig. 8a, Colour plate 8). The roundworms, *Ascaridia galli* and *Heterakis gallinarum,* are the most common species found in the intestines of poultry. Eggs of both species have a long survival rate in the environment, and, thus, a high infection potential. Because of these biological peculiarities and because of the establishment rates of *A. galli* that strongly decrease with increasing infection doses (Permin et al. 1997), preventive measures, effective in other species, are less efficient in poultry.

Free-range hens are also at an increased risk of being infected by the protozoan, *Histomonas meleagridis* (blackhead disease), which is transmitted by *H. gallinarum*. At present, no feed additives against blackhead disease are registered (EC 2002). Helminth control is therefore probably the most effective measure against these protozoan infections.

The chicken is host to seven species of *Eimeria spp*. These coccidian parasites vary in their pathogenicity, but generally, coccidiosis is a major economically significant intestinal disease of commercially reared poultry. Good management contributes to the prevention of coccidiosis, but additional measures are needed under indoor and outdoor conditions (McDonald and Shirley 2009).

8.3 Methods of Control

8.3.1 Hygiene: Reducing the Environmental Sources of Infection

The general objective is to reduce the contact between the host and the infective parasitic stages. Under indoor conditions, this is usually based on the principles of disinfection and hygiene. The main target is protozoan infections, although helminths are also found in poultry. However, methods relying on the overall principles of hygiene to reduce pasture contamination have also been developed.

Management of Indoor Areas

There are four possible sources of coccidian infections in ruminants: (i) oocysts from previous faecal contamination that survive in the environment; (ii) fresh oocysts passed by the ewes; (iii) fresh oocysts passed by the lambs; and (iv) contaminated ewe fleeces and udders (Pout 1973). The management system, hygiene status and nutrition play an important role in the development of subclinical vs. clinical coccidiosis (Gregory and Catchpole 1990). Weaning of lambs/kids in combination with other stress factors such as high stocking densities, multiple lambing with lower colostrum and milk intake, cold and wet weather and depressed immunity predispose animals to disease (Catchpole et al. 1990). However, the use of pastures does not always mean a lack of coccidiosis risk. For example, calves on organic dairy farms have also been found to be heavily exposed to coccidia when they are group-housed in deep-litter pens indoors and later introduced into permanent pastures (Nielsen et al. 2003).

In pigs, continuous housing in permanent deep-litter systems may lead to increasing levels of parasite infections (Holmgren and Nilsson 1998). Similarly, the introduction of enriched environments like sprinkler systems may be associated with the increased risk of helminth transmission (Roepstorff et al. 2011). This can be partly overcome by the ample provision of new straw bedding, but the only rational approach is all-in-all-out systems. In a recent study, some sanitary and management protocols aimed at lowering oocyst uptake and excretion from infected piglets have been described. These include farrowing rooms, no cross-fostering or fostering during the first 24 h after farrowing, plastic flooring in the farrowing pens, farrowing rooms with fewer pens, and measures to prevent caretakers from entering the farrowing pens (Skampardonis et al. 2012).

In poultry, good hygiene of hen houses is also essential to prevent accumulation of the long-lasting infective parasite eggs and/or oocysts over time. Thorough cleaning of the hen house is only possible between flocks. Maurer et al. (2009) found similar helminth egg concentrations at different litter management regimes (replacement or addition of litter during flocks) and no effects of egg density in litter on worm burdens of the layers.

Outdoor Management

Under outdoor conditions, the runs and pastures cannot be "disinfected" to destroy infectious stages. The use of possible chemicals on a large scale has proved to be inefficient under field conditions and is not acceptable according to OF rules. Therefore, different strategies of grazing management have been developed. They were initially designed for ruminants in the late 1960s and mainly aimed at controlling infections with GI nematodes. Their pros and cons were then evaluated under a wide range of epidemiological conditions (Michel 1976; Barger 1999). These different grazing management strategies aim at reducing the parasitic risk over time and/or space

- by reducing the larval density by applying low stocking rates to dilute the risk;
- by taking advantage of the natural death rate of infective larvae by pasture rotation systems;
- by accelerating larval mortality through biological control or co-grazing between different hosts.

Due to the different epidemiological patterns of the parasite species involved, preventive strategies are less effective in monogastrics than in ruminants, although hygiene and proper management of runs and pastures are also the basis for preventing helminth infections in pigs and poultry.

Dilutions

Many studies on GIN infections in ruminants have shown the existence of a relationship between the stocking rate and the level of host infection. However, this phenomenon does not appear to be linear (Thamsborg et al. 1996). The limitation of animal outdoor stocking rates, which is one of the cornerstones of the OF rules, is based on common sense with regard to parasitic control in ruminants. In pigs, the effects of nose rings (to avoid the destruction of the sward) and stocking rate on the uptake of infections are not clearly described (Thomsen et al. 2001). Similarly, stocking rates within the hen house and in the hen run had no effect on *A. galli* infections (Permin et al. 1998). Moreover, Heckendorn et al. (2009) reported that the stocking rate of hens in the outdoor run did not influence the transmission patterns of *A. galli* and *H. gallinarum,* and repeated moving of runs did not reduce helminth infections. Lower stocking rates, however, led to a substantial improvement of the run vegetation.

Rotation of Pastures

Because the survival of infective stages of parasites on pastures is limited in time, systems of pasture rotations aimed at introducing animals in paddocks after the risk related to the infective stages has been substantially reduced because of the natural death rate. This concept has been applied in ruminants and monogastrics but the efficiency of the methods highly depends on (i) the biology of parasitic stages, and (ii) the local/regional climatic and epidemiological conditions.

For example in ruminants, more efficient results have generally been obtained with rotation methods under tropical compared to temperate conditions because of the peculiarities of the biology of *Haemonchus contortus,* which is a highly prevalent nematode in the tropics (Torres Acosta and Hoste 2008).

In pigs, pasture rotation is also likely to play a central role in the control of some helminth species. Infective larvae of *Oesophagostomum* spp. and *H. rubidus* can survive for a maximum of 1 year on pasture and have a poor over-winter survival rate in northern temperate climates (Thomsen et al. 2001; Mejer 2006).

Consequently, even pastures heavily contaminated in autumn may be totally clean the following spring, and *Oesophagostomum* spp. do not constitute a problem in strictly outdoor sow herds (Carstensen et al. 2002).

In contrast, the most common helminths (*A. suum* and *T. suis*) are characterised by hard-shelled eggs with a sustained longevity on pasture of up to 10 years (Roepstorff 2003), despite initial high death rates of eggs within the first 6–12 months (Larsen and Roepstorff 1999), indicating that a shorter rest period may still serve to reduce the transmission. Because of this long persistence in the environment of these worm species, it is still unknown as to whether pigs should be moved to clean pasture one or two times per year, and when it is safe to be back to a previously contaminated pasture. Ongoing experiments using naive pigs (never previously exposed to parasites) to trace the levels of contamination on pastures after initial deposition of eggs, have yielded two results: (1) transmission levels increased the first 2 years, indicating an unexpectedly slow development to infectivity in colder climates; and (2) infection levels did not markedly decrease after 4 years (Helena Mejer 2010, unpublished data). Moreover, Carstensen et al. (2002) described a farm that produced organically for 8 years with a stringent 3-year pasture rotation. *Ascaris* was found with a prevalence >90 % in the weaners and fatteners. These results may indicate either a failure in the management of the scheme or simply that 3 years are not sufficient to reduce pasture contamination with both parasites.

Last, in some poultry production systems (e.g., free ranging broilers or layer flocks in mobile systems), an all-in-all-out system, where new areas are provided for each batch of animals, is feasible (Thamsborg et al. 1999). Where hens are kept in solid hen houses with surrounding runs, a rotation scheme with sufficient resting time between flocks (>1 year) is nearly impossible to achieve. In a 2-year on-farm experiment, a rotation scheme where the hen flock returned to the same area during one season helped to maintain the sward, but it was not effective to reduce *A. galli* and *H. gallinarum* burdens (Maurer et al. 2013).

For the coccidian parasites, including *Isospora suis* in pigs, efficient control seems achievable (if pastures ungrazed the previous year are provided at turn-out), in particular in herds where farrowing huts are routinely moved before farrowing (Roepstorff et al. 1992). Isosporosis is thus not a clinical problem in organic pig production.

Co-grazing Between Different Host Species

Coccidia are highly specific to their hosts. For helminths, with the exception of trematodes and a few nematode species (e.g., *Trichostrongylus axei*) that are ubiquists, most nematodes and cestodes are specific to their hosts, although the phenomenon is not fully exclusive between large and small ruminants. This parasite specificity means that when different animal species graze together in the same pasture, each host contributes to reducing the pasture-related risk for the second animal species by "destroying" the parasitic stages specific to this second host.

Most studies on these co-grazing methods have been performed in ruminants, generally between cattle and sheep. The results obtained under various epidemiological conditions and in situations of either simultaneous or alternate grazing have usually shown a reduced level of infection in sheep, especially with the pathogenic species *H. contortus* (Hoste et al. 2009). The effects on cattle were more limited. These practices have usually been associated with better resilience (see Box 1) of the animals, including a better use of the grazing rejects (see review by Barger 1999).

When sows have nose rings, it is possible to have them graze together with ruminants. In a 3-year study with heifers and dry sows, mixed grazing has been shown to increase the performance of the sows, although the most important effect was to reduce *Ostertagia* infections of the heifers and to increase their weight gain (Thamsborg et al. 1999). The effect was attributed to better utilisation of pasture, as well as disruption of cattle faecal pats and the surrounding tussocks. However, the use of nose rings is now an animal welfare issue.

As suggested by Barger (1999), there are also potential risks in mixed systems. One is that previously host-specific helminths adapt to other hosts. For example, cross contaminations with sheep nematodes, leading to clinical diseases, have been reported in calves (Armour et al. 1988). Patent *Ascaris* infections have also been described in lambs grazing in pastures previously used by pigs. Reciprocally, the ruminant species, *Teladorsagia* spp and *Trichostrongylus vitrinus,* were also reported in pigs. Last, the increased risk related to trematodes has to be more closely monitored.

Biocontrol and Use of Nematophagous Fungi

Under laboratory conditions, a range of various biological agents (e.g., *Bacillus thuringiensis*, or "cannibal" nematodes) have been shown to affect the biology of environmental stages of helminths, particularly nematodes (Torres-Acosta and Hoste 2008). However, under farm conditions, the most outstanding results of biocontrol to decrease pasture infectivity were obtained with some nematophagous fungi that can kill nematode larvae in faeces. Their effects have been studied within a wide range of host species and/or epidemiological conditions, in particular for *Duddingtonia flagrans.* This fungus species was of particular interest because of the ability of the spores to develop and to trap the nematode larvae in faeces after a digestive passage (Larsen 2000). In ruminants and horses, convincing results have also been repeatedly obtained under both laboratory and farm conditions. However, no commercial product of this promising, innovative option has yet been launched.

D. flagrans was also shown to significantly reduce *Oesophagostomum dentatum* infections in grazing pigs (Larsen 2000). In contrast, the main parasitic nematodes of monogastric animals have egg-dwelling and not larval infective stages. Therefore, until now, no promising biocontrol agents have been identified for use against these species.

8.3.2 Improving the Host Response: Immunostimulation

Vaccination

Conceptually, vaccination represents a highly attractive solution to control any infectious disease and is usually well accepted in OF. Therefore, this solution is clearly included in the general "basket of options" explored to address the issues related to parasitism in OF. However, antiparasitic vaccines still face several limits today in terms of their implementation under field/farm conditions because of considerations related to either (i) ethics (GMO vaccines are not acceptable in OF systems; use of live vaccines with potential risks of diffusion of diseases after mutation); (ii) technology (the difficulties to develop vaccines based on recombinant proteins against helminths); or (iii) economy (multivalent vaccines are often demanded).

For gastrointestinal helminths, no commercialised vaccines are currently available despite considerable research efforts to produce a vaccine against *H. contortus* based on some hidden antigens of the nematode GI tract (Smith and Zarlenga 2006). However, a live vaccine with partly inactivated larvae of the cattle lungworm *Dictyocaulus viviparus* (Dictol®) has long been commercialised and widely used in some countries where lungworms are endemic (Benitez Usher et al. 1976). For protozoa, the use of live (attenuated or non-attenuated) vaccines (consisting of oocysts of different *Eimeria* species) to control coccidiosis due to various *Eimeria* infections in layer and broiler breeder chickens is well established (Chapman et al. 2002; Williams 2002). Vaccination with a reduced number of *Eimeria* species is also increasingly applied in one-day old organic broiler chickens (Williams 2002), but conventional broiler production still largely relies on anticoccidial drugs (McDonald and Shirley 2009).

Selective Breeding for Resistance and/or Resilience to Parasites

The preference for local livestock breeds, is one of the cornerstones of the OF philosophy to address sanitary problems and to reduce the reliance on chemical drugs. This recommendation relies on the concept of coevolution between parasite and hosts, which, after natural selection, led to hosts that were more adapted to parasite infections depending on the regional/epidemiological conditions. In the case of GI parasite infections, early results that support the hypothesis that some genetic components are involved in differences in resistance to helminths were obtained in sheep because, until recently, helminth challenges remained limited in monogastrics under conventional production systems.

In ruminants, differences in the level of gastrointestinal infections between breeds have been thoroughly documented in sheep. However, most of the studies compared imported vs. local breeds under tropical conditions (Bishop and Morris 2007). In contrast, less data is available comparing local rustic breeds to more intensively selected breeds in temperate conditions.

The existence of possible individual differences in resistance against GI nematodes within a breed has also been widely studied (Bishop and Morris 2007; Vagenas et al. 2002). The level of nematode egg excretion has been the main phenotypical criterion used for selection. Heritabilities (h^2) of faecal egg counts (FECs) in sheep range between 0.08 and 0.43 (Gasbarre and Miller 2000). This has promoted programmes (e.g., the commercial DNA-marker WORMSTAR programme for sheep in New Zealand) to select animals with a better response to GINs and responsible for a lower pasture contamination because of significantly decreased FECs after several generations (Hunt et al. 2008; Vagenas et al. 2002). Resilience selection programmes aimed at better animal productivity under parasitic challenges have been another option that was explored to counteract the negative effects of parasites on their hosts. However, this option has received less attention than selection for resistance (Bisset et al. 1996).

Such genetic differences in resistance are progressively assessed in monogastrics. Some Danish results (Schou et al. 2003) indicated that the epidemiology of *A. galli* infections in chicken may similarly be influenced by a genetic component. Abdelquader et al. (2007) found that there is not only a variation in the genetic background of the hens, but also that *A. galli* isolates from different geographic areas differ in their ability to infect different chicken genotypes. Gauly et al. (2008) observed significantly higher faecal egg outputs of helminths in white laying hens than in brown hens. They estimated sufficiently high h^2 of FECs of *A. galli* (0.13–0.19 for white hens) as well as for *H. gallinarum* worm burdens (0.31–0.41) to allow selection for helminth resistance. However, parasite resistance is not a seriously considered criterion in poultry breeding at present because genotypes are mainly selected for best performance under indoor conditions where the parasitic challenges are of minor importance.

In pigs, studies based on examination of 200 offspring of known matings revealed h^2 of FECs of 0.3–0.4 for *A. suum,* and of 0.4–0.7 for *T. suis* (Nejsum et al. 2009). For *T. suis,* the h^2 depended on time in relation to the onset of infection: during the early expulsion phase, h^2 were highest, probably indicating close genetic control of the onset of immunity. For *Ascaris,* other parameters such as worm burden, total egg output and antibody levels were also heritable, whereas this was not the case for the size and fecundity of the worms (Peter Nejsum, personal communication). It is obvious that breeding for increased host resistance is also an option within the pig industry and may be highly relevant in free-ranging systems, although more studies must to be performed.

8.3.3 Treatments Affecting Parasite Biology

When treatments are required to cure animals and to improve their welfare, the general OF recommendations are to promote alternative medicines (phytotherapy, homeopathy) rather than synthetic chemical (allopathic) drugs. However, in many circumstances, the efficiency of alternative medicines remains to be fully established.

Herbal Drugs

The use of natural (herbal) remedies (phytotherapy, including the use of essential oils) to cure and/or to prevent diseases is not a novel concept (Githiori et al. 2006; Waller et al. 2001). According to both time and geographical scales, chemical drugs remain the norm. It is worth recalling that more than 70% of the drugs currently used worldwide are natural and that many of the so called "chemical drugs" are derived from natural products (Wilcox et al. 2001). For example, halofuginone, a quinazolinone alkaloid from *Dichroa febrifuga*, has been used as a coccidiostat because febrifugine, the original plant extract, possesses antiprotozoan activity (Youn and Noh 2001). Natural (mainly plant) materials used as an alternative to control parasitic diseases can be broken down into two different categories: phytotherapy remedies vs. nutraceuticals. In both cases, the observed antiparasitic activity is usually linked to the presence of plant secondary metabolites (PSMs) in sufficient concentration, including condensed tannins and flavonoids, sesquiterpens, proteinases, etc. (Rochfort et al. 2008).

Herbal drugs are preparations of plants and/or plant extracts that aim at curing infected animals after a short-term administration. Their general therapeutic recommendations are close to those of chemical drugs, except that the active compound(s) are usually not well identified and measurable. Herbal remedies are often a mixture of plants and/or plant extracts obtained by various physical or chemical processes. Accordingly, they are usually composed of a high number of biochemical components. This complexity is a general characteristic of drugs derived from plants. It has consequences on the definition and standardisation of the products and, consequently, on the validation of their therapeutic efficacy as a function of their variations.

A variety of plants worldwide have been shown to affect survival and/or reproduction of helminths of chicken or ruminants *in vitro* or *in vivo*. However, in some cases, severe side effects on the host have been observed after use of plant products (e.g., Javed et al. 1994; Akhtar and Riffat 1985). In other studies (e.g., Chota et al. 2009), positive effects of a plant preparation (*Carica papaya*) were confused with the effects of better nutrition during the experiments. The effect of these herbal drugs was tested by Maurer et al. (unpublished data) under controlled conditions in eight series of hens and chickens artificially infected with *A. galli*. Although significant effects of some plant preparations were observed in a particular series, none of the plant extracts tested resulted in reductions of egg counts or of worm burdens in a reproducible manner (unpublished data). Similar difficulties have been found in a series of studies comparing the efficacy of plant preparations against helminths in lambs (Bouilhol et al. 2003; Hördegen et al. 2003) and pigs (van Krimpen et al. 2010).

Research on alternative methods for controlling *Eimeria* spp. with plants showed that the effect on coccidia differs from the one on nematodes. Some plants and plant products such as *Azadirachta indica* (Tipu et al. 2002) and artemisinin from *Artemisia annua* (Almeida et al. 2012) reduced the excretion of oocysts of some *Eimeria* spp. and/or decreased mortality and intestinal lesion rates. Youn and Noh (2001) tested extracts of 15 plants against *E. tenella* in chickens and found extracts of *Sophora flavescens* to be the most efficient in terms of survival rates, bloody diarrhoea, lesion scores, body weight gains and oocyst excretion.

The use of antioxidant compounds in the management of coccidiosis has been shown to be effective because they decrease the degree of intestinal lipid peroxidation, which is associated with coccidian parasite-induced host cell destruction. One of the most potent veterinary anticoccidials, toltrazuril, is believed to achieve some, if not all, of its beneficial effects by limiting the degree of lipid peroxidation (Eraslan et al. 2004).

In support of this hypothesis, *in vivo* studies in South Africa have included treatment with extracts from *Tulbaghia violacea, Vitis vinifera* and *Artemisia afra* (Naidoo et al. 2008), and in China with proanthocyanidin extracts from grape seeds (Wang et al. 2008). Both improved the performance of broiler chickens and relieved the clinical symptoms caused by avian coccidian infection.

Nutritional Approaches and Nutraceuticals

The term "nutraceutical" is defined as "any substance that may be considered as a food or part of a food which provides health benefits, including the prevention and treatment of disease" (Andlauer and Furst 2002). Nutraceuticals are a feed resource, used either fresh or conserved, but the main reason for using them is linked to their potential benefits on animal health. Compared to herbal remedies, they are administered for a longer term (at least a few days), and the first objective is to prevent or limit the level of infections and, consequently, to reduce the reliance on chemotherapy to control parasites. In contrast to allopathic treatments that aim at eliminating worms, nutraceuticals act more by "slowing down" the biology of parasites and the dynamics of infection. Since the end of the 1990s, results on the use of nutraceuticals to control GINs have been obtained either in ruminants or pigs, illustrating how this option might be relevant in various OF systems (Waller and Thamsborg 2004).

In small ruminants (Fig. 8b, Colour plate 8), most current data on the potential represented by nutraceuticals have been obtained on the use of tannin-rich (TR) legume fodders, (e.g. sulla, *Sericea lespedeza*, big and birdsfoot trefoils, and sainfoin). Overall, the first effect associated with the distribution of tanniniferous legumes is a reduction of GIN egg excretion due to either a reduction in worm number or the reduced fertility of female worms (Hoste et al. 2006; Athanasiadou et al. 2001; Heckendorn et al. 2007; Manolaraki 2011). Reductions in the establishment of infective larvae have also been described (Hoste et al. 2006). Less consistent results have been obtained on possible reduced development from eggs to infective larvae in the environment (Niezen et al. 2002).

In addition to TR legumes, many studies in sheep have focused on chicory (*Cichorium intybus*) when used as forage. Its consumption has often been associated with some favourable effects on worm biology. For chicory, the suspected bioactive compounds are not tannins but sesquiterpene lactones (Marley et al. 2003; Athanasiadou et al. 2007).

In pigs, the addition of easily fermentable carbohydrates to the diet was shown to significantly diminish *O. dentatum* numbers and female fecundity (Petkevičius et al.

2003). This promising principle is currently being investigated using inulin-rich diets for sows in OF. Chicory roots and lupin seeds are rich in fermentable carbohydrates, particularly fructans (inulin). In pigs, an almost complete reduction of the *Oesophagostomum* egg output was obtained by adding purified inulin (Petkevičius et al. 2003) or dried chicory roots to the diet (Mejer 2006). High reductions in worm counts have been observed in some, but not all studies (Petkevičius et al. 2003; Mejer 2006). Incomplete elimination of worms may explain why depression of egg excretion was partially reversible since egg counts were shown to increase when the carbohydrates were withdrawn from the diet. Since the fermentable carbohydrates are only partially degraded in the small intestine, the action mechanism is probably related to the production of short-chain fatty acids by fermentation in the large intestine (Petkevičius et al. 2004). These fatty acids could directly or indirectly cause adverse conditions for the residing nematodes just as there is a shift in microbial composition. Consequently, *T. suis*, another inhabitant of the large intestine, is moderately affected, but results are inconsistent (Thomsen et al. 2007). Furthermore, penetration of early larval stages of *A. suum* in the large intestine before the migratory liver phase, and the establishment of incoming infections may be affected (Mejer 2006) but not adult established infections (Mejer, personal communication). Because *A. suum* and *T. suis* are the major targets of nematode control in outdoor pig production, these findings require further validation if they are to be considered to be of practical relevance.

Allopathic Treatments

Effective alternative methods for parasite control in poultry are lacking, and the use of conventional anthelmintics is the rule on organic as well as on conventional farms, although the extent of their use may vary. The situation is particularly delicate in the case of laying hens because only one anthelmintic (flubendazole) is registered. This issue is a major problem for organic egg production.

In ruminants, methods aimed at a "tailor-made" (more adapted) use of chemical drugs are also currently being explored in conventional systems because of the increasing challenge represented by the development of resistance to antiparasitic drugs (Kaplan 2004). These approaches might also be of benefit to the OF systems. For an overview of recent results and approaches used for these Targeted Selective Treatments (TST), the reader is referred to a special issue of Veterinary Parasitology (2009) on "Novel Approaches for the Sustainable Control of Nematodes in Ruminants" that summarises the main data from the PARASOL EU project.

8.4 Discussion/Conclusions

Parasite infections are also described as "long term interactions". Therefore, they probably represent a paradigm to illustrate how the control of any livestock disease can be challenged in OF systems whose general aim is sustainable production.

Moreover, the issues related to the control of parasitic infections in livestock illustrate some key features of both the societal and scientific questions raised by the development of OF systems. Obviously, one main keyword underlying the concept of OF disease control is “variability”.

Consequently, a whole range of questions are emerging, for example: (i) How can we challenge the differences between the consequences of this variability in parasitic risks with the consumer’s image of OF production? (ii) How can we analyse the variability of situations and host-parasite interactions? (iii) How can this variability in OF production systems be integrated into the general framework, which includes the three main principles of disease control? (iv) Do we have the appropriate methodological tools to measure the effects of the different methods of control?

Promoting OF rules in livestock is usually linked to improved animal welfare and better product quality by the consumer public. However, when considering the parasitic challenge, several pros and cons must be taken into account. Indoor breeding is usually associated with higher animal concentrations, less possibility of expressing social behaviour and, consequently, a lower status of animal welfare. However, outdoor breeding, which is promoted by OF, also means a higher risk of parasitic infections, degradation of animal health and welfare and, in some cases, can raise questions about the safety of livestock products (e.g., increased risk of zoonoses related to possible wildlife reservoirs).

Throughout this chapter, we aimed at illustrating the robustness of the three well-defined principles for controlling all pathogenic processes (Torres-Acosta and Hoste 2008): (1) by reducing the contact between the host and the infective agents [including the disinfection/elimination of pathogens in the environment (hygiene)]; (2) by improving the host response to the parasites (immunostimulation); and (3) by eliminating the parasites in the hosts (treatment). These principles still remain the cornerstones of actions against any pathogen, including parasites. They have generic implications, as illustrated for the different parasitic agents in the different livestock species under both conventional (CF) and OF systems.

The main difference between the two modes of production is probably linked to the order of priorities given to the different actions. In CF, the reliance on curative and supposedly fully suppressive synthetic antiparasitic drugs has been the first and nearly exclusive part of the tripod because of the difficulty to develop vaccines against parasitic agents. According to OF rules, disease prevention based on hygienic principles relies first on the stimulation of the host response as a result of genetic, nutritional factors and/or use of homeopathy, whereas curative synthetic drugs are only used as a last-ditch option when phytotherapy or natural substances have been proven inefficient (EC 2002).

The variability of interactions with the environment makes it necessary to adapt the solutions derived from these three main principles of control. This can be illustrated by two examples.

In pigs and poultry, the difficulty in applying genetic selection for resistance to parasites from genotypes selected primarily for performance under indoor conditions has been reported. A second option is to better evaluate and possibly to use local breeds that have been traditionally maintained outdoors. This corresponds to one of the general recommendations in OF rules. However, the reduced size of the remaining populations is usually a limiting factor.

Compared to the well-standardised synthetic antiparasitic drugs, the variability in resources is also a key factor to consider when assessing the efficacy of nutraceuticals. The amount and/or quality of PSMs, which are usually thought to be responsible for the activity, vary with genetic, environmental and even technological factors (Manolaraki 2011). This leads to the need for research to identify the main active compounds and to develop simple, affordable methods to measure biological and/or biochemical markers of activity. The situation is even more complex with phytotherapy and herbal remedies that often correspond to mixtures of plants. Last, for homeopathy, which is often assumed to have some long-term effects, the methodology to demonstrate the activity in experimental studies needs to be adapted to livestock species.

Specific studies performed under OF conditions to evaluate adapted solutions for the control of digestive parasitism have existed for nearly 10 years, although differences in progress still exist in the different livestock species (Lund and Algers 2003; Hovi et al. 2003). Because conventional production systems in ruminants have involved extensive use of grazing for many years, some agronomical and technical solutions have been widely explored in cattle, sheep and goats, compared to pig and poultry production. One of the objectives of this review was to illustrate how results acquired on ruminants can be of benefit to monogastrics since the principles for controlling parasitic diseases remain the same, regardless of the host species. It is also worth emphasizing that many of the solutions explored in OF can be of benefit to conventional systems of production. For example, because of the widespread diffusion of resistances to antiparasitic drugs, the sole reliance on chemotherapy in conventional systems is increasingly recognised as being unsustainable, and the need for alternative options is evident. Another reason is the implementation of more restrictive regulations on the use of chemical treatments in livestock, sparking a major interest in alternative approaches. A more restrictive use will delay the diffusion of resistance to antiparasitic drugs in both protozoa and helminths in any system. In both OF and conventional systems, the general goal today is to promote integrated control that relies on a "basket of options" adapted to on-farm situations. In this respect, the current development of software models to integrate abiotic and/or biotic factors at the farm level in the analysis of the parasitic risk of GIN infection in cattle and to propose adapted control measures seems promising (Chauvin 2009).

Acknowledgments The authors wish to express their gratitude for a Marie Curie project entitled "*Healthy Hay*" (contract: MRTN- CT- 2006-035805), as well as for the CAPARA FA0805 COST Action project (*Goat–Parasite Interactions: from Knowledge to Control*) for its financial support. The authors are also grateful for the financial participation of the European Community under the Seventh Framework Programme for Research, Technological Development and Demonstration Activities, for the Integrated Project, LowInputBreedsFP7-CP-IP 222623.

The views expressed in this publication are the sole responsibility of the author(s) and do not necessarily reflect the views of the European Commission.

Neither the European Commission nor any person acting on behalf of the Commission is responsible for the use that might be made of the information contained herein.

References

Abdelquader A, Gauly M, Wollny CBA (2007) Response of two breeds of chickens to *Ascaridia galli* infections from two geographic sources. Vet Parasitol 145:176–180

Akhtar MS, Riffat S (1985) Evaluation of *Melia Azedarach* Linn. Fruit (Bakain) against a *Ascaridia galli* infection in chickens. Pak Vet J 5:34–37

Almeida GF de, Horsted H, Thamsborg SM, Kyvsgaard NC, Ferreira JFS, Hermansen JE (2012) Use of *Artemisia annua* as a natural coccidiostat in free-range broilers and its effects on infection dynamics and performance. Vet Parasitol 186:178–187

Andlauer W, Furst P (2002) Nutraceuticals: a piece of history, present status and outlook. Food Res Int 35:171–176

Armour J, Bairden K, Dalgliesh R, Ibarra Silva AM, Salman SK (1988) Clinical nematodiriasis in calves due to *Nematodirus battus* infection. Vet Rec 123:230–231

Athanasiadou S, Kyriazakis I, Jackson F, Coop RL (2001) Direct anthelmintic effects of condensed tannins towards different gastrointestinal nematodes of sheep: in vitro and in vivo studies. Vet Parasitol 99:205–219

Athanasiadou S, Gray D, Younie D, Tzamaloukas O, Jackson F, Kyriazakis I (2007) The use of chicory for parasite control in organic ewes and their lambs. Parasitology 134:299–307

Barger IA (1999) The role of epidemiological knowledge and grazing management for helminth control in small ruminants. Int J Parasitol 29:41–48

Benitez-Usher C, Armour J, Urquhart GM (1976) Studies on immunisation of suckling calves with dictol. Vet Parasitol 2:209–222

Bishop SC, Morris CA (2007) Genetic of disease resistance in sheep and goats. Small Rumin Res 70:48–59

Bisset SA, Morris CA, Squire DR, Hickey SM (1996) Genetics of resilience to nematode parasites in young Romney sheep; use of weight gain under challenge to assess individual anthelmintic treatments. N. Z. J Agric Res 39:314–323

Bouilhol M, Mage C, Archimbaud E (2003) Activité antiparasitaire de produits phytothérapiques. CR 2013206. Institut de l'Elevage, 3–4

Cabaret J, Bouilhol M, Mage C (2002) Managing helminths of ruminants in organic farming. Vet Res 33:625–640

Carstensen L, Vaarst M, Roepstorff A (2002) Helminth infections in Danish organic swine herds. Vet Parasitol 106:253–264

Catchpole J, Norton CC, Gregory MW (1990) Immunisation of lambs against coccidiosis. Vet Rec 132:56–59

Chapman HD, Cherry TE, Danforth HD, Richards G, Shirley MW, Williams RB (2002) Sustainable coccidiosis control in poultry production: the role of live vaccines. Int J Parasitol 32:617–629

Chapman HD, Jeffers TK, Williams RB (2010) Forty years of monensin for the control of coccidiosis in poultry. Poult Sci 89:1788–1801

Chauvin A (2009) Parasitic risks linked to grazing and their control. Fourrages 199:255–264

Chota A, Sikasunge CS, Phiri AM, Musukwa MN, Haazele F, Phiri IK (2009) A comparative study of the efficacy of piperazine and *Carica papaya* for the control of helminth parasites in village chickens in Zambia. Trop Anim Health Prod 42:315–318

Colditz IG (2002) Effects of the immune system on metabolism: implications for production and disease resistance in livestock. Livest Prod Sci 75:257–268

EC, 2002, Council Regulation (EC) No 1756/2002 of 23 September (2002) Amending Directive 70/524/EEC concerning additives in feeding stuffs as regards withdrawal of the authorisation of an additive and amending Commission Regulation (EC) No 2430/1999 Official Journal of the European Communities, L 265/261–L 265/262

Eraslan G, Cam Y, Eren M, Cem Liman B (2004) Changes in malondialdehyde level and catalase activity and effect of toltrazuril on these parameters in chicks infected with *Eimeria tenella*. Bull Vet Inst Pulawy 48:251–254

Gasbarre LC, Miller JE (2000) Genetics of Helminth Resistance. In: Axford RFE, Bishop SC, Nicholas FW, Owen JB (eds) Breeding for disease resistance in farm animals. CAB International, pp 129–152

Gauly M, Kanan A, Brandt H, Weigend S, Moors E, Erhardt G (2008) Genetic resistance to *Heterakis gallinarum* in two chicken layer lines following a single dose infection. Vet Parasitol 155:74–79

Githiori JB, Athanasiadou S, Thamsborg SM (2006) Use of plants in novel approaches for control of gastrointestinal helminths in livestock with emphasis on small ruminants. Vet Parasitol 139:308–320

Gregory MW, Catchpole J (1990) Ovine coccidiosis: the pathology of *Eimeria crandallis* infection. Int J Parasitol 20:849–860

Heckendorn F, Haring DA, Maurer V, Senn M, Hertzberg H (2007) Individual administration of three tanniferous forage plants to lambs artificially infected with *Haemonchus contortus* and *Cooperia curticei*. Vet Parasitol 146:123–134

Heckendorn F, Häring DA, Amsler Z, Maurer V (2009) Do stocking rate and a simple run management practice influence the infection of laying hens with gastrointestinal helminths? Vet Parasitol 159:60–68

Holmgren N, Nilsson O (1998) Inverkan av produktionsplanering och skabbsanering på tarmnematoder hos bis-grisar (Influence of production planning and mange control on nematodes of pigs). Project report from Köttböndernas Forskningsprogram project No. G37 97, Svenska Djurhälsovården

Hördegen P, Hertzberg H, Langhans W, Maurer V (2003) The anthelmintic efficacy of five plant products against gastrointestinal trichostrongylids in artificially infected lambs. Vet Parasitol 117:51–60

Hoste H, Jackson F, Athanasiadou S, Thamsborg SM, Hoskin SO (2006) The effects of tannin rich plants on parasitic nematodes in ruminants. Trends Parasitol 22:253–261

Hoste H, Cabaret J, Grosmond G, Guitard JP (2009) Alternatives aux traitements anthelminthiques en élevage biologique des ruminants. INRA Productions Animales 22:245–254

Hovi M, Sundrum A, Thamsborg SM (2003) Animal health and welfare in organic livestock production in Europe: current state and future challenges. Livest Prod Sci 80:41–53

Hunt PW, McEwan JC, Miller JE (2008) Future perspectives for the implementationm of genetic markers for parasite resistance in sheep. In: Proc.of 5th International Workshop: Novel Approaches to the Control of Helminth Parasites of Livestock. Trop Biomed 25(Supplement 1):18–33

Javed I, Akhtar MS, Rahman ZU, Khaliq T, Ahmad M (1994) Comparative anthelmintic efficacy and safety of *Caesalpinia crista* seed and piperazine adipate in chickens with artificially induced *Ascaridia galli* infection. Acta Vet Hung 42:103–109

Kaplan RM (2004) Drug resistance in nematodes of veterinary importance: a status report. Trends Parasitol 20:477–481

van Krimpen MM, Binnendijk GP, Borgsteede FHM, Gaasenbeek CPH (2010) Anthelmintic effects of phytogenic feed additives in *Ascaris suum* inoculated pigs. Vet Parasitol 168:269–277

Larsen M (2000) Prospects for controlling animal parasitic nematodes by predacious fungi. Parasitology 120:S121–S131

Larsen MN, Roepstorff A (1999) Seasonal variation in development and survival of *Ascaris suum* and *Trichuris suis* eggs on pastures. Parasitology 119:209–220

Lindgren K, Lindahl C, Höglund J, Roepstorff AK (2008) Occurrence of intestinal helminths in two organic pig production systems. I: Proceedings of the 16th IFOAM Organic World Congress. s.1–4

Lund V, Algers B (2003) Research on animal health and welfare in organic farming—a literature review. Livest Prod Sci 80:55–68

Manolaraki F (2011) Propriétés anthelminthiques du sainfoin (*Onobrychis viciifoliae*): Analyse des facteurs de variations et du rôle des composés phénoliques impliqués. PhD thesis, INP Toulouse, 185 pages

Marley CL, Cook R, Keatinge R, Barrett J, Lampkin NH (2003) The effect of birdsfoot trefoil (*Lotus corniculatus*) and chicory (*Cychorium intybus*) on parasite intensities and performance of lambs naturally infected with helminth parasites. Vet Parasitol 112:147–155

Maurer V, Amsler Z, Perler E, Heckendorn F (2009) Poultry litter as a source of gastrointestinal helminth infections. Vet Parasitol 161:255–260

Maurer V, Hertzberg H, Heckendorn F, Hördegen P, Koller M (2013) Effects of paddock management on vegetation, nutrient accumulation, and internal parasites in laying hens. J Appl Poult Res 22:334–343

McDonald V, Shirley MW (2009) Past and future: vaccination against Eimeria. Parasitology 136:1477–1489

Mejer H (2006) Transmission, infection dynamics and alternative control of helminths in organic swine. PhD. Thesis, Copenhagen University

Michel JF (1976) The epidemiology and control of some nematode infections in grazing animals. Adv Parasitol 14:355–397

Naidoo V, McGaw LJ, Bisschop SPR, Duncan N, Eloff JN (2008) The value of plant extracts with antioxidant activity in attenuating coccidiosis in broiler chickens. Vet Parasitol 153:214–219

Nejsum P, Roepstorff A, Jørgensen CB, Fredholm M, Göring HHH, Anderson TJC, Thamsborg SM (2009) High heritability for Ascaris and Trichuris infection levels in pigs. Heredity 102:357–364

Nielsen BK, Thamsborg ST, Kristensen T (2003) Feed supplement for young dairy bred calves after turn out to pasture: effect on weight gain and subclinical coccidiosis in organic production systems. Acta Agric Scand Sect A 5:1–10

Niezen JH, Charleston WAG, Robertson HA, Shelton ID, Waghorn G, Green R (2002) The effect of feeding sulla (*Hedysarum coronarium*) or lucerne (*Medicago sativa*) on lamb parasite burdens and development of immunity to gastrointestinal nematodes. Vet Parasitol 105:229–245

Permin A, Bisgaard M, Frandsen F, Pearman M, Kold J, Nansen P (1999) Prevalence of gastrointestinal helminths in different poultry production systems. Br Poult Sci 40:439–443

Permin A, Bojesen M, Nansen P, Bisgaard M, Frandsen F, Pearman M (1997) *Ascaridia galli* populations in chickens following single infections with different dose levels. Parasitol Res 83:614–617

Permin A, Nansen P, Bisgaard M, Frandsen F, Pearman M (1998) Studies on *Ascaridia galli* in chickens kept at different stocking rates. Avian Pathol 27:382–389

Petkevičius S, Bach Knudsen KE, Murrell KD, Wachmann H (2003) The effect of inulin and sugar beet fibre on *Oesophagostomum dentatum* infection in pigs. Parasitology 127:61–68

Petkevičius S, Murrell KD, Bach Knudsen KE, Jørgensen H, Roepstorff A, Laue A, Wachmann H (2004) Effects of short-chain fatty acids and lactic acids on survival of *Oesophagostomum dentatum* in pigs. Vet Parasitol 122:293–301

Pout DD (1973) Coccidiosis of lambs. I. Observations on the naturally acquired infection. Br Vet J 129:555–567

Rochfort S, Parker AJ, Dunshea FR (2008) Plant bioactives for ruminant health and productivity. Phytochemistry 69:299–322

Roepstorff A (2003) *Ascaris suum* in pigs: population biology and epidemiology. Doctorate Thesis, The Royal Veterinary and Agricultural University, Copenhagen 2003, p 227

Roepstorff A, Jørgensen RJ, Nansen P, Henriksen SA, Pedersen JS, Andreasen M (1992) Parasitter hos økologiske svin. Report paper from a project, financed by the Danish Ministry of Agriculture, pp 36

Roepstorff A, Mejer H, Nejsum P, Thamsborg SM (2011) Helminth parasites in pigs: new challenges in pig production and current research highlights. Vet Parasitol 180:72–81

Schou T, Permin A, Roepstorff A, Sorensen P, Kjaer J (2003) Comparative genetic resistance to *Ascaridia galli* infections of 4 different commercial layer-lines. Br Poult Sci 44:182–185

Skampardonis V, Sotiraki S, Kostoulas P, Leontides L (2012) Factors associated with the occurrence and level of Isospora suis oocyst excretion in nursing piglets of Greek farrow-to-finish herds. BMC Vet Res 8

Smith WD, Zarlenga DS (2006) Development and hurdles in generating vaccines for controlling helminth parasites of grazing ruminants. Vet Parasitol 139:347–359

Steenhard NR, Jungersen G, Kokotovic B, Beshah E, Dawson HD, Urban Jr JF, Roepstorff A, Thamsborg SM (2009) *Ascaris suum* infection negatively affects the response to a *Mycoplasma hyopneumoniae* vaccination and subsequent challenge infection in pigs.Vaccine 27:5161–5169
Taylor MA, Coop RL, Wall RL (2007) Veterinary Parasitology, 3rd Revised edition. Wiley-Blackwell Editors, p 904
Thamsborg SM, Roepstorff A (2003) Parasite problems in organic livestock production systems and options for control. J Parasitol 89:S277–S284 (Suppl)
Thamsborg SM, Jorgensen RJ, Waller PJ, Nansen P (1996) The influence of stocking rate on gastrointestinal nematode infections of sheep over a two-year grazing period. Vet Parasitol 67:207–224
Thamsborg SM, Roepstorff A, Larsen M (1999) Integrated and biological control of parasites in organic and conventional production systems. Vet Parasitol 84:169–186
Thomsen LE, Mejer H, Wendt S, Roepstorff A, Hindsbo O (2001) The influence of stocking rate on transmission of helminth parasites in pigs on permanent pasture during two consecutive summers. Vet Parasitol 99:129–146
Thomsen LE, Knudsen KEB, Jensen TK, Christensen AS, Møller K, Roepstorff A (2007) The effect of fermentable carbohydrates on experimental swine dysentery and whip worm infections in pigs. Vet Microbiol 119:152–163
Tipu MA, Pasha TN, Zulfiqar A (2002) Comparative efficacy of salinomycin sodium and neem fruit (*Azadirachta indica*) as feed additive anticoccidials in broilers. Int J Poult Sci 1:91–93
Torres-Acosta JFJ, Hoste H (2008) Alternative or improved methods to limit gastro-intestinal parasitism in grazing/browsing sheep and goats. Small Rumin Res 77:159–173
Trujillo GR, Mata C (2000) The Dehesa: an extensive livestock system in the Iberian Peninsula. In: Proceedings of the 2nd NAHWOA Workshop. Cordoba, 8–11 January, 2000, pp 50–61
Vagenas D, Jackson F, Merchant M, Wright IA, Bishop SC (2002) Genetic control of resistance to gastro-intestinal parasites in crossbred cashmere-producing goats: responses to selection, genetic parameters and relationships with production traits. Anim Sci 74:199–208
Waller PJ, Thamsborg SM (2004) Nematode control in "green" ruminant production systems. Trends Parasitol 20:493–497
Waller PJ, Bernes G, Thamsborg SM, Sukura A, Richter SH, Ingebrigtsen K, Hoglund J (2001) Plants as deworming agents of livestock in the nordic countries: historical perspective, popular beliefs and prospects for the future. Acta Vet Scand 42:31–44
Wang ML, Suo X, Gu JH, Zhang WW, Fang Q, Wang X (2008) Influence of grape seed proanthocyanidin extract in broiler chickens: effect on chicken coccidiosis and antioxidant status. Poult Sci 87:2273–2280
Wilcox ML, Cosentino MJ, Pink R, Bodeker G, Wayling S (2001) Natural products for the treatment of tropical diseases. Trends Parasitol 20:493–499
Williams RB (2002) Anticoccidial vaccines for broiler chickens: pathways to success. Avian Pathol 31:317–353
Youn HJ, Noh JW (2001) Screening of the anticoccidial effects of herb extracts against *Eimeria tenella*. Vet Parasitol 96:257–263

Chapter 9
Animal Healthcare Strategies in Organic and Conventional Farming

Christian Nicourt and Jacques Cabaret

Abstract The sociology of animal health is a relatively new subject, but one that is central to the effective management of disease control in farm animals. Understanding how and why farmers select certain health strategies is paramount for the effective dissemination of technical proposals. These strategies may be influenced by a variety of factors such as whether the farms follow conventional practices or organic guidelines, the type of animal farmed, the production aim (e.g., egg, milk, meat), the differences in pathocoenosis between these farms or the individual farmers' beliefs. This chapter will consider each of these complexities in turn to reach a conclusion on why farmers do what they do.

Keywords Pathocoenosis · Disease · Farm animal · Treatment · Organic

9.1 Introduction

Originating in the post World War II period, medical sociology was the forerunner of veterinary sociology. Similarly veterinary medicine followed the tracks of biomedicine.The first characteristic of biomedicine is the wide adoption of reductivism: "biomedicine assumes that health and diseases are natural phenomena which exist in the individual body rather than in the interaction of the individual and the social world" (Annandale 1998). The human body is considered as a complex machine apart from the mind, a conceptualisation that has its origin in the work of the seventeenth century philosopher, Descartes, who conceived the mind and the body as distinct entities. The second characteristic of biomedicine is the doctrine of specific aetiology, which is a corollary of the reductivist approach. This doctrine was first described by Dubos (1960) to depict the change that occurred at the end

C. Nicourt (✉) · J. Cabaret
INRA, UR1216 RiTME, 65 boulevard de Brandebourg,
94205 Ivry, France
e-mail: christian.nicourt@ivry.inra.fr

J. Cabaret
INRA-Université de Tours, UMR1282 Infectiologie et Santé Publique,
37380 Nouzilly, France

S. Bellon, S. Penvern (eds.), *Organic Farming, Prototype for Sustainable Agricultures,*
DOI 10.1007/978-94-007-7927-3_9,

of the 19th century where "disease could be produced at will by the mere artifice of introducing a single specific factor—a virulent micro-organism—into a healthy animal". The third characteristic of the biomedical model is the claim of scientific neutrality: "medicine can be rational, objective and value free, treating each individual according to their need and irrespective of any sense of moral worth" (Annendale 1998). In the 1970's, health and social psychologists developed a different view and produced the Health Belief Model (HBM) for humans (Abraham and Sheeran 2005). The HBM focuses on two aspects of how health is represented to an individual: threat perception and behavioural evaluation. Threat perception is the perceived susceptibility to illness and its anticipated severity. Behavioural evaluation consists of two sets of beliefs: the efficacy of a recommended health behaviour and the cost of enacting the behaviour.

The sociology of veterinary health is still in its infancy. Despite its compelling role as "something of a silent player in the development of rural sociology" (Enticott 2009), there is a decided lack of knowledge in the area since animal health has been seen principally as an adjunct to the competitiveness of modern agriculture, whereas contemporary rural sociology has been shaped more by aspects of agricultural sustainability. Enticott (2008) believes that "rural sociology has the potential to make a fundamental contribution to understanding the impact of animal disease upon rural societies, advise policy makers how they might prevent future outbreaks, and how the consequences of outbreaks may be alleviated". Of the few studies that have been carried out in the area, the majority focused upon crises in animal health and their repercussions on the consumer. Fewer still have focused on how decisions are reached on farms (Cabaret and Nicourt 2009b; Saddiqui et al. 2012 Cabaret et al. 2011). Perhaps the least studied is the sociology of veterinary health in relation to organic farming (Nicourt and Cabaret 2010), yet there is no doubt that the ideals and regulations of organic health clash with those used in biomedicine. Organic husbandry (i) views animals as living beings as opposed to complex machines; (ii) accepts the functional/equilibrium theory (Laplantine 1992), and thus promotes disease prevention and is opposed to the doctrine of the specific aetiology theory (Nicourt and Cabaret 2010); and (iii) targets the ideal of a value-orientated medicine that promotes the reduction of synthetic chemical drugs and the use of alternative medicines in contrast to value-free medicine (Kaltoft 1999).

Animal health strategies are constructed on the basic knowledge of what causes disease in a particular situation with the intention of preventing (by means of adequate practices) or alleviating target problems. In comparing the health strategies used in conventional versus organic farming, two important points should be considered. Firstly, if diseases differ across farms, it is logical that the adopted health strategies would also differ in a bid to control distinct targets. Secondly, if the health problems are similar among the farms, then the different health strategies are due to different therapeutic choices based on organic regulations and farmers' personal values. This chapter (i) will evaluate if the occurrence of disease is comparable across conventional and organic farms. The diseases considered include either those that are independent within a system or communities of interacting diseases (pathocoenosis); and (ii) compare and contrast strategies for managing health in homogenous groups of farms. Such a sociological study could unveil the reasons why farmers do what they do and answer the question: how are decisions made on the farm?

9.2 Do Organic Farms have Pathocoenoses that are Distinct from Conventional Ones?

9.2.1 Identifying Pathocoenosis

Grmek (1969a, b) developed his concept of pathocoenosis (derived from the term of biocoenosis used by ecologists) as "the ensemble of pathological states present in a specific population at a given moment in time. It consists of a system with precise structural properties that should be studied so as to determine its nosological parameters in qualitative and quantitative terms." He considered the virulence of a pathogen isolate, the susceptibility of the host and the probability of encounter, as well as the existence of other diseases. He recognised the fact that the occurrence of a disease is not independent of the occurrence of other diseases, making it reasonable to study a set of diseases rather than only one disease within a site. Conventional pig farmers established the concept of "farm microbism", which also corresponds to pathocoenosis: each farm has its own set of pathogens that remains under a delicate equilibrium and that could transform into a pathological episode when an unknown microbe is introduced onto the farm (Christian Nicourt, unpublished observation, 2011). Biodynamical farmers do not accept the eradication of a disease (e.g., botfly eradication programme in France) or temporary suppression of pathogens (like internal parasites) since they believe that their presence has a meaning and contributes to the general equilibrium of health, and therefore apply the pathocoenosis concept (Cabaret, personal observations, 2011). Standard organic farmers consider health as the result of an equilibrium between many different forces such as feeding, housing and interactions between pathogens. The concept of pathocoenosis is then closely linked to organic farmers' beliefs but is never used among them. Obtaining a clear understanding of which diseases are present within a farm is a subjective process. It primarily relies on the farmers' description of disease, often inaccurately diagnosed. The diagnosis in goats was effective for mastitis that had fairly obvious symptoms, and therefore led to homogeneous reported prevalences (16 to 22 %) between equivalent farms, but was poor for gastrointestinal parasite infections (13 to 36 %) for which clinical expression covers a variety of symptoms such as decreased appetite, anaemia, diarrhoea, low production, etc. (Cabaret 2003). The farmer's accuracy range is much better for those diseases with obvious symptoms such as tetanus or rabies that occur infrequently, as opposed to diseases that occur frequently but with less discernible symptoms such as diarrhoea caused by gastrointestinal parasites or coughing resulting from lungworms. The diagnostic expert may also inadvertently over-represent the diseases he views to be more significant and immediately disturbing, while over-looking the more mild, chronic diseases considered to be of limited importance. It is thus difficult to construct pathocoenoses on the declaration of farmers alone since they reflect a mixture of easily and poorly diagnosed diseases, with very different degrees of accuracy in establishing prevalences. We need more exact records of the diseases and syndromes that occur on-farm and this could be achieved with farmers' health plans (Vaarst et al. 2012) or formally established interactions between farmers and veterinarians (Robinet 2011). In the latter case, a contract between the farmers and the vets exists, and the obligation of the vets is to elucidate and solve husbandry problems with the aid and cooperation of the farmer. One of the oldest structures founded in 1979 in the centre of France (AVEM)[1] still exists (Robinet 2011). The agreement implies that the vet will perform two kinds of tasks: systematic visits to the farm to monitor flock health, and instruction concerning animal health for the farmer. The vet can also ensure emergency visits. The farmer should accept to follow the veterinarian's recommendations, even if it means additional work or the absence of therapeutic solutions. His interest is to achieve the overall control of a health problem on his farm. This practice is similar to experiential science that aims at integrating the implicit and reflected knowledge of the practitioner (farmer) and the expert (vet) in organic farming (Baars 2011).

9.2.2 Pathocoenosis in Different Organic Productions

A previous survey of conventional and organic farms (cattle, sheep, pig and poultry) showed that similar pathocoenoses existed within the same production types (e.g., egg, milk or meat) (Thamsborg et al. 2004; Berg 2001). Although useful, these surveys are potentially flawed, as mentioned above, and should be critically examined. A tentative qualitative meta-analysis of the diseases recorded in organic productions is shown in Table 9.1. On average, it appears that each production type has its own pathocoenoses and that there is little difference between conventional and organic farming. In Table 9.1, for diseases recorded for all types of production ($n=33$), 54% of the diseases (17/33) are equally prevalent in organic and conventional farms, 31% are more prevalent and 15% are less prevalent in organic compared to conventional farms. However, since diseases are subject to strong individual and breed modifications, pathocoenoses will not be complete. Therefore, even if parasitism is the main problem in meat sheep production, it will not appear on every farm (Cabaret et al. 2009). Based on these results, we put forward the hypothesis that animal health strategies differ between conventional and organic farms as a result of formal rules (i.e., organic regulations) and the values of the individual farmers.

[1] Association Vétérinaires Eleveurs du Millavois: Association of veterinarians and breeders of Millavois.

Table 9.1. Diseases in organic compared to conventional animal husbandry (a meta-analysis of Thamsborg et al. 2004; Berg 2001; Cabaret and Nicourt 2009a, Cabaret et al. 1986; Benoit et al. 2009; Prunier and Cabaret 2010) and putative pathocoenoses

Production	Mastitis	Reproduction diseases	Metabolic diseases	Parasitic diseases	Lameness/ foot rot/Pica	Pathocoenosis
Milk cattle	+	+	–	+	+	Mastitis, reproduction, parasites*, lameness
Meat cattle	?	=	?	+	=	Parasites, reproduction, lameness
Milk sheep	+	=	–	=	=	Reproduction, footrot, parasites
Meat sheep	?	=	–	+	=	Parasites, reproduction, footrot
Dairy goats	=	=	=	=	=	Mastitis, parasites, reproduction, lameness
Pigs	?	–	=	+	=	Reproduction, mastitis, parasites, lameness
Broilers	NA	NA	+	+	=	Metabolic disorders, parasites, pica
Layers	NA	=	+	+	=	Metabolic disorders, parasites, pica, reproduction

Symbols:+ more frequent in organic,= equivalent in organic and conventional, –less frequent in organic farms,? not known, NA not applicable

*Parasites are generally gastrointestinal nematodes

9.3 The Views of Farmers for Constructing a Strategy to Control Diseases

We chose the term of "views" rather than more precise words like "beliefs" or "ideals" in order to encompass the reality of farmers.

9.3.1 The Contrasting Views of Animal Health Caretakers

While scientists can evaluate the risk of a particular disease with probabilities, it seems that farmers are tied to their own individual experiences. In a study of 37 Australian farms (beef, cow milk or sheep producers), Palmer et al. (2009) found that "*farmers clearly adopt a constructionist stance that a risk is never entirely an objective reality but is influenced by pre-existing knowledge*". Such idiosyncratic practices extend to all the various animal health caretakers, i.e., farmers, vets, veterinary assistants, traditional practitioners. A study in an arid zone of Pakistan (Saddiqi et al. 2012) showed widely contrasting ideas among the different animal health caretakers on how to treat parasitic diseases. The epistemic experts such as vets (they know why they treat) referred to a universal strategy for the control of parasitosis using synthetic drugs at putative risk periods. The farmers and traditional practitioners were pragmatic experts (they know how to treat) who used local solutions to cope with problems in a diversity of situations. Such an approach is based on the old Greek concept of mètis: adaptive practical intelligence (Cabaret and Nicourt 2009b). Ultimately, farmers decide on what advice to accept or reject and on which therapeutic means they will apply on their farms. They are the cornerstone for the dissemination of technological knowledge, which makes understanding how they make their decisions of primary importance.

9.3.2 Contrasting Views Among Farmers: A French Meat Sheep Example

Regional differences and economic constraints influence the strategies used in disease control. In resource-poor countries, the cost of a treatment alone can be the deciding factor on whether it is used (see Saddiqi et al. (2012), for Pakistan, and Ouzir et al. (2011) for Morocco). However, in developed countries, the efficacy of the treatment to reduce the occurrence of the disease is the driving force. The cost of treatment may vary widely in developed countries from one farm to another. We used the data of Benoit and Laignel (2002), established for a network of 49 meat sheep farms in the centre of France, in a semi-mountainous region. These farms are considered as sustainable since they were engaged in the same activity for several years. The average veterinarian costs in conventional farms were € 6 per ewe and per year with limited variations. The organic farms had veterinarian costs that were highly variable (from € 4 to a maximum of € 21 per ewe and per year), which could be linked to their general treatment strategy (see below). This is indicative of the fact that management of animal health on organic farms is highly dependent on the individual farm. Internal parasitism is considered to be one of the greatest problems in sheep farming (Cabaret et al. 2009, 2011). The symptoms of such infections are not often clearly expressed and farmers must often rely on what they believe rather than on objective signs when diagnosing them. Conventional sheep farmers in France do this by cross-referencing a checklist provided by vets and animal technicians that contain suggestions for actions, suggestions that are not always applied (Cabaret 2003). We found that these farmers have a limited interest in developing their own sanitary norms (Nicourt and Cabaret 2010). In contrast, organic farmers target a stable sanitary equilibrium in line with organic regulations, with a 'prevention is better than cure' ideal. They broadly follow Canguilhem's (1966) proposal to "precisely determine the content of norms within which life stabilises, without precluding the possibility or impossibility of an eventual correction of these norms". Thus, the construction of norms among individual farmers results in different health care and husbandry strategies within organic farms and between conventional farms. The organic meat sheep farmers could be divided into two categories (Nicourt and Cabaret 2010): (i) isolated, autonomous "self-made" farmers who aimed for a health equilibrium using a large set of practices; and (ii) "creative" farmers orientated towards the correction of health disorders with the use of organic drugs they have experimented with themselves.

9.3.3 Self-Made Organic Autonomous Farmers

The beliefs of these farmers constitute a real breach from those of conventional farming in that they drastically limit the use of any therapeutics. They also believe that disease results from a rupture in the equilibrium of their animals and should be treated with a restoration of their integrity. This directly contradicts the aetiology concept prevalent in biomedicine and explains why veterinarians are not part of their main advisory resource pool (Cabaret et al. 2011). Their ideal is a demographically stable flock, without any sanitary disorders of importance in a process of self-regulation. Such a stable flock is the result of accomplished animal health care, a kind of masterpiece for the farmer. It is also related to the concept of naturality (living under conditions comparable to those that they may be subjected to when living wild in nature): when animals are bred under good conditions, they have a stronger capacity to react to and control bioagressors without external help.

9.3.4 Creative Organic Farmers

Creative organic farmers do not strictly exclude any potential health strategies. Although they prioritise disease prevention by carefully monitoring their animals and adhering to good hygiene practices and high quality feed, they also accept that drugs can cure disease. They do not, however, dispose of the wide array of available drugs in the way conventional farms do. Instead, they communicate with other stakeholders in organic farming and with conventional farmers, and consider experimenting with prospective new drugs on their farms. This may partially explain the high veterinary costs of some organic farmers. As such, creative farmers may be regarded as altruistic by testing strategies on their own farms for the good of the organic community. Their ideal is a flock maintained under their own sanitary control and in line with organic regulations

9.3.5 Innovative Internal Sheep Parasite Management and Farmers' Views

Targeted Selective Treatment (TST) was a strategy developed to reduce the need for drug treatments against internal parasites in sheep. It selectively targets only those animals in need of drugs opposed to the traditional mass treatment approach. The animals to be treated are identified by physiological indicators of the infection, the presence of anaemia and diarrhoea. Despite its success on both private and experimental (Benoit et al. 2009) farms, conventional farmers show no interest in adopting TST. They deem that the cost of time needed to detect the target animals outweighs the low financial cost of treating all the animals. Due to the limitations of drug use in organic regulations, organic farmers found TST a more worthwhile strategy, which was in accordance with their beliefs. Thus, technically valid health strategies such as TST (Ouzir et al. 2011) are accepted or rejected on the basis of a farmer's beliefs and values.

9.4 Conclusion

Pathocoenoses between conventional and organic farms do not substantially differ, although the practices are very different. This could be due to the fact that the evaluation of pathocoenoses is not sensitive and has many biases. The farmer has practical expertise on what happens in the everyday life of the flock, whereas the vet has academic knowledge of diseases that is only occasionally applied on a farm. Pathocoenosis can only be properly assessed if farmers and vets cooperate to construct valuable databases that illustrate the recent history of diseases on each farm. Innovations regarding health developed by researchers are not 'naturally' transferred into everyday practices. The personal dimension and willingness of the key actor, the farmer, has not been adequately taken into consideration. The Health Belief Model established for human diseases (a set of events like the risk of acquiring the disease, the dangerousness of the disease and the difficulties to control the disease) could be a conceptual aid for promoting a constructed view of diseases for farmers.

Acknowledgments The financial help of the European Union (Parasol programme) and the ANR Pan project is gratefully acknowledged, as are the comments of Caroline Chylinski (INRA) on an earlier version of the text.

References

Abraham C, Sheeran P (2005) The health belief model pp 28–80. In: Conner M, Norman P (eds) Predicting health behaviour. Open University Press, Maidenhead, p 385

Annendale E (1998) The sociology of health and medicine. A critical introduction. Polity Press, Cambridge, p 334

Baars T (2011) Experential science: towards an integration of implicit and reflected practioner-expert knowledge in the scientific development of organic farming. J Agric Environ Ethics 24:601–628

Benoit M, Laignel G (2002) Constraints under organic farming on French sheepmeat production: a legal and economic point of view with an emphasis on farming systems and veterinary aspects. Vet Res 33:613–624

Benoit M, Tournadre H, Dulphy JP, Laignel G, Prache S, Cabaret J (2009) Is intensification of reproduction rhythm sustainable in an organic sheep production system? A 4-year interdisciplinary study. Animal 3:753–763

Berg C (2001) Health and welfare in organic poultry production. Acta Vet Scand 95:37–45
Cabaret J (2003) Animal health problems in organic farming: subjective and objective assessments and farmers' actions. Livestock Production Science 80:99–108
Cabaret J, Nicourt C (2009a) Les problèmes sanitaires en élevage biologique: réalités, conceptions et pratiques. INRA Prod Anim 22:235–244
Cabaret J, Nicourt C (2009b). Experts et novices en médicine vétérinaire: de l'épistemè à la mètis pour la construction des décisions de soins. 3d Congress of the Society of Philosophy of Science, Paris, 12–14 November, p 42
Cabaret J, Anjorand N, Leclerc C (1986) Dairy goat farms in Touraine. I. Management, parasitism and estimation of disease in adult goats. Recueil de Médecine Vétérinaire 162:575–585
Cabaret J, Benoit M, Laignel G, Nicourt C (2009) Current management of farms and internal parasites by conventional and organic meat sheep French farmers and acceptance of targeted selective treatments. Vet Parasitol 164:21–29
Cabaret J, Benoit M, Laignel G, Nicourt C (2011). Health advisors in organic meat sheep farms: the role of the veterinarians. Open Vet Sci J 5 (Suppl 1: M3):7–11
Canguilhem G (1966) Le normal et le pathologique. PUF, Paris, ed. Quadrige 1993:232
Dubos R (1960) Mirage of health. George Allen and Unwin, London
Enticott G (2008) The ecological paradox: social and natural consequences of the geographies of animal health promotion. Trans Inst Br Geogr 33:433–446
Enticott G (2009) Introduction to the theme section: rural sociology and animal disease. Sociol Ruralis 49 doi:10.1111/j.1467-9523.2009.00498.x
Grmek MD (1969a) Discussion on medicine and culture. In: Poynter FNL (ed) Medicine and culture. Welcome Institute, London, p 119–120
Grmek MD (1969b) Préliminaires d'une étude historique des maladies. Annales ESC 24:899–900
Kaltoft P (1999) Values about nature in organic farming practice and knowledge. Sociol Ruralis 39:39–53
Laplantine F (1992) Anthropologie de la maladie. Payot, Paris, p 411
Nicourt C, Cabaret J (2010) Création de normes, innovation sanitaire et éthique des éleveurs ovin *bio*. In: P. Béguin, B. Dedieu et E. Sabourin (eds) Le travail en agriculture dans les sciences pour l'action. L'Harmattan, Paris, pp 59–69
Ouzir M, Berrag B, Benjouad A (2011) Use of pathophysiological indicators for decision of anthelmintic treatment of ewes against gastro-intestinal nematodes in Morocco. Vet Parasitol 180:372–377
Palmer S, Fozdar F, Sully M (2009) The effect of trust on West Australian farmers' responses to infectious livestock diseases. Sociol Ruralis 49:360–374
Prunier A, Cabaret J (2010) Les facteurs de risque d'infestation par les parasites internes chez les porcs élevés en Agriculture Biologique. Prévention de la santé des porcs. Résultat du programme de recherches Corepig. Actes du colloques, 14 décembre 2010, 46–54. http://www.interbiobretagne.asso.fr/upload/File/Recherche/Elevage/Corepig/Actes_Corepig_2010.pdf. Accessed 24 Nov 2013
Robinet A (2011) Larzac-Millau-Grands Causses. Elevage et partage des savoirs. Graveurs de mémoire, L'Harmattan, Paris, p 271
Saddiqi HA, Jabbar A, Iqbal Z, Babar W, Sarwar M, Cabaret J (2012) Contrasting views of animal healthcare providers on worm control practices for sheep and goats in an arid area. Parasite 19:53–61
Thamsborg SM, Roderick S, Sundrum A (2004) Animal health and diseases in organic farming: an overview. In: Vaarst M, Roderick S, Lund V, Lockeretz W (eds) Animal Health and welfare in agriculture. Cabi Publishing, Wallingford, pp 227–252
Vaarst M, Winckler C, Roderick S, Smolders G, Ivemeyer S, Brinkmann J, Mejdell CM, Whistance LK, Nicholas P, Walkenhorst M, Leeb C, March S, Henriksen BIF, Stöger E, Gratzer E, Hansen B, Huber J (2012) Animal health and welfare planning in organic dairy cattle farms. Open Vet Sci J 5:19–25

Chapter 10
Optimisation of Breeding Systems and Land use to Maximise Feed Self-Sufficiency and Economic Outcomes in Organic Sheep-for-Meat Production

Marc Benoit and Gabriel Laignel

Abstract With the development of organic agriculture in France, sheep-for-meat farmers are converting to this way of production in order to meet consumer demand. However, constraints, technical but mainly economic, are considerable, given the high use of concentrates and their cost. Analysis of a network of 42 sheep farms including 13 under organic farming (OF) showed that the gross margin per ewe is 24% lower in OF in upland areas due to high feed costs, and comparable in lowland areas where levels of fodder self-sufficiency are potentially higher given the options for rotating temporary pastures and increasing the share of pasture grazing in the livestock diet. Furthermore, cereal production can increase feed self-sufficiency and keep farms less dependent on outside feed sources. In upland areas, system logics are more difficult to define, and the allied technical management roadmaps are relatively subsector-specific. Our analysis of four demonstration farms showed that the different context settings require specific livestock management strategies depending on whether the farm has tillable land. If the proportion of tillable land is limited, lambings are split equally across spring and autumn in order to maximise forage self-sufficiency while optimising ewe productivity and diversifying the sales windows. When there are options for crop production, lambings are focused on autumn with good added value on lamb sales. In a context of inflationary concentrate prices, the economic viability of both organic and conventional sheep-for-meat farming systems is contingent on high levels of forage and feed self-sufficiency.

Keywords Sheep-for-meat · Farming system · Organic · Economics · Feeding · Self-sufficiency

---

M. Benoit (✉) · G. Laignel
INRA, UR1213 Herbivores, 63122 Saint-Genès-Champanelle, France
e-mail: marc.benoit@clermont.inra.fr

S. Bellon, S. Penvern (eds.), *Organic Farming, Prototype for Sustainable Agricultures*,
DOI 10.1007/978-94-007-7927-3_10,

10.1 Introduction

In 2007, organic sheep-for-meat production accounted for 1.8 % of sheep meat in France, a smaller proportion than for sheep milk (3.9 %) but on the same order as that observed for dairy or suckler cattle (1.5 %) (Agence Bio 2007). This small proportion, which even declined between 2006 and 2007 (−6 %), can be explained by: (i) strong constraints related to animal feeding (quantity and high cost of concentrates); (ii) a sanitary control, considered to be difficult, especially in terms of endoparasites such as helminths (Hovi et al. 2003); and (iii) a lack of market structure (Nardone et al. 2004).

Nevertheless, the market is growing regularly, with + 72 % of preparers and distributors between 2005 and 2009 (Agence Bio 2010). Regarding sanitary issues, the importance of infectious and parasitic diseases is fairly similar in organic and conventional husbandry (Thamsborg et al. 2004; Cabaret and Nicourt 2009), but there is much less use of synthetic drugs and we may expect that on the longer term, health problems could be solved with more adequate management or production facilities (Thamsborg et al. 2004).

Thus, feeding remains the major issue as far as the differential price remains persistently very high between organic and conventional feeding.

Sheep-for-meat production in France is geographically centred on essentially grassland or pastureland regions, often in far-from-ideal conditions, in disadvantaged lowland and upland areas. Studies underline that profitability on conventional sheep farms hinges primarily on ewe productivity (Benoit et al. 1999; Bellet and Morin 2005). However, this kind of system will tend to draw heavily on concentrate inputs. As a rule, lambs are fattened in sheep stalls, generally on a ration system that offers free access to concentrate.

In this setting, where does organic sheep farming stand, given the limits to added value on lamb prices (nearly 15 % during the period 2008–2010, i.e., higher than previous years) and the sky-high costs entailed in purchasing concentrate (price per kg + 42 % in 2008 to + 76 % in 2010, compared to conventional)? What level of ewe productivity is achieved, and through what resource use? Do organic sheep farmers apply the same type of strategies as conventional sheep farmers, and what kind of revenue do they ultimately generate?

The concept of 'land-related stock farming' is a prominent feature of organic farming from the standpoint of organic principles and regulations (proportion of feed produced on-farm) and the standpoint of simple economics (costs incurred by purchasing feed). How does this translate into farms? Hormone treatments are widely used in conventional sheep farming (above all, in plain breeds) to allow out-of-season reproduction (lambings in autumn) in order to provide lambs throughout the year for the sheep industry. How do organic farms deal with this issue?

To address these questions, long-term follow-up data on a network of conventional sheep farms (Laignel and Benoit 2004) was completed with data on organic farms from the mid-2000s. This made it possible to cross-compare results between systems, highlighting the factors affecting revenues and the degree of system

Table 10.1 Average farm structures in organic and conventional farming, in upland and lowland areas

2006 data	Upland			Lowland	
	Conv.	OF		Conv.	OF
Number of farms (*n*)	21	8	4[a]	8	5
UAA (Ha)	77	51	60	132	88
% PFA (Princ.Forage.Area)	96	85	82	71	88
Ewes (number)	494	252	270	611	492
FO (Farm operator)	1.62	NA	1.03	1.58	1.38
Equivalent LU[b]/FH	54	NA	54	74	73
Stocking rate (LU[b]/Ha PFA)	1.07	0.98	0.91	1.06	1.24

NA data not available
a Without agricultural college or experimental farms
b LU = Livestock Unit (for example, one ewe = 0.14 LU)

cohesiveness in both lowland and upland areas, i.e., the relationship between the technical production level, the stocking rate level, feeding resources needed for the flock, and the type of feeding resources available on the farm. These reference datasets were enhanced by four experimental or demonstration farms (located in upland areas) to broaden the field of observation and, more importantly still, to highlight two specific flock management policies (based on seasonally-cycled production) built to fit specific feed resources available in the local landscape environment.

10.2 Material and Methods

This study network included 42 farms (Table 10.1) located in the ten administrative departments (*departements*) of the Massif Central in France (farms between 500 to 1,100 m above sea level on subsoil granite, in general, sometimes volcanic, but excluding the southern xeric habitats) and adjacent lowlands. The study was conducted using 2006 data to cross-compare 13 organic sheep farms (an 8–5 upland-lowland split with all farms beyond the 5-year conversion window) with 29 conventional sheep farms (a 21–8 upland-lowland split). The conventional farms are part of a long-term tracking scheme used by our research unit. They were recruited as specialised sheep farms that provided reliable data and that—on paper—had a sufficiently large UAA to make them economically viable. As a rule, these farms boast technical performances that are above average for the region, tied to the fact that they often use an accelerated lambing system (three lambings per ewe every 2 years). Flock size averaged 305 ewes per farm operator in upland farms and 387 in lowland farms. These figures are above the national average of 274 ewes per farm operator (AGRESTE 2006). Income, at € 14,500 and 14,300 per farm operator in upland and lowland farms, respectively, is also well above the 2006 national average calculated at € 9,149 per farm operator. However, the region covered counts too few organic farms to build a significantly large population sample, and there is no national-scale barometer to serve as a benchmark. Five of the 13 organic sheep farms sell sheep meat directly to consumers, an activity that represents 10–50% of total sales (in number of lambs). These 13 organic farms raise ten different breeds.

More in-depth analysis was carried out on the four experimental (INRA) or demonstration farms (operated by agricultural colleges) (see Box 10.1). Although they operate in different contexts, all four share the same core objective to develop production systems that draw added value from local assets (agronomic factors, breeds) to meet market needs as profitable businesses. They use various strategies within a framework that targets natural-to-high animal productivity compatible with a good level of feed self-sufficiency.

Box 10.1: Characteristics of the four experimental and demonstration farms

The first three farms are run by agricultural colleges for teaching and demonstration purposes, with the overriding goal of achieving profitability. The Redon farm is run by INRA, which sets up experiments designed to assess the viability and performances of farm systems (technicity, economics, sustainability).

***The Cambon farm, St. Affrique Agricultural College (Aveyron)** (organic since 2000)

The Cambon farm covers 50 ha used by the Lacaune breed sheep (130 ewes; high prolificacy, good potential for counter-season reproduction, rather good lamb conformation) and cattle (feedlots of 20 Aubrac × Charolais cross-breed heifers). The farm is located at 350 m a.s.l., under irregular rainfall (average precipitation: 850 mm/year) in strong decline over the last 5 years, on good-quality alluvial soil and relatively unfarmable red sandstone. The stocking density is 0.89 Livestock Unit/ha. Given the summer season fodder deficit, breed characteristics and market features (good added value on winter slaughter lambs), the ewes are lambed in November and then again in January–February ('re-mating' non-pregnant ewes and ewe lambs). Cereals and alfalfa hay produced on-farm provides a cheaper way of providing ewes with a diet complement and fattening the lambs. This is consistent with the farm's objective to optimise the mixed sheep-plus-cattle system in terms of forage use and health management (parasitism).

***The Charriol farm, Brioude Bonnefond Agricultural College (Haute-Loire)** (organic since 1998)

This farm is located at 500 m a.s.l., in a fast-drying zone (precipitation: 500 mm; soil over granite), counts 57 ha and 430 Bizet-breed ewes (high potential for reproduction in counter-season, high hardiness), with a high stocking density of 1.44 LU/ha. The principal forage area is largely made 'viable' by a significant share of tillable farmland (featuring temporary grassland and cropland). Lambing mainly takes place in autumn (starting late-August), since viable productivity fares better than in spring (lower lamb mortality rate) and autumn lambing opens better market options for higher-priced lambs. A share of the concentrate used (with lactating mothers) is produced on-farm.

***The Prades farm, Rochefort-Montagne Agricultural College (Puy-de-Dôme)** (organic since late 2001)

Located on a volcanic bedrock (altitude: 800 m a.s.l.; precipitation: 1000 mm), this farm spans 40 ha of permanent grassland and counts 270 Rava-breed ewes (high potential for reproduction in counter-season, high hardiness) at a stocking density of 0.96 LU/ha. As part of a strategy to optimise ewe productivity and draw added value from available grass, lambing is done over two periods (March/April and September/November), an attempt to combine high ewe requirements with high-quality pasture grass. The Rava breed's hardiness and maternal qualities are sufficiently adapted to this context to give relatively good flock performances with relatively little concentrate or manpower input.

***The Redon farm (INRA Clermont-Ferrand Theix)** (organic since early 2002) The farm's 50 ha (altitude: 850 m a.s.l.; rainfall: 750 mm; surface soil over granite) are used by 200 Limousin-breed ewes, split into two 'systems' (good potential for reproduction in counter-season, better lamb conformation than Bizet or Rava breeds). This study focused on system 1: the "grassland system". In addition to specificational compliance, this system aims to comply with the principles of organic farming (soil sustainability, minimal non-accelerated reproductive demand on ewes). Heavy reliance on forage and the natural landscape environment dictate the low stocking density, at just 0.7 LU/ha. In order to maximise forage self-sufficiency (little farmable cropland hectarage), lambings are split 50/50 between March and November (in a rationale similar to that of the Prades farm). Spring lambs are grass-fattened, while concentrate is kept down to 40 % or 50 % of the ration for autumn lambs.

The second "accelerated lambing system" was set up to assess the potential benefits and limits of the '3/2' system in an organic farming setting. Discontinued in 2003, this system is described and cross-compared with the "grassland system" in Benoit et al. (2009).

Analysis was carried out according to the method described in Benoit and Laignel (2006). The main analytical criteria addressed farm structure, flock performances (including ewe productivity: number per year of live lambs per ewe aged >12 months), characteristics of the end product sold to market, overhead, and income per farm operator. In relation to the notion of 'land-related stock farming', our study also integrated self-sufficiency criteria: feed self-sufficiency, defined as the fraction of flock requirements (in forage units [FU], INRA system) sourced on-farm, and forage self-sufficiency, defined as the fraction of flock requirements sourced from the forage area. The fact that hormone treatments are not allowed in organic farming standards makes the proportion of out-of-season lambings critically important, given the sector's year-round demand. In order to obtain a more fine-grained assessment of this proportion, we use the "out-of-season lambing index" criterion, which weights lambings per fortnight according to a coefficient running between the extreme bound values of 100 (lambing around late-August to September) to 0 (lambing in spring) (Benoit and Laignel 2006). The cost of cereals produced on-farm and transferred to the sheep stock (internal transfer) is evaluated based on market prices.

10.3 Results

10.3.1 Cross-Comparisons of Organic vs. Conventional Farm Performances

Structural data was compiled solely from private-sector-run farms. Flock system and performance data was compiled integrating data from the experimental farms.

Farm Structure

Organic farms were smaller-scale with a lower labour force (Table 10.1) than conventional-system farms both in upland and in lowland areas. LU per farm operator was ultimately similar between organic and conventional farms in both upland and lowland settings. The proportion of cropland was significantly higher on organic upland farms (15% of UAA vs. 4%), tied to the objective of enhancing feed self-sufficiency. Lowland organic farms sidelined 12% of UAA as cropland, all for the flock, compared to 29% for conventional farms that sell off part of their cereal harvest. Stocking density on the principal forage area (PFA) was 15% lower on upland organic farms, linked to the lower degree of forage area intensification, but higher on lowland organic farms. This inverse pattern may be explained by the geographic location of the farms, since lowland organic farms were essentially clustered within the Allier department, whereas lowland conventional farms were clustered in the 'Brandes' rangeland of the southern Vienne department where the best lands are devoted to cash crops and the poorest ones to breeding.

Flock Management and Flock Performance

Variance in ewe productivity (EP) was essentially tied to differences in lambing regimes, which varied strongly between farms (Fig. 10.1). There was very little difference in EP between lowland-based organic and conventional farms (Table 10.2). However, in upland-based farms, the lambing rate was clearly lower in organic farms than conventional farms due to the fact that organic farms refuse to accelerate lambing since it could lead to heavy demand on the ewes, which in turn would use

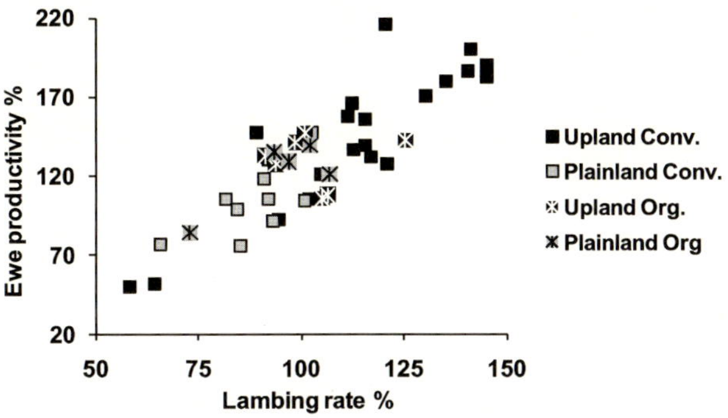

Fig. 10.1 Relationship between *lambing rate* (number of lambings per ewe and per year) and *ewe productivity* (number of lambs alive per ewe and per year)—42 farms

more concentrate (Benoit et al. 2009). Upland organic farms registered a lower EP than conventional farms, which is largely explained by the lower lambing rate and a 3.5 point higher new-born lamb mortality rate.

Upland organic farms have a higher proportion of out-of-season lambings, tied to their use of hardy breeds that quickly and comfortably adapt to out-of-season lambing. The out-of-season index evens out between organic and conventional lowland-based farms since the organic farmers use specific techniques such as the 'ram effect' (Tournadre et al. 2002) and/or techniques well-adapted to natural out of season lambing genotypes for a part of the flock (since hormonal treatments are forbidden).

Flock Diet

In upland-based farms, the organic systems appeared to make heavy use of concentrate input compared to conventional systems (Table 10.2). The cumulative impacts of successive droughts appear to have medium-term repercussions, with a decline in the quality of permanent grassland (loss of legume crops) and forage nitrogen depletion (as observed on the Redon experimental farm), as well as a chronic lack of stocks since overly expensive forage purchases had to be kept to a minimum. However, on-farm crop growth helped offset part of this forage deficit. The conventional systems experienced equally strong impacts, but adding nitrogen inputs made it easier to replenish the stocks, while forage deficits were systematically absorbed by purchasing feed and forage. Organic farms followed the same lamb diet patterns as conventional farms, with grass-fattening remaining rare. In lowland-based farms, organic systems were much better positioned since they used far less concentrate inputs by grass-fattening their lambs. In organic upland-based farms, the cost of forage produced and feed purchased accounted for 51 % of the animal gross product, compared to only 30 % in the other three settings (conventional upland, conventional lowland and organic lowland). Forage self-sufficiency was 66 % in organic upland areas vs. 70 % in conventional upland areas. However, on-farm cereal crop production meant that feed self-sufficiency in organic upland farms was

Table 10.2 Technical and economic results for organic and conventional farms studied, upland and lowland, 2006

		Upland			Lowland	
		Conv.	OF		Conv.	OF
	Number=	21	8	4[a]	8	5 (3[b])
Reproduction	Prolificacy %	145	148	147	142	ND
	Lambing rate	1.15	1.07	1.06	0.92	0.89
	Lamb mortality %	14.1	17.6	18.1	19.5	ND
	Ewe productivity	1.43	1.30	1.27	1.05	1.07
	Out-of-season lambing index[c]	45	39	27	27	24
Feeding	Concentrate/ewe (kg)	157	167	185	177	122
	Price (€/kg)	0.19	0.26	0.26	0.17	0.25
	Forage self-sufficiency (%)	72	66	61	70	ND
	Feed self-sufficiency (%)	76	77	70	84	ND
Economic results	Carcass weight (kg/head)	16.8	16.4	15.8	18.7	20.0
	Carcass price (€/kg)	4.95	5.46	5.68	5.35	5.21
	Gross margin/ewe (€)	66	50	44	59	67 (66)
	Net income/farm operator (€)	14,500	ND	15,200	14,300	9,200 (13,700)
	Overhead (€/Equivalent LU)	451	ND	370	442	539 (444)

NA: data not available
a Without agricultural college or experimental farms
b Without two farms with very high levels of overhead
c Used to qualify the amount of out-of season lambing (autumn) (maximum = 100)

virtually identical to upland conventional farms. Hence, the proportion of feed self-sufficiency accounted for by cereal crops is 11 % for organic farmers compared to just 4 % for conventional-system farmers.

Gaining Added Value From Lambs

In upland farms, organic lambs weighed less than conventionally-reared lambs, with a 10 % higher added value per kilo (Table 10.2). In lowland farms, organic lamb carcasses were much heavier, but sold at little less price per kilogram than conventionally-reared lambs. Looking at the overall picture for the last 5 years, upland-reared organic lamb had a mean added value of € 0.5/kg, whereas the price of lowland-reared organic lamb was not significantly different from that of conventional lamb, part of which goes to official quality label schemes that keep sale prices relatively high, especially in the off-season. Official label schemes are quite diverse in terms of breeding constraints, added value and the proportion of lambs sold in the pool of farms studied, and one farm may even have several labels. Generally speaking, certain labels such as the "Red Label" (*Label Rouge*) generate a higher added value but have an impact on only a small part of the production. They are based on the age at slaughter, carcass conformation, fat and meat colour. The level of the added value often depends on the season, which is a way to better adapt the demand.

Veterinary Expenses and Mineral and Vitamin Costs

Veterinary expenses (drugs and veterinarians' fees) at organic farms are 8 and 16% lower in upland and lowland areas, respectively. On upland farms, the two systems make comparable use of minerals, vitamins and diet supplements, whereas on lowland farms, the organic system uses three-fold more (€ 4.9/ewe vs. 1.6), with three farms claiming that major mineral and vitamin supplement outlays were needed to ensure good flock performances (argument unchecked). Internal parasite control in young sheep hinges on very limited use of anthelmintics (one or two only, against lungworms and tapeworms), possibly opting instead for alternative medicines (plant sources) but most importantly by preventive integrated management (clean pastures, good-quality nutritional management). It is consequently possible in upland farms to avoid using anthelmintics in the majority of grass-fattened lambs (Benoit et al. 2009). Since upland farms rarely pasture-graze weaned lambs, they are exposed to far less risk of parasitism, but are consequently reliant on a mainly concentrate-based diet. This pen-based management system is the result of a range of constraints (paddocks far from farm buildings, strongly-contrasting climate swings), together with traditional practices.

Revenue Performance of the Sheep Stock

In upland-based farms, the added value on organic lamb sales partly offsets the lower ewe productivity—profit per ewe is ultimately 5% less than for conventional-farmed sheep (€ 117 vs. 123). However, per-ewe operational costs are € 10/ewe higher (feed costs: +18%), meaning that the gross margin per ewe is € 16/ewe lower (Table 10.2), i.e., −24%. In lowland-based organic farms, the marginally higher EP means that the profit per ewe is 6% higher (€ 123/ewe), with slightly better control over costs (−5%). Overall then, the margin per ewe is 14% higher for organically-farmed sheep than conventionally-farmed sheep. It should, however, be underlined that the benchmark conventional system figures appear to be mediocre in lowland farms, whereas in upland farms, there were several '3/2'-system farms that showed good economic returns as a result of their high level of ewe productivity. Like the conventional system, organic farming also posted a broad range of economic figures (Fig. 10.2), mainly in relation to the variability in ewe productivity where the observed range of gross margins ran from € 30 to 93/ewe in lowland farms and from € 37 to 75/ewe in upland farms. The best lowland organic results appeared in farms where ewe productivity was 1.2–1.4 at 100–110 kg of concentrate per ewe, compared to upland organic farms where the best results were linked

Table 10.3 Technical and economic results of the four agricultural college and experimental farms (mean of 5 years 2002–2006)

	Farms	Cambon	Charriol	Prades	Redon
Reproduction	Prolificacy %	195	142	154	170
	Lambing rate	1.03	1.01	0.98	1.00
	Lamb mortality %	25	20	12	12
	Ewe productivity	1.51	1.15	1.33	1.51
	Out-of-season lambing index[a]	72.5	60.6	44.5	31.7
Sales	% sales on organic	89	85	87	65
	Carcass weight (kg/head)	17.7	16.5	17.0	15.7
	Price (€/kg)	5.51	5.48	5.00	4.91
Concentrates	Concentrate/ewe (kg)	238	167	109	111
	—Purchased (kg)	78	98	109	84
	—On-farm (kg (% total))	160 (67%)	69 (41%)	0 (0%)	27 (24%)
	Mean price (€/kg)	0.223	0.288	0.341	0.341
	Concentrate/kg carcass (€)	2.3	3.2	2.1	1.9
Self-suffic.	Forage (%)	62	66	79	79
	Feed (%)	89	81	79	83
	Tied to crops (%)	27	15	0	4
Economic Results	Gross product/ewe (€)	142	119	109	120
	Operational costs/ewe (€)	80	73	49	60
	Gross margin/ewe (€)	62	46	60	60
	Gross margin/ha used for flock (€)	439	540	434	294

a Used to qualify the amount of out-of season lambings (autumn) (maximum = 100)

Conversely, the Charriol farm and especially the Cambon farm derive part of their feed resources from rotational grassland and grass leys (particularly alfalfa) and cereal crops that provide rations adapted to animals with high energy needs, whether stalled indoors, out of season (autumn-lactating ewes), or overwintering (lambs for fattening). Furthermore, these lambs generally have higher added value than spring-born lambs.

Consequences in Terms of Value Gain on Lambs, Concentrate Input, and Feed and Forage Self-Sufficiency

Different seasonally-cycled production systems have impacts on lamb market value: both farms that opted for out-of-season lambing (Charriol and Cambon) posted 10% higher added value per kilo, at € 5.5/kg, compared to € 4.9–5.0 (Table 10.3). Redon posted a lower share of sales on the organic market due to the fact that lamb quality did not meet organic label demand (weight, conformation, finishing stage).

High concentrate inputs were correlated with the existence of on-farm cereal crop production and reached 238 kg/ewe at Cambon. At Prades and Redon—where

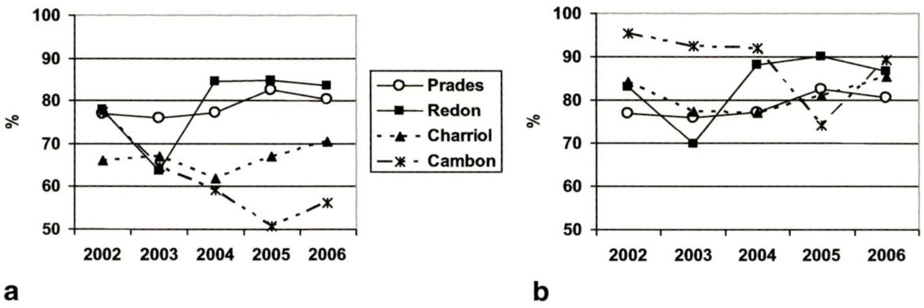

Fig 10.5 Change in the forage self-sufficiency (**a**), and feed self-sufficiency (**b**) of the four agricultural college and experimental farms, from 2002 to 2006

concentrate input was mainly brought in—it approached 110 kg/ewe, but at a much higher cost (€ 0.34/kg compared to 0.26). Concentrate input costs fell in the € 1.9–2.3 per kilo carcass weight range for three farms (Redon, Prades, Cambon), but was virtually 60 % higher at Charriol (€ 3.2/kg carcass weight) where concentrate was used to offset the lack of available forage under high-density stocking. Faced with little flock feed other than forage, Redon and Prades reached almost 80 % forage self-sufficiency. However, Cambon and Charriol were able to take advantage of cereal crop production to reach 89 and 81 % feed self-sufficiency, respectively, which was comparable if not better than Redon and Prades. All four farms, which work within radically different soil-climate settings, experienced strong climatic fluctuations over the five-year period. Figures 5a, b illustrate variations and convergences in feed self-sufficiency (around 80–90 %) via the various strategies outlined above at the four farms, for 2006.

Technical and Economic Performances

Ewe productivity (EP), shown to be a determining factor governing gross margin per ewe in conventional farms (Benoit et al. 1999), was a major factor here as well. Differences in EP stem from mother prolificacy and lamb mortality, since lambing rates were comparable across the four flocks. High prolificacy rates enabled Cambon and Redon to reach a good 1.51 EP level. Charriol came in last at 1.15 EP due to a mediocre prolificacy rate and a high mortality rate (20 %). Farms with little on-farm cereal production had lower operating costs due to lower concentrate inputs. Overall, two strategies culminated in roughly the same gross margin per ewe (€ 60), whether it was high profit and high outlays like at Cambon, or lower profit and lower outlays like at Redon and Prades. The Charriol farm posted a 23 % lower margin per ewe (€ 46), with very high feed costs, given its production output.

10.4 Discussion

Given the small sample population and strong within-sample heterogeneity, particularly among organic sheep farms, the emphasis should be placed less on cross-comparing organic farms with conventional farms and more on investigating the factors that shape technical and economic performances under these contrasted systems.

10.4.1 Feed Self-Sufficiency: A Decisive Factor With Between-Region Differences

Performances on **upland** organic farms are exposed to heavy constraints. Economic factors are particularly decisive since feed self-sufficiency is often low, driving up production costs. On average, around 41 % of the lamb sale value has to cover feed inputs that cannot be produced on-farm; when market pressures drive cereal prices upwards (as in 2008 or 2010), this figure can rise up to 50 %. The fact that the use of concentrate remains rather high on organic farms partly explains the lower gross margin per ewe: € 50 on organic farms compared to € 66 on conventional farms. Heavily reducing infrastructure costs and increasing earnings (mainly through direct-to-consumer sales) is the only way to keep income per farm operator comparable to conventional-system farms. Stronger land-related stock farming should lead to at least 80 % feed sufficiency in grazing land systems, or at least 90 % on farms with the resources for producing cereal crops. This hinges on identifying the optimal stocking density (Thériez et al. 1997), as well as integrating the years exposed to potentially severe drought events, which can have greater financial impacts on organic farmers since: (i) compensatory organic feed purchases will be more expensive; and (ii) organic farmers are intrinsically resistant to purchasing forage and concentrate, which can weaken medium-term technical performances. Several issues need to be resolved: where growing cereal crops is a possibility, how can they be optimised into added value (ewes vs. lambs)? Upland farm systems based on a high proportion of permanent rangeland need to determine what type of flock management system to adopt in order to sustainably secure legume crops or promote legume crop regrowth following years marked by harsh climate events (Doyle and Topp 2004). Permanent pastures are fundamental to the forage self-sufficiency equation, and if they cannot be transformed into rotational grassland, other solutions need to be implemented to maintain or improve production potential (overseeding, grazing-driven vegetation composition). Lastly, although wholly absent from upland farms, is grass-fattening of lambs a conceivable option? Although possible, it requires a high level of technical skill (Thériez et al. 1997; Prache et al. 1986) to optimally control forage quality and parasitic infectivity (Cabaret 2004).

Average income on lowland organic farms is comparable to conventional farms when infrastructure costs are brought under control. With better feed and forage self-sufficiency figures than conventional systems, they are able to maintain a re-

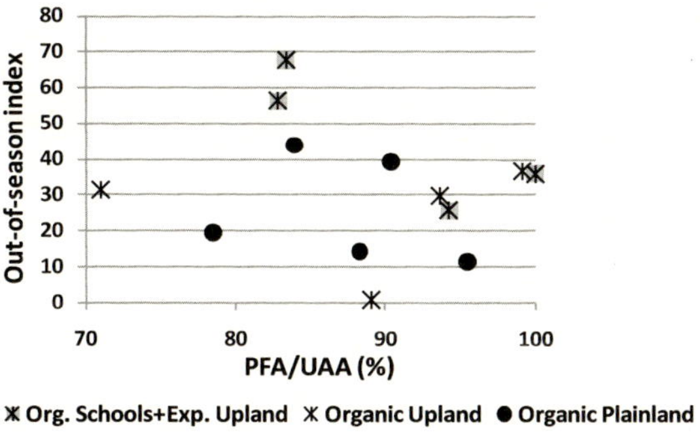

Fig. 10.6 Proportion of the principal forage area (PFA) in the UAA and out-of season index (organic farms), 2006

spectable margin per ewe at comparable flock productivity and lamb value-gain levels. Lowland organic farms are able to take advantage of not just quality forage cut from temporary pastures (such as for the grass-fattening of lambs) but farm-produced concentrate that meets the bulk of flock needs as well—including out-of-season lamb finishing. Furthermore, natural conditions may ensure a much longer annual grazing season than in the uplands (meaning less stocks to build up and better-quality pasture stock), which in turn extends the possibilities for grass-fattening lambs (Bellet 2010).

In upland settings, system logics are more difficult to define, and the allied technical management roadmaps are relatively 'subsector-specific', especially in settings where there are no options for growing cereal crops and few options for setting up temporary pastures.

10.4.2 Cereal Crops: Farm System Viability and Economic Impacts

This study of four agricultural college and experimental farms reveals the links between possibilities for producing cereal crops and the major fraction of out-of-season lambings tied to high quantities of concentrate inputs. Figure 10.6 shows that this system logic does not apply for private-sector upland farms, none of which drew heavily on out-of-season breeding (out-of-season lambing index <40). However, concentrate input per ewe did appear correlated to options for cereal crop production (Fig. 10.7). No trend emerged in lowland settings (Figs. 10.6 and 10.7). The size of our sample combined with the broad diversity among organic farms and the observed variability in technical control preclude any reliable generalisation of the observations from these experimental farms. In particular, the relatively significant use of direct-to-consumer sales as a production outlet by private-sector farms could translate into decreased follow-up of flock management, which sometimes leads to relatively poor technical performances as farmers focus more on monetising their products.

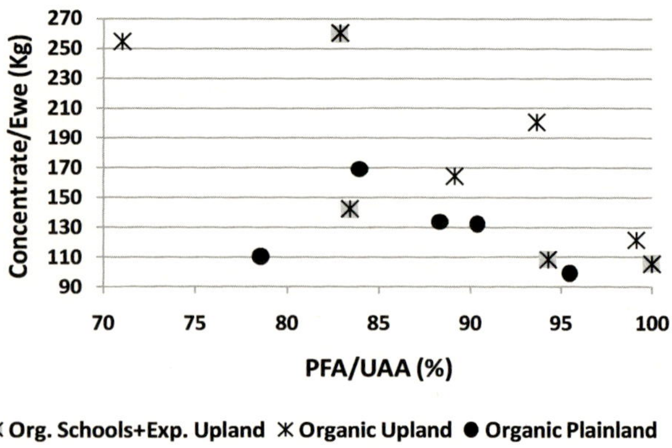

Fig. 10.7 Proportion of the principal forage area (PFA) in the UAA and concentrate consumption per ewe (organic farms)

The method routinely used for calculating margins on sheep stock accounts for on-farm cereal crop consumption as 'sheep' stock expenditure and 'cereal' stock profit. It thus follows that, especially in organic systems, the (high) price of between-stock cereal transfers can lead to a low margin per ewe and a high margin per ha of cereal crops. Farmers themselves tend to picture the economic utility of a system in terms of the overall profit margin, which can be expressed as a margin per ha of farm area used by the flock. Consequently, as shown in Table 10.3, although the Charriol farm has a gross margin per ewe of € 46, which is 23 % lower than at the other three experimental sites, its margin per ha used by the flock (forage plus cropland) was the highest at € 540, driven by the high stocking rate achieved through on-farm cereals that have a very high margin per ha. The options for growing on-farm crops (and the potential offered by the local environment) therefore play a major role in margin per ha used (which was low at Prades and Redon). The between-farm variability in agronomically-driven potential highlights the limitations inherent to working with per-ha ratios in agronomically-contrasted settings.

Brought to the forefront by organic farming, soil sustainability via cereal crops has a strong positive effect at the farm scale. Furthermore, the consequently lower reliance on outside sources provides a resilient buffer against market risks. In cereals, market conditions such as those that marked 2008 could be decisive for farm earnings, not just on upland farms, for which concentrate prices could be extremely high, but on lowland farms as well where farmers could gain room to maneuver and sell off a share of the crops harvested. Finally, growing cereal crops leads to measurable cuts in straw costs.

10.4.3 Towards Technical Optimisation Research

Obtaining a good margin per ewe hinges on pinpointing the technical optimum that combines a high level of productivity with the lowest possible level of related inputs (concentrate included). From the forage management perspective, this entails reducing the stocking rate so that forage production (quantity and quality) empowers the farmer to become non-dependent on concentrate use (optimal efficiency). However, as a measure to avoid losing headage and turnover, the best way forward is to upsize UAA. Thériez et al. (1997) showed that reducing the stocking rate from 1.2 to 0.80 LU per ha (in a conventional upland-based farm on soil over granite), it was possible to cut concentrate inputs by 26% without losing flock-wide ewe productivity, yielding a 27% increase in the gross margin per ewe. From the animal perspective, there is also an optimality process to look at since an accelerated rate of reproduction leads not just to higher inputs (especially feed costs), but to the excessive demand on the ewes, which fragilises the system: significant variability in out-of-season fertility, increased vulnerability to various health problems, sharper impacts of exceptional climate conditions and quality of forage available (Benoit et al. 2009). The cohesiveness of the sheep stock system is illustrated in Fig. 10.2. On organic farms, at comparable EP levels (1.2–1.3), the gross margin per ewe varies 1–3-fold (from € 30 to 90), in relation to worse situations with very high concentrate inputs. Our observations on (upland-based) experimental farms show that under certain sets of conditions, it is possible to achieve EP levels of 1.3–1.5 with 70–80 kg of concentrate per ewe. These combinations were also observed in two lowland-based organic sheep farms that posted a gross margin per ewe close to € 90 with an EP of 1.2–1.35 and concentrate intakes at around 100 kg per ewe. These findings overlap with the analysis carried out on conventional lowland-based farms ('self-sufficient grassland' systems), which although highly seasonally-cycled with 75% of lambings taking place in late winter, still managed to obtain very high gross margins at a baseline 100 kg concentrate per ewe and 1.5 EP (Benoit et al. 1999).

Cross-comparison of four contrasted sheep farming systems highlighted the degree of variability in farm system strategies. Strategy success essentially hinges on fitting objectives to local context (soil-climate setting, labour resources) by drawing added value from the potentialities of the genotype farmed. The four breeds chosen, all anchored to their local geography, shared similar hardiness (maternal qualities, roaming ability, and gaining value from different types of feed resources) and out-of-season breeding characteristics. This makes it possible to optimise reproduction performance (non-fertile ewes quickly recycled back into mating bands, managing ewe-lamb reproduction), as well as to produce lambs in periods of potential shortage (winter). However, the lower conformation of the lambs may require farmers to identify specific market niches or to complexify flock management by introducing terminal meat rams.

10.4.4 Features Specifically Tied to the French Context

The conclusions drawn above are to be gauged in relation to the study setting (central France), where the trend is towards a generalised drop in grazing intensity on the PFA of organic farms (vs. conventional intensified farming with chemical fertilisers), concentrate inputs used as flock feed, and options for growing on-farm crops.

These conclusions may be modulated or nuanced for other settings, especially in conventional systems that are relatively unintensive or exclusively grass-based. A comparative study led in a New Zealand-based experimental domain (Richardson and Richardson 2006) reported similar conclusions on animal performances to those reported here, with a slight drop in ewe productivity (6 %) and lamb weight (− 1 kg carcass weight) in organic vs. conventional farms that are offset by added value at sale, resulting in a 19 % increase in profit. With the PFA stocking rate held constant, costs—when no concentrate inputs are used—are reduced (− 18 %), particularly veterinary expenditures through the use of parasite-resistant foundation stock.

However, mirroring our conclusions once again, economic success hinges on minimising losses in flock performance, gaining significant added value at sale, reducing operating costs, and making outstanding feed-value gains on grass.

10.5 Conclusion

These findings highlight the limits to economic profitability in organic sheep-for-meat farms in the contexts presented, and the need to fine-tune adjustments in technical management roadmaps on ewe productivity—feed costs, lamb monetisation triangle—the latter two factors being even more important in comparison to conventional sheep farm systems. In 2008, inflationary cereal prices strongly challenged the economic profitability of sheep farms. Within this market context, self-sufficiency (for both forage and feed) has major impacts, as discussed in this study, and becomes the pivotal make-or-break condition governing respectable profitability in organic sheep-for-meat farming. If these self-sufficiency thresholds can be met, organic sheep meat production has perspectives for expansion in upland and disadvantaged lowland areas, provided that options for pen-fattening lambs are also maintained on early spring or late autumn.

The predominant issue of feed costs is fairly specific to sheep-for-meat farming, and possibly dairy cattle farming as well. For suckler cattle farming, other constraints may come into play such as product 'finish', which often leads to a wholesale reorganisation of the farm system (Veysset et al. 2009).

Globally speaking, the limits to organically-farmed sheep-for-meat systems are less centred on the ability to achieve good ewe productivity and good control of animal health factors than on securing a substantial added value on lambs at sale through organised subsector channels, and on keeping production costs under control. Our research into forage resource quality and forage strategies to be used in terms of reorganising farm systems will continue in this direction.

Deeper insight and progress on these points could encourage farmers to convert to organic farming, against a background where the mid-term appraisal of the CAP (Common Agricultural Policy) has sent promising signs to the organic farming sector, with specific economic support and recognition (Barnier 2007, 2010). Indeed, after a stagnation in the number of farms from 2003 to 2007, the increase of farms on OF in France was 55 % between 2008 and 2010, for all types of farms.

Given the projected changes in economic context, particularly the outlook for raw material prices, the same challenges organic farmers have to face in terms of maximising feed self-sufficiency may well resurface in exactly the same way for conventional farmers tomorrow, in a repeat of 2008 (Hovi et al. 2003). The reflexive analysis and approaches implemented by these organic farms over the last few years could provide conventional farmers with a platform of strategies that mirror those described here to improve feed self-sufficiency performance as a defensive measure at a new economic turning point in the cereals market.

Acknowledgments This study was carried out through a partnership with the Chambers of Agriculture of the Allier, Puy-de-Dôme and Haute-Vienne departments, with the financial support of FNADT and the Auvergne Region. We also thank the farm managers and farmers who ran the four experimental and agricultural college farms studied (Hervé Tournadre and Marcel Verdier, Jean-Paul Guittard, Jean-Pierre Chaput and Jean-Claude Giraudet) and their staff.

References

Agence Bio (2007) L'agriculture biologique, Chiffres clés. Edition 2007. Agence Française pour le Développement et la Promotion de l'Agriculture Biologique. p 168

Agence Bio (2010) L'agriculture biologique, Chiffres clés. Edition 2010. Agence Française pour le Développement et la Promotion de l'Agriculture Biologique. ISBN 2-9519133-9-7. p 236

AGRESTE (2006) http://agreste.maapar.lbn.fr/ReportFolders/ReportFolders.aspx. Accessed 25 Nov 2010

Barnier M (2007) Agriculture biologique horizon 2012. Grand Conseil d'orientation de l'Agence Bio. 12 septembre 2007. Ministère de l'Agriculture et de la Pêche. p 14

Barnier M (2010) Bilan de santé de la PAC: mise en œuvre pour une PAC préventive, juste et durable. Ministère de l'Agriculture et de la Pêche, p 4. http://agriculture.gouv.fr/IMG/pdf/Bilandesantefrance.pdf. Accessed 25 Nov 2010

Bellet V (2010) Convertibilité comparée des systèmes ovins viande français. Mars 2010. Institut de l'élevage. Compte rendu n°001055001. ISBN 978-2-84148-834-6. p 44

Bellet V, Morin E (2005) Approche des coûts de production et des déterminants du revenu en élevage ovin viande. Compte-rendu final 11 05 50 020. Éditions Institut de l'Élevage. p 73

Benoit M, Laignel G (2006) Méthodologie d'élaboration de résultats technico-économiques en élevage ovin allaitant. Illustration en France, en zone de plaine et de montagne, Séminaires Méditerranéens: L'évolution des Systèmes de Production Ovine et Caprine: Avenir des Systèmes Extensifs face aux changements de la Société 70:57–65

Benoit M, Laignel G, Liénard G (1999) Facteurs techniques, cohérence de fonctionnement et rentabilité en élevage ovin allaitant. Exemples du Massif Central Nord et du Montmorillonnais. Renc Rech Ruminants 6:19–22

Benoit M, Tournadre H, Dulphy JP, Laignel G, Prache S, Cabaret J (2009) Is intensification of reproduction rhythm sustainable in an organic sheep production system? A four-year interdisciplinary study. Animal 3(5):753–763

Cabaret J (2004) Parasitisme helminthique en élevage biologique ovin: réalités et moyens de contrôle. INRA Prod Anim 17(2):145–154

Cabaret J, Nicourt C (2009) Les problèmes sanitaires en élevage biologique : réalités, conceptions et pratiques. INRA Prod Anim 22(3):235–244

Doyle CJ, Topp CFE (2004) The economic opportunities for increasing the use of forage legumes in north European livestock systems under both conventional and organic management. Renew Agr Food Syst 19(1):15–22 (8)

Hovi M, Martini A, Padel S (2003) Socio-economic aspects of animal health and food safety in organic farming systems. Proceedings of the 1st SAFO Workshop, Florence, Italy, 5–7 September 2003

Laignel G, Benoit M (2004) Production de viande ovine en agriculture biologique comparée à l'élevage conventionnel: Résultats technico-économiques d'exploitations de plaine et de Montagne du Nord Massif Central. INRA Productions Animales 17:133–143

Nardone A, Zervas G, Ronchi B. (2004) Sustainability of small ruminant organic systems of production. Livest Prod Sci 90:27–39

Prache S, Brelurut A, Thériez M (1986) L'élevage de l'agneau à l'herbe. Effets de l'âge au sevrage sur les performances d'agneaux élevés à l'herbe puis engraissés en bergerie. Annales Zootechnie 35:231–254

Richardson A, Richardson S (2006) A comparison of the organic and conventional livestock farming systems of Avalon farming in West Otago. Proc New Zeal Grassl Assoc 68:127–132

Thamsborg SM, Roderick S, Sundrum A (2004) Animal health and diseases in organic farming: an overview. In: Vaarst M, Roderick S, Lund V, Lockeretz W (eds) Animal health and welfare in agriculture. Cabi Publishing, Wallingford, pp 227–252

Thériez M, Brelurut A, Pailleux JY, Benoit M, Lienard G, Louault F, De Montard FX (1997) Extensification en élevage ovin viande par agrandissement des surfaces fourragères. Résultats zootechniques et économiques de 5 ans d'expérience dans le Massif Central Nord. INRA Productions Animales 10:141–152

Tournadre H, Bocquier F, Petit M, Thimonier J, Benoit M (2002) Efficacité de l'effet bélier chez la brebis Limousine à différents moments de l'anoestrus saisonnier et selon la durée de l'intervalle tarissement-mise en lutte. Rencontres Recherches Ruminants 9:143–146

Veysset P, Bécherel F, Bébin D (2009) Élevage biologique de bovins allaitants dans le Massif Central: résultats technico-économiques et identifications des principaux verrous. Inra Prod Anim 22(3):189–196

Chapter 11
Experiencing Organic Mixed Crop Dairy Systems: A Step-by-Step Design Centred on a Long-term Experiment

Xavier Coquil, Jean-Louis Fiorelli, André Blouet and Catherine Mignolet

Abstract To provide more sustainability for their farming activity, farmers may attempt to redesign it according to their perception of their global environment. This paper focuses on the step-by-step design approach. The main objective of this approach is to produce resources to empower farmers to develop a more sustainable farming activity. We postulate that self-sufficient agricultural systems designed on the basis of natural land properties will generate sustainable farming activity. The methodology developed is anthropocentric and centred on a long-term experiment. Two organic and self-sufficient mixed-crop dairy systems were designed in 2004 at the INRA ASTER-Mirecourt Experimental Station. Research scientists designed and redesigned the systems, step-by-step, according to their perception of the natural properties of their agro-ecological environment. Since 2005, the two systems have been evolving by repairing system malfunctions or by improving their self-sufficiency. Step-by-step design is an approach based on methodologies that create experience in situations. This approach has proven its relevance to create knowledge for (i) the transition of farming systems towards more self-sufficient forms of agricultural activity, and (ii) the adaptation of systems to environmental fluctuations.

Keywords Design · Long-term experiment · Experience · Organic · Self-sufficiency · Mixed crop dairy systems · Grazing system

11.1 Introduction: Designing Sustainable Agricultural Systems

Sustainable development considerations make it necessary to redesign agricultural activity in rural territories (Godard and Hubert 2002). Sustainable agriculture is faced with the challenge of a new type of agricultural activity development that gives equal consideration to social, environmental and economic developments. At the farming system level, these challenges can be interpreted and taken up by farmers. They can act or react according to their perception of changes in their

X. Coquil (✉) · J.-L. Fiorelli · A. Blouet · C. Mignolet
INRA, UR055 SAD ASTER-Mirecourt, 662 avenue Louis Buffet, 88500, Mirecourt, France
e-mail: coquil@mirecourt.inra.fr

S. Bellon, S. Penvern (eds.), *Organic Farming, Prototype for Sustainable Agricultures*,
DOI 10.1007/978-94-007-7927-3_11, © Springer Science+Business Media Dordrecht 2014

environment and thus redesign their farming activity in order to improve its sustainability. In this context, one of the questions that arises is how to empower farmers in their search for sustainability in their farming activity.

We postulate that agricultural systems designed on the basis of natural land properties with no use of chemical inputs will be sustainable, at least in ecological terms. Designing systems on the basis of natural land properties is a real change of paradigm in the agronomic sciences. Natural properties are then seen as production factors for agricultural activity (Auricoste et al. 1985), whereas the currently dominant paradigm in agronomy considers the natural properties (the ecological environment) as limiting or a matrix on which farming activity must limit negative externalities (Legrand et al. 2002; Lopez-Ridaura et al. 2009). This new paradigm targets the expression of the diversity of the natural properties of the land by limiting its artificialisation. Preserving the functional integrity of the land (Thompson 1997) by minimising input use becomes a condition and a means of managing natural resources. For the total expression of natural land properties, we decided to design low input systems. French specifications for organic agriculture are seen as a good framework to test ecological farming systems. Forbidding the use of chemicals inputs and limiting the use of allopathic specifications (stricter than European specifications), a framework is created that: (i) prevents many ecological problems as a result of the precaution principle; (ii) prevents shortcuts since the commitment is official; and (iii) stimulates agro-ecological functioning of farming systems since we choose to use no alternative inputs.

Second, a farming system has been defined by Osty (1978) as a biotechnical and a human sub-system connected by information flows and interacting with the global environment (ecological, political, economic, etc.). We thus postulate that changes in farming systems are based on two types of dynamics (Lamine and Bellon 2008): biotechnical dynamics such as changes in weed cycles and in soil functioning during conversion to organic farming, as well as human dynamics such as the acquisition of new know-how and experience while changing farming systems. The theoretical option that we developed is quite original in the field of farming system design: it aims at designing self-sufficient farming activity focused on the resources required by experimenters (including researchers and all the technical staff) to develop them, whereas mainstream farming system design is focused on biotechnical systems (Vereijken 1997; Dogliotti et al. 2003; Sadok et al. 2009; Blazy et al. 2010). Inspired by Béguin (2010), resources are instruments that have been useful for the experimenters to develop their activity in the systems. By development here, we mean the ways in which experimenters have been stimulated to think and to act that would have been impossible when they were dealing with conventional farming. Development relies on the hypothesis that small changes can lead to large changes for the farmer in relation to his farming system. Design and redesign is therefore dedicated to innovation for the subject in his specific system.

We propose an anthropocentric approach in this paper: the two sub-systems interacting with their environment are considered through the perception of humans, in our case, experimenters of the INRA ASTER-Mirecourt Experimental Station (ES). We decided to build some propositions to empower farmers looking for sustainability on their farms. We have been designing two organic and self-sufficient mixed crop dairy systems since 2004. Designing self-sufficient systems with the objective to produce resources to empower farmers has given rise to methodological questions that can be summed up as follows: *what kind of resources do we need to design self-sufficient systems? How can we produce them?* The methodological approach developed at the ES is thus based on a change of posture considering the role of experimental stations in agronomy and giving a real place to experimenters considered as learning subjects. ES might be used to improve the experimenters' know-how and not only, as was the case in the past, to collect quantitative data in order to understand biotechnical and ecological processes. Breaking down our epistemic position at the ES level, we postulate that we are experiencing systems and not making an experiment because the design approach is centred on experimenters' experiences: (i) the experience and know-how of the experimenters is based on the evolution of resources that they mobilise to act in the systems; (ii) this experience is contextualised and therefore singular; and (iii) formalising the construction of this experience can create resources for the construction of the experience of other farmers (Sève 1987, cited by Schwartz 1992).

We first present the step-by-step design approach. The methods and principles mobilised to design the systems are explained and the contributions of these methods to the step-by-step design approach are then illustrated. We then discuss the interest of the methods used in the step-by-step approach. We also discuss the step-by-step approach considering the main design methods used in the agronomical sciences, as well as the need to design dynamic farming systems to be able to address future challenges.

11.2 A Step-by-Step Design

All experimental methods are not equivalent in terms of their contribution to the design of farming systems. In this section, after presenting the global approach to the design of the systems used at the INRA ASTER-Mirecourt Experimental Station, we will explain the methods already mobilised in the long-term trial and their contribution to the step-by-step design approach.

11.2.1 A Design Approach Centred on an ES: From the Biotechnical Properties of the System to the Development of Experimenters' Activities

The experiment presented in this article takes place at the ES of the INRA ASTER-Mirecourt research unit, located in the Vosges plain in north-eastern France (Coquil et al. 2009a). A group of 21 experimenters composed of five researcher scientists and engineers and 16 technicians are running a long-term experiment at the system scale. The ES extends over 240 ha and includes 100 Holstein and Montbeliard dairy cows and replacement heifers. The soils have a predominantly clayey texture and are located on limestone plateaux and clayey plains. The climate is of the semi-continental type.

In 2004, we configured the territory and the systems according to the natural properties of the land. Since then, experimenters' activities have been designed following a step-by-step process, as a central part of the system design.

In 2004, experimenters defined field potentialities from their knowledge of local realities backed, for some of them, by more than 20 years of experience. The territory was configured on the basis of two imperatives: (i) to cultivate all the fields recognised as being suitable for cultivation; and (ii) to enable two dairy systems to be designed using just one milking parlour. The main criteria used to define field potentialities were: (i) the agronomic and geographical characteristics of the fields (type of soil, area and shape of fields, slope, hydromorphy); and (ii) the logistical constraints (accessibility for the dairy cows). Fields judged suitable for cropping were allocated to crop rotations that were differentiated according to (i) their capacity to grow alfalfa, not requiring hydromorphic soils, and (ii) their capacity to cultivate spring cereals, requiring a good bearing capacity at the end of winter. The fields judged unsuitable for cropping were allocated to permanent grassland, identifying those, in particular, whose distance from the buildings made them accessible to the dairy cows (less than 2 km from the milking parlour). On the basis of the configuration of land occupation, experimenters decided to build a Grazing System (GS) and a Mixed Crop Dairy System (MCDS) in order to: (i) design two different kinds of systems with potentially different operating modes; (ii) design systems with seasonal but complementary milk production over the year; and (iii) be able to be self-sufficient at the system as well as at the small regional scale by allowing small but equivalent exchanges between the systems.

Since the beginning of the trial in 2004, the systems were considered to be in a period of learning and instability due to the dynamics of the biological resources composing the systems (animals, plants, etc.) and the experimenters' inexperience in managing organic systems. System management had to evolve (i) to ensure the continuity of the systems, and (ii) to improve the degree of achievement of their self-sufficiency objectives. We therefore implemented a pragmatic approach for a step-by-step design. To build this pragmatic approach, we tested methods consisting in analysing agricultural practices to re-design, and other methods consisting in creating innovative knowledge to re-design. Experimenters decided to act and to make decisions in a more collective manner in order to share their new experiences that were acquired while working at the station (Fig. 11.1). Collective decision-making was also a way to negotiate decisions and to formalise evolutions in the know-how of the experimenters since the conversion to organic agriculture.

Our results consist of territory and system configurations. We also present methods that we used to design the systems step-by-step. We have attempted below to provide a few answers to the question of the type of resources required by experimenters to build self-sufficient systems.

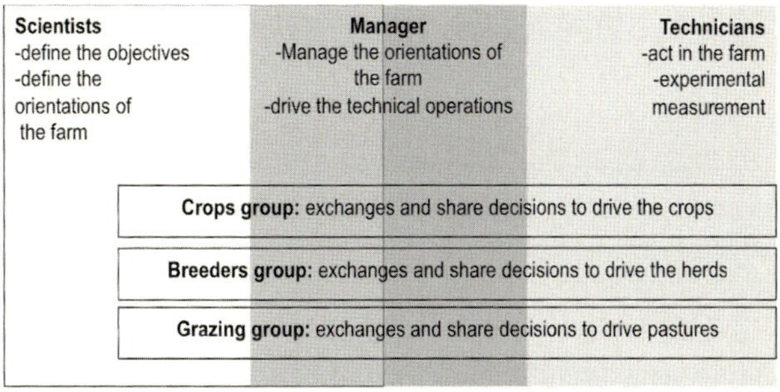

Fig. 11.1 Towards a collective decision-making organisation: governance of the INRA ASTER-Mirecourt ES since 2005

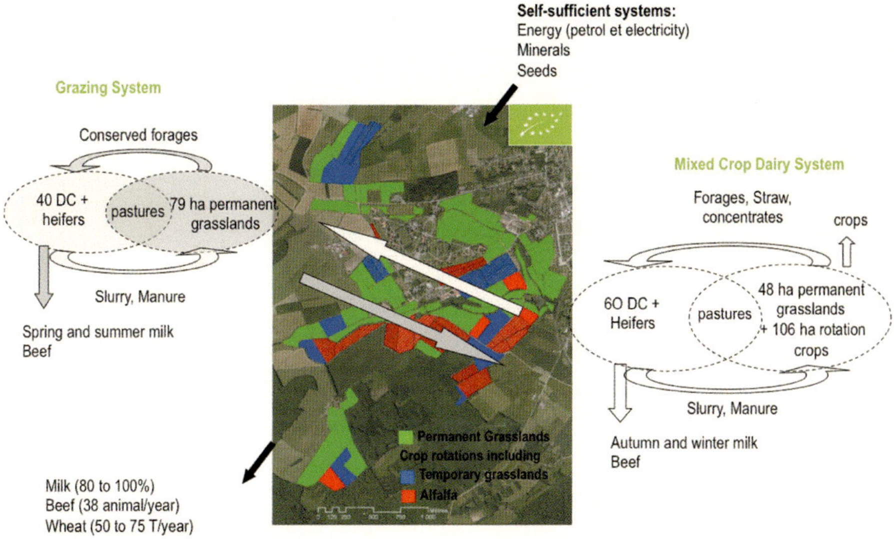

Fig. 11.2 Configuration of the ES of the INRA ASTER-Mirecourt unit, based on the definition of natural land properties by local experimenters: design of a Mixed Crop Dairy System and a Grazing System

11.2.2 Systems Designed on the Basis of the Natural Properties of the Land

Two production systems were configured: a Mixed Crop Dairy System (MCDS) and a Grazing System (GS) whose field patterns interlock within the territory (Fig. 11.2) (Coquil et al. 2009a). The MCDS is responsible for self-sufficiency in straw and concentrates of both systems, and the GS pays it back in manure. These exchanges are intended to limit the transfer of fertility between the two systems. Each system is self-sufficient in forage. The GS aims at managing the seasonality of herd requirements to maximise grazing. The MCDS aims at structuring the diversity of crops and livestock to complete the cycle of materials.

11.2.3 Pragmatic Approach to Deal with the Lack of Experience of Experimenters: A Step-by-Step Design

We first focus on methods mobilised to analyse the malfunctions of systems in order to redesign them. We then present methods mobilised to improve system self-sufficiency.

Analysing and Formalising Malfunctions of Systems as a Source of Redesign

The method used consists in formalising and analysing the practices implemented within the systems tested in order to, firstly, understand the sources of malfunction and, secondly, to attempt to compensate for them. The formalisation is carried out using graphic and/or statistical methods from data taken from the agronomic assessment.

Improving Reproduction Performances of GS cows

In the perspective of extreme self-sufficiency in terms of feed, the dairy cows and heifers in the GS are managed according to a feeding strategy that maximises grazing. This system is managed without the use of concentrates except for the calves. In order to maximise grazed grass in the diet: (i) calvings were grouped over a 3-months period in late winter in order to synchronise the cows' needs with grass growth; and (ii) the grazing period was lengthened by early turnout to pasture in the spring (between 20 March and early April) and a late return indoors in autumn (last ten days of November). This led to average grazing times of 242 days/year (215 days + nights/year) from 2005 to 2008. Consumption of stored forage rose to only 1.9 and 2.1 T DM/cow/year over the years 2007 and 2008, respectively. During the same period, the milk production of the herd was 5403 kg/cow/year for the Holstein cows and 4887 kg/cow/year for the Montbeliard cows. However, in 2005, the reproduction results compromised the survival of the herd: in 2006, only 55 % of the breeding females (i.e., 33 % of the dairy cows and 94 % of the heifers) were successfully incalved during the reproduction period of 101 days.

Reproduction performances of the dairy cows in the grassland system were analysed over the key breeding period, i.e., from calving to the end of the insemination period, i.e., the end of winter and the spring for this herd. Over this period, we formalised the feeding and breeding practices of the dairy cows and analysed their reproduction performances from the viewpoint of the evolution of their milk performances and their body condition scores (indicator of their energy balance) (Gouttenoire et al. 2010).

Since the sustainability of the GS herd was not ensured in 2006, we decided to keep 12 dairy cows that were not pregnant (7 Hn and 5 Mo) for breeding during the following farm year (2007). The lactations of these cows were therefore prolonged (579 days on average), applying the normal drying off rule of the system. Even so, this option made it possible to express average productions of 10388 kg of milk/cow/lactation for the Holstein and 8790 kg of milk/cow/lactation for the Montbeliard, i.e., 96% and 88%, respectively, of the level of production of two lactations that were not prolonged. One of the suspected causes for the deterioration in herd fecundity was a negative energy balance during the breeding period. In fact, the cows' state of fattening was particularly low during their breeding period (from May to August). A precise analysis of 2005 and 2006 data revealed greater success in reproduction for the cows that had started their lactation before the grazing period. The breeding period was brought forward by a month for spring 2007 (Gouttenoire et al. 2010). The reproduction performances were then greatly improved (at least 75% of females incalved in 2007 and 2008), ensuring the long-term survival of the herd. However, the longevity of the cows did not increase in the herd.

Improving Calves' Health in the GS and the MCDS

From 2005 to 2009, some diseases frequently occurred in calves aged 0 to 8 months in the GS and MCDS herds: diarrhoea (frequent occurrences in 2005–2006 and in 2009) and infections of the navel and keratitis. In order to understand the difficulties in treating these illnesses, we analysed the curative care practices and formalised their evolutions over the period under consideration.

The successes, failures and lessons learned in the matter of alternative treatments administered to the calves aged 0–8 months were analysed for all of the health care given to the calves raised in the two systems over the period 2005 to 2009. Over this period, we formalised the occurrence of illnesses and alternative treatments administered to the calves according to timelines. This formalisation was used as a basis for discussion with the ES experimenters for the collection of determinants of practices *a posteriori*.

On the basis of the analysis of practices, we observed factors that accounted for treatments administered to calves aged 0 to 8 months in the GS and the MCDS. These included the systemic character of the management of animal health, the diversity of use of treatments not correlated with the pressure and diversity of the diseases encountered, and the multiplicity of explanatory factors of use of a variety of treatments. The choice of treatments used by the experimenters depends on many factors, including work organisation and decision-making within the calf unit, and the know-how, experience, affinities and power to act of each one. This preponderant place of work organisation and empowerment (responsibility, continuity in the management of the calves, sufficient working time to carry out tasks and address uncertainties, etc.) of those involved in the sound management of a unit was previously demonstrated by Vaarst and Sorensen (2009). For the livestock in the Mirecourt ES, treatments were chosen according to the calf/illness pair. This way to act seems, from the viewpoint of the experimenters involved, consistent with "the philosophy" of use of alternative treatments (homeopathy, aromatherapy and phytotherapy). The long presence or repeated occurrences of the same illness

Fig. 11.3 Step-by-step design centred on long-term experimentation implemented at the INRA ASTER-Mirecourt Experimental Station

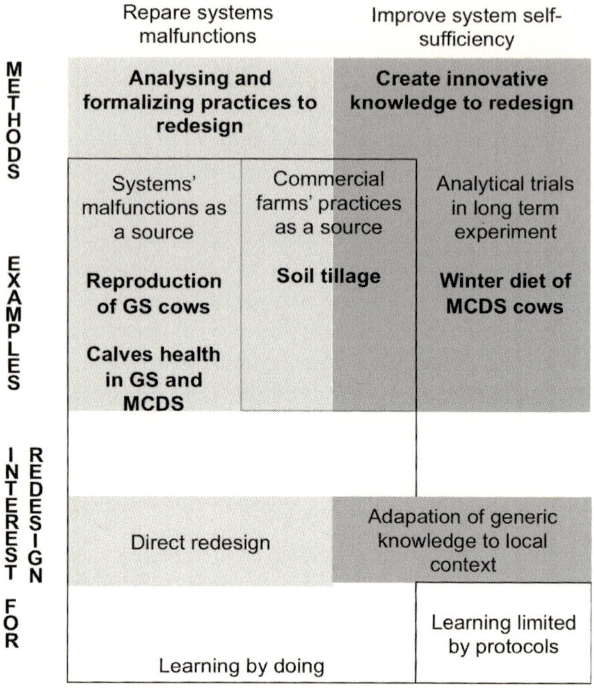

in a calf seems to be a situation that stimulates learning about the use of alternative treatments. Indeed, such a situation leads the experimenters to question the effectiveness of the treatments used and thus the test of new treatments. Nevertheless, it seems that learning about the effectiveness of a treatment is more or less difficult and slow, depending on the illness concerned (e.g., diarrhoea = risk of death for the animal), as well as on the diversity of treatments available on the market to care for the same illness.

The analysis of practices that cause malfunctions within systems can shed light on the adjustments of practices and the modifications of the organisation of systems to be operated to eliminate these malfunctions, promoting the redesign of the systems involved. It can stimulate the creation of technical alternatives (e.g., prolonged lactations) and learning by trial and error in the situation, i.e., consistent (Schwartz 2009) with the system in which these innovations are mobilised. These assessment methods for redesign are particularly interesting in the case of acquiring know-how for the experimenters (Fig. 11.3).

Integrating External Knowledge to Improve System Self-Sufficiency

The methods consist in creating knowledge through analytical trials or studies in commercial farms and attempt to integrate this knowledge into the redesign of the systems in order to improve their self-sufficiency.

Choosing Cereal/High-Protein Plant Mixtures by Carrying Out Analytical Trials in the MCDS

In order to stabilise the type of concentrates in the dairy cows' diets, we wanted to determine the zootechnical interests of several cereal and high-protein plant mixtures (oats/horse beans, triticale/peas, barley/lupine) recognised as being of interest for their agronomic properties in the crop rotations.

The method tested consists of setting up analytical trials aimed at comparing management methods that could potentially be mobilised in the MCDS. These trials are constructed according to experimental trials that are relevant at the statistical level (e.g., Latin square), being careful to test modalities that are not too different in order to limit the risks of backtracking on systems at medium or long time steps. Zootechnical use of three cereal/high-protein plant mixtures in the form of crushed grains (oats/horse beans, barley/lupine and triticale/peas) was tested as a supplement to a diet of alfalfa/orchard grass hay and permanent grassland hay. They were compared during an analytical trial on dairy cows in middle lactation within the MCDS (Coquil et al. 2009b). This trial was carried out in a Latin square composed of three batches of eight dairy cows, balanced according to the breed (50 % Holstein, 50 % Montbeliard) and according to the parity. The animals received 4 kg/cow/day of one of the cereal/high-protein plant mixtures, 8 kg DM/cow/day of alfalfa/orchard grass hay and hay from permanent grassland *ad libitum*. The assessment was centred on the individual performances of the dairy cows: intake, milk production, milk quality and weight of the animals.

The tested diets presented a good nitrogen/energy balance: this ratio significantly varied according to the diet (101 to 106 g PDIN/UFL). The milk production (20.3 kg of milk/cow/day) and the fat content (41.3 g/kg of milk) did not significantly vary according to the diet. The dairy cows that received the oats/horse bean mixture produced milk with significantly lower protein content than the other two mixtures (−0.6 g of protein content/kg of milk). The concentrates slightly contributed to the supply of nitrogen in the diet (22.6 % of PDIN inputs). We thus concluded that in organic mixed crop dairy systems with nitrogen-rich forage such as alfalfa hay, the choice of cultivating associations of cereals and high-protein plants and the choice of the composition of these associations can, as a priority, be rationalised initially on the basis of their agronomic interests.

The redesign of systems by the injection of scientific knowledge from analytical trials seems difficult. Analytical trials, interesting at the level of the creation of scientific knowledge, do not appear to be very favourable for the creation of pragmatic concepts (Pastré 2009) and knowledge that can be mobilised in action. The modalities tested are not adjustable by the experimenters, regardless of the biotechnical and practical consequences observed. Moreover, in the MCDS, the balance between forage, cereal and straw crops and their use by the herds fluctuates depending on the production of the system's territory, making permanent adjustments to the diets of different batches of animals necessary.

Table 11.1 Fuel consumption (l/Ha) and grain yield (t/Ha) per crop for each of the strategies tested

		Yield in grain at 15 % humidity (t/Ha)			Fuel consumption (l/Ha)		
Year	Crops	PPC	NP	AP	PPC	NP	AP
2006–2007	Winter wheat	0.23	1.16	1.08	39	76	65
2007–2008	Winter rye	0.41	2.50	2.06	35	84	93
2008–2009	Spring oats—horse beans	4.23	1.14	3.95	136	52	92
2009–2010	Winter wheat	1.05	0.80	3.29	32	89	100
	Average for 2007–2010	1.48	1.40	2.59	61	75	88

PPC permanent plant cover, *NP* non-ploughing, *AP* agronomic ploughing

Rethinking Cropping Practices by Analysing Pioneer Practices on Commercial Farms

The progressive self-sufficiency in inputs of the MCDS led us to work on the design of farming activity that limits fuel consumption. The main energy consumptions of the MCDS come from cropping interventions within the cropping systems. According to Bochu et al. (2008), soil tilling practices are the primary consumers of fuel in organic cropping systems. To date, very little knowledge can be mobilised to set up lower fuel soil tillage strategies within organic cropping systems.

The method tested consists of collecting, analysing and formalising practices of interest implemented on commercial farms. It consists of adopting innovative practices to encourage the achievement of the objectives of the systems tested at the Mirecourt ES. This approach was mobilised to think about the implementation of low-energy soil tillage practices in the cropping systems of the organic MCDS tested. The data were collected by surveys of 12 organic farmers who were socially recognised for their soil tillage practices without ploughing in predominantly cereal cropping systems. The analysis and formalisation of the information were carried out according to a typology of soil tillage strategies centred on their energy performances. The strategies of interest for the systems of the Mirecourt ES were selected and tested in conjunction with experts from Lorraine and experimenters from the ES (Coquil et al. submitted).

This method made it possible for us to acquire a certain amount of practical knowledge for the implementation of three soil tillage strategies: sowing under permanent plant cover, non-ploughing with systematic intercropping and agronomic ploughing. The lack of experience of the experimenters in the management of "sowing under permanent plant cover" and "non-ploughing" methods leads us to be careful about the biotechnical results. The "non-ploughing" and "sowing under permanent cover" soil tillage strategies tested in the framework of this approach were more economical in fossil energy consumption than the soil tillage strategies with ploughing (Table 11.1), but they were not as productive. This approach came up against the filter of technical paradigms of the different players called on during the collection of interesting practices, the design of tillage strategies by experts and then by the experimenters of the ES. There were difficulties in linking the tillage strategies designed and tested in the MCDS cropping system trial to the technical paradigms on which these strategies are founded and implemented by farmers. These difficulties led us to the conclusion that this approach limits the range of innovation to the technical paradigms of the experts and experimenters, regardless of the innovative character of the strategies mobilised as a source of inspiration (Coquil et al. submitted).

The integration of innovations from practices collected from farmers and tested in systemic trials seems relevant for stimulating the experience of the experimenters of the systems tested (Kummer et al. 2008). Nevertheless, the development of the activity of the experimenters and the discovery of new practices from outside depends on their capacity to grasp practices that come within a variety of technical paradigms.

11.3 Discussion: Step-by-Step Design Seen as a Transition

In this paper, we present a step-by-step approach to the design of self-sufficient mixed crop dairy systems. The step-by-step approach integrates the practical feasibility of the systems. We focused on resources mobilised by experimenters during the progressive design of the systems. Design is seen as a transition where operation and emerging properties of the systems are part of the approach of step-by-step design. This approach, centred here on a long-term experiment, makes the creation of original experimental methods necessary. First, we discuss the contribution of this approach to the scientific community involved in design in agronomical sciences. Second, we discuss the methodological considerations when using a step-by-step approach to design. Third, we discuss the interests of this step-by-step approach for designing adaptive systems.

11.3.1 Step-by-Step Design: Creation of Resources for Transition Towards Self-Sufficiency

A large part of research in the field of design in agronomy aims at proposing stabilised technical solutions (Vereijken 1997; Dogliotti et al. 2003; Sadok et al. 2009; Blazy et al. 2010). This is based on the implicit hypothesis that rules implemented in the designed systems might be applied in farmers' situations. Most design approaches in agronomy are based on the idea that the farmer is a rational actor as defined by Simon (1978) and applied to farming systems by Sebillotte and Soler (1990). The farmer then acts according to a plan that may be designed by a researcher (or by himself), and a designed farming system or technical solution might be a good way to propose innovation to farmers. In methodological terms, most of the design methods in agronomy are based on mathematical modelling or expert knowledge to design virtual systems and to select the best systems to be tested and assess their performances. This is a form of prototyping as defined by Vereijken (1997). In the same field, Meynard (2008) and Mischler et al. (2009) propose an alternative, trying to design evolving technical solutions and systems. Their proposition relies on a step-by-step design considering design as a continuous and iterative improvement of the "plan" by confronting the designed plan to agronomical realities in an experiment or a farm.

In our text, step-by-step design relies on a different epistemic position about the relationship between research and people potentially interested in the pragmatic use of research. Step-by-step design also relies on a different representation of the way a farmer might act. This step-by-step approach considers biotechnical specificities of situations and the singularity of subjects that select technical options in their activity according to their individual feelings. This approach focuses on resources (i.e, prolonged lactations in the GS, change of reproduction period in the GS, work organisation in the calf unit, etc.) mobilised by experimenters and their learning during the progressive design of the systems. It is an anthropocentric approach to designing technical systems. The systems designed are supports considered to be specific to the natural and agronomical situation, and to the experimenters. The aim consists in formalising how experimenters can progressively discover significant information and know-how to progress from a situation (in our case, conventional agriculture) to another (in our case, self-sufficient and organic agriculture). This is based on the hypothesis that formalisation of the progressive discovery of the system by the experimenters can allow farmers to discover their farming systems while changing to more self-sufficient systems. In the step-by-step design, the farmer is considered as an autonomous actor, able to be creative during action (Joas 1999). Step-by-step design aims to develop resources to empower farmers in their search for a new system. From a methodological point of view, the step-by-step approach leaves no room for virtual systems. Systems are designed according to pragmatic considerations that integrate the experimenter's perceptions about his specific system (performances, practical feasibility, etc.).

11.3.2 Methods for the Step-by-Step Design Approach

The approach presented in this article is centred on a long-term experiment at the scale of production systems, referred to as "system experimentation". System experimentation can be conducted in order to assess the technical, environmental and economic performances of farming systems managed according to methods fixed over periods greater than the farming year (Reau et al. 1996, Verloop et al. 2006, Benoit et al. 2009; Delaby et al. 2009). System experimentation can also be a special experimental framework aimed at defining and validating the strategic and operational management of production systems using a multiannual test (Dedieu et al. 2002). In the step-by-step design defined in this paper, we focus on a period of changes for experimenters. To change, they are attentive to the dynamics of their environments and have to improvise and be creative to act on a daily basis, distancing themselves from their past know-how and values that were efficient in conventional farming.

All experimental methods do not stimulate the know-how of experimenters in the step-by-step design in the same way. Analysing practices to explain system malfunctions is particularly interesting to redesign systems and stimulate know-how for the experimenters. This method is efficient to shed light on practices to adjust and redesign systems in order to repair system malfunctions. It stimulates learning in a given situation. Integrating interesting practices into experimental systems by analysing practices on commercial farms (e.g., tillage practices with no ploughing in crop rotations) and testing them in the systems might be interesting to stimulate experimenters' learning. However, the capacity to integrate practices into the system not only depends on the capacity to adapt these practices to a new agronomic situation, but it also depends on the capacity of experimenters to consider practices that rely on different technical paradigms.

It is much more difficult to redesign or to adjust experimental systems by mobilising scientific knowledge resulting from analytical trials. Analytical trials, interesting for creating scientific knowledge, are not relevant for creating new pragmatic concepts that might empower experimenters to act in the system. This could be partly explained by the fact that in an analytical trial, compared modalities must not be adjusted, even if the modality goes wrong from a practical or a biological point of view. Thus, the trial provides information on what to avoid, but does not inform the experimenter as to what he might do in that kind of situation. Moreover, fixed modalities are not really relevant in self-sufficient systems based on permanent adjustments.

In the step-by-step design, creation of efficient know-how relies on methods that enhance the creation of experience that are consistent with the working activity of experimenters (Schwartz 2009)

11.3.3 A Relevant Approach for the Adaptation of Systems to Environmental Changes

The step-by-step design takes evolutions and the contingency of technical systems subject to environmental fluctuations and the evolution of experimenters' experience into consideration.

The approach highlights the resources that can be mobilised by the players at the technical level, according to the identified fluctuations of the system and of its environment. Experimentation over the long-term makes it possible to assess the fluctuations of the ecological and human environments of the system in real time, as well as the biotechnical adaptations built in response. The fluctuations in the performance of dairy cows in the GS are one example of this. Lack of clarity in the distribution of responsibilities to the people working in the calf unit and difficulties in capitalising on experience in alternative treatment methods are another.

The step-by-step design makes it possible to determine where there are gaps in knowledge or systemic problems, and calls for pragmatism in order to resolve these malfunctions and to not compromise the survival of the systems tested. Thus, it is a resource that is relevant to scientific questioning but also to the creation of knowledge relevant for action. For example, the interest of prolonged lactations to compensate for reproduction problems in the GS appeared to solve, on the short-term, systemic problems in 2006.

Nevertheless, this very anthropocentric approach leads us to wonder about how to integrate new objectives and new indicators into system management. On the basis of the past years of experimentation, it seems important to us to transform new objectives into questions and precaution principles that are significant in agronomical and even in agricultural terms. For example, considering the problem of nitrate losses in organic farming rotations, the identification of sources of nitrate leaching (e.g., the turning of temporary grassland, etc.) makes it possible to transform an environmental question, “How can we preserve water quality?”, into an agronomic question, “How can we limit the turnover of grassland on the same catchment area during a farming year?”, and then into a farming question, “How can we reconcile the turnover of grasslands per catchment area with the functioning of the farm?”.

11.4 Conclusion: A Step-by-Step Design as an Anthropocentric Approach to Design Evolving Systems

This paper is a methodological proposition to designing farming systems by considering agricultural activity at the farm scale. Farmer’s activity is considered by developing an anthropocentric approach to designing systems: we assume that farmers can, like the experimenters in this paper, reconsider their farming system step-by-step to reach a more sustainable type of agriculture. The organic agriculture framework is used as a precaution principle to develop self-sufficient systems from the natural properties of the milieu. This legal framework is a pragmatic way to remain self-sufficient.

This experience-based approach aims at producing operating and contextualised knowledge for the design and management of mixed crop dairy systems progressively evolving towards self-sufficiency in the sense of systems that use low inputs and empowerment of the stakeholders involved. Knowledge produced for the design must be consistent with the systemic context in which it will be mobilised and according to methods that generate pragmatic learning for the experimenters who implement it. The analysis of system malfunctions with a view to proposing technical alternatives is particularly effective in this regard. The introduction of new knowledge in order to refine the degree of achievement of the objectives of the systems is more delicate. The difficulties in appropriating and implementing this scientific or expert knowledge in a singular systemic context raise questions and pose difficulties for the creation of knowledge for action. New methodologies that introduce innovative knowledge to develop these systems must be devised to generate pragmatic learning in the research group in terms of technical alternatives that do not come within their own technical paradigm (Coquil et al. submitted).

Nevertheless, the knowledge produced within the framework of this step-by-step design approach needs to be more effectively contextualised. We need to have a better definition of the experimenters' design work space (Broberg 2010). Thus, the step-by-step approach offers an alternative proposition to the generic programming of knowledge. Moreover, the step-by-step approach proposed in this paper recognises qualities in singular knowledge, provided that the context justifying these singularities is made explicit. We propose to make these singularities of the design context explicit by focusing on the processes at play in the evolution of the systems and the resources mobilised to make these systems evolve during the design. We therefore propose to analyse the design approach described in this paper by more deeply characterising the learning and the resources mobilised for learning of the group that controls and operates within these systems. Identified resources relevant to creating experimenters' experience in self-sufficiency might be a good foundation for focusing on exchanges with farmers interested in self-sufficient systems.

Acknowledgments The authors would like to thank the collective of the INRA ASTER-Mirecourt Experimental Station for its active participation in this work. The authors thank Benoît Dedieu, Pascal Béguin and Cécile Fiorelli for their advices during the writing of this paper. The authors would also like to thank two ANR projects for their financial support: DISCOTECH (ANR-05-PADD-04-01) and O_2LA (ANR-09-STRA-09).

References

Auricoste C, Deffontaines JP, Fiorelli JL, Osty PL (1985) Points de vue d'agronomes sur l'évaluation des potentialités des terrains en friches ou en parcours. Bulletin Technique d'Information, 301–306

Béguin P (2010) Conduite de projet et fabrication collective du travail: une approche développementale. Sciences sociales: société, santé, décision. Habilitation à Diriger la Recherche. Université Victor Segalen Bordeaux 2, p 188

Benoit M, Tournadre H, Dulphy JP, Laignel G, Prache S, Cabaret J (2009) Is intensification of ewe reproduction rhythm sustainable in an organically managed sheep production system? A 4-year interdisciplinary study. Animal 3:753–763

Blazy JM, Tixier P, Thomas A, Ozier-Lafontaine H, Salmon F, Wery J (2010) BANAD: A farm model for ex ante assessment of agro-ecological innovations and its application to banana farms in Guadeloupe. Agricult Sys 103(4):221–232

Bochu JL, Risoud B, Mousset J (2008) Consommation d'énergie et émissions de GES des exploitations en agriculture biologique: synthèse des résultats PLANETE 2006. Colloque international: Agriculture biologique et changement climatique. Contribution de l'agriculture biologique et de nos choix alimentaires à l'effet de serre, Lempdes (FRA), p 8

Broberg O (2010) Workspace design: a case study applying participatory design principles of healthy workplaces in an industrial setting. Int J Technol Manag 51(1):39–56

Coquil X, Blouet A, Fiorelli JL, Bazard C, Trommenschlager JM (2009a) Conception de systèmes laitiers en agriculture biologique: une entrée agronomique. Productions Animales 22:221–234

Coquil X, Trommenschlager JM, Bazard C, Despres S, Delaby L (2009b) Valorisation par les vaches laitières de mélanges céréales/protéagineux fermiers dans une ration hivernale. Innovations Agronomiques 4:145–149

Coquil X, Gerber M, Etienne C, Peigné J (submitted). Pioneering by experiencing: agronomic innovation restricted by professional values of designers. The case of tillage strategies in organic agriculture

Dedieu B, Louault F, Tournadre H, Benoît M, De Montard FX, Thériez M, Brelurut A, Toporenko G, Pailleux JY, Liénard G (2002) Conception de systèmes d'élevage intégrant des préoccupations environnementales: contribution d'une expérimentation système en production ovin viande. In: agronomique, I.N.-I.n.d.l.r., Institut de, l.E (eds) 9. Rencontres Recherches Ruminants. Institut de l'Elevage, Paris (FRA), pp 391–394

Delaby L, Faverdin P, Michel G, Disenhaus C, Peyraud JL (2009) Effect of different feeding strategies on lactation performance of Holstein and Normande dairy cows. Animal 3:891–905

Dogliotti S, Rossing WAH, van Ittersum MK (2003) ROTAT, a tool for systematically generating crop rotations. Eur J Agronomy 19(2):239–250

Godard O, Hubert B (2002) Le développement durable et la recherche scientifique à l'INRA. Rapport à Madame la directrice générale de l'INRA. Décembre 2002, p 58

Gouttenoire L, Fiorelli JL, Trommenschlager, JM, Coquil X, Cournut S (2010) Understanding the reproductive performance of a dairy cattle herd by using both analytical and systemic approaches: a case study based on a system experiment. Animal 4:827–841

Joas H (1999) La créativité de l'agir. les éditions du Cerf, Paris, 306

Kummer S, Leitgeb F, Vogl CR (2008) Changes as triggers and as results of farmers' experiments: examples of organic farmers in Austria. 8th European IFSA Symposium, Clermont-Ferrand (France), 413–422

Lamine C, Bellon S (2008) Conversion to organic farming: a multidimensional research object at the crossroads of agricultural and social sciences. A review. Agron Sustain Dev 29:97–112

Legrand P, Fraval A, Laurent C (2002) INRA faced with sustainable development: landmarks for the Johannesburg conference. Dossiers de l'Environnement de l'INRA, p 212

Lopez-Ridaura S, Van Der Werf H, Paillat JM, Le Bris B (2009) Environmental evaluation of transfer and treatment of excess pig slurry by life cycle assessment. J Environ Manage 90:1296–1304

Meynard JM (2008) Produire autrement: réinventer les systèmes de culture. In: Reau R, Doré T (eds) Systèmes de cultures innovants et durables: quelles méthodes pour les mettre au point et les évaluer? Educagri éditions

Mischler P, Lheureux S, Dumoulin F, Menu P, Sene O, Hopquin JP, Cariolle M, Reau R, Munier-Jolain N, Faloya V, Boizard H, Meynard JM (2009) Huit fermes de grande culture engagées en production intégrée réduisent les pesticides sans baisse de marge. Courrier de l'environnement de l'INRA 57:73–91

Osty PL (1978) L'exploitation agricole vue comme un système: diffusion de l'innovation et contribution au développement. BTI 326:43–49

Pastré P (2009) Didactique professionnelle et conceptualisation dans l'action. In: Barbier JM, Bourgeois E, Chapelle G, Ruano-orba JC (eds) PUF Encyclopédie de la formation, p 1236

Reau R, Meynard JM, Robert D, Gitton C (1996) Des essais factoriels aux essais "conduite de culture". Comité potentialités-ACTA-Ministère de l'Agriculture, de la Pêche et de l'Alimentation (DERF)-10 janvier 1996

Sadok W, Angevin F, Bergez JE, Bockstaller C, Colomb B, Guichard L, Reau R, Messean A, Dore T (2009) MASC, a qualitative multi-attribute decision model for ex ante assessment of the sustainability of cropping systems. Agron Sustain Dev 29:447–461

Schwartz Y (2009) Produire des savoirs entre adhérence et désadhérence. In: Beguin P, Cerf M (eds) Dynamique des savoirs, dynamique des changements. Octares, Toulouse, pp 15–28

Sebillotte M, Soler LG (1990) Les processus de décision des agriculteurs. In: Brossier J, Vissac B, Le Moigne, JL (eds) Modélisation systémique et système agraire: Décision et organisation: Actes du séminaire du département de recherches sur les systèmes agraires et le développement (SAD). INRA, St-Maximin, pp 3–117

Sève L (1987) La personnalité en gestation. In: Je, sur l'Individualité, Messidor, Editions Sociales. Cité par Schwartz Y (1992) Une science du sujet singulier est-elle possible? In: Travail et Philosophie. Convocations mutuelles. Première edition (1992) Octares edition, Toulouse

Simon HA (1978) Rationality as process and as product of thought. Am Econ Rev 65:1–16

Thompson PB (1997) The variety of sustainability in livestock farming. Proc. 4th Int. Symp. Liv. Farming Systems, Foulum, EAAP, 5–15

Vaarst M, Sorensen JT (2009) Danish dairy farmers' perceptions and attitudes related to calf-management in situations of high versus no calf mortality. Prev Vet Med 89:128–133

Vereijken P (1997) A methodical way of prototyping integrated and ecological arable farming systems (I/EAFS) in interaction with pilot farms. Eur J Agron 7:235–250

Verloop J, Boumans LJM, Keulen H van, Oenema J, Hilhorst GJ, Aarts HFM, Sebek LBJ (2006) Reducing nitrate leaching to groundwater in an intensive dairy farming system. Nutr Cycl Agroecosys 74:59–74

Part II
Organic Performances

Chapter 12
Using Life Cycle Analysis to Analyse the Environmental Performances of Organic and Non-organic Apple Orchards

Aude Alaphilippe, Sylvaine Simon and Frank Hayer

Abstract Although the conventional farming system (CV) for apple production remains the common practice worldwide, the organic farming system (OF) is becoming increasingly important. Few global assessments of the environmental impacts of organic orchard systems are currently available. In this work, we analyse the weak and strong points of the environmental performance of the growing phase of two organic and one conventional apple orchard, using a pluri-annual dataset from experimental orchard systems located in the Middle Rhone valley in France, with life cycle analysis (LCA). LCA, also referred to as cradle-to-grave analysis, allows a quantitative and global evaluation of an orchard's environmental performance. The analysis was performed using the SALCA (Swiss Agricultural Life Cycle Assessment) method (SALCA-Crop V3.1, adapted for pome fruit) and included relevant impact categories based on characterisation models derived mainly from the EDIP97 and CML01 methods, as well as those developed by Agroscope (ART).

Seven impact categories that included ecotoxicity and human toxicity, as well as energy consumption and other environmental impact categories, were calculated and are discussed here. The OF systems appeared to have less of an impact than the conventional system, considering the surface-based functional unit (ha/year). However, the basic substitution of conventional with organic inputs or mechanised activities was not sufficient to radically improve the overall environmental performance of the orchard systems. These results need several years of full production to be validated.

Keywords Life cycle analysis · Orchard system experiment · Environmental performance · Multi-criteria evaluation

---

A. Alaphilippe (✉) · S. Simon
INRA, UE695 UERI Unité Expérimentale Recherches Intégrées,
Domaine de Gotheron, 26320, Saint Marcel les Valence, France
e-mail: aude.alaphilippe@avignon.inra.fr

F. Hayer
Agroscope Reckenholz Tanikon ART, 8046, Zurich, Switzerland

S. Bellon, S. Penvern (eds.), *Organic Farming, Prototype for Sustainable Agricultures*,
DOI 10.1007/978-94-007-7927-3_12,

12.1 Introduction

In addition to the multi-functional and multi-dimensional aspects of agricultural and food systems (Rickerl and Francis 2004), numerous and various issues have already been addressed in terms of the environmental performances of farming systems. Indeed, concerning agricultural practices, organic farming systems do not only differ from conventional farming systems by the choice of non-synthetic pesticides and fertilisers, but also by the use of a range of alternative methods, e.g., mechanisation. Thus, to estimate the environmental performance of organic production systems, methodologies that evaluate only the ecotoxicity and human toxicity are not sufficient since organic practices may affect environmental categories such as energy use and global warming potential.

Regarding the two studies on the environmental performances of different apple production systems that include organic farming (OF), the results are inconsistent. In Simon et al. (2011), OF had the greatest environmental effects, whereas in Reganold et al. (2001), it was the most sustainable. However, agricultural practices and indicators were not similar in those studies, which may explain such discrepancies in the results. Compared with the indicators used in these two studies, the strength of the Life Cycle Analysis (LCA) methodology is that it has been standardised, since LCA procedures are part of the ISO 14000, ISO 14040:2006 and ISO 14044:2006 environmental management standards. The development of international standards for LCA is an important step for consolidating LCA procedures and methods via their harmonisation, which facilitates its international acceptance as an environmental indicator and allows comparison. Another advantage of the LCA method is that it includes the supply chain and its impacts and therefore makes it possible to quantify possible burden shifts. This point is especially important when comparing conventional and organic systems because, for example, replacing synthetic pesticides by alternative mechanical methods might decrease ecotoxicity but increase energy consumption. Thus, to estimate the environmental performance of organic production systems, methodologies that evaluate only ecotoxicity and human toxicity are not sufficient since organic practices may affect environmental categories such as energy use and global warming potential. To avoid any pollution transfer from one environmental problem to another (Rebitzer et al. 2004), a more global methodology such as LCA that includes different impact categories is required.

This methodology, also referred to as cradle-to-grave analysis, allows an objective and general comparison of the systems analysed (Milà et al. 2006; Mouron et al. 2006). LCA evaluates not only the direct inputs and field emissions, but the emissions from the supply chain of all inputs and the disposal or recycling of the outputs as well, and is suitable for comparing farming systems (Haas et al. 2001). It was originally developed for industrial processes, but was adapted to agriculture and, in particular, to arable cropping systems and animal production several years ago (Cowell and Clift 1997). In the last few years, not only cereals but specialised crops such as apple and tomato as well have been analysed with LCA (Anton et al. 2004; Milà et al. 2006; Mouron et al. 2006). A recent study showed the suitability of LCA to assess and compare the environmental effects of agricultural systems (Xavier and Caldeira-Pires 2004; Bockstaller et al. 2009). The strong points of LCA represented by the SALCA software tool, compared to other methods for assessing the environmental impact of agricultural systems, were the coverage of environmental issues, the inclusion of production factors, the depth of environmental analysis and the avoidance of incorrect conclusions. The main disadvantages concerned feasibility issues such as user friendliness and similarities to existing farming software (Bockstaller et al. 2009). Moreover, when the LCA methodology is used to evaluate different impact categories, it does not give a single result but instead calculates a list of indicators, making it difficult to draw conclusions (Kägi et al. 2008a). This last aspect is often criticised.

The aim of this paper is to give a first insight into the environmental performances of two organic and one conventional apple orchard production system using a pluri-annual dataset (2006–2008) derived from experimental orchard systems still in their first years of fruit setting (planting year: 2005). The comparison will focus on the differences in plant protection practices and related activities, with a holistic approach for seven impact categories (i.e., demand for non-renewable energy resources, global warming potential over 100 years, eutrophication potential, acidification potential, terrestrial and aquatic ecotoxicity potential, and human toxicity potential). This paper addresses several questions:

- How do organic vs. conventional apple orchards compare in terms of environmental performance?
- What are the hot spots for each system?
- Is LCA methodology suitable for such an evaluation?

This paper only gives a first insight into the environmental impact assessment of orchards since the dataset only covers the installation phase of the orchards studied. Evaluation of the full production years is underway and results will be presented later.

12.2 Material and Methods

12.2.1 Life Cycle Inventory Assessment

The design of an LCA study has been outlined in four parts (ISO 14044) and its principles are described in Jolliet et al. (2004). In this section, these four parts will be detailed with emphasis on the specific aspects of apple production.

Goal and Scope

The goal of this LCA study was to compare the environmental impacts of two cultivars under organic-certified management with a conventional production system in their first years of production. The pluri-annual dataset was surveyed in a field trial

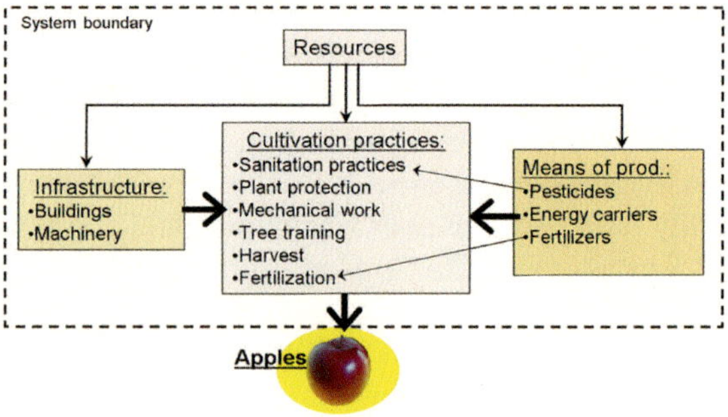

Fig. 12.1 System boundary including all cultural practices and related resources that were assessed for their environmental impacts

located in the Middle Rhone Valley in France. The results of the present LCA were expressed using the functional unit (FU), hectare (ha)/year, reflecting the social perspective of preserving landscape and of sustaining land use. The functional units, kg and €, representing the production and generation of revenues, were not taken into account in this study since the orchards were still not fully productive and their yield was, therefore, not representative.

System Definition and Boundaries

Since our aim was to evaluate the production system only, the system boundary was set at the field gate. As already mentioned, the LCA methodology is a cradle-to-grave analysis that evaluates not only the direct inputs and field emissions, but the emissions from the supply chain of all of the inputs (including transportation) and the disposal or recycling of the outputs and processes required to produce apples and deliver them at the field gate as well (Fig. 12.1). The foreground system included all field operations, fertilisers and pesticides, as well as the direct field emissions and the background system, including the production and maintenance of agricultural inputs, e.g., fertilisers, machinery and pesticides (Appendix 1). Data collected in the experimental design (Sect. 12.2.2) were used to describe the foreground system. The background system was characterised using inventories from the Ecoinvent database v2.01 and the SALCA database (Nemecek and Erzinger 2005), which is the basis for life cycle inventories created by ART and adapted, when necessary, by an INRA/ART joint research unit, but not included in the Ecoinvent database.

Only the field operations that individualise systems were taken into account. Thus, irrigation, orchard planting and removal were not considered here. All processes including storage and activities related to apple commercialisation were also not taken into account in the present environmental evaluation. In this study, the environmental impact of manual labour was not considered at all, as is the case for most LCA studies.

The cycle began after the harvest in the previous year and ended with the harvest in the current production year. The post-harvest activities in autumn were attributed to the next harvest.

This study covered the first 3 years of production corresponding to the harvests of years 2006, 2007 and 2008. An LCA was performed for each production year, but the mean from 2006–2008 is presented here.

Direct Field Emissions

The direct field emissions (NH_4, N_2O, phosphorus, NO_3^-, heavy metals and pesticides) were estimated using the SALCA method (Nemecek and Erzinger 2005; Nemecek et al. 2008; Gaillard and Nemecek 2009). An adapted INRA-ART version of ART's SALCA-crop v3.1 LCA calculation tool was used. The tool consisted of modules programmed in Microsoft EXCEL® and a system implemented in TEAM™ software (version 4.0) (PriceWaterHouse Coopers/Ecobilan, Paris, France). The nitrate-leaching module from SALCA was adapted to apple production in order to better represent the nitrogen (N) uptake by this perennial crop.

Impact Assessment

The SALCA method developed within ART's Life Cycle Group (Gaillard and Nemecek 2009) includes relevant impact categories and mid-point impact assessment methods mainly derived from the EDIP97 (Hauschild and Wenzel 1998) and the CML01 (Guinée et al. 2001) methods, as well as those developed by ART. As in most agricultural LCAs (Hayashi et al. 2006), a mid-point approach was chosen because the main questions in this study raised in the introduction required a separate interpretation of single impact categories. The assessment methods were chosen according to their scientific soundness and their applicability to agricultural LCAs.

The following impact categories were considered:

- Demand for non-renewable energy resources (NRE, in MJ-Eq. ha^{-1} $year^{-1}$) including: direct (diesel and electricity) and indirect (coal, oil gas, uranium) non-renewable energy (Hischier et al. 2009).
- Global warming potential over 100 years (GWP, in kg CO_2-Eq. ha^{-1} $year^{-1}$): the main emissions considered in agriculture are carbon dioxide, methane, nitrous oxides in the air compartment at a global scale, stemming from the use of energy resources, and nitrogen fertilisers (IPCC 2007).
- Eutrophication potential or nutrient enrichment potential (NEP, in kg N-Eq. ha^{-1} $year^{-1}$): enrichment of the nutrients phosphorus and nitrogen in aquatic and terrestrial ecosystems (EDIP97).
- Acidification potential (in kg SO_2-Eq. ha^{-1} $year^{-1}$): aerial emissions of acidifying substances (mainly SO_2, NH_3 and nitrous oxides) (EDIP97).
- Terrestrial and aquatic ecotoxicity potential (TEP and AEP, in 1,4-DCB kg-Eq. ha^{-1} $year^{-1}$): toxic impacts to ecosystems are mainly caused by pesticides and heavy metals (CML01).
- Human toxicity potential (HTP, in 1,4-DCB kg-Eq. ha^{-1} $year^{-1}$): impact of toxic pollutants on human health through aerial emissions (CML01).

The missing characterisation factors for chemical compounds were calculated using the SYNOPS database (Gutsche and Rossberg 1997; Gutsche and Strassemeyer 2007) as a reference, and FOOTPRINT PPDB[1] for data gaps. The pesticide emissions were assumed to be flows into agricultural soil.

12.2.2 Orchard Design and Cultural Practices

Experimental Orchards

Data from an experimental design of an apple orchard were used to perform the present LCA analysis and to describe the foreground system. This experimental design has already been described by Simon et al. (2011). Briefly, this orchard is located in the Middle Rhone Valley in France. It was planted in 2005 over an area of 3.3 ha. Planting density is 1000 trees/ha, with a grass cover between rows. Three plots of 0.37 ha each were assessed:

- The reference conventional plot (CV): Integrated Pest Management guidelines of the French National Apple Board were used to manage the Golden Delicious type scab-susceptible cultivar Golden D. 2832 T®, referred to here as Golden D.
- Two plots under OF management: according to EEC rules (EEC rules 2092/91; 834/2007 (2009)) and French regulations, two cultivars were planted, Golden D. and Melrose, a low-susceptible cultivar to scab (*Venturia inaequalis*) and powdery mildew (*Podosphaera leucotricha*).

The Golden Delicious type can be considered as a reference for the CV production system, with more than 30 % of the French apple orchard surface area planted with this cultivar. Low-susceptible cultivars are relatively more frequently planted in organic compared to conventional or integrated production systems (Sauphanor et al. 2009).

The orchard was assumed to have no drainage, and no slope was considered. The climate was continental with relatively dry weather during the summer due to Mediterranean influences, and approximately 850 $L.m^{-2}$ $year^{-1}$ of precipitation. The soil contains 2 % humus, 15 % clay and the rooting depth is 0.4 m.

Description of Cultural Practices and Activities Within the Orchards

Activities described here are fertilisation, plant protection and within-row management.

Fertilisation. Organic fertilisers (i.e., on farm compost and feather meal) were applied in the OF systems, whereas mineral fertilisers were applied in the CV system. The mean total yearly available nitrogen was 45 kg ha^{-1} and the mean K_2O supply

[1] http://www.eu-footprint.org/ (10.02.2010)

Table 12.1 Plant protection practices: mean number of active ingredients and machinery use of the 3 years (2006–2008)

		CV Golden D.	OF Golden D.	OF Melrose
Active ingredients (except pheromones and biocontrol agents)	Number of applied compounds	24	5.3	5.3
	Insecticides	8	3.3	3.3
	Herbicides	5	–	–
	Fungicides	10	2	2
	Other (thinning, etc.)	1	–	–
	Number of pesticide applications	34.7	28.3	16
Machinery use	Number of herbicide applications	3	–	–
	Number of mechanical weedings	–	6	6

CV conventional, *OF* organic farming

was approximately 65 kg ha^{-1} $year^{-1}$ in all plots. P_2O_5 supply was 30 kg ha^{-1} $year^{-1}$ in OF orchards and 50 kg ha^{-1} in the CV orchard (Simon et al. 2011).

Management of weeds, pests and diseases (plant protection, Table 12.1) was defined by sets of decision rules that were related to the farming system and the associated guidelines, and could vary according to the cultivar, especially for disease management constrained by cultivar susceptibility. The applied compounds are listed in Appendix 1 with their corresponding characterisation coefficients for calculating human toxicity and aquatic and terrestrial ecotoxicity.

In all three systems, more than half of the total pesticide applications targeted scab. The protection against arthropods represented the other half of the total annual chemical applications, mainly against codling moth (*Cydia pomonella* (L.)) and aphids.

Sanitation practices (such as leaf litter shredding or ploughing in for scab management) were performed once to twice a year in the OF systems but not in the CV system.

Within-row and between-row management. Whereas weeds under the trees (within-row) were controlled by herbicides only in the CV system, mechanical and/or occasional manual labour was used in the OF systems. An inter-row disc (Ommas: four tilling and three earth turning discs) was used for mechanical understorey management. Ommas machinery was also used for sanitation practices in the OF systems (to plough in the leaf litter in order to decrease scab inoculum) and to plough in organic fertilisers. Only mechanical labour was used (mostly mulching) between rows.

12.3 Results and Discussion

12.3.1 *Toxicity and Ecotoxicity Impacts*

For both aquatic and terrestrial ecotoxicity, as well as for human toxicity, the two organic systems had a lower impact compared with the CV system used as a reference (Tables 12.2 and 12.3). The steepest decrease was observed for terrestrial ecotoxicity with a decrease of about 93 % in the OF Melrose system compared to CV Golden D. For both human toxicity and aquatic ecotoxicity, the decrease was greater for the orchard planted with the low scab-susceptible cultivar, Melrose, compared to the susceptible one, Golden D.

Table 12.2 Potential aquatic and terrestrial ecotoxicity (1,4-DCB kg-Eq. ha^{-1} $year^{-1}$) of non-pesticide origin (e.g., inputs and machinery use) and pesticide origin (Compounds applied in the three orchards causing either aquatic or terrestrial ecotoxicity) according to target pest for the CV Golden D., OF Melrose and OF Golden D. systems (3-year mean value) and percentage of variation compared to the conventional CV Golden D. farming system

		CV Golden D.	OF Golden D.	OF Melrose
Water ecotoxicity	Non-pesticide	29.5	83.6	80.8
	Herbicides	56	0.0	0.0
	Fungicides	90.9	195.7	142.8
	Insecticides	235.7	0.1	0.1
	TOTAL	412.1	279.4 (−32 %)	223.6 (−46 %)
Terrestrial ecotoxicity	Non-pesticide	1.2	4.1	4.0
	Herbicides	44.3	0.0	0.0
	Fungicides	3.7	7.8	4.6
	Insecticides	79.4	0.2	0.2
	TOTAL	128.6	11.6 (−91 %)	8.8 (−93 %)

CV conventional, *OF* organic farming

Table 12.3 Potential human toxicity (1,4-DCB kg-Eq. ha^{-1} $year^{-1}$) of non-pesticide origin[2] and pesticide origin[3] for the CV Golden D., OF Melrose and OF Golden D. systems (3-year mean value) and percentage of variation compared to the conventional CV Golden D. farming system

		CV Golden D.	OF Golden D.	OF Melrose
Human toxicity	Non-pesticide	521.2	695.2	583
	Herbicides	210.5	0.0	0.0
	Fungicides	17.0	32.0	24.0
	Insecticides	227.7	1.3	1.3
	Other (thinning, etc.)	0.1	0.0	0.0
	TOTAL	976.4	728.5 (−25 %)	608.4 (−38 %)

CV conventional, *OF* organic farming

[2] e.g. input productions and machinery use

[3] compounds applied in the three management systems of apple orchards causing either aquatic or terrestrial ecotoxicity

Table 12.4 Contribution analysis of non-renewable energy (NRE) (in MJ-Eq. ha^{-1} $year^{-1}$) for the CV Golden D., OF Melrose and OF Golden D. systems (3-year mean value) and percentage of variation compared to the conventional CV Golden D. farming system

	CV Golden D.	OF Golden D.	OF Melrose
1 Tree training and harvest	*1647*	*1256*	*1913*
2 Between-row mechanical work	*1829*	*1600*	*1486*
3 Fertiliser production	4605	1305	1305
4 Fertiliser application	*398*	*1055*	*1055*
5 Sanitation practices	*0*	*1037*	*952*
6 Within-row mechanical work	*0*	*1515*	*1515*
7 Pesticide production	2373	533	293
8 Pesticide application (treatment)	*6764*	*5572*	*3147*
Sum of plant protection activities (5,6,7 and 8)	9137	8657 (−5 %)	5907 (−35 %)
Sum of fertilising activities (3 and 4)	5003	2360 (−53 %)	2360 (−53 %)
Sum of machinery contribution (in italics)	*10638*	*12035 (+13 %)*	*10068 (−5 %)*
TOTAL Non-Renewable Energy used (1–8)	17616	13873 (−21 %)	11666 (−34 %)

CV conventional, *OF* organic farming

Tables 12.2 and 12.3 indicate the contribution of pesticide and non-pesticide inputs related to ecotoxicity. The impact for both ecotoxicity categories was mainly caused by direct pesticide field emissions (according to their target: fungicides, herbicides, insecticides and others such as thinning chemicals; see list of compounds in Appendix 1).

The aquatic ecotoxicity (Table 12.2) of the CV system was the highest because of the recurrent applications of highly toxic compounds such as chlorpyrifos-ethyl insecticides. In the OF systems, the use of mineral fungicides such as sulphur and copper also had a strong impact, depending on the amount applied. The scab-susceptible cultivar, Golden D., which received around 100 kg $year^{-1}$ of sulphur (3-year mean value) vs. 38 kg $year^{-1}$ of sulphur in the low-susceptible Melrose cultivar, displayed the highest impact.

Concerning the terrestrial ecosystem, the CV system also had the highest impact once again due to the organo-phosphate insecticide applications.

Unlike aquatic and terrestrial ecotoxicity, human toxicity (Table 12.3) was mainly caused by non-pesticide emissions that occurred in the production and use of buildings and machinery (and not just chemical products).

The difference between OF Melrose and OF Golden D. was mainly due to a higher number of treatments and thus to an increased use of the sprayer with associated emissions. Indeed, the number of treatments varied from 16 (OF Melrose) up to 34.7 (CV Golden D.) (see Table 12.1).

12.3.2 Non-Renewable Energy (NRE) Resources

The contribution in terms of energy use of each cultural practice and input were calculated for each system (Table 12.4). The most demanding input of this impact category was the use of machinery, mostly due to diesel consumption but also to the energy used for the machinery construction, ranging from 10068 to 13873 MJ-Eq. ha^{-1} $year^{-1}$. Plant protection (i.e., pesticide production and application, mechanical practices) was the activity with the highest energy consumption due to either pesticide applications in the CV system or to pesticide applications combined with other mechanical plant protection measures in the OF systems. Crop protection was also the activity that differentiated the modalities according to the system and to the cultivar. Indeed, Golden D. had more of an impact than Melrose since the number of pesticide applications varied from 16 (Melrose OF) to 34.7 (Golden D. CV), corresponding to 3147 and 6764 MJ-Eq. ha^{-1} $year^{-1}$, respectively. In the OF systems, herbicide applications were replaced by mechanical understorey management with an inter-row disc (Ommas). This Ommas was used up to seven times per year and consumed a large amount of energy. Although sanitation practices were not performed in the CV system, overall comparison of the plant protection activities revealed that the energy consumed for the within-row mechanical work was counterbalanced by the higher number of pesticide applications and the amount of active ingredients applied in the CV Golden D. system. Thus, OF systems use less energy for weed control for both within and between-row management.

Fertilisation also differentiated the production systems, with 5003 Eq. ha^{-1}. $year^{-1}$for the CV systems, compared to only 2360 MJ-Eq. ha^{-1}.$year^{-1}$for the OF systems (53 % less energy consumption in the OF systems). The production of mineral fertilisers, a high energy consuming process, represented a large part of the energy consumption for fertilising. The OF systems had a lower energy demand since compost instead of mineral fertiliser was applied, although compost application in the field had a greater impact.

Concerning tree training and harvest, a self-propelled elevator was used, explaining the energy consumption observed.

12.3.3 Global Warming Potential (GWP)

Field operations and emissions from fertiliser production and application are the main sources of global warming potential (Table 12.5). The emissions from fieldwork activities accounted for 52 to 58 % of the total GWP, and fertilisation was responsible for between 39 and 47 %. Although the global warming potential was similar for all systems, the nature of emissions was different between production systems (Fig. 12.2): there was a specific methane emission in the OF system related to the use of organic fertiliser (compost).

Compared to the CV system, the level of CO_2 emissions from plant protection was lower in the OF systems due to a decrease in pesticide use. However, the use of sanitation practices (leaf litter management to decrease winter scab inoculum) and the substitution of chemical treatments by mechanical weeding completely counterbalanced this lower emission. CV and OF systems therefore globally displayed similar levels of CO_2 emissions, but Golden D. had a slightly higher GWP than Melrose.

Table 12.5 Contribution analysis of global warming potential (GWP) (in kg CO2-Eq. ha^{-1} $year^{-1}$) for the CV Golden D., OF Melrose and OF Golden D. systems (3-year mean) and percentage of variation compared with the conventional CV Golden D. farming system

	CV Golden D.	OF Golden D.	OF Melrose
1 Tree training and harvest	*106*	*81*	*124*
2 Between-row mechanical work	*106*	*92*	*92*
3 Fertiliser production	441	443	501
4 Fertiliser application	*25*	*67*	*67*
5 Sanitation practices	*0*	*65*	*60*
6 Within-row mechanical work	*0*	*95*	*95*
7 Pesticide production	80	32	17
8 Pesticide application (treatment)	*415*	*342*	*193*
Sum of plant protection activities (5,6,7 & 8)	494	374 (−24%)	210 (−57%)
Sum of fertilising activities (3 & 4)	466	510 (+9%)	568 (+22%)
Sum of machinery contribution (in italics)	*652*	*742 (+14%)*	*630 (−3%)*
TOTAL Global warming potential (1–8)	1173	1217 (+4%)	1148 (−2%)

CV conventional, *OF* organic farming

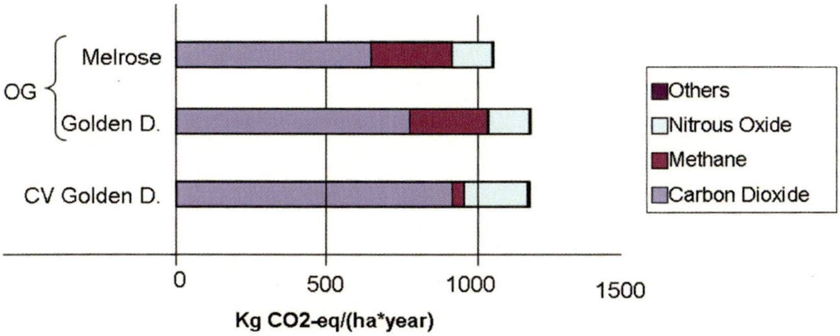

Fig. 12.2 Global warming potential emissions (kg CO_2-Eq. ha^{-1} $year^{-1}$) in the three apple orchard management systems (3-year mean; *CV* conventional, *OF* organic farming)

This scab-susceptible cultivar required more fungicide applications and, therefore, an increased use of the sprayer, inducing more CO_2 emissions (Table 12.5).

12.3.4 Eutrophication Potential

The eutrophication potential was two-fold lower in the OF systems compared to the CV system (Table 12.6). Eutrophication, also referred to as nutrient enrichment potential, was mainly due to fertilising activities: either to direct field emissions for the CV Golden D. system or to both sources of emissions, the production process (maturation) of the compost applied and its associated field emissions for the two OF systems.

Table 12.6 Contribution analysis of nutrient enrichment potential (in kg N-Eq. ha^{-1} $year^{-1}$) for the CV Golden D., OF Melrose and OF Golden D. systems (3-year mean) and percentage of variation compared to the conventional CV Golden D. farming system

	CV Golden D.	OF Golden D.	OF Melrose
1 Tree training, understorey management and harvest	0.47	0.40	0.50
2 Fertiliser production	4.42	2.18	2.18
3 Fertiliser field emissions	8.69	2.94	2.94
4 Fertiliser application	0.08	0.14	0.14
5 Sum of fertilising activities (2–4)	13.18	5.26 (−60%)	5.26 (−60%)
6 All plant protection activities	1.15	1.37 (+20%)	0.96 (−17%)
TOTAL Eutrophication potential (1; 5; 6)	14.80	7.03 (−52%)	6.72 (−55%)

CV conventional, *OF* organic farming

Table 12.7 Contribution analysis of air acidification potential (in kg SO_2-Eq. ha^{-1} $year^{-1}$) for the CV Golden D., OF Melrose and OF Golden D. systems (3-year mean) and percentage of variation compared to the conventional CV Golden D. farming system

	CV Golden D.	OF Golden D.	OF Melrose
1 Tree training, understorey management and harvest	1.38	1.15	1.44
2 Fertiliser production	*3.54*	*5.50*	*5.50*
3 Fertiliser field emissions	*0.01*	*2.27*	*2.27*
4 Fertiliser application	*0.21*	*0.39*	*0.39*
5 Sum of fertilising activities (2–4)	3.77	8.16 (+116%)	8.16 (+116%)
6 All plant protection activities	3.79	6.66 (+76%)	3.92 (+3%)
7 All machinery emissions (including diesel consumption)	*4.78*	*5.37 (+12%)*	*4.52 (−5%)*
TOTAL acidification potential (1, 5, 6)	8.94	15.97 (+79%)	13.52 (+51%)

CV conventional, *OF* organic farming

This impact was mainly due to direct NO_3 field emissions for CV Golden D., and NH_3 for both OF systems. NO_x and phosphorus emissions were also responsible for this effect.

12.3.5 Acidification Potential

The air acidification potential (Table 12.7) was higher in OF orchards since the use of organic fertilisers generally induced emissions during the production processes (corresponding to the maturation phase of compost preparation with a high level of emissions of ammonia, among other compounds) since the compost is ploughed in and only small air emissions occur in the field. Besides, the use of machinery associated with fuel combustion emitted large amounts of nitrous oxides, which strongly determined this impact category. Thus, the increased number of pesticide applications on the scab-susceptible Golden D. cultivar was associated with an increased

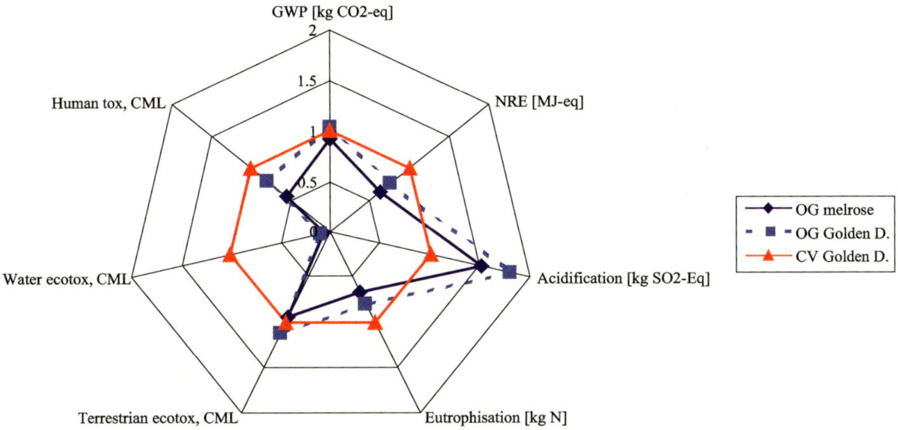

Fig. 12.3 Seven impact categories (3-year mean) of the two organic farming systems (OF Golden D. and OF Melrose) compared to CV Golden D. impacts, which were set to the value of 1 for each calculated impact category (functional unit: ha $year^{-1}$)

impact of the OF Golden D. system compared to the OF system planted with a less susceptible cultivar, i.e., Melrose.

Globally, fertilisation is the main activity responsible for the impact, especially fertiliser production.

12.3.6 Overall Environmental Performance

When globally comparing the orchard-growing phase, the OF systems revealed lower impacts than the CV system (Fig. 12.3), except for the acidification potential category. For all of the six other impact categories, the OF systems had less impact per ha. Apart from the eutrophication potential, the major contributors to the six other impact categories were the construction and use of machinery for field activities, together with pesticide and fertiliser production. The minor contributors were the application of fertilisers, followed by sanitation practices and the activities of tree training, understorey management and harvest.

The introduction of the low-susceptible cultivar, Melrose, improved the overall environmental performance of the organic orchard, compared with the OF system planted with the scab-susceptible Golden D. cultivar. Indeed, the OF Melrose system had much lower toxicity, ecotoxicity and non–renewable energy consumption for plant protection activities compared with the other two systems as a result of a decrease in the number of applications and the amount of active ingredients applied in the orchard. Conversely, the scab-susceptible Golden D. cultivar was intensively treated in both the CV and OF management systems.

12.4 Global Overview and Conclusions

12.4.1 Environmental Performances and Hot Spots for Each System Studied

In our study, the overall environmental performance of the orchard system was not drastically improved when changing from conventional production practices (CV Golden D.) to organic practices with no re-design of the orchard (OF Golden D.). Indeed, the basic substitution of conventional inputs (CV Golden D.) by organic inputs and/or mechanical labour (OF Golden D.) only slightly improved the overall environmental performance. In contrast, the OF Melrose orchard design that combines various means to achieve pest control (including low-susceptible cultivars) reduced the impacts per ha by 93 % (terrestrial ecotoxicity), except for the acidification potential that is increased by 51 %.

This result highlights the important role of the cultivar in orchard design with the aim of developing more environmentally-friendly apple production systems.

Concerning the hot-spots for each production system, we can conclude that, first, the organic farming system decreases the toxicity and ecotoxicity. However, the impact categories related to ecotoxicity are not as low as we could expect because of the large amounts of copper and sulphur applied in the OF systems and the potential aquatic ecotoxicity of these compounds. Second, concerning the substitution of chemical treatments by alternative methods potentially involving increased mechanisation, the results here showed that the higher energy demand due to mechanisation and the increased emissions due to fuel combustion are both counterbalanced by less pesticide application and a smaller amount of applied compounds. Last, concerning nutrient resource management, the use of organic fertilisers increased the acidification potential (compost maturation process), but decreased the eutrophication potential (less risk of nitrate leaching), thus inducing a pollution transfer from one impact category to another one.

12.4.2 Result Perspectives

The orchard system experiment, which reproduces commercial farming conditions in Southern France, will continue to provide data and thus permit a long-term evaluation. Years of full production will be analysed and allow a combined view of each system covering all of the aspects of a multifunctional agriculture, which are: (i) land management (ii) production and (iii) the generation of revenues as described in (Nemecek et al. 2011). Production and revenues are not regarded in this study due to the facts that the orchards were still not fully productive and yields were not representative. As a first estimation in the ECOPHYTO R&D report (Sauphanor et al. 2009), the yield is twice as high in conventional compared with organic systems but, in contrast, the commercial value is more than doubled in the organic farming system compared with the conventional one. Thus, on this basis, it is likely that the

CV system will have less of an impact per kg fresh fruit and will be equivalent to the results per ha of the organic farming systems at the financial level. Further analysis of experimental data will cast light on this conclusion.

12.4.3 Limits of this Study and Suitability of the LCA Methodology

Although the experimental design used for this LCA was not at full production, the management systems for plant protection were similar for the standard conventional and organic systems (Simon et al. 2011) and for older orchards as well because fruit damage had to be taken into account. Two previous studies using LCA to evaluate and compare different apple production systems are available (Milà et al. 2006; Kägi et al. 2008b). The present results are consistent with both papers in which orchards were at full production. The value range of the potential impact corresponded to the one published by Mila et al. (2006). Concerning the comparison between organic and integrated apple production systems, the Endure project reported the same conclusions with an overall advantage of the organic production systems over the integrated production systems when considering the functional unit per ha (i.e., considering the landscape protection function) (Kägi et al. 2008b). Other functions and thus functional units might be used in the future to give a complete picture of the environmental performances of organic farming systems. The results are consistent throughout the 3 years evaluated, but the variability throughout the years must also be analysed, compiling data from the following years.

The results of the first 3 years could therefore provide an initial insight into the environmental performance of these three different plant protection systems. The comparison of orchard systems by means of LCA permitted the identification of some strong and weak points of organic farming strategies, confirming the suitability of the LCA methodology for such evaluations. On this basis, possible improvements to create orchard systems with a lower environmental impact could be proposed, including systems that combine low-susceptible cultivars with alternative methodologies (e.g., sanitation practices, substitution of chemical applications with mechanised activities, etc.).

However, this methodology still requires some adaptation to specific types of productions such as apple orchards. There is a lack of studies and work on apple production, leading to a lack of specific inventories for specific tools, as well as a lack of knowledge to estimate some of the direct emissions such as specific nitrate leaching models or even to calculate some direct potential impacts such as those on biodiversity and on soil quality. Specific models or methods have been developed within the framework of the SALCA method to calculate these direct emissions and impact categories, but are not applicable to orchard farming systems. However, this missing information raises scientific questions to be addressed concerning these specific needs, and might also point out possible optimisation of the farming systems that could be obtained by working and developing knowledge and tools concerning this missing information.

Appendix 1: List of Pesticides and CML Toxicity Coefficient Used for Calculation

Reference: SYNOPS database, see description in material and methods.

Active ingredient	Pesticide class	CML aquatic tox	CML terrestrial tox	CML human tox
Acetamiprid	Cyclic N-compounds	4.17E−03	9.79E−04	1.53E+00
Acetic acid	Pesticide unspecified	9.61E−01	8.17E−02	7.04E+00
Aminotriazole	Cyclic N-compounds	1.32E+00	4.82E−02	9.27E+01
Ammonium thiocyanate	Pesticide unspecified	7.90E+01	5.18E+01	7.04E+00
Azinphos-methyl	Organophosphorus-compounds	3.07E+01	2.04E+01	2.08E+01
Beta-cyfluthrin	Pyretroid-compounds	8.88E+00	1.35E+01	1.35E+01
Bupirimate	Cyclic N-compounds	2.27E+00	2.32E−01	1.26E+01
Bacillus thurengensis	Pesticide unspecified	0.00E+00	0.00E+00	0.00E+00
Calcium chloride	Pesticide unspecified	9.61E−01	8.17E−02	7.04E+00
Captan	Phtalamide-compounds	2.83E−03	1.44E−02	1.72E−01
Carbaryl	[Thio] Carbamate-compounds	1.63E−01	6.87E−02	5.15E+01
Chlorpyrifos-methyl	Organophosphorus-compounds	6.17E+01	2.64E+01	5.02E−01
Copper	Pesticide unspecified	5.55E+01	1.48E+00	9.61E+00
Difenoconazole	Cyclic N-compounds	6.36E+01	2.90E+00	5.59E+01
Dithianon	Nitrile-compounds	4.24E−03	7.42E−03	3.00E−02
Diquat	Bipyridylium-compounds	4.57E+02	5.06E+02	2.06E+03
Dodine	Pesticide unspecified	9,61E−01	8,17E−02	5,04E−04
E, e−8, 10-dodecadiene−1-ol	Pesticide unspecified	0,00E+00	0,00E+00	0,00E+00
Ethephon	Organophosphorus-compounds	1,17E−02	5,86E−03	9,35E−02
Flonicamid	Cyclic N-compounds	1,07E−01	6,57E−03	8,08E+00
Fluazifop-P-butyl	Diphenylether-compounds	7,64E−03	5,49E−03	8,10E+00
Fludioxonil	Nitrile-compounds	4.02E−02	1.17E−02	3.73E−02
Glufosinate-ammonium	Pesticide unspecified	1.55E−03	2.46E−04	1.26E+00
Glyphosate	Organophosphorus-compounds	7.75E−05	9.91E−06	8.90E−03
Granulosevirus	Pesticide unspecified	0.00E+00	0.00E+00	0.00E+00
Isoxaben	Acetamide-anillide-compounds	1.48E+00	8.70E−02	1.24E+01
Kresoxim-methyl	Pesticide unspecified	1.76E+01	9.20E−01	2.62E−01
Mancozeb	Dithiocarbamate-compounds	6.38E−04	9.30E−04	1.08E−02
Oryzalin	Dinitroaniline-compounds	1.45E−1	2.69E−02	1.09E+1
Oxydemeton-methyl	Organophosphorus-compounds	2.42E+01	2.40E+00	1.29E+03
Paraffin (C18–C30)	Pesticide unspecified	4.02E−4	5.2143E−05	0.00E+00
Paraffin (C11–C25)	Pesticide unspecified	2.33E−3	6.33E−3	0.00E+00
Pyridaben	Pesticide unspecified	8.96E+1	1.71E+2	1.05E+2

Active ingredient	Pesticide class	CML aquatic tox	CML terrestrial tox	CML human tox
Phosmet	Organophosphorus-compounds	2.34E−02	6.45E−02	8.91E−02
Pyrimethanil	Acetamide-anillide-compounds	1.03E+02	4.03E+00	8.46E+00
Rotenone	Pesticide unspecified	4.50E−02	3.69E−02	7.04E+00
Spinosad	Pesticide unspecified	3.65E−01	3.58E−02	2.20E−01
Sulfur	Pesticide unspecified	1.04E−1	2.32E−2	0.00E+00
Thiacloprid	Nitrile-compounds	4.64E−03	1.00E−03	1.37E+00
Thiophanate-methyl	[Thio] Carbamate-compounds	2.63E−01	2.44E−02	2.15E+00

References

Anton A, Castells F, Montero JI, Huijbregts M (2004) Comparison of toxicological impacts of integrated and chemical pest management in Mediterranean greenhouses. Chemosphere 54:1225–1235

Bockstaller C, Guichard L, Keichinger O, Girardin P, Galan MB, Gaillard G (2009) Comparison of methods to assess the sustainability of agricultural systems. A review. Agro Sustain Develop 29:223–235

Cowell SJ, Clift R (1997) Impact assessment for LCAs Life Cycle Assessments involving agricultural production. International J Life Cycle Assess 2:99–103

Gaillard G, Nemecek T (2009) Swiss Agricultural Life Cycle Assessment (SALCA): an integrated environmental assessment concept for agriculture. Proceedings of the Conference on integrated assessment of agriculture and sustainable development: Setting the Agenda for Science and Policy (AgSAP 2009), Hotel Zuiderduin, Egmond aan Zee, The Netherlands, 10–12 March 2009, pp. 134–135

Guinée JB, Gorrée M, Heijungs R, Huppes G, Kleijn R, de Koning A, van Oers L, Wegener Sleeswijk A, Suh S, Udo de Haes HA, de Bruijn H, van Duin R, Huijbregts MAJ, Lindeijer E, Roorda AAH, Weidema BP (2001) Life cycle assessment—an operational guide to the ISO standards. Ministry of Housing, Spatial Planning and Environment (VROM) and Centre of Environmental Science (CML), Den Haag and Leiden, Netherlands

Gutsche V, Rossberg D (1997) SYNOPS 1.1: a model to assess and to compare the environmental risk potential of active ingredients in plant protection products. Agr Ecosyst Environ 64:181–188

Gutsche V, Strassemeyer J (2007) SYNOPS—a model to assess the environmental risk potential of pesticides. Nachrichtenblatt des Deutschen Pflanzenschutzdienstes 59:197–210

Haas G, Wetterich F, Kopke U (2001) Comparing intensive, extensified and organic grassland farming in southern Germany by process life cycle assessment. Agr Ecosyst Environ 83:43–53

Hauschild M, Wenzel H (1998) Environmental assessment of products: scientific background. Chapman & Hall, London

Hayashi K, Nakagawa A, Itsubo N, Inaba A (2006) Expanded damage function of stratospheric ozone depletion to cover major endpoints regarding life cycle impact assessment. Int J Life Cycle Assess 11:150–161

Hischier R, Weidema B, Althaus HJ, Bauer C, Doka G, Dones R, Frischknecht R, Hellweg S, Humbert S, Jungbluth N, Köllner T, Loerinick Y, Margni M, Nemecek T (2009) Implementation of life cycle impact assessment methods In: Inventories S.C.f.L.C. (ed), Ecoinvent report, Dübendorf

Jolliet O, Muller-Wenk R, Bare J, Brent A, Goedkoop M, Heijungs R, Itsubo N, Pena C, Pennington D, Potting J, Rebitzer G, Stewart M, de Haes HU, Weidema B (2004) The LCIA midpoint-damage framework of the UNEP/SETAC life cycle initiative. Int J Life Cycle Assess 9:394–404

Kägi T, Bockstaller C, Gaillard G, Hayer F, Mamy L, Strassemer J (2008a) Multicriteria comparison of RA and LCA toxicity methods with focus on pesticide application strategies. 6th International Conference on Life Cycle Assessment in the Agri-Food Sector, Zurich, p. 414

Kägi T, Casado D, Delval P, Gaillard G, Otto S, Strassemeyer J (2008b) Life cycle assessment of integrated and organic apple production systems in Europe. In: Network E (ed), Endure network meeting

Milà i Canals L, Burnip GM, Cowell SJ (2006) Evaluation of the environmental impacts of apple production using Life Cycle Assessment (LCA): case study in New Zealand. Agr Ecosyst Environ 114:226–238

Mouron P, Scholz RW, Nemecek T, Weber O (2006) Life cycle management on Swiss fruit farms: relating environmental and income indicators for apple-growing. Ecol Eco 58:561–578

Nemecek T, Erzinger S (2005) Modelling representative life cycle inventories for Swiss arable crops. Int J Life Cycle Assess 10:68–76

Nemecek T, von Richthofen JS, Dubois G, Casta P, Charles R., Pahl H (2008) Environmental impacts of introducing grain legumes into European crop rotations. Euro J Agro 28:380–393

Nemecek T, Dubois D, Huguenin-Elie O, Gaillard G (2011) Life cycle assessment of Swiss farming systems: I. Integrated and organic farming. Agr Syst 104:217–232

Rebitzer G, Ekvall T, Frischknecht R, Hunkeler D, Norris G, Rydberg T, Schmidt WP, Suh S, Weidema BP, Pennington DW (2004) Life cycle assessment Part 1: Framework, goal and scope definition, inventory analysis, and applications. Environ Int 30:701–720

Reganold JP, Glover JD, Andrews PK, Hinman HR (2001) Sustainability of three apple production systems. Nature 410:926–930

Rickerl D, Francis C (2004) Multidimentional thinking: a prerequisite to agroecology. In: Al-Amoodi L, Barbarick KA, Roberts CA, Dick WA (eds), Agroecosystems analysis. American Society of Agronomy, Inc. Crop Science Society of America, Inc., Soil Science Society of America, Inc., Madison, Wisconsi, USA, pp. 1–17

Sauphanor B, Dirwimmer C, Boutin S, Chaussabel A-L, Dupont N, Fauriel J, Gallia V, Lambert N, Navarro E, Parisi L, Plenet D, Ricaud V, Sagnes J-L, Sauvaitre D, Simon S, Speich P, Zavagli, F (2009) ECOPHYTO R & D vers des systèmes de culture économes en produits phytosanitaires. INRA, p. 68

Simon S, Brun L, Guinaudeau, J., Sauphanor B (2011) Pesticide use in current and innovativbe apple orchard systems. Agro Sustain Develop 31:541–555

Xavier JHV, Caldeira-Pires A (2004) Potencial use of Life Cycle Assessment (LCA) methodology for characterization of agriculture environmental impacts. CC & T, Cadernos de Ciencia & Tecnologia 21, 311–341

Chapter 13
The Potential of Organic Agriculture to Mitigate the Influence of Agriculture on Global Warming—A Review

Adrian Muller and Claude Aubert

Abstract The biggest mitigation potential of agriculture lies in soil carbon sequestration. The most promising practices for this, such as the use of legume leys and organic fertilisers, are common to organic agriculture, thus suggesting considerable mitigation potential for this farming system. However, mitigation in agriculture needs to be assessed beyond the level of single farming practices. This is best illustrated with the issues of fertility management and animal husbandry. Optimisation of fertility management necessitates optimisation of the soil-fertiliser system as a whole and, thus, the assessment of the links between crop rotations, fertiliser types, tillage and soil carbon sequestration. Optimisation of animal husbandry requires a global view, accounting for life-cycle emissions of feed production. Feeding roughage leads to lower life-cycle emissions. Most effective, however, is a drastic reduction of the number of animals, which necessitates consideration of aspects beyond agriculture. Wider societal changes such as dietary changes to reduce meat consumption or behavioural changes to reduce wastage are necessary. Organic agriculture is well positioned to mitigate climate change in such a systemic context. However, addressing mitigation in agriculture leads to some change of thought in conventional agriculture. By acknowledging the essential role of soil carbon sequestration, for example, systemic ideas have gained increasing importance in conventional agriculture. This development should be used to move towards a global approach to sustainable systemic and multifunctional agriculture.

Keywords Carbon dioxide · Carbon sequestration · Climate change · Global warming · Greenhouse gas · Methane · Mitigation · Nitrous oxide · Organic agriculture

A. Muller (✉)
Research Institute of Organic Agriculture (FiBL), Ackerstrasse 113,
Postfach 219, 5070 Frick, Switzerland
e-mail: adrian.mueller@fibl.org

C. Aubert
Lempdes, France
e-mail: aubertcl@wanadoo.fr

S. Bellon, S. Penvern (eds.), *Organic Farming, Prototype for Sustainable Agricultures*,
DOI 10.1007/978-94-007-7927-3_13,

13.1 Climate Change Mitigation and Agriculture

Globally, greenhouse gas (GHG) emissions have to be cut or offset by more than 50% of the 1990 levels by 2050, i.e., more than by 60% from the 2009 levels. Given that the largest share of these reductions should be borne by industrial countries, this amounts to reductions of 80–95% from the 1990 levels by 2050 for those countries in question (Rogelj et al. 2010a, b; den Elzen et al. 2010; ENB 2010; Meinshausen et al. 2009; UNFCCC 2009, IPCC 2007; Schellnhuber et al. 2006; EC 2005). These reduction goals are motivated by the necessity to avoid "dangerous" climate change, which is largely defined as irreversible processes such as the meltdown of the polar ice caps, a shutdown of the Gulf Stream, high sea-level rises and other gradual effects with detrimental consequences for many nations. Avoiding dangerous climate change requires preventing an increase in average global temperatures of over 2 °C by 2100. The emission reduction targets presented above would still lead to only a 50% chance of keeping an increase in global warming to below 2 °C. This illustrates the truly drastic emission reduction efforts needed to achieve at least some results in mitigating climate change.

Agriculture accounts for 10–12% of total global greenhouse gas emissions (Bellarby et al. 2008). When including emissions from land-use change such as deforestation to gain additional cropland, this percentage rises to more than 30%. Between 1990 and 2005, global agricultural nitrous oxide (N_2O) and methane (CH_4) emissions increased by 17% and are expected to considerably rise further in a business-as-usual scenario (50% by 2030) (Smith et al. 2007). The largest contributions are therefore N_2O emissions from fertilised soils (2.1 Gt CO_2 equivalents (CO_2e)/yr), CH_4 emissions from cattle enteric fermentation (1.8 Gt CO_2e/yr) and CO_2 emissions due to soil carbon losses from deforestation and other land-use change (5.9 ± 2.9 Gt CO_2e/yr). Further contributions stem from the burning of biomass waste and crop residues (0.7 Gt CO_2e/yr, N_2O and CH_4) and rice cultivation (0.6 Gt CO_2e/yr, methane). Finally, there are the N_2O and CO_2 emissions due to synthetic fertiliser production (0.4 Gt CO_2e/yr, including energy use), CH_4 and N_2O emissions from manure management (0.4 Gt CO_2e/yr) and CO_2 emissions from fuel use for irrigation and farm machinery (0.5 Gt CO_2e/yr). All estimates are from Bellarby et al. (2008) and refer to data modelled for 2005, based on 2000 and older data (USEPA 2006). These are also the ones used in the fourth assessment report of the Intergovernmental Panel on Climate Change (IPCC) (Smith et al. 2007). Framed differently, agriculture accounts for about 60% of total global N_2O emissions, about 50% of CH_4 emissions and 90% of non-fossil fuel-related CO_2 emissions (including land-use change and deforestation) (USEPA 2006). Thus, emissions from agriculture (including land-use change and deforestation) are second to emissions from fossil energy use (the latter accounting for about 60% of total global emissions) (USEPA 2006). Combining this data on the contribution of agriculture to global warming and the mitigation measures necessary to curb climate change to a tolerable level clearly shows that significant mitigation efforts in agriculture are unavoidable.

In this review, we focus on the mitigation potential of organic agriculture and how it may be improved whenever possible. Despite the importance of climate change mitigation and the dominance of this issue in the political debate, it has to be kept in mind that agriculture is not primarily concerned with mitigation. Agriculture is concerned with food production and food security. Furthermore, conserving production capacity, fertile soils, ecosystem health, water bodies, etc., is also of key importance. Socio-economic aspects are of similar importance, particularly in the Global South where a large percentage of the population directly depends on agriculture for their livelihoods. Agriculture thus needs to be concerned with sustainable food production in its entirety.

In approaching this issue, we first (Sect. 2) provide details on the mitigation potential of agriculture and organic agriculture, in particular, and discuss how organic agriculture may further improve its mitigation potential. In Sect. 3, we point out the importance of a systemic approach to mitigation in agriculture. Sect. 4 draws some general conclusions, emphasizing the importance of going beyond sectorial boundaries when addressing mitigation in agriculture in order to account for the multifunctionality of agriculture and to rethink the framing of organic agriculture and its positioning in relation to conventional agriculture.

13.2 Mitigation of Greenhouse Gas Emissions and Soil Carbon Sequestration in Agriculture

Mitigation in agriculture is usually linked to certain practices that can be employed in conventional and organic systems. Those practices are summarised in Table 13.1 below. Assessments of the whole farming system are thus based on a presence or absence of these practices, and many of them are core aspects of organic agriculture. However, they can also be used in conventional agriculture. We thus often only illustrate the mitigation potential of a certain practice and not the potential of organic in comparison to conventional agriculture. We structure this discussion by differentiating between the various GHGs and then discuss soil organic carbon (SOC) sequestration. A general overview on mitigation in agriculture is provided by Smith et al. (2008). An overview that specifically addresses mitigation in sustainable agriculture, including policy aspects, is provided by Muller et al. (2011). Recent reviews on organic agriculture and climate change are provided by El-Hage Scialabba and Müller-Lindenlauf (2010) and Lynch et al. (2011), the latter focusing on energy use. We also point out that many of the mitigation needs identified for agriculture in the IPCC Fourth Assessment Report (2007) can be met by common practices of organic agriculture as outlined in Niggli et al. (2009).

Table 13.1 Mitigation measures and their mitigation potential, as discussed in the text; we also refer the reader to the assessment of mitigation options in Smith et al. (2008)

Measure		Mitigation effect	Source	Notes
Fertilizer and biomass management	Avoided use and production of synthetic fertilizers	1–10 kg CO_2e per kg N	Wood and Cowie 2004; Snyder et al. 2007	Production emissions from organic fertilizers have to be accounted for e.g. compost production 0.1–0.2 t CO_2e/t waste, IPCC 2006, vol. 5, Chap. 4
	Reduced fertilizer N input	1–2 % N_2O-N from N	Bouwman et al. 2002; Alluvione et al. 2010; IPCC 2006, vol. 4, Chap. 11	Large variations in dependence on site characteristics, fertilizer type, etc. in particular urea: 0.7 t CO_2e per t urea applied,IPCC 2006, vol. 4, Chap. 11
	Optimised manure management	− 1/3 to 3/4 emissions	Pattey et al. 2005; Vanotti 2008	
	Optimised compost production	Reduced N_2O	Dias et al. 2010	By addition of bulking material
	Avoided burning of biomass residues	0.08 t CO_2e/t residue	IPCC 2006, vol. 4, Chap. 2	
	Biogas production (methane capture)	No emissions besides physical leakage		
Soil management	Use organic fertiizers	Increases soil organic carbon	Leifeld and Fuhrer 2010, (Gattinger et al. 2012)	General notes: higher Soil-C levels correlate with higher N_2O emissions (Bouwman et al. 2002, Li et al. 2005) and reduced energy use for tillage (less dense soil structure) and irrigation (higher water holding capacity)
	Optimized crop rotations	Incr. soil org. C: 0.8 t CO_2e/ha/y	Smith et al. 2008; West and Post 2002	More research is needed, but the tendency seems robust (Gattinger et al. 2012)
	Use of legumes	Increases soil organic carbon	Smith et al. 2008	
	Reduced tillage	Increases soil organic carbon	Smith et al. 2008	
	No tillage	Incr. soil org. C: 2 t CO_2e/ha/y	West and Post 2002	The effects of no tillage on N_2O emissions are unclear, Smith et al. 2008. The interaction between reduced tillage and optimized crop rotations seem to further increase soil organic carbon, West and Post 2002 reduced/no tillage is a viable option for organic agriculture, if weeds and diseases can be controlled, Peigne et al. 2007

Table 13.1 (continued)

Measure		Mitigation effect	Source	Notes
	Avoided soil compaction	Reduces N_2O emissions	Bhandral 2007; Bouwman et al. 2002	
	Agroforestry	Incr. soil org. C: 3–8 t CO_2e/ha/y	Mutuo et al. 2005, Albrecht and Kandji 2003	
	Plantation of hedges	Increases soil organic carbon		
	Permanent grass cover (e.g. in vineyards and orchards)	Increases soil organic carbon		
	Pasture instead of cropland	Increases soil organic carbon	Guo and Gifford 2002	Has to be seen in a larger context of changed production patterns (e.g. fewer animals, cf. below, and those on pastures without concentrate feed)
	Biochar	Increases soil organic carbon	Sohi et al. 2009	More research is needed
Animal husbandry	4–5 % of lipids as feed additives	Methane emission reduction by 15–20 % or more	Martin et al. 2010	May has adverse effects on animal health and welfare
	High concentrate instead of roughage	− 1/3 methane emissions	Shibata and Terada 2010	May has adverse effects on animal health and welfare
	Avoided use of concentrate feed	Avoids deforestation/land use change and corresponding soil carbon losses		
	Increased longevity of dairy cows	− 13 % emissions by doubling the number of lactations	O'Mara 2004	
	Increased productivity: higher milk yields per animal			Has adverse effects on animal health and welfare

Table 13.1 (continued)

Measure		Mitigation effect	Source	Notes
	Increased productivity: faster growth of meat animals			Has adverse effects on animal health and welfare
Energy use	No heated greenhouses	Reduces emissions		Consumer aspects need to be accounted for (seasonal consumption)
	Energy efficient machinery	Reduces emissions		
	Optimised machinery use	Reduces emissions		
	No use of synthetic biocides	Reduces emissions		Account for production emissions of other biocides
	Pest-resistant varieties with less spray cycles	Reduces emissions	Consumer aspects need to be accounted for acceptance of other varieties)	
	Provision/use of bioenergy	Reduces emissions		
Systemic	Well-managed combined animal-grassland systems	Climate neutral	Soussana et al. 2010	
	Reducing number of ruminants	Reduces emissions	Carlsson-Kanyama and Gonzalez 2009	Consumer aspects need to be accounted for (eating less meat)
	Switch from ruminants to monogastric animals (Pigs, poultry)	2–5 more efficient feed protein in meat protein conversion		
	Switch to organic	Incr. soil org. C: 1-1.7 tCO_2e/ha/yr	Gattinger et al. 2012	Low sequestration rates are for closed systems without import of organic matter; these sequestration rates show a saturation dynamic and decrease over time.

13.2.1 Nitrous Oxide Emissions

Quantifying N_2O emissions from agriculture is notoriously difficult. We can, however, have the most confidence in estimates of GHG emission reductions due to *avoidance of synthetic fertiliser use* in organic agriculture. Synthetic fertiliser production is based on standardised industrial processes with correspondingly standardised emissions and quantification. Nevertheless, emission factors from fertiliser production can vary considerably, even for the same chemicals (due to different plant efficiencies and abatement measures). Not using synthetic fertilisers thus contributes to a reduction of between 3 and 7.1 kg CO_2e per kg N for ammonium nitrate and about a third of this for urea (Wood and Cowie 2004). This translates into per hectare numbers of synthetic fertiliser use of 0.05–0.35 t CO_2e (for an application of 50 kg N/ha) to 0.25–1.75 t CO_2e (for 250 kg N/ha). Emission factors for other fertilisers have ranges within the same order of magnitude. Transport emissions from these fertilisers need to be taken into account as well, but quantification is difficult. Snyder et al. (2007) report somewhat higher values when transport is included (e.g., 9.7 kg CO_2e per kg N for ammonium nitrate).

Avoiding the burning of biomass avoids the corresponding N_2O (and methane) emissions from this incomplete and inefficient burning process. In organic agriculture, burning of biomass waste and agricultural residues is prohibited, whereas it is common practice in conventional agriculture (e.g., pre-harvest burning of sugar cane and on-field burning of crop residues after harvest). Some quantification can be based on IPCC default values (0.06 and 0.02 t CO_2e per t agricultural residues for N_2O and methane, respectively) (IPCC 2006, vol. 4, Chap. 2). Besides its mitigation potential, avoiding the burning of biomass and agricultural waste has additional benefits regarding nutrient recycling since it can be used for compost production or mulching, thus replacing synthetic fertilisers.

The greatest contribution of N_2O to overall GHG emissions is via N_2O *emissions from fertilised soils* (Van Groenigen et al. 2010). Those emissions are the most difficult to quantify since they are highly dependent on local soil and weather conditions (temperature, precipitation, soil humidity and changes thereof, etc.), and optimal quantification would require continuous measurement (Bouwman et al. 2002). In addition, these emissions are highly dependent on the type of fertiliser applied, on soil characteristics, on crops planted and crop rotation characteristics, etc. Detailed information on how these parameters influence emissions, however, is lacking for many of these parameters, and default values are by far too coarse to adequately illustrate the situation within a specific site. Depending on variations in these parameters, these emissions can vary by factors of 2 to 4 (Bouwman et al. 2002).

An important result is the fact that N_2O emissions generally appear to correlate with *nitrogen input.* Reduced nitrogen input thus leads to reduced N_2O emissions from fertilised soils. The meta-analysis of Bouwman et al. (2002) reveals that this effect is particularly pronounced when shifting from high nitrogen inputs to medium levels. Rough translation of their emission values to CO_2e results in 0.5 t CO_2e/ha at 50–150 kg N input/ha to 2 t CO_2e/ha at more than 250 kg N input/ha (with 1 t CO_2e/ha at 200 kg N input/ha, this corresponds to 1–2 % N_2O-N from N inputs). The relationship is therefore not linear. The correlation between N input and N_2O emissions is also the basis for the IPCC recommendation for a simplified global emission factor that does not differentiate between fertiliser type or any other characteristic and that is set at 1 % of the total nitrogen applied if more detailed information is lacking (IPCC 2006, vol. 4, Chap. 11; however, this factor also applies to N from legumes where the fixation process itself does not generate emissions). An average value of 1–2 % N_2O-N from N inputs thus seems to be a reasonable emission factor from fertilised soils (see Petersen et al. 2006), but the spread and the dependence on other parameters are huge.

Since organic agriculture has lower nitrogen input levels than conventional agriculture, it also has correspondingly lower N_2O emission levels per ha. On the other hand, synthetic fertiliser application can also be optimised regarding the quantity applied, timing, type of fertiliser used and type of application (e.g., the "4R" approach: right source, rate, time, place) (IFA 2009). Due to the potential to control the nutrient release dynamics of synthetic fertilisers, such optimisation has considerable potential to lower the nitrogen applied in conventional agriculture. Canada and the US, for example, base N_2O emission reduction strategies on this (GoA 2010; MSU-EPRI 2010). Finally, we point out that reducing N inputs correlates with yield reductions. Assessing emissions on a per unit output basis, there is an optimum N input level where those emissions are lowest and yield reductions that further offset emission reductions from lower N inputs if calculated on a per output basis (Van Groenigen et al. 2010). It is another question, however, whether such an assessment per unit output is the adequate approach for comparing sustainable agricultural production systems (see Sect. 4).

Regarding fertiliser types, organic and ammonium-based synthetic fertilisers seem to have higher N_2O emissions than nitrate-based synthetic fertilisers (Bouwman et al. 2002). Of particular interest for organic agriculture are emissions from compost and legumes. A recent case study from northern Italy (Alluvione et al. 2010), for example, indicates that emissions from compost use are by far lower (by a factor of more than 20) than emissions from legumes or synthetic fertiliser, for which they report similar emissions. The high emissions of legumes can be due to fast decomposition of the fresh plant residues incorporated in the soil. On the other hand, this study reports high CO_2 emissions from compost application, which can be treated as zero since they are emissions resulting from renewable biomass use. It is, nevertheless, not clear how much N_2O emission occurs during the composting process since this is highly dependent on the source material and process management. Adding bulking material such as biochar or sawdust, for example, can reduce N_2O emissions from optimal compost production (see, e.g., Dias et al. 2010). Global default emission values are available from the IPCC for composting (0.2 t CO_2e/t waste treated on a dry weight basis and 0.1 t CO_2e/t waste treated on a wet weight basis) (IPCC 2006, vol. 5, Chap. 4). However, for soil emissions, these values are subject to a high degree of uncertainty.

N_2O emissions from soils are also influenced by *management techniques*. Heavy machinery and corresponding soil compaction increases N_2O emissions (Bhandral

2007; Bouwman et al. 2002). In organic agriculture, because of the generally high organic matter content and biological activity, the soils are, in most cases, less compacted than in conventional agriculture. On the other hand, high soil carbon contents seem to correlate with higher N_2O emissions (Bouwman et al. 2002; Li et al. 2005), but a high degree of uncertainty still exists (Smith et al. 2008).

Finally, somewhat tangentially, we point out that some of the other emissions related to fertiliser application, namely CO_2 emissions from urea application, may be quantifiable with less uncertainty than soil N_2O emissions. They are given by the chemical characteristics of urea and amount to 0.7 t CO_2e per t urea applied (IPCC 2006, vol. 4, Chap. 11). Nevertheless, these emissions also depend, to some extent, on soil dynamics.

A general pattern governing emissions from organic matter such as compost (as well as manure and slurry) and in soils is the amount of available oxygen. For biomass under anaerobic conditions, nitrous emissions are negligible and potentially large methane emissions occur, whereas the opposite tends to be true under aerobic conditions. Aeration decreases and even eliminates methane emissions, while it tends to increase N_2O emissions. This trade-off is relevant for manure and slurry management, in particular, where methane-avoidance practices can lead to increased N_2O emissions (see Sect. 2.2. below). It is not important for manure application, although when methane emissions are low, creating an aerobic situation, techniques such as the use of drag hoses can considerably reduce N_2O emissions (Lovanh et al. 2010).

To sum up, we emphasize the following points. Avoided the burning of biomass waste in organic systems clearly reduces N_2O emissions, while biomass burning is widely used in conventional systems in the Global South. The avoidance of synthetic fertilisers in organic agriculture also leads to reduced emissions. On the other hand, the emissions from organic fertilisers and legume incorporation should instead be quantified on a life-cycle basis for an encompassing emission assessment for all fertiliser types. Reduced nitrogen inputs clearly lead to lower emissions from soils, and organic farms tend to have lower nitrogen inputs and, thus, additional mitigation benefits. Nitrogen application on conventional farms is, however, continuously optimised with corresponding emission reductions. In consequence, the relative advantage of organic agriculture with regard to N_2O emissions from fertile soils tends to be reductive.

13.2.2 Methane Emissions

The processes that lead to methane emissions in agriculture are usually somewhat simpler to describe and less dependent on local and short-term variations than those for N_2O emissions. As for N_2O, burning of biomass waste and crop residues is a source of methane emissions and by avoiding it, we also avoid the corresponding emissions (see above).

Most important, however, is *enteric fermentation in ruminants.* Reduction of these emissions is possible by using feed additives, by optimising the composition and type of feed, by increasing the longevity and productivity of the animals, or by reducing the number of animals. Feed additives also require further research regarding their effectiveness and, in particular, regarding their effects on animal health.

There are some technically promising feed additives such as fatty acids. Adding 4–5 % of lipids to the feed, for example, can reduce emissions by 15–20 % and even considerably more (Martin et al. 2010). Another positive effect of this addition is to increase the content in omega-3 fatty acid of dairy products and meat. Such additions may reduce feed digestibility and thus productivity, and the economic viability, as well as the long-term effects on animal welfare, still need to be investigated (Sejian et al. 2010). Feed additives that have the characteristics of antibiotics or other drugs are clearly problematic. For a short overview of some options, see Smith et al. (2008).

The type of feed provided (concentrate feed vs. roughage) has a considerable effect on emissions (Shibata and Terada 2010), resulting in a third less emissions from high concentrate than from roughage-rich feed, which seemingly puts organic agriculture at a disadvantage. However, this issue has to be addressed on a global scale. The production of some concentrate feed, mainly soy cake, goes hand-in-hand with heavy land-use change and deforestation in the tropics for the production of soy, thus leading to huge carbon losses from the corresponding soils and forests that can offset the reductions in methane emissions from enteric fermentation due to concentrate feed (Hörtenhuber et al. 2010).

Another mitigation approach is based on specific breeding goals. An option is to increase the longevity of dairy animals, thus shortening the unproductive in relation to the productive phase and reducing the emissions per litre of milk. As the estimate of O'Mara (2004) shows, this can reduce emissions by 13 % following a doubling of the average number of lactations from 2.5 to 5. There is also the option to use other cattle races that can deliver both milk and meat (so called dual-purpose breeds), thus reducing emissions per kg output by considerably increasing output per animal (since both meat and milk can be used). Both these options can be carried out without adverse effects on animal welfare, and they ideally fit organic production systems. However, for both options, further research is needed. Another option is to further increase productivity of the animals. This would clearly lead to emission reductions per litre of milk and per kg of meat, but would also lead to further increased health problems for the animals, which are already bred to a level of specialisation that considerably interferes with animal welfare.

The most effective and sustainable measure thus remains a reduction in the number of ruminant animals. This clearly means a reduction in consumption of meat and dairy products and thus touches on considerations beyond the agricultural production sector, necessitating changing consumption patterns. Nevertheless, given the huge reduction potential of this strategy (e.g., the emission of 1 kg CO_2e allows the production of 160 g whole wheat protein but of only 10 g beef protein) (Carlsson-Kanyama and Gonzalez 2009), the absence of problems related to animal welfare and the effects it would have on land use and deforestation, this issue must be addressed in terms of mitigation in agriculture. A reduction of animal numbers is very well possible in organic agriculture where stocking rates per ha tend to be lower than in conventional agriculture. Optimised mixed farming systems that combine crop production and grazing with animal husbandry have considerable potential for closed nutrient cycles. Specifically, it has been found that a well-managed combined animal-grassland system can be essentially climate-neutral (Soussana et al. 2010; Lynch et al. 2005).

In contrast to ruminants,*monogastric animals* (mainly pigs and poultry) do not emit methane by enteric fermentation but only from the fermentation of manure. Emissions per kg of meat are therefore considerably lower (Gonzalez et al. 2011). Moreover, they are much more efficient than ruminants for transforming plant protein into animal protein. Ruminants have to eat about 25 kg of plant protein to produce 1 kg of meat protein, whereas pigs need about 10 kg and poultry around 5 kg (Smil 2002). On the other hand, monogastric animals cannot digest roughage and mainly eat concentrate feed like grain and soy cake, which leads to N_2O emissions during production and potential soil carbon losses if deforestation for land-use change is involved.

As already mentioned above, there is some mitigation potential in *manure management* regarding collection, storage and application. Methane emissions mainly stem from the storage phase. The trade-off mentioned above between aeration as a treatment to reduce methane emissions and N_2O emissions that increase with aeration is seen in solid manure, for example, where composted, aerated manure has lower emissions than stacked manure (less methane by a factor of 5 but 1.5 times more N_2O, leading to a total reduction of 1/3) (Pattey et al. 2005). This pattern also applies to liquid manure where methane emissions can be drastically reduced by aerobic treatment, whereas this can strongly increase the emission of N_2O (IPCC 2006, vol. 4, Chap. 10). Optimised systems can, however, considerably reduce both methane and N_2O. Vanotti et al. (2008), for example, describe pit storage of liquid swine manure in comparison to separation, composting of the solid part and sequential aeration of the liquid part, which reduces methane emissions by almost 100% and N_2O emissions by 75%. When assessing such emission factors, it also has to be kept in mind that the processes in manure are highly complex and depend on many specific situation parameters such as temperature, etc. (IPCC 2006, vol. 4, Chap. 10).

Techniques using methane capture with subsequent flaring or biogas use are most promising to reduce these emissions. Since anaerobic conditions are maintained, no additional N_2O emissions occur.

Finally, we mention *wet rice* as a specific single important crop for methane emissions. Rice is usually grown in flooded conditions due to weed management advantages. Permanently flooded rice leads to methane emissions through the anaerobic microbiological decomposition of soil organic matter from the flooded soils. This is a particular challenge for organic agriculture where only organic fertilisers are applied. An option to reduce these emissions is the switch to temporary flooding where the reliable quantification of the mitigation potential requires additional measurements and research, particularly since N_2O emissions will increase in partially flooded systems (Wassmann et al. 2000; Wassmann and Dobermann 2006). More research is also needed on how such partially flooded systems perform in organic agriculture and, in particular, on optimal pest and weed management.

The above can be summed up as follows. Avoiding biomass burning reduces methane emissions. The potential for optimised manure management, storage and application is still subject to uncertainty, in particular due to the trade-offs between methane and N_2O emissions. Optimised systems with net reductions nevertheless seem possible. A reduction is clearly achieved by capturing methane and using it as biogas. For direct animal emissions, a clear reduction per unit land area is achieved by reducing the number of animals. Some other mitigation options such as certain feed additives and productivity increases are not applicable in part in organic farming or of only limited effect if interference with animal welfare is avoided. Other additives seem technically promising, but more research is needed, in particular, on their economic viability. When adopting a global view or life cycle analysis (LCA) approach that includes not only direct methane emissions but land-use change and deforestation as well, the use of roughage for feed clearly reduces emissions compared to concentrate feed.

13.2.3 CO_2 Emissions

Most important are CO_2 emissions from land-use change. Due to their relation to soil carbon sequestration, they will be covered in the next subsection. The other direct CO_2 emissions mainly accrue from fossil fuel consumption for machinery use and irrigation and heated greenhouses, and constitute about 10% of total direct agricultural emissions (Bellarby et al. 2008). Besides the emissions due to heated greenhouses, which are restricted in some organic standards (e.g., BioSuisse or KRAV, under discussion in the EU organic standard), these emissions are similar for both organic and conventional agriculture and can be reduced through increased energy efficiency of the machines, their optimised use (e.g., adequate engine power, lower speed on the field, optimised tyre pressure, etc.) and conservation tillage. Higher organic matter contents in organically managed soils increase water holding capacity and can thus lead to reduced irrigation needs. Soils also tend to be less heavy with corresponding lower energy requirements for tillage. Mondelaers et al. (2009), Gomiero et al. (2011) and Schader et al. (2011) present recent detailed compilations of the environmental performance of organic farming.

On the other hand, less effective pest control can increase machinery use in organic farming in the event that more spray cycles are needed than for a chemical pesticide (e.g., for non-resistant apples or grapes). However, increased machinery use for pest control may be avoided by cropping pest-resistant varieties. Clearly, energy use is highly dependent on the specific aspects of the cropping system (crops, intensity, etc.).

Some mitigation potential can also be achieved by renewable energy provision where this seems to be a promising option (e.g., biogas, solar heat and electricity from large roof areas, windmills, etc.), and by increasing the energy efficiency of buildings (e.g., through improved insulation).

As with N_2O, the absence of synthetic fertilisers in organic agriculture does away with the corresponding emissions from their production and, in the case of urea, from their application (see above). CO_2 emissions from organic fertilisers can be treated as zero since they are sourced from renewable biomass. Synthetic pesticides and herbicides are not used and the corresponding emissions from the production process therefore do not arise. For a thorough assessment, life-cycle emissions of the products used in organic agriculture (e.g., aeration and shifting of compost windrows) would, however, have to be weighed against those avoided emissions.

13.2.4 Soil Carbon Sequestration

The most important contribution to mitigation in agriculture stems from soil carbon sequestration. The global biophysical potential is estimated at 5–5.5 Gt CO_2e/y by 2030. It would be possible to achieve 3.5–4 Gt CO_2e/y at a cost of up to \$ 100/t CO_2e (based on Smith et al. 2008), but a high degree of uncertainty exists. Soil carbon sequestration is directly linked to the build-up of soil organic matter and the avoidance of losses through erosion. It can be promoted by various agricultural practices such as application of organic instead of chemical fertilisers, use of optimised crop rotations, use of legumes and reduced tillage (Smith et al. 2008). The soil carbon levels that can be reached also depend on land use, prior history of management and a soil's carbon saturation capacity. Pastures store much more carbon in the soil than croplands, for example (Guo and Gifford 2002), and agroforestry systems also have a high potential for soil—and biomass—carbon sequestration per ha and year (e.g., on the order of 3–8 t CO_2e/ha/y in soil carbon, as reported in Mutuo et al. 2005 or Albrecht and Kandji 2003).

In principle, soil carbon can be increased both in organic and in conventional agriculture. Due to the underlying processes and given the advantageous practices just listed, maintaining and increasing soil carbon is an integral part of organic agriculture with its focus on soil fertility and soil organic matter build-up and where these practices just listed are common. It is, by far, less in line with conventional agriculture where synthetic fertiliser use, soil compaction, erosion and monocultures instead reduce soil fertility and soil carbon stocks over the years (Matson et al. 1997; for statistics on soil carbon losses, see Lal 2004).

When comparing soil carbon sequestration figures, it is important to note that the processes involved follow a saturation dynamic. Soil carbon sequestration does not continue forever but levels off at an equilibrium after some decades. Thus, the mitigation from soil carbon sequestration primarily helps to gain time for other mitigation measures, e.g., those related to fossil energy use. Without changes in global emissions, the annual mitigation effect from soil carbon sequestration will be lost after the new, saturated equilibrium of carbon stocks in soils is reached. In addition, soil carbon sequestration is a reversible process, and changing management practices after some decades can lead to fast release of all the carbon previously sequestered. This directly links soil carbon sequestration to the CO_2 emissions from land-use change and deforestation. A huge potential of soil carbon sequestration lies in the reversion of these losses through optimised land-use change, reduced husbandry. In particular, as a result of soil carbon sequestration, systemic ideas have thus gained increasing importance in conventional agriculture. This development should be used to slowly move towards a global approach to sustainable systemic agriculture beyond the organic vs. conventional divide. Clearly, organic agriculture must not compromise many of its core aspects, and within the context of a unified sustainable agriculture, conventional agriculture would most likely have to move much further towards organic agriculture than the other way round. Similarly, however, since conventional agriculture acknowledges the potential of several aspects of organic agriculture in the context of climate change mitigation (Smith et al. 2007, 2008), organic agriculture may also adopt some aspects of conventional agriculture. Examples could be limited to the use of chemical fertilisers in nutrient-deficient contexts or the moderate use of certain conventional biocides.

Such suggestions clearly challenge the basic understanding of organic agriculture, but some modernisation of organic agriculture is needed and climate change mitigation is an ideal approach for this. Organic agriculture was originally developed to address certain environmental and ethical problems raised by industrialised agriculture and was optimised to resolve these problems. Climate change mitigation was not a concern when organic agriculture was at the developmental stage. Now, however, climate change mitigation is a key global concern but is not related to this defining basis of organic agriculture. It forces both organic and conventional agriculture to provide answers—and neither system is currently optimised for mitigation. Due to its principles, organic agriculture has some intrinsic advantages regarding mitigation, but conventional agriculture takes up some of these principles and improves its performance in such a way that the differences between organic and conventional agriculture diminish when compared in terms of their mitigation benefits. There is nothing "exclusively organic" about mitigation in agriculture (see Table 13.1), and the mitigation potential is also only one of the many parameters that determine a sustainable multifunctional agriculture.

Acknowledgments We thank the anonymous reviewers and the editors for their very thoughtful comments and criticisms that considerably improved this paper. AM: Financial support from the Mercator Foundation Switzerland is gratefully acknowledged.

References

Albrecht A, Kandji ST (2003) Carbon sequestration in tropical agroforestry systems. Agr Ecosyst Enviro 99:15–27

Alluvione F, Bertora C, Zavattaro L, Grignani C (2010) Nitrous oxide and carbon dioxide emissions following green manure and compost fertilization in corn. Soil Sci Soc Am J 74:384–395

Badgley C, Moghtader J, Quintero E, Zakem E, Jahi Chappell M, Aviles-Vazquez K, Samulon A, Perfecto I (2007) Organic agriculture and the global food supply, Renew Agr Food Syst 22(2):86–108. For a criticism of this study and response from the authors see the same journal, 22(4):321–329

Bellarby J, Foereid B, Hastings A, Smith P (2008) Cool farming: climate impacts of agriculture and mitigation potential. Greenpeace International, Amsterdam, The Netherlands

Bhandral R, Saggar S, Bolan NS, Hedley MJ (2007) Transformation of nitrogen and nitrous oxide emission from grassland soils as affected by compaction. Soil Till Res 94:482–492

Bouwman A, Boumans L, Batjes N (2002) Emissions of N_2O and NO from fertilized soils: summary of available measurement data. Global Biogeochem Cy 16(4)

Carlsson-Kanayama A, Gonzalez AD (2009) Potential contributions of food consumption patterns to climate change. Am J Clin Nutr 89(5):1704S–1709S

Den Elzen MGJ, van Vuuren DP, van Vliet J (2010) Postponing emission reductions from 2020 to 2030 increases climate risks and long-term costs. Clim Change 99:313–320

Dias B, Silva C, Higashikawa F, Roig A, Sanchez-Monedero M (2010) Use of biochar as bulking agent for the composting of poultry manure: effect on organic matter degradation and humification. Bioresour Technol 101:1239–1246

EC (2005) Council of the European Union. Presidency Conclusions—Brussels, 22/23 March 2005 (European Commission 2005). Avoiding dangerous climate change (Cambridge Univ. Press 2006)

El-Hage Scialabba N, Müller-Lindenlauf M (2010) Organic agriculture and climate change. Renew Agr Food Syst 25(2):158–169

ENB (2010) Summary of the Bonn climate change talks: 31 May–11 June 2010. Earth Negot Bull 12(472):4

Eshel G, Martin P (2006) Diet, energy, and global warming. Earth Interact 10(9):1–17

FAO (2007) Report on the International Conference on Organic Agriculture and Food security, Rome 3–5 May 2007. Food and Agriculture Organisation of the United Nations FAO, Rome

Gattinger A, Jawtusch J, Muller A, Mäder P (2011) No-till agriculture—a climate smart solution? Climate Change and Agriculture Report No. 2, MISEREOR

Gattinger A, Muller A, Häni M, Skinner C, Fliessbach A, Buchmann N, Mäder P, Stolze M, Smith P, El-Hage Scialabba N, Niggli U (2012) Enhanced top soil carbon stocks under organic farming. Proceedings of the National Academy of Sciences of the United States of America PNAS, 109(44):18226–18231

Godfray H, Beddington J, Crute I, Haddad L, Lawrence D, Muir J, Pretty J, Robinson S, Thomas S, Toulmin C (2010) Food security: the challenge of feeding 9 billion people. Science 327:6

GoA (2010) Quantification protocol for agricultural nitrous oxide emissions reductions. Government of Alberta

Gomiero T, Paoletti M, Pimentel D (2011) Environmental impact of different agricultural management practices: conventional vs. organic agriculture. Crit Rev Plant Sci 30:95–124

Gonzalez A, Frostell B, Carllsson-Kanyama A (2011) Protein efficiency per unit energy and per unit greenhouse gas emissions: potential contribution of diet choices to climate change mitigation. Food Pol 36:562–570

Guo L, Gifford M (2002) Soil carbon stocks and land use change: a meta analysis. Glob Change Biol 8:345–360

Hörtenhuber S, Lindenthal T, Amon B, Markut T, Kirner L, Zollitsch W (2010) Greenhouse gas emissions from selected Austrian dairy production systems—model calculations considering the effects of land use change. Renew Agr Food Syst 25(4):316–329

IFA (2009) The global "4R" nutrient stewardship framework. International Fertilizer Industry Association IFA

IPCC (2007) Pachauri RK & Reisinger A (eds) Climate change 2007: synthesis report (Intergovernmental Panel on Climate Change, Cambridge, UK 2007)

IPCC (2006) IPCC guidelines for national greenhouse gas inventories. Intergovernmental panel on climate change IPCC. http://www.ipcc-nggip.iges.or.jp/public/2006gl/index.html

Lal R (2004) Soil carbon sequestration impacts on global climate change and food security. Science 304(5677):1623–1627

Leifeld J, Fuhrer J (2010) Organic farming and soil carbon sequestration: what do we really know about the benefits? AMBIO 39(8):585–599

Li C, Frolking S, Butterbach-Bahl K (2005) Carbon sequestration in arable soils is likely to increase nitrous oxide emissions, offsetting reductions in climate radiative forcing. Clim Change72:321–338

Lovanh N, Warren J, Sistani K (2010) Determination of ammonia and greenhouse gas emissions from land application of swine slurry: a comparison of three application methods. Bioresour Technol 101:1662–1667

Lynch D, Cohen R, Fredeen A, Patterson G, Martin R (2005) Management of Canadian prairie region grazed grasslands: Soil C sequestration, livestock productivity and profitability. Can J Soil Sci 85:183–192

Lynch D, MacRae R, Martin R (2011) The carbon and global warming potential impacts of organic farming: does it have a significant role in an energy constrained world? Sustainability 3:322–362

MacRae R, Lynch D, Martin C (2010) Improving energy efficiency and GHG mitigation potentials in Canadian organic farming systems. J Sustain Agr 34(5):549–580

Martin C, Morgavi D, Doreau M (2010) Methane mitigation in ruminants: from microbe to the farm scale. Animal 4(3):351–365

Matson P, Parton W, Power A, Swift M (1997) Agricultural intensification and ecosystem properties. Science 277:504–509

Meinshausen M, Meinshausen N, Hare W, Raper SCB, Frieler K, Knutti R, Frame DJ, Allen MR (2009) Greenhouse-gas emission targets for limiting global warming to 2°C. Nature 458:1158–1162

Mondelaers K, Aertsens J, Van Huylenbroeck G (2009) A meta-analysis of the differences in environmental impacts between organic and conventional farming. Brit Food J 111(10):1098–1119

MSU-EPRI (2010) Quantifying N_2O emissions reductions in US agricultural crops through N fertilizer rate reduction, methodology proposal submitted to the voluntary carbon standard VCS, Michigan State University MSU and Electric Power Research Institute EPRI

Muller A, Jawtusch J, Gattinger A (2011) Mitigating greenhouse gases in agriculture—a challenge and opportunity for agricultural policies. Report commissioned by „Brot für die Welt“ (Germany), „Brot für alle“ (Switzerland), DanChurchAid (Denmark) and Church of Sweden. With case study contributions from Friedhelm Gölthenboth and Joergen Olesen

Mutuo PK, Cadisch G, Albrecht A, Palm CA, Verchot L (2005) Potential of agroforestry for carbon sequestration and mitigation of greenhouse gas emissions from soils in the tropics. Nutr Cycl Agroecosyst 71:43–54

Niggli U, Fließbach A, Hepperly P, Scialabba N (2009) Low Greenhouse Gas Agriculture: Mitigation and Adaptation Potential of Sustainable Farming Systems. Food and Agriculture Organization of the United Nations FAO, Rev. 2

O'Mara F (2004) Greenhouse gas production from dairying: Reducing methane production. Advances in Dairy Technology 16:295–309

Pattey E, Trzcinski MK, Desjardins RL (2005) Quantifying the Reduction of Geenhouse Gas Emissions as a Result of Composting Dairy and Beef Cattle Manure. Nutr Cycl Agroecosyst 72(2):173–187

Peigne J, Ball B, Roger-Estrade J, David C (2007) Is conservation tillage suitable for organic farming? A review. Soil Use Manag 23(2):129–144

Petersen S, Regina K, Pöllinger A, Rigler E, Valli L, Yamulki S, Esala M, Fabbri C, Syväsalo E, Vinther F (2006) Nitrous oxide emissions from organic and conventional crop rotations in five European countries. Agric. Ecos. Env 112:200–206

Pretty JN, Noble AD, Bossio D, Dixon J, Hine RE, Penning deVFWT, Morison JIL (2006) Resource-Conserving Agriculture Increases Yields in Developing Countries. Environmental Science & Technology 40

Rogelj J, Nabel J, Chen C, Hare W, Markmann K, Meinshausen M, Schaeffer M, Macey K, Hohne N (2010a) Copenhagen Accord pledges are paltry. Nature 464:1126–1128

Rogelj J, Chen C, Nabel J, Macey K, Hare W, Schaeffer M, Markmann K, Höhne N, Krogh Andersen K, Meinshausen M (2010b) Analysis of the Copenhagen Accord pledges and its global climatic impacts—a snapshot of dissonant ambitions, Environ. Res. Lett 5(034013):1–9

Schader C, Stolze M, Gattinger A (2011) 'Environmental performance of organic agriculture', in Boye J, Arcand Y (eds.), Green Technologies in Food Production and Processing, New York, Springer.

Schellnhuber JS, Cramer W, Nakicenovic N, Wigley TML, Yohe G (2006) Avoiding Dangerous Climate Change, Cambridge Univ. Press

Sejian V, Lakritz J, Ezeji T, Lal R (2010) Forage and flax seed impact on enteric methane emission in dairy cows, Research Journal of Veterinary Sciences

Shibata M, Terada F (2010) Factors affecting methane production and mitigation in ruminants. Anim Sci J 82:2–10

Smil V (2002) Eating meat: Evolution, Patterns, and Consequences. Popul Devel Rev 28(4):599–639

Smith P, Martino D, Cai Z, Gwary D, Janzen H, Kumar P, McCarl B, Ogle S, O'Mara F, Rice C, Scholes B, Sirotenko O (2007) Agriculture. In Climate Change (2007) Mitigation. Contribution of Working Group III to the Fourth Assessment Report of the Intergovernmental Panel on Climate Change [Metz B, Davidson OR, Bosch PR, Dave R, Meyer LA (eds)], Cambridge University Press, Cambridge, UK, and New York, NY, USA. http://www.ipcc.ch/pdf/assessment-report/ar4/wg3/ar4-wg3-chapter8.pdf

Smith P, Martino D, Cai Z, Gwary D, Janzen HH, Kumar P, McCarl B, Ogle S, O'Mara F, Rice C, Scholes RJ, Sirotenko O, Howden M, McAllister T, Pan G, Romanenkov V, Schneider U, Towprayoon S, Wattenbach M, Smith JU (2008) Greenhouse gas mitigation in agriculture. Philosophical Transactions of the Royal Society of London Series B Biological Sciences 363: 789–813

Snyder CS, Bruulsema TW, Jensen TL, (2007) Greenhouse Gas Emissions from Cropping Systems and the Influence of Fertilizer Management, A literature review, International Plant Nutrition Institute, Norcross, Georgia, U.S.A

Sohi S, Loez-Capel E, Krull E, Bol R (2009) Biochar's roles in soil and climate change: A review of research needs. CSIRO Land and Water Science Report 05/09

Soil Association (2009) Soil carbon and organic farming, Soil Association

Soussana J, Tallec T, Blanfort V (2010) Mitigating the greenhouse gas balance of ruminant production systems through carbon sequestration in grasslands. Animal 4(3):334–350

UNEP-UNCTAD (2008) Organic Agriculture and Food Security in Africa. United Nations, New York and Geneva

UNFCCC (2009) UNFCCC Decision 2/CP.15 Copenhagen Accord, United Nations Framework Convention on Climate Change UNFCCC

USEPA (2006) Global Anthropogenic Non-CO_2 Greenhouse Gas Emissions: 1990–2020. U.S. Environmental Protection Agency USEPA, Washington, DC

Van Groenigen J, Velthof G, Oenema O, van Groenigen K, van Kessel C (2010) Towards an agronomic assessment of N_2O emissions: a case study for arable crops. Eur. J. Soil Sci 61:903–913

Vanotti MB, Szogi AA, Vives CA (2008) Greenhouse gas emission reduction and environmental quality improvement from implementation of aerobic waste treatment in swine farms. Waste Manage 28(4):759–766

Wassmann R, Dobermann A (2006) Greenhouse gas emissions from rice fields: what do we know and where should we head for. 2nd Joint International Conference on Sustainable Energy and Environment (SEE 2006), November 2006, *Thailand*

Wassmann R, Neue HU, Lantin RS, Buendia LV, Rennenberg H (2000) Characterization of methane emissions from rice fields in Asia. I. Comparison among field sites in five countries. Nutr. Cycling Agroecosyst. 58: pp. 1–12

West T, Post W (2002) Soil organic carbon sequestration rates by tillage and crop rotation: a global data analysis. Soil Sci Soc Am J 66:1930–1946

Wood S, Cowie A (2004) A review of greenhouse gas emission factors for fertilizer production. Input for IEA Bioenergy Task 38

Chapter 14
The Freedoms and Capabilities of Farm Animals: How Can Organic Husbandry Fulfill Them?

Jacques Cabaret, Caroline Chylinski and Mette Vaarst

Abstract Organic farming promotes animal husbandry practices that consider the welfare of the animals on the farm. The concept of animal welfare and the standards that should encompass this concept have in many cases been largely generalised in practice, which leaves relevant aspects of animal freedom or capabilities insufficiently addressed. This chapter puts forth the prospect that the capabilities approach offers an appropriate practical platform by which to improve welfare in farm animals by meeting a wider range of their natural needs and abilities. The capabilities approach coupled with effective health planning could foster organic husbandry towards a more acceptable production system for farmers and consumers alike.

Keywords Welfare · Animal liberty · Animal capabilities · Organic husbandry

14.1 Introduction

Animal welfare is a much debated, and often highly emotive topic. Many different views can be taken when considering how to best provide our animals with ethically-just living conditions to meet their needs. In some ways, it is our immediate sympathy and compassion for nonhuman animals that has driven the notion of creating just relations between humans and animals, and ensuring that animals live lives that are worth living from their own perspective. On the other hand, we cannot rely on sympathy alone to bring about conscientious change. With the increased industrialisation of farming, research and scientific facts are needed to support ethical

J. Cabaret (✉) · C. Chylinski
INRA-Université de Tours, UMR1282 Infectiologie et Santé Publique, 37380 Nouzilly, France
e-mail: jacques.cabaret@tours.inra.fr

J. Cabaret
Université Paris I—Sorbonne, 13 rue du Four, 75005 Paris, France

M. Vaarst
Department of Animal Science, Aarhus University, PO-Box 50, 8830 Tjele, Denmark

S. Bellon, S. Penvern (eds.), *Organic Farming, Prototype for Sustainable Agricultures*,
DOI 10.1007/978-94-007-7927-3_14,

decision-making. The aim of this chapter is to discuss the reality of animal welfare on organic farms today, as a function of evidence and knowledge.

Deepening our scientific knowledge of animal welfare provides greater support for the deserved moral status of the animals. Several practical methods for assessing animal welfare have already been developed and implemented in the field (see Veissier et al. 2012). It is beyond the scope of this chapter to discuss them in detail. Instead, this chapter evaluates the extent to which current organic principles, as proposed by the IFOAM (International Federation of Organic Movements 2012), meet the animal welfare standards as laid out in two different concepts, the five freedoms and the capabilities of animals.

14.2 The Five Freedoms: Ideal States for Welfare?

The five freedoms were developed to account for both the physical and psychological well-being of the animal which, taken together, define animal welfare. These were proposed as 'ideal states' to work towards rather than standards of acceptable (minimum) levels of welfare. They were intended to provide a framework for welfare analysis for animals on the farm, in transit, at market or at the place of slaughter (Farm animal Welfare Council: fawc.org accessed 03/04/12). The five freedoms include:

1. Freedom from Hunger and Thirst—by sufficient access to fresh water and a diet to maintain full health and vigour.
2. Freedom from Discomfort—by providing an appropriate environment including shelter and a comfortable resting area.
3. Freedom from Pain, Injury or Disease—by health promotion, disease prevention and relevant immediate intervention in the event of any condition that causes pain, injury or disease.
4. Freedom to Express Normal Behaviour—by providing sufficient space, proper facilities and company of the animal's own kind.
5. Freedom from Fear and Distress—by ensuring conditions and handling that minimise mental or emotional suffering.

Certain elements of the five freedoms may be at odds with organic animal husbandry. Firstly, the five freedoms most likely challenge the cultural values or personal opinions of the farmer or their respective society (Tague 2010). Farmers values concerning nature (Kaltoft 1999) are particularly important as they translate into choices made in husbandry (i.e., whether to be an organic producer, whether to sell their products directly to consumers or not). Secondly, fulfilling the five freedoms may not be conducive to the most economically profitable options in the short term. And lastly, the five freedoms may present a conflict of interest within the organic system itself. For example, under European organic legislation, the use of chemical substances is limited, with one exception, anthelmintics. However, the same legislation further states that responsible action should be taken to prevent animal suffering in cases of disease where the use of efficient drugs is then recommended. Yet, according to American organic legislation, the animals cannot be sold as organic if they have been treated with antibiotics, which thereby discourages the use of medical treatment and instead encourages health promotion on a very radical level. It is paramount that animals that require treatment should receive it in a qualified way. It is also important that animals are not only treated for disease alone, but that better health is also promoted, e.g., through supportive feed, adequate rest and comfort, etc. Efforts to minimise antibiotics should be based solely on the promotion of animal health and welfare to thereby remove the need for treatment. Research has shown that regulations and goals focusing on minimising antibiotic treatment can support the emphasis on health and welfare promotion (Ivemeyer et al. 2012; Bennedsgaard et al. 2010), which will also minimise the risk of suffering. This will be discussed below.

14.3 Organic Agriculture Principles in Relation to Welfare

Organic agriculture is based on four principles (IFOAM 2012): health, ecology, fairness and care.

> Health is the wholeness and integrity of living systems. It is not simply the absence of illness, but the maintenance of physical, mental, social and ecological well-being. Immunity, resilience and regeneration are key characteristics of health. Organic agriculture should avoid the use of fertilisers, pesticides, animal drugs and food additives that may have adverse health effects.

This health principle is directly linked to the five freedoms, especially the freedoms of avoiding pain, suffering, disease, and discomfort: it is at all times important to keep the animal in a healthy state that enables it to resist diseases and be supported mentally, emotionally and physically.

The principle of ecology emphasizes that 'Organic Agriculture should be based on living ecological systems and cycles, to work with them, to emulate them and to help to sustain them'. Animals are part of farming systems and they should contribute to a well-balanced system where feed and manure circulate, and where space and production is in harmony with what the soil and the rest of the farm can produce in a way that is sustainable from an environmental, social, institutional and economic point of view.

The principle of fairness promotes a type of agriculture that is built on 'relationships that ensure fairness with regards to the common environment and life opportunities'. This principle stresses that animals should be provided with the conditions and opportunities of life that accord with their physiology, natural behaviour and well-being. This means that the animal should not be pushed to exceed its capabilities in any way, physically, physiologically, mentally or emotionally. It emphasizes the need to ensure that the animals are given an environment in which they can live in dignity and in accordance with their needs. An example to illustrate a lack of fairness is the farming of bees, which are maintained in difficult monocultural landscapes, with minimal flowering plants for a relatively short period, thus pushing the bees to exceed their physical and physiological capabilities.

The principle of care stipulates that 'Organic Agriculture should be managed in a responsible manner to protect the health and well-being of current and future generations and the environment'. It also directs us to not bring animals into life situations that cannot be adequately managed. This principle may describe most directly the human role towards organic animals. It addresses the responsibility of the farmers and of those who take responsibility for animals, to provide a care system that allows the animals to live their lives as close to their respective 'nature' as possible. This refers to the animals' ability to perform natural behaviours, to access food that meets their physiological needs and in all ways, to be the animals that nature intended, as far as can be done under farming conditions. Achieving this principle relies upon human understanding of the animals and their natural needs, as well as sound judgment on when human intervention is necessary and what the appropriate measures to employ would be (Vaarst and Alroe 2012). This requires skill, knowledge and careful observation. It is important to stress that in the case of disease, relevant action can include the provision of special care and support as well as administration of the necessary and appropriate treatment. Lund et al. (2004) described this as a mutual ethical contract between human (care) and animal (production), which eventually comes to an unavoidable 'natural end' at slaughter.

The human responsibility to manage animals well and to ensure their health and welfare can incorporate expertise from varying points of view. Science can play an effective and supportive role in making ethical judgements. In turn, this can influence the future development of organic livestock farming to continuously strive towards improved standards of animal health, consumer safety and ecological and environmental sustainability. Yet, these ethical decisions cannot be made on the basis of science alone. Valuable solutions can also be found in the accumulated wisdom of farmers, agricultural workers and veterinarians, as well as in traditions and indigenous knowledge tested and experienced over time. For the animals to benefit from this combined knowledge, continual exchange and development is necessary. Theoretically, the implementation of these four principles should indirectly ensure that the five freedoms are met. Animal welfare and the ethics of organic husbandry have been described (Verhoog et al. 2004) and discussed at length in numerous articles and books. However, many of these have also referred to elements of discord that can arise from applying ethical approaches to everyday farming, often resulting in compromises. Porcher (2014, in this book) highlighted the difficulty that many farmers have in maintaining their organic standards:

> This applies mainly to the transport and slaughtering of animals, the choice of breeds, the specifications and, more broadly, the utilitarian and economic paradigm underlying organic animal husbandry.

Porcher expresses her concern that the animals, as sentient beings, have become the forgotten partners in the contract of organic agriculture. Thus, although the ideals of organic animal welfare and the ideals of the five freedoms first appear aligned, the application and guidance provided by these two 'sets of principles' cannot be expected to—in any way—be a guarantee for good animal welfare.

Welfare limitations can be viewed in the light of animal rights. We will thereby explore to what extent animal rights are covered in the five freedoms, and if they are met in an organic animal husbandry. We will also examine how the five freedoms hold up in relation to the capabilities approach, a concept that goes beyond basic survival to characterise those aspects required to attain a 'quality of life' as initially described for humans by Sen (2005), and others. To note, we assume that the freedoms relating to 'no hunger or thirst' and 'provide shelter and a comfortable resting area' are such evident basic needs that we do not pay specific attention to them.

14.4 Animal Rights and Welfare

The rationale behind animal welfare stems from findings that animals are sentient beings, with the ability to experience both pleasure and pain. This broke with the previously held view that humans were above animals and, therefore, have special inherent 'rights' (speciesism). When all human and nonhuman animals are capable of suffering, all should be worthy of equal consideration and rights. A central argument in Singer's book, 'Animal Liberation', published in 1975, is an expansion of the utilitarian idea in that "the greatest good of the greatest number" is the only measure of welfare and, therefore, the only valid guide to ethical behaviour. Although Singer belonged to the movement that promoted humaneness and respect for animal life and welfare, he does not raise specific concern about eating animals or using animals for work, insofar as they are raised and killed in a way that does not involve fear, pain and suffering. Animal rights defenders are less focused on this perspective, but rather more so on a complete ban of the use of animals altogether (see Jeangène Vilmer (2008) for the detailed position of Regan or Francoine).

14.5 Freedoms in Relation to 'Animals' Capabilities'

It is our opinion that the five freedoms should not be the sole matrix upon which animal welfare is based and analysed. Having originally been defined and intended as guidelines, they may be subject to substantial interpretation among the various care providers (from farmers to vets). We propose examining animal welfare in light of the 'capabilities approach' as outlined by Nussbaum (2001). This approach emphasizes the fact that animals have a moral status as defined by Warren in 1997:

> To have moral status is to be morally considerable, or to have moral standing. It is to be an entity toward which moral agents have, or can have, moral obligations. If an entity has moral status, then we may not treat it in just any way we please; we are morally obliged to give weight in our deliberations to its needs, interests, or well-being". Although Nussbaum's approach was originally designed to capture the liberties that encompass a 'quality of life' in humans (Sen 2005), they have since been adapted to animals. Anand and coauthors (2005) stated: "Sen defines capabilities as what people are able to do or able to be—the opportunity they have to achieve various lifestyles and as a result, the ability to

live a good life. He differentiates this from what he calls functionings—the things a person actually does and experiences. Functionings in humans may vary from the elementary, such as being adequately nourished and being free from avoidable disease, to complex activities or personal states, such as taking part in the life of the community and having self-respect.

This approach contributes towards understanding the relationship between an animal's inherent nature (what they are able to do or to be) and suitable welfare. We find that this capabilities approach adds another dimension to the issue of welfare by presenting interesting and relevant perspectives that, in addition, have the merit of being applicable to organic animal husbandry situations. The 10 capabilities proposed by Nussbaum (2001) for animals are specified and discussed below.

1. *Life*. Being able to live to the end of a life of normal length; not dying prematurely, or before one's life is so reduced that it is not worth living.
2. *Bodily Health*. Being able to have good health, including reproductive health; to be adequately nourished; to have adequate shelter.
3. *Bodily Integrity*. Being able to move freely from place to place; to be secure against violent assault, including sexual assault and domestic violence; having opportunities for sexual satisfaction and reproductive choices.
4. *Senses, Imagination, and Thought*. Being able to have an adequate education. Being able to have pleasurable experiences and to avoid non-beneficial pain.
5. *Emotions*. Being able to attach to things and people outside ourselves; to love those who love and care for us, to grieve at their absence; in general, to love, to grieve, to experience longing, gratitude, and justified anger. Not having one's emotional development blighted by fear and anxiety.
6. *Practical Reason*. Being able to plan one's life. Nussbaum (2001) discussed that this is not easily applicable for farm animals, although some animals may be engaged in "projects".
7. *Affiliation*. Being able to live with others, to engage in various forms of social interaction.
8. *Other Species*. Being able to live in relation to animals, plants, and the world of nature.
9. *Play*. Being able to play.
10. *Control Over One's Environment*. Being able to hold "property" (for example, a place to sleep or to be milked).

The extent to which these capabilities can be applied to welfare under organic animal husbandry conditions will now be discussed in relation to those of conventional farms.

14.6 A First Set of Capabilities in Organic Animal Husbandry: Life, Bodily Health and Integrity

We suggest that the three capabilities of 'life', 'bodily health' and 'integrity' relate directly to the five freedoms. We therefore choose to focus on them by considering the extent to which these capabilities can be applied to animal welfare under organic farming conditions. This will be discussed in relation to cases with free-range farms (as all organic farms are) and contrasted with conventional farms. These cases serve as examples to bring the discussion into a practical framework.

14.6.1 Life

Somewhat ironically, death is the measure by which the capability of life is measured. Although this is a crude evaluation of the capability and does not account for a reduced quality of life, it remains the best available indicator. Specifically, this relates to the age of animals at culling and mortality. A study by Benoit and Laignel (2009) comparing 25 conventional and nine organic grassland-based sheep production systems in the Centre region of France, provides a practical example of this. Lamb mortality rates were found to be 16.2 and 16.5 % in conventional and organic farms, respectively, and ewe mortality rates were 5.9 and 5.3 %. The culling percentage was also very similar at 20.4 and 20.6 %. None of these variables were significantly different between the farms. A similar study that investigated 152 conventional and 22 organic beef cattle farms in the same area (Veysset et al. 2009) found slightly greater differences. The calf mortality rates were 6 and 7 % for conventional and organic farms, respectively, whereas the culling rate was 20 and 23 %. A study in Wisconsin dairy herds did not show any difference in culling rates: 18 vs. 17.2 % (Sato et al. 2005). A greater difference is seen among mono-gastric animals that are largely maintained indoors on conventional farms, but on pasture in organic farms. The mortality rate of organic piglets from birth to weaning is highly variable among European countries (Prunier 2010), ranging from 15 % in Italy to 35 % in France. This range was smaller in Austria, Denmark, Germany and Sweden at 28–30 %. These figures are nearly 25 % higher than those observed in conventional pig husbandry. Based on a five-year study in Western France, broiler chicken mortality was found to be 4, 3 and 5 % in organic, free-range and conventional husbandry, respectively. Mortality in laying hens is slightly higher in organic than in non-organic free-range farms, and is nearly double that observed in conventional production (Magdelaine 2006). The reasons for greater mortality in organic husbandry of mono-gastric animals may be related to how suitable the currently used breeds are, e.g., for outdoor production.

One of the greatest challenges to the life capability is the ending of life, namely how the animals are transported and slaughtered. No differences exist in these practices between organic and conventional management systems. Farmers become increasingly dissatisfied with these practices according to Porcher (2003) and she has consequently proposed, and tested, the concept of mobile abattoir facilities. The idea is to reduce stress associated with transport and waiting time before slaughter (Porcher and Daru 2005).

Irrespective of the welfare standards attained in relation to the methods of transport and slaughter, the death of the animal is somewhat difficult to tie into the capability of life described by Nussbaum (2001). Reconciling the death with the life capability is overcome by some cultures that transform the act of killing into a sacrifice as seen in the practices of Halal and Kosher. However, the validity of these arguments is under much debate. The human conscience can also accommodate the act of death by establishing strong dissociations between the animals and the slaughterhouse (farmers) or animals and the food (consumers).

14.6.2 Bodily Health

In both conventional and organic farms, the farmer is assumed to offer adequate food and shelter to his animals to meet the ethical expectations of consumers and citizens, and to meet legislative and market requirements (sanitary and organoleptic quality of the produced meat). On organic farms, the animals are fed with organic feed that, whenever possible, is local in origin and grown without pesticides or synthetic fertilisers that may interfere with the health status of the animals, although no data has confirmed this (Zollitsch et al. 2004). Rather surprisingly, Nussbaum (2001, 2006a, b) did not clearly identify diseases and appropriate disease control methods in relation to her bodily health capability. Instead, she assumed that if good conditions are provided for the animals, then disease will disappear or not appear at all. As a consequence of this, we chose to include animal health and welfare planning under bodily integrity in our attempt to loyally follow the capability approach of Nussbaum.

Health is much more than the mere ‘absence of disease’, and adequate health promotion practices can improve the overall health of the animals through access to fresh air, exercise, high quality feed and clean water. The maintenance of good health clearly includes disease prevention practices and strategies such as the use of drying-off to prevent mastitis in dairy cows. It also comprises immediate and relevant intervention where there are signs of disease. Clearly, organic animal husbandry operates under a ‘health promotion and disease prevention is better than cure’ ideal, and attempts are made where possible to avoid dependency on veterinary medicinal inputs, including vaccines, through adapted management practices. The use of vaccines is considered acceptable as long as the animals on a particular farm are at high risk of a particular disease. Organic broilers and laying hens are a good example of the intensive use of vaccinations, mostly against viral diseases. A survey in Sweden showed that the 56 laying hen farms vaccinated against Marek’s disease, infectious bronchitis, avian encephalitis and coccidiosis (Berg 2001). In Switzerland, the following protocol recommended by the FIBL (Research Institute of Organic Agriculture) is applied for organic laying hens at different ages: day 1: Marek’s disease; day 9: coccidiosis (eight coccidia species); week 3: Gomboro disease; week 5: infectious bronchitis; week 7: Gomboro disease; week 9: infectious bronchitis; week 12: avian encephalitis: week 15: infectious bronchitis. The vaccination protocols in organic and conventional poultry herds remain largely similar.

According to the current organic EU regulation, alternative medicines with demonstrated efficacy should be selected over allopathic medicine. The complementary and alternative medicines (CAM) are a group of diverse medical and health care systems, practices and products, most of which are not considered as a part of biomedicine ('conventional medicine'/'school medicine'). Phytotherapy and homoeopathy are among the most commonly used alternative medicines in organic husbandry. Cabaret (1986) described the use of phytotherapy in organic and, to a lesser extent, conventional farms, and Vaarst et al. (2004) described the use of homeopathy in organic herds. The efficacy of these therapeutics is often difficult to assess using natural scientific methods, and only a few randomised controlled trials have been done (Wynn and Wolpe 2005). Veterinarians have a moral and ethical obligation to provide therapies with proven or experienced efficacy, and they have the obligation to respect the farmer's wishes and beliefs (Wynn and Wolpe 2005). Vaarst and co-authors (2004) stated that:

> It has been suggested that the restriction on medicine use may lead to unnecessary suffering if animals are left without treatment. This appears to be a general concern within the veterinary profession...avoidance of suffering clearly overrules any limits on the use of medicine.

The choice of treatment and the type of medicine selected will often be the result of a compromise between the farmer's wishes versus those of the veterinarians (Cabaret et al. 2012). In any case, it is important to give the animal the sufficient care, attention and support, no matter which treatment it has received. Disease patterns in conventional and organic farms do not appear to be very different and, indeed, it has been shown that the difference between herds within each production system are bigger than a systematic difference between organic and conventional (Thamsborg et al. 2004; Cabaret et al. 2012). This highlights the importance of management and husbandry choices on a farm level.

Organic husbandry practices could be improved through animal health and welfare planning programmes (Hovi et al. 2004; Vaarst et al. 2012). These programmes supplement the Nussbaum capabilities, which remain limited in the area of disease. The health plans can be directed at acute problem-solving, goal-oriented efforts to avoid particular diseases, or long-term health planning, based on the farmer's goals for the herd or flock. An example of these planning programmes can be seen in the Farmer Field Schools, a concept for learning, knowledge exchange and empowerment that is being developed and used in some countries. In Denmark, the concept has been adapted to Danish conditions and is referred to as 'Stable Schools'. The first four Stable Schools were established in 2004 with the aim of phasing out the use of antibiotics in Danish organic dairy herds (Vaarst et al. 2007). In Germany, the same approach was adapted and the first results seem very promising (Brinkmann et al. 2012). In Stable Schools, problems are identified and solutions proposed based on the farmer's individual wishes and goals for the farm. The successful implementation of a plan combines the farmers' knowledge with facts about actual events and conditions in the herd. This information is collected in a systematic way and the views of external persons such as other farmers or health advisors (Vaarst et al. 2012). The European CORE-organic project ANIPLAN highlighted a number of principles that are important when creating lasting changes on a farm (see Box 14.1 below) (Nicholas and Vaarst 2011). This project aimed to develop a

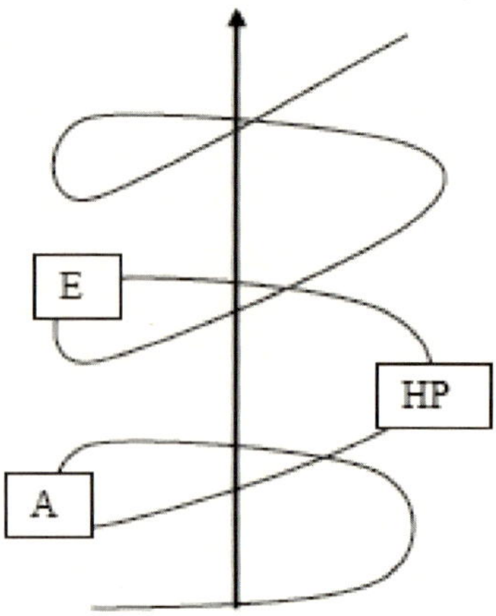

Fig. 14.1 Representation of animal health and welfare planning as a continuous process based on assessment (*A*), planning (*HP*) and evaluation (*E*)

model for animal health and welfare planning that can be implemented in different types of farming environments, e. g., large-scale dairy husbandry as well as alpine, smallholder and diverse farming systems. The principles were developed through discussion, which catalyses this process and which is necessary in order to achieve a balance between farmers' needs, animals' needs and the wider societal perception of health and welfare. At the same time, the multiple objectives of organic animal husbandry should also be fulfilled. The first of the key principles is illustrated in Fig. 14.1 and focuses on how animal health and welfare planning should be seen as a continuous process.

Box 14.1. The nine principles characterised by the European CORE-organic project ANIPLAN as a good and appropriate animal health and welfare planning process.

1. A health planning process should aim at continuous development and improvement and should incorporate the promotion of health and disease handling based on a strategy that incorporates (as described in Fig. 14.1 above) a learning cycle among the involved persons, including: assessing the situation, evaluating, taking action and reviewing the development, etc.
2. Farm specific: all planning should be based on the specific farm.
3. Farmer ownership: the farmer should lead the way, and not the advisor.
4. External person(s) should be involved: planning should be based on dialogue between the owner and somebody seeing the situation from the outside.
5. External knowledge: the knowledge that provides background for decisions should be partly based on an outside 'view of the farm', including its data, seen by somebody else).

6. Organic principles framework (systems approach): the organic principles should always guide planning on an organic farm, which implies taking a systems approach.
7. Written: common memory is necessary; written minutes are crucial and should be based on what the farmer commits him/herself to do, and not recommendations alone.
8. Acknowledge good aspects: don't just focus on problem areas, but also remember to include a description of positive developments.
9. Involve all relevant persons in the process: all those with responsibilities and tasks to do within the herd should be involved in the process to ensure the exchange of knowledge and common understanding, as well as joint action.

The Danish Stable Schools demonstrated that farmer groups that shared a common goal to phase out antibiotic use were able to decrease antibiotic use by 50 % in one year with no negative side effects in terms of disease or production (Bennedsgaard et al. 2010). One major factor for this was the focus on the promotion of health and welfare rather than disease handling in terms of disease prevention and disease treatment.

14.6.3 Bodily Integrity

The ability of animals to freely move from place to place is limited in both organic and conventional livestock systems. This capability remains unfulfilled for conventional animals reared intensively indoors, which is the case for the vast majority of pig and poultry husbandry practices. Many pasture management choices to optimise herbage production, such as combining grazing and hay making, may restrict the opportunities for animals to graze and move freely on pasture. In Europe, there are only a few extensive animal husbandry systems that permit a greater freedom of movement for the animals, and they are more common in organic than conventional practices. The bodily integrity capability encompasses sexual and other types of violence, which often differs only very slightly between organic and conventional livestock production. Some organisations and researchers have debated the issues surrounding the value of natural reproduction versus artificial insemination (Piccardi et al. 2011), although the choice will always be made by the farmers in the end. The genetic selection of animals has been developed over many years and much progress has been made through artificial insemination and breeding programs in the different breed registries, often with the focus on maximising production. Organic animal husbandry may be more inclined to include different breeding objectives.

> Regardless of species, the primary breeding objectives for organic farming are likely to include disease resistance and longevity. Another area of importance is increased reliance

> on forage in ruminant diets. Good mothering abilities are also an important selection criterion in pigs and sheep. (Pryce et al. 2004)

These breeding objectives are achieved by means of a selection index, but in organic systems, the task may be complicated by the multiplicity of goals to be achieved and the relatively low number of animals involved in the production. The maintenance of genetic diversity, which may be of interest in organic animal husbandry, is difficult to cultivate in a selection index. Selective breeding means that the animals will not have any reproductive freedom, and the use of artificial insemination will preclude any sexual satisfaction. In that case, the 'animal capability' can be said to be restricted in organic as well as in conventional systems. Cross-breeding is justified by the increase in productivity through the heterosis phenomenon, and the F1 generation resulting from the two breeds or two lines are used both in conventional and organic pig and bird breeding. In that case, not only is the sexual capability of the animals impaired but, in addition, the farmers themselves are not the ones to make the decisions regarding reproduction. They fully rely on breeding companies to obtain hybrid lines, and this accounts for the near totality of poultry farmers, conventional as well as organic (Guéméné et al. 2009). To restore "genetic" independence, the farmers would have to turn to local breeds that are adapted to the specific environmental conditions in which they are reared, or to kin-selection on the farm.

Additional dilemmas exist regarding the bodily integrity of animals in both conventional and organic systems (Menke et al. 2004). For example, mulching (the removal of skin near the lambs' anus without any anaesthesia) is a practice widely used in Australian sheep to protect them against highly detrimental fly strike that may result in death following long periods of suffering. In this instance, the farmers largely perceive the ends (prevention of fly strikes) to justify the means (suffering during the rapid operation). Furthermore, they estimate that the use of anaesthetics would only serve to slow down the total operation time since it is performed on thousands of animals at a time. A similar situation exists for the castration of piglets, which has been widely practiced in organic as well as in conventional systems to prevent the undesirable odour ('boar smell') of the meat (banned by EU regulations as of 2015). This too is practiced without anaesthesia for time-saving reasons. The nose-ringing of sows is another issue. It is carried out to prevent rooting, a natural behaviour for pigs, but destructive to pastures and, hence, environmentally damaging, especially in large farms. No agreement has been reached on whether it should be allowed within the EU. It is currently banned in Sweden and the UK, but permitted in Denmark. Reaching an agreement on these issues is complicated by other concerns such as environmental pollution to accommodate the needs of animals and their welfare, which contradicts the principles of organic husbandry all together. Another example of this reverse husbandry is the practice of beak trimming in poultry. Intended to reduce feather pecking between individuals, this behaviour is only the result of high animal stocking densities. Thus, it is carried out to enable greater stocking densities, not for the benefit of the birds. This practice is prohibited in organic poultry farming and is therefore more respectful of animal welfare.

14.7 A Second Set of Capabilities in Organic Animal Husbandry: A Social Life with Emotions

We include the following capabilities here: relating to members of their own species, as well as the senses and imagination, emotions and play. These capabilities largely correspond to the freedom to express normal behaviour and as such, they can be viewed as a key to attaining a life of quality.

14.7.1 To Live in Relation with Other Species in Nature

This means that animals will have access to the external world and that they will not be confined in buildings except under special circumstances. This meets consumers' expectations and is generally found in organic herds. Much has been proposed and debated on how species-specific behaviour can be maintained, mostly based on observations of wild groups (Waiblinger et al. 2004). Most data focus on social structure within a flock or herd, e.g., on intra-species interactions. There is a lack of knowledge about relationships with other species, except with humans (Waiblinger et al. 2006). These relationships need to be established between humans and each of the animals, and therefore require a certain number of humans per number of animals. The organic principles for animal husbandry emphasizes a framework that allows a life with 'naturalness' as much as possible, and human care and intervention whenever necessary. If this condition is fulfilled, sufficient human involvement will be ensured and will necessitate more people in larger farms.

In some countries, interactions between different animal species are frequent in organic husbandry. This may include mixed grazing to maximise grass production or to reduce internal parasitism (from the tropics (Giudici et al. 1999) to the subarctic (Sormunen-Cristian et al. 2008)). The potential benefits of inter-species interactions on welfare have not yet been studied although their impact on farm pathocoenosis (pathogen dynamics) has been suggested (Nicourt and Cabaret 2014, Chap. 9). The interaction of wild animals has not been studied either, or if so, only from the negative viewpoint of predation, particularly in free-range and organic poultry. These multispecies interactions are part of natural life and as such, add 'naturalness' to the animals' lives and may even alter the range of pathogens on a farm.

14.7.2 To Experience Education and Emotions

Play and social interaction is an important element in animal life. Play among piglets, for example, begins within the first few days after birth, peaking between 2 and 6 weeks of age (Waiblinger et al. 2004). Social models have been shown to play an important role in the behaviour and diet selection of young animals (Thorhalsdottir et al. 1987). They serve to enhance learning efficiency by reducing the need for animals to rediscover foraging information through trial and error. Such learning may be transmitted from the experienced mother to the offspring. Studies have shown that lambs guided by their mothers were able to distinguish between different species of grass and legumes that were either safe or toxic to consume (Ginane and Dumont 2011). Orphaned lambs were unable to do the same. Studies have also shown that there is a transmission of self-medicative behaviour from mother to offspring (Sanga et al. 2011). The transmission of information from one generation to another in non-human animals relies on the memorisation of multi-sensorial cues (Nowak et al. 2011). Very early weaning will thus interrupt the mother-offspring relationship and reduce the share of education and emotions it provides. The cognitive abilities of animals have been substantially overlooked, but an increasing number of studies are showing that farm animals can perform 'executive' cognitive tasks that have typically only been equated with primate intelligence (Morton and Avanzo 2011). This 'executive' function refers to the ability to react adaptively: to learn associations between stimuli, actions and outcomes, and to then adapt their behaviour to changes in the environment. Such skills would be necessary for their survival in nature, which makes them integral aspects of welfare. Studies in sheep have shown that they are able to discriminate between different colours (Morton and Avanzo 2011), recognise and remember faces of different people (Kendrick 1991) and other sheep (Kendrick et al. 2001), and adapt their behaviour to other sheep as part of a social hierarchy.

Farm animals could use these cognitive abilities when maintained in a variable environment. The availability of pasture or paddocks may be a source of variable environments. The intensive production environment, however, particularly those for pigs and poultry, would certainly not provide this opportunity for the animals.

14.7.3 Conclusions and Perspectives

Farm animals' conditions have significantly changed over the last 50 years. These have included positive effects such as more adequate quality of feed and better health management. But there have also been negative impacts such as those brought about by a one-sided ambition to increase profitability. This has led to farms with increased concentrations of animals, and with practices catered towards meeting this goal rather than maintaining animal health and welfare. In one respect, farm animals are increasingly invisible and instead seen as numbers, amounts, units and subjects for trade. Yet, there is also a movement that recognises farm animals as living sentient beings that should be treated with dignity and fairness. The five freedoms for animals provided the first step in recognising, and improving, animal health and welfare. It is our opinion that Nussbaum's framework of capabilities can now act as the much needed next step in providing a strong framework to elevate the lives of farm animals. Where the five freedoms were more based in principle, the capabilities approach supplies practical goals with clear direction in fulfilling them. The capabilities approach has the added advantage of being able to be understood within the same framework as human capabilities and may act as a universal grid for justice for animals and for those people working with them. Many of the capabilities are clearly better fulfilled in organic husbandry, but there is still definite room for improvement, specifically pertaining to the capacities of bodily health and integrity. We should thus be mindful that the increasing organic production continues to build upon principles that emphasize good animal health and welfare, and are not based on the goals and practices that resemble conventional animal husbandry. The capabilities approach can provide a good point of reference in ensuring the ethical development of future farming.

Acknowledgments Caroline Chylinski is a grateful recipient of a EU Marie-Curie PhD grant, "NematodeSystemHealth". Several sets of data presented here were obtained from research funded by the French Alteravi and ANR Pan, as well as the EU Aniplan programmes. Discussions with Nicolas Delon (Paris 1-Sorbonne) on Martha Nussbaum's framework of capabilities were enjoyable and useful for the present review. Mette Vaarst thanks the partners in the ANIPLAN project as well as the Danish partners who contributed to many research and development projects for having shared their knowledge and insight.

References

Anand P, Hunter G, Smith R (2005) Capabilities and well-being: evidence based on the Sen-Nussbaum approach to welfare. Soc Indic Res 74:9–55

Bennedsgaard TW, Klaas IC, Vaarst M (2010) Reducing use of antimicrobials—experiences from an intervention study in organic dairy herds in Denmark. Livest Sci 131:183–192. doi: 10.1016/j.livsci.2010.03.018

Benoit M, Laignel G (2009) Performances techniques et économiques en élevage biologique d'ovins viande: observation en réseau d'élevage et fermes expérimentales. INRA Prod Anim 22:197–206

Berg C (2001) Health and welfare in organic poultry production. Acta Vet. Scan Supp 95:37–45

Brinkmann J, March S, Winckler C (2012) 'Stable Schools' to promote animal health in organic dairy farming—first results of a pilot study in Germany. Agr Forest Res, Special Issue No. 362 (Braunschweig 2012). http://literatur.vti.bund.de/digbib_extern/dn050757.pdf

Cabaret J (1986) 156 plantes pour soigner les animaux. Phytothérapie vétérinaire. Ed. Point vétérinaire, Maisons-Alfort, France, p 192

Cabaret J, Benoit M, Laignel G, Nicourt C (2012) Health advisors in organic meet sheep farms: the role of the veterinarians. Open Vet Sci J 5:7–11

Farm Animal Welfare Council (2012). www.defra.gov.uk/fawc. Accessed 20 Feb 2012

Ginane C, Dumont B (2011) Do sheep (*Ovis aries*) categorize plant species according to botanical family? Anim Cogn 14:369–376

Giudici C, Aumont G, Mahieu M, Saulai M, Cabaret J (1999) Changes in gastro-intestinal helminth species diversity in lambs under mixed grazing on irrigated pastures in the tropics (French West Indies). Vet Res 30:573–581

Guémené D, Germain K, Aubert C, Bouvarel I, Cabaret J, Chapuis H, Corson M, Jondreville C, Juin H, Lessire M, Lubac S, Magdelaine P, Leroyer J (2009) Les productions avicoles biologiques en France: état des lieux, verrous, atouts et perspectives. INRA Prod Anim 22:161–178

Hovi M, Gray D, Vaarst M, Striezel A, Walkenhorst M, Roderick S (2004) Promoting health and welfare through planning. In: Vaarst M, Roderick S, Lund V, Lockeretz W (eds) Animal health and welfare in agriculture. Cabi Publishing, Wallingford, pp 253–277

IFOAM (Internal Federation of Organic Movements) (2012). http://www.ifoam.org/about_ifoam/principles/index.html. Accessed 20 Feb 2012

Ivemeyer S, Smolders G, Brinkmann J, Gratzer E, Hansen B, Henriksen BIF, Huber J, Leeb C, March S, Mejdell C, Nicholas P, Roderick S, Stoger E, Vaarst M, Whistance LK, Winckler C, Walkenhorst M (2012) Impact of animal health and welfare planning on medicine use, herd health and production in European organic dairy farms. Livest Sci 145:63–72. doi: 10.1016/j.livsci.2011.12.023

Jeangène Vilmer JB (2008) Ethique animale. PUF, Paris, p 304

Kaltoft P (1999) Values about nature in organic farming practice and knowledge. Sociol Ruralis 39:39–53

Kendrick KM (1991) How the sheep's brain controls the visual recognition of animals and humans. J Anim Sci 69:5008–5016

Kendrick KM, da Costa AP, Leigh AE, Hinton MR, Peirce JW (2001) Sheep don't forget a face. Nature 414:165–166

Lund V, Anthony R, Röcklingsberg H (2004) The ethical contract as a tool in organic animal husbandry. J Agr Env Ethics 17:23–49

Magdelaine P (2006) Marché Français et Européen des œufs de consommation. Journée Nationale des professionnels de la pondeuse et de l'œuf de consommation. 5 Décembre, Ploufragan, France, p 12

Menke C, Waiblinger S, Studnitz M, Bestman M (2004) Mutilations in organic husbandry dilemmas involving animal welfare, humans and environmental protection. In: Vaarst M, Roderick S, Lund V, Lockeretz W (eds) Animal health and welfare in agriculture. Cabi Publishing, Wallingford, pp 163–188

Morton AJ, Avanzo L (2011) Executive decision-making in the domestic sheep. PLOS One 6:1–8

Nicholas P, Vaarst M (2011) From plan to planning. In: Vaarst M, Roderick (eds) The process of minimising medicine use through dialogue based animal health and welfare planning: workshop report—core organic project nr.: 1903—ANIPLAN. 2011, 22–25. http://orgprints.org/18404/1/18404.pdf. Accessed 18 Nov 2013

Nicourt C, Cabaret J (2014) Animal healthcare strategies in organic and conventional farming. In: Bellon S, Penvern S (eds) Organic farming, prototype for sustainable agricultures. Springer, Chap. 9, pp 171–179

Nowak R, Keller M, Levy F (2011) Mother-young relationships in sheep: a model for a multidisciplinary approach of the study of attachment in mammals. J Neuroendocrinol 23:1042–1053

Nussbaum MC (2001) Symposium on Amartya Sen's Philosophy: 5. adaptive preferences and women's options. Econ Philos 17:67–88

Nussbaum MC (2006a) The moral status of animal. Chronical of higher education, February 3:1–6

Nussbaum MC (2006b) Frontiers of justice: disability, nationality, species membership. The Belknap Press, Cambridge

Piccardi M, Capitaine Funes A, Bo GA, Balzarini M (2011) Impacto del nivel de produccion, estacion de parto y el tipo de servicio sobre la tasa de prenez acumulada a 100 dias en vacas lecheras en la Argentina. AgriScientia 28:127–135

Porcher J (2003) La mort n'est pas notre métier (Death is not our job). Editions de l'Aube

Porcher J (2014) Breaking with animal production paradigm: a major issue for organic animal husbandry. In: Bellon S, Penvern S (eds) Organic farming, prototype for sustainable agricultures. Springer, Chap. 15, pp 279–293

Porcher J, Daru E (2005) Concevoir des alternatives à l'organisation industrielle du travail. Un camion pour le transport et l'abattage des animaux (Designing alternatives to industrial organization of work. A truck for transport and slaughter of animals in farm). FaçSADe (INRA-SAD), n°2005-23. http://www.inra.fr/sciences_action_developpement/publications/facsade/concevoir_des_alternatives_a_l_organisation_industrielle_du_travail_en_elevage. Accessed 18 Nov 2013

Prunier A (2010) Caractéristiques des élevages enquêtés dans le projet Corepig. Prévention de la santé des porcs biologiques. Actes du colloque Corepig, 14 décembre, Rennes. http://www.

interbiobretagne.asso.fr/upload/File/Recherche/Elevage/Corepig/Actes_Corepig_2010.pdf. Accessed 24 Feb 2012
Pryce JE, Conington J, Sørensen P, Kelly HRC, Ryhdmer L (2004) Breeding strategies for organic livestock. In: Vaarst M, Roderick S, Lund V, Lockeretz W (eds) Animal health and welfare in agriculture, Cabi Publishing, Wallingford, pp 357–388
Sormunen-Cristian R, Manninen M, Jauhiainen L, Oksanen A (2008) Effect of mixed grazing of suckler cows and lambs on faecal egg counts and animal performance. In: Hopkins A, Gustafsson T, Bertilsson J et al (eds) Biodiversity and animal feed: future challenges for grassland production. pp 861–863. Proceedings of the 22nd General Meeting of the European Grassland Federation, Uppsala 9–12 June 2008
Sanga U, Provenza FD, Villalba JJ (2011) Transmission of self-medicative behaviour from mother to offspring in sheep. Anim Behav 82:219–227
Sato K, Bartlett PC, Erskine RJ, Kaneene JB (2005) A comparison of production and management between Wisconsin organic and conventional dairy herds. Livest Prod Sci 93:105–115
Sen A (2005) Human rights and capabilities. J Hum Dev 6:151–166
Singer P (1975) Animal liberation. A new ethics for our treatment of animals. Random House, New York
Tague IH (2010) Companions, servants, or slaves? Considering animals in eighteenth-century Britain. Stud Eighteenth Cent 39:111–130
Thamsborg SM, Roderick S, Sundrum A (2004) Animal health and diseases in organic farming: an overview. In: Vaarst M, Roderick S, Lund V, Lockeretz W (eds) Animal health and welfare in agriculture, Cabi Publishing, Wallingford, pp 227–252
Thorhalsdottir AG, Provenza FD, Balph DF (1987) Food aversion learning in lambs with or without a mother: discrimination, novelty and persistence. Appl Anim Behav Sci 18:327–340
Vaarst M, Alroe HF (2012) Concepts of animal health and welfare in organic livestock systems. J Agr Environ Ethic 25(3):333–347
Vaarst M, Nissen TB, Østergaard S, Klaas IC, Bennedsgaard TW, Christensen J (2007) Danish stable schools for experiential common learning in groups of organic dairy farmers. J Dairy Sci 90:2543–2554
Vaarst M, Martini A, Bennedsgaard T, Hektoen L (2004) Approaches to the treatment of diseased animals. In: Vaarst M, Roderick S, Lund V, Lockeretz W (eds) Animal health and welfare in agriculture. Cabi Publishing, Wallingford, pp 279–307
Vaarst M, Smolders G, Leeb C, Henriksen BIF, Roderick S, Walkenhorst M, Winckler C, Gratzer E, Stöger E., Brinkmann J, March S, Ivemeyer S, Mejdel C, Nicholas P, Whistance LK (2011) Education and advisor systems related to dairy organic farming in the participating countries In: Vaarst M, Roderick (eds) The process of minimising medicine use through dialogue based animal health and welfare planning: Workshop report—Core Organic project nr.: 1903—ANIPLAN. 2011, 53–63. http://orgprints.org/18404/1/18404.pdf
Vaarst M, Winckler C, Roderick S, Smolders G, Ivemeyer S, Brinkmann J, Mejdell CM, Whistance LK, Nicholas P, Walkenhorst M, Leeb C, March S, Henriksen BIF, Stöger E, Gratzer E, Hansen B, Huber J (2012) Animal health and welfare planning in organic dairy cattle farms. Open Vet Sci J 5:19–25
Verhoog H, Lund V, Alroe HJ (2004) Animal welfare, ethics and organic farming. In: Vaarst M, Roderick M, Lund V, Lockeretz W (eds) Animal health and welfare in Organic agriculture. CABI Publishing, Wallingford, p 73–94
Veissier I, Botreau R, Perny P (2012) Multicriteria evaluation applied to farm animal welfare: difficulties and solutions from the Welfare Quality (R) project. Prod Anim 23:269–283
Veysset P, Bécherel F, Bébin D (2009) Elevage biologique de bovins allaitants dans le Massif Central: résultats technico-économiques et identifications des principaux verrous. INRA Prod Anim 22:189–196
Waiblinger S, Baumgartner J, Kiley-Wothington M, Niebuhr K (2004) Applied ethology: the basis for improved animal welfare in organic farming. In: Vaarst M, Roderick M, Lund V, Lockeretz W (eds) Animal health and welfare in organic agriculture. CABI Publishing, Wallingford, pp 117–161

Waiblinger S, Boivin X, Pedersen V, Tosi MV, Janczak AM, Visser EK, Jone RB (2006) Assessing the human-animal relationship in farmed species: A critical review. Appl Anim Behav Sci 101:185–242

Warren MA (1997) Moral status: obligations to persons and other living things. Oxford University Press, Oxford, p 274

Wynn SG, Wolpe PR (2005) The majority view of ethics and professionalism in alternative medicine. JAVMA 226:516–520

Zollitsch W, Kristensen T, Krutzinna C, MacNaeihde F, Younie D (2004) Feeding for health and welfare: the challenge of formulating well-balanced rations in organic livestock production. In: Vaarst M, Roderick S, Lund V, Lockeretz W (eds) Animal health and welfare in agriculture. Cabi Publishing, Wallingford, pp 329–356

Chapter 15
Breaking with the Animal Production Paradigm: A Major Issue for Organic Husbandry

Jocelyne Porcher

Abstract Can organic animal husbandry become a prototype for future animal farming? If so, under what conditions? In this chapter, it is shown that organic farmers are currently finding it difficult to distinguish their practices from those imposed in animal production. This applies mainly to the transport and slaughtering of animals, the choice of breeds or genotypes, the specifications and, more broadly, the utilitarian and economic paradigm underlying organic animal husbandry as reflected in the ambiguous term, "organic animal production". Apart from issues of conceptual definition, which are indeed crucial, the actual challenges of organic animal husbandry—between "animal welfare" and biotechnologies—are huge. Data are based on the results of numerous interviews with animal farmers and their employees in France, Belgium, Portugal and Quebec, as well as specific studies on 30 organic animal farmers. The archetypal example of an organic pig farmer attests to the lack of support that organic farmers receive and to the solitude in which they have to deal with ethical issues stemming from the collective organisation of work modelled on the "conventional" approach. The sustainability of animal husbandry and the potential added value of organic methods depend on the capacity of organic husbandry to break with the animal production paradigm, to give up its utilitarian approach and to take the meaning of work for farmers and for their animals into account.

Keywords Animal husbandry · Animal production · Farm animals · Farmers · Work · Organic farming · Utilitarianism · Gift theory · Ethics · *In vitro* meat

In industrialised countries, animal husbandry, in its predominantly animal production form, is currently subject to severe criticism. It is accused of damaging the environment (FAO 2006; Pelletier and Tyedmers 2010), undermining animal welfare,

J. Porcher (✉)
INRA-SupAgro-CIRAD, UMR951 Innovation, Campus SupAgro—2 place Viala,
34060 Montpellier Cedex 2, France
e-mail: jocelyne.porcher@supagro.inra.fr

S. Bellon, S. Penvern (eds.), *Organic Farming, Prototype for Sustainable Agricultures*, 279
DOI 10.1007/978-94-007-7927-3_15,

threatening the health of the public and farm worker, and failing to meet consumer demands in developing countries. Consequently, increased industrialisation (Steinfeld et al. 2006), more severe regulations related to animal welfare, the increasing use of grain-based diets, vegetarianism (Goodland 1997; Holm and Jokkala 2008) and consumption of "*in vitro* meat[1]" (Edelman et al. 2005; Hopkins and Dacey 2008) are now often regarded as the only viable ways of addressing these problems. However, such strategies overlook a third way, one that breaks with animal production and restores a sounder approach to animal husbandry. It is with respect to this third way that organic farmers could play a major role. But how could they achieve this? This is discussed in the present chapter.

I postulate that organic animal farming as currently practiced is not very different from conventional animal production in that it often gives priority to technical and economic outcomes and, in so doing, does not depart from the agriculture industrialisation paradigm and the search for productivity and profit only. My data in this article was taken from numerous interviews that I conducted with animal farmers and livestock workers in France, Belgium, Portugal and Quebec and, more specifically, from interviews focused on human-animal relationships at work with 30 organic animal farmers in these countries. These interviews were held during the period from 2005 to 2007 within the framework of research on pleasure and suffering in animal production (France, Quebec), as well as on the human-animal relationship (France, Belgium, Portugal, Quebec).

I will first review the theoretical and conceptual issues at stake today as a result of the work relationship with farm animals. I will then discuss challenges faced by organic farmers in practicing their chosen form of animal husbandry, e.g., the lack of availability of better alternatives to standard forms of animal transport and slaughter, the economic pressure to work with modern industrial breeds, especially in pig farming, the problems of organic specifications (in the USA, for example, where the organic label is refused to animals who have received any antibiotic treatment whatsoever, even in the case of an emergency), the prevalence of a technical-economic rationale and, more broadly, the repression of emotional bonds with farm animals (Porcher 2006). I will examine the new alternatives to organic farming, namely "animal welfare" and, on the medium-term, the production of "*in vitro* meat", and will investigate the role of animal farmers and their animals in the development of organic farming, based on the particular but archetypal experience of Eric Simon, a farmer in the Hérault department in France. I then conclude by discussing the conditions that would be required for organic animal husbandry to be a paradigm for other forms of animal husbandry, in view of the current expansion of industrial organic production.

[1] The production of cultured meat *in vitro* is based on muscle tissue engineering techniques.

15.1 Behind Words: Contrasted Theories, Concepts, Practices and Issues

The modernisation of agriculture or, in fact, its industrialisation since the 18th and 19th centuries, was accomplished with violence like in other sectors (textiles, for example), and was met with resistance by a large number of subsistence farmers. In the name of science and social progress, animals were reduced to machines as science and technology replaced knowledge based on a balanced relationship with animals and their well-being. That is why organic agriculture was born in the early 20th century in Europe—to build an alternative type of agriculture for farmers and consumers.

My analysis of the international bibliography on organic animal farming shows that the question of the meaning of work, when it is raised, and of the values that it conveys for individuals is formulated in terms of the values of organic agriculture (Kaltoft 1999) in general, and rarely refers to a moral relationship with farm animals as working partners. It is via the question of animal welfare that ethical questions arise and even then, primarily, from a theoretical point of view (Mepham 2006). The most widely used theoretical frameworks are those of utilitarianism, which refer to the notion of a contract (Singer 1975; Larrère and Larrère 2000; Lund et al. 2004). However, what appears in farmer's discourses regarding their work and their relationships with animals is much more the notion of gift than of contract. Gift theory (Mauss 1990; Godbout et al. 1998) enables us to understand why animal farmers and organic animal farmers suffer from the inability to give their animals what they would like to, notably, a good life, a life worth living (FAWC 2009), that could justify their eventual slaughter (Porcher 2002b, 2003; Mouret 2009). As stressed by Budiansky (1992) and Haraway (2003), domestication is a co-evolution process. From the beginning of domestication, we, humans and animals, have been engaged in a triple obligation: to give, to receive and to reciprocate (Porcher 2002b, 2011b). A gift is at once self-interested and disinterested, and involves both freedom and obligation (Caillé 1994; Osteen 2002). The problem today is to reconsider the debt that we owe to animals. From the point of view of farmers involved in animal husbandry and their employees, animal production takes everything from animals and gives nothing back to them, as a result of industrialisation. That is why it is necessary to understand the differences between animal production and animal husbandry and what organic farming can really do for farm animals and not only for consumers or the environment.

Several terms exist to refer to the working relationship with farm animals: breeding, animal rearing, animal husbandry, animal farming, animal production, livestock production, and so on[2]. In the FAO report, "Livestock's Long Shadow", published in 2006, which had considerable media aftereffects like many other reports and articles, these terms are used interchangeably without any specific precision.

[2] I would like to thank Ben Mepham, who I first asked to write this paper with me, for his help in clarifying some terms in English.

Organic animal farmers have to deal with many difficulties because the particularities of animal husbandry and of organic animal husbandry are not taken into account (Aroe 2001). Organic specifications refer to animal production as if animal production and animal husbandry were synonymous. For example, European organic specifications stipulate that "livestock farming is a land-related activity" (Art. 16 RCE 834/2007), which means that what is consumed and excreted remains within the same farm (or region). However, regardless of the type of production, *animal production* is not linked to the land. Instead, it is linked to international trade, unlike *animal husbandry* that is indeed linked to the land.

I use the term *animal husbandry*, which denotes the historical working relationship that we have always had with farm animals. By historical, I mean anchored in history, including in its actualisation and not just belonging to the past. Animal husbandry relates to multiple rationales, the main one being relational (Porcher 2002a, 2004). I use the term *animal production* to denote that or what we also refer to as "livestock production" and intensified, industrial production (also known as "animal factories" or "intensive livestock production"), where the prevailing rationale underlying the organisation of work is the search to maximise profit and productivity. I use the term animal farming or organic animal farming in a more general sense.

Apart from the need for semantic clarification—because the multiplication of terms clearly does not facilitate an understanding of the issues—the possibility of a future development of organic farming also depends on the anchorage of the meaning conveyed by words. Should the term "organic" be added to animal husbandry or to animal production? What value can "organic" add to animal husbandry?

15.2 A Gap Between Ideal and Real Work

Agencies for organic farming use the image of farm animals and their welfare to promote organic agriculture by providing evidence of its difference from animal production. Most of the animal farmers we met have an ideal image of their work and aspire to interact with animals in accordance with their values. This is the case for animal farmers and the majority of organic animal farmers as well (Porcher 2002a; Wilkie 2010).

Yet, as Allen and Kovach (2000) point out, "organic labelling is simply not enough to create an agri-food system that provides real value" because organic animal farming is becoming an industry based on the "conventional agriculture" paradigm. This trend partly results from the lack of real interest that organic agencies have expressed in relation to animals. The basis of common representations seems to be that animal husbandry is a necessary evil rather than a key component of organic farming because of the bond with animals. The role of animals is secondary compared to the other concerns of the promoters of organic farming, as are the questions of work and, more broadly, social issues (Herman 2008). As regards animal husbandry, organically-produced meat is promoted as "good for the environment" before being good for the animals[3], and the criteria put forward to emphasize the "good life of animals on organic farms" are sometimes deceptive. Hence, the argument advanced by the promoters that "they grow at their own pace to provide quality meat" is out of touch with the reality of the actual life expectancy of pigs and poultry.

This argument has two implications. First, animals must live peacefully before being slaughtered; second, the quality of their meat is better. Indeed, the quality of meat for consumers[4] is highly dependent on the age of animals at slaughter. In the case of pigs, the slaughtering age for most industrial breeds is increasingly the same as that in industrial farming (5–6 months); in the case of poultry, chickens are slaughtered later (81 days instead of 40 or less) but, as in pig farming, the problem of breed or genotypes is real. Most organic chickens and laying hens are hybrid animals produced by international corporations specialised in genetics. Furthermore, many chickens and laying hens are raised under intensive conditions because the farmers who raise them work for big groups[5] and have a work organisation that leaves them with little autonomy.

What does this gap signify? It encompasses various elements that weaken the emotional bond between farmers and animals, and raises the question of the moral commitment to animals. Life expectancy is not a detail. It is part of what Sen and Nussbaum (Terenschenko 2010) refer to as "capabilities" and that, in reference to Marx (Marx and Engels 1975), can be linked to work and the opportunity for farm animals to live a life that can offer the possible expression of the animal's potentialities (Porcher 2011b).

15.3 Is This Organic?

15.3.1 Industrial Animal Breeds (the Case of Pigs)

The industrialisation of pig farming has led to the disappearance of dozens of pig breeds. In France, only six old breeds remain, with a total of less than two thousand sows[6]—compared to 1.3 million industrial sows. Boars and sows produced by cross-breeding for industrial production (Large White, Landrace, Piétrain, Duroc and hyper-prolific Chinese breeds) are the property of private companies, unlike old breeds that historically belong to nobody and therefore to everybody. Breed

[3] For example, "Good for nature, good for us", as can be seen on a poster for organic meat. (http://www.produitslaitiersetviandebio.com/).

[4] Meat quality is not a concept. There are several qualities of meats that are sometimes conflictual: for consumers (taste, tenderness), for meat transformers (pH, glycogen, etc.), for butchers (average fat content, conformation, etc.).

[5] For example, "Douce France", part of the group, "Gastronome" (annual turnover: M€ 745) itself part of the group, "Terrena" (annual turnover: € 3.9 billion).

[6] Cul noir du Limousin (136 sows), Pie noir du Pays Basque (448), Bayeux (229), Gascon (871), Blanc de l'Ouest (115), Corse (150).

selection decisions concerning industrial breeds depend on company strategies and on the interests of the hog industry. For example, the selection criterion, "number of piglets born", concerns industry because profits depend on the tonnage of meat produced. That is why industrial sows can produce 18, 20, 22 or more piglets per litter, even if some of them are stillborn or so weak that workers have to kill them (Porcher 2010a, 2011a). With regard to the disappearance of old animal breeds, industrial breeders argue that their genes are stored at –196 °C for posterity's sake. However, farmers are interested in animals, not genes:

> I don't think there is any point in keeping genes in the freezer, assuming that if we need them in 100 years, we'll just take them out. There will be no more farmers able to rear a pig and no consumers able to taste its meat (Lauvie 2007, p. 250).

The genetic origin of animals remains an issue when implementing organic specifications. Organic processing firms request meat characteristics that are much the same as in industry, i.e., above all, no fat and standard conformation or size for chicken! That is why most organic pig farmers work with industrial breeds, except those who process and sell the meat themselves. This has important consequences for animals and for the sustainability of pig farming since a significant part of the work organisation is determined by the breed. The life expectancy of animals is similar to that in industry because after a certain age, the quality of the meat of these industrial breeds declines. Hence, meat quality for consumers may not be very different from that of industrial meat. That is why, in France, the "red label" for pigs is associated with local production, whereas the organic label for pigs is more commonly associated with environmental protection[7].

If old pig breeds disappear—and they will if no more farmers are able to work with them—how will pig farmers be able to raise organic pigs, considering their dependence on industrial systems? How will they be able to develop their specific values if they are unable to give a good life to their animals?

15.3.2 Slaughterhouses

> I don't usually go to the slaughterhouse, but I sometimes have to. They want us to take the cows to the slaughterhouse the day before they are killed, but we would rather they didn't have to spend the night since we think they are afraid. We have nothing against the animals and no reason to mistreat them (a French animal farmer).

Legislation requires the killing of animals to take place in a slaughterhouse. However, this slaughterhouse centralisation process has eliminated most small public facilities and forced farmers to take their animals to distant slaughterhouse factories—even if they have only a few animals to slaughter. Instead of taking cows in the early morning to a small abattoir where the farmer knows the workers and can trust them, he has to take them the previous day or entrust them to a trucker, letting them go without any possible control over the procedures and no farewell. This procedure is the cause of intense suffering for farmers and, we can assume, for the animals as well.

---

[7] Ecocert. "Le guide pratique d'élevage porcin".

It is a cause of suffering for farmers because they know what work in factory slaughterhouses is, and organic or not, cows and pigs will follow the very same industrial process, at least in France[8]. Farmers are emotionally involved with their animals and therefore feel that they are abandoning them and breaking their commitment towards both the animals and consumers (Porcher 2003; Wilkie 2010). As mentioned above, when the human-animal relationship in animal husbandry is considered in terms of the gift theory, a good life, a life worth living, is seen as a gesture of gratitude for the gift that animals give us of their presence, their work and their death. A good life necessarily includes a good death, i.e., a minimum of respect, presence and good sense. Since the slaughterhouse factory cannot provide this, farmers feel guilty. They feel that they are letting their animals down, not fulfilling their commitments. Moreover, they have the impression of lying to consumers who believe that differences between industrial production and organic farming include the animal's death.

Delegation is a cause a suffering for animals because transport to the industrial slaughterhouse is a complete rupture with the good life that most organic farmers endeavour to give them. Instead of meadows, relationships with other animals and with humans, friendly spoken words, caring looks and contacts, they are confronted with an incomprehensible environment: unknown people in a hurry who do not look at or speak to them, unfamiliar noises and smells, frightened congeners, huge buildings, etc.

Following our publication that proposed a concept of a truck-slaughterhouse to slaughter animals at the farm (Porcher and Daru 2005), I received, and still receive, many e-mails, letters and phone calls from organic and non-organic farmers throughout France, asking me where this truck could be found or when it will be produced and commercialised. In 2004, the French Ministry of Agriculture informed me that slaughter *in situ* was strictly forbidden by French regulations[9]. That meant that my proposal was irrelevant. Since then, groups of animal farmers have worked on the subject and it appears that the ban could be lifted. It is the necessary first step and the first condition for a truck manufacturer to be interested. Otherwise, many farmers or groups of farmers are attempting to maintain their small local slaughterhouses, but it is a difficult battle considering that big industrial groups need to increase their production to make their investments profitable.

15.3.3 Organic Specification Problems

Some organic specifications entail problems. In pig farming, sows must suckle their piglets for at least 40 days. In industrial systems, the age of weaning is supposed to be 21 days. In fact, it is often less (15 or 7) because of the constant pursuit of productivity: reducing the production cycle is necessary to increase productivity. In organic farming, the age of weaning is around 60–80 days for

[8] Ecocert. Guide pratique, abattoirs, ateliers de découpe et boucheries. ID SC190, 23/11/2010.

[9] Except pigs for family consumption.

an old pig breed because the sows have 8–10 piglets and because piglets grow slowly. Gradually, the sow pushes away her piglets. This does not apply to industrial breeds where sows are tired and weaken before 40 days. Hence, this organic specification that is good for the piglet is not so good for the sow because of the genetic choices of the industrial system, even if industrial sows have outdoor behaviours that are very similar to those of old breeds and even to those of wild boars (Chartier and Porcher 2009).

We met organic pig farmers in Quebec who work for the US market and therefore have to follow US organic specifications that prohibit any antibiotics at any stage, even in the case of serious disease. Farmers are forced to leave their sick animals to die or else to kill them, which is not the case in Europe. The US organic production that considers the health of consumers, with no consideration whatsoever for the animals, is difficult for farmers because it conflicts with their moral values and their duties towards animals. Justification of these specifications refers to a sort of neo-Darwinism: animals that are not strong enough should die; that is the law of nature. However, animal husbandry does not refer to nature but to work. That is why this justification is not strong enough to prevent farmers from suffering. Note that, from an economical and utilitarianism point of view, these animals cannot be sold to conventional markets overnight.

15.3.4 For Farmers, Animals are the Forgotten Partners of the Organic Movement

Organic animal farming today, like organic agriculture, seems to be reduced to no pesticides, no chemical fertilisers, no drugs. The theoretical underpinnings of work are supposed to revolve around "modernisation", based on measurements, figures, diagrams, etc. The technical-economic rationale imposes its way of considering work and overlooks the fact that work is part of life, i.e., sensitivity and subjectivity (Dejours 2009). This is contrary to what makes sense for farmers: a sensible and sensitive bond with animals based on the inter-subjectivity of the work relationship.

Many animal farmers are aware that the aim of their work with animals is not profit alone. Farmers must earn their living, but money is not the main reason why they choose to work with domestic animals. Above all, they want to live with animals and working with them is the best way to do so (Porcher 2011b). Hence, the primary purpose of their work concerns relationships with animals. It therefore has an emotional dimension, as well as identity, moral and philosophical dimensions (Porcher 2002a; Mouret 2009).

> Anyway, when animals are well, it makes those who work with them happy. This, I can tell you. It has its difficulties but it provides happiness in return. If it makes you live well, fine, but we don't live only for money and I understand that more and more. Finally, at the end of your life, when you die, does it matter whether you have 50 € or 500 million? (a French animal farmer).

> I don't ask for any subsidies. I don't receive a single euro. The way I watch my animals, count them, watch a calf being born, if they are healthy, is a source of well-being. [...] I am here to take care of the animals... their calm, their lack of concern, and the way they consider life... Actually I would like to live like them. But it is not possible (a Portuguese animal farmer).

As in animal production, the management of organic animal farming overlooks the fact that animal farming is done by farmers and by animals, and that animals matter. For farmers, they are not just things or machines but working partners. How can organic farming take the contribution of animals to work into account?

15.4 New Paradigms?

15.4.1 Animal Welfare

The question of animal welfare enables us to highlight the issue of animals' living conditions in organic farming. Although some animal production systems take animal welfare into consideration[10], and would have no difficulty qualifying for the label, what welfare can organic animal production, as it is currently defined, claim to ensure[11]? Why will consumers concerned about animal welfare buy organic meat if conventional products are supposed to guarantee that products have an animal welfare added value?

The creation of a welfare quality label implies the ability to quantify welfare and to reduce it to a set of simple measures (Botreau 2008). There is nothing new about this endeavour to reduce animal welfare to biological and behavioural measurements, all other things being equal, that is, without calling the organisation of work in animal production into question. It has underlain the issue of animal welfare since the 1980s (Porcher 2005) and involves the training and formatting of workers in industrial and intensified systems (Hemsworth et al. 2002). I think that animal welfare scientists as well as some lobbyists are totally in the dark as to what animal husbandry actually is, and this ignorance is reflected in European legislations, as well as initiatives such as "Welfare Quality", which are set up and run without animal farmers and their employees, and without the animals.

Breaking with the industrial organisation of work with farm animals is the basic condition for animal welfare because the industrial world is irreconcilable with animal husbandry.

[10] See, for example, the Freedom Food label in the UK that supplies McDonalds. http://www.freedomfood.co.uk/news/2013/04/mcdonalds.

[11] See the winners of the Animal Welfare Awards given by the CIWF (Compassion In World Farming). For example: McDonalds, Subway, Sodexo, Marks & Spencer, Coca-Cola, Lesieur, Barilla, etc.

15.4.2 In Vitro *Meat*

On the longer term, and in addition to what was discussed above, organic animal production, like all animal production, runs the risk of being replaced by bio-technological innovations such as *in vitro* meat production[12] (Hopkins and Dacey 2008; Porcher 2010b). Since animal husbandry has become the *bête noire* of animal rights activists and some ecologists, *in vitro* meat production is acclaimed as the ultimate innovation in favour of animal welfare. Agri-food firms support this type of initiative since no animals would be required to produce pork, poultry or bovine "matter"[13] for fast foods (sausages, hamburgers, nuggets, etc.) or mass consumption. Animals are a huge constraint for these firms because they are alive, sensitive and communicative and, as a result, bonds exist between them and their farmers—bonds that slow down and complicate work (Porcher 2006).

Can organic animal production provide an alternative to *in vitro* meat that is defended both by agri-food firms and by animal rights activists and ecologists? Probably not, because for the same quality, that is, if organic farming is content simply with animal production, consumers are likely to prefer *in vitro* chicken or pork that is guaranteed to be free of animal suffering and death, rather than chicken or pork that is "good for the environment"—which *in vitro* meat production also claims to be (Tuomisto 2011). Indeed, without animal production, there is no greenhouse gas effect, no deforestation, no water pollution, etc.

15.5 Organic Animal Husbandry?

We see that the challenges in organic animal husbandry are huge. Based on a recent experience with Eric Simon[14]—which is quite common among organic animal husbandry farmers in France—and from his own point of view, I would like to show the limits of current organic farming for an animal husbandry farmer, and the hopes that he had for organic animal husbandry.

15.5.1 Breeding Pigs Organically: Eric Simon's Experience

When I met Eric Simon, his pig farming system included 80 sows and 70 ha of "causses" (small limestone plateaux in southern France) with pastures and woods.

[12] See Sorente (2012) for technical details.

[13] The term "minerais" (ore) is used in French in animal production supply chains to refer to meat and, thus, to highlight representations of work with animals as an activity consisting of the extraction of animal matter.

[14] Eric Simon has just published a handbook on issues related to animal well-being, environmental protection and food safety: Simon 2013. Une vraie vie de cochons, élever des truies en agriculture biologique. Educagri Editions, 90 pages.

All of the animals lived outdoors. Eric Simon has been farming pigs since 2002 in France. A former lecturer in animal production science, he decided to take up farming and experiment with free-range pigs, both to test his theoretical knowledge and to provide his sows and himself with living conditions in keeping with his values. As he put it: "I had to learn to work to suit the sows", i.e., he first had to establish a trusting relationship that enabled him to work with animals who were free to move about as they wished—which means free to come to you or not. This freedom, which Ervin Straus (2000) links to the "symbiotic understanding" that binds humans and animals, is often described by animal husbandry farmers as a *sine qua non* condition for a good life for animals and goes hand-in-hand with the right to pastures as a must for all species (Porcher 2002a).

The crisis in the pork industry and, above all, the obligation to castrate piglets hung heavily over Eric Simon. He decided to stop castrating and instead sold his young animals directly to consumers. Apart from a workload that was far too heavy, Eric realised that although many consumers called for animal welfare, few were actually prepared to support farmers' initiatives in this respect. For example, even though Eric was awarded the OABA[15] prize in 2004 for the best animal husbandry farmer, this had no effect on his sales. He consequently switched to organic farming and sold his piglets in the organic supply chain. He then entered a partnership with another pig farmer to birth and fatten non-castrated pigs before selling them to the meat-salting industry. Everything was done free-range and provided the animals with living conditions conducive to their welfare. Moreover, his farm was promoted by an animal rights organisation (PMAF[16]). However, faced with mediocre technical results and a difficult context for pork production, he decided to give up fattening pigs and to revert to selling young pigs directly to restaurants. Today, Eric describes himself as being "free of the supply chains", but he is also very divided as to what organic farming actually contributed to his farming practice.

From a positive point of view, Eric Simon considers that organic methods enabled him to continue free-range farming. He explains, however, that paradoxically, organic farming makes it economically difficult to work with old pig breeds. Given the price of organic feed and the slow growth of these breeds, the production costs are much higher. The selling price must therefore also be higher. Eric emphasizes, "activists for organic farming don't eat meat and those who do want it to be good, whether it's organic or not". As far as animal welfare is concerned, he says that organic farming is not a solution. The piglets are also castrated and this practice is not challenged any more than is the use of industrial breeds[17].

Eric Simon believes that the potentially fast growth of the organic market is leading conventional firms towards organic products. Technicians and engineers working in industrial production systems then impose their ways of producing and

[15] Association for Assistance to Slaughterhouse Animals (*Œuvre d'Assistance aux Bêtes d'Abattoirs*).

[16] *Protection Mondiale des Animaux de Ferme*, affiliated with Compassion In World Farming.

[17] The Improvac vaccination (Pfizer) against boar taint will make the problem even more complex. Is Improvac compatible with organic regulations?

thus influence the specifications. This, he adds, can lead many animal husbandry farmers to fail and also to discredit organic farming due to the ignorance of these technicians and engineers concerning the practical conditions in which organic animal husbandry is practiced. The experience of organic farmers, at least in pig farming, is not taken into consideration: "it is neither sought after nor valued". Eric explains that the situation is the same downstream: "the animal resource is treated like an inanimate natural resource like iron ore". The comparison even extends to the animal production supply chain, using words such as "minerals" and "extraction". The credo, "make organic accessible to everyone", conceals the same logic everywhere in the global market: produce more at a lower cost. Eric says: "The animal is still the mineral, the farmer the miner".

In the final analysis, Eric concludes with: "organic animal husbandry means, above all, feeding the animals with organic cereals. But is this really what being an animal husbandry farmer is all about?"

15.5.2 An Immense Amount of Know-how

How can farmers live and work sustainably and intelligently with their animals? In what terms can we formulate the ethical question of our relationships with farm animals and of their slaughter?

The animal farmers—both organic and not—that I have met have personal answers to these questions, and have adapted their farming systems accordingly. Eric Simon has developed an animal husbandry system that enables him to avoid castrating his piglets because castration was emotionally and morally problematic for him. This is not the case of many other farmers. Some, like Thierry Schweitzer in Alsace, have first chosen to anesthetize the piglets then, instead, to use Improvac®. The same applies to the new rules applying to the slaughtering of animals: everyone deals with the constraints and events at work in their own way.

That is why the animal husbandry system reflects the farmer or the collective work that developed it. It is also why these systems can differ so widely, even when they concern the same species. This is one of the richest aspects of animal husbandry that industrialisation has destroyed by standardising practices. An immense amount of know-how has thus been lost by farmers and their animals.

Farmers who claim to practice animal husbandry all have in common their concern to give their animals a good life in which the legitimacy of their relations with them is grounded. Giving—receiving—giving back: these are the three aspects of a gift. Giving life/giving work/giving a good life/giving recognition. It is life that circulates between animal husbandry farmers and their animals, life itself, transcending individuals, rooted in animal and human genealogies.

> We were educated on our parent's farm, a little farm with ten cows that we milked every day. Our parents milked of course too. Us, we were used to being in a relationship with the cows. We knew all of our cows by their name—it was a direct relationship. Each child had its own cow. Our relative, it was our cow. With the Blanc Bleu, it was over. We didn't have time to bond because they left the farm so quickly (a Belgian farmer).

15.6 An Organic Animal Husbandry Design?

Animal husbandry is disappearing, engulfed by animal production reduced to its simplest expression, as if we were beings with no past. What organic farming can do for animal husbandry is, above all, enable it to be perpetuated: first, by formalising and actualising its existence, along with that of animal farmers and their animals, and by recognising animal husbandry as an essential part of agriculture and, second, by recognising that farm animals have a legitimate role in society. This means taking the social question seriously: the living conditions at work not only of the farmers and their employees but also of the animals.

Moreover, it would be necessary not to train animal husbandry farmers like in industrial agriculture, but to help them to construct and to make sustainable the animal husbandry system that suits them and that suits consumers. As Eric Simon emphasizes, organic animal husbandry implies giving up cheap meat, but the price that consumers do not pay is paid by others—by the animals and by their owners, usually in terms of suffering.

That is why research is needed on the following two issues: (1) why and how organic animal farming could support animal husbandry; (2) the role of animals in the construction of breeding systems, with respect for their needs and contributions (Porcher and Schmitt 2012).

15.7 Conclusion

Making animal husbandry a common good and farm animals a common link, changing our utilitarian relationship with animals and renouncing our quest for profit at all costs would enable us to hope for a future better than one in which only the fittest can survive. This is what organic farming can contribute to animal husbandry, *with* farmers and their animals and not without them or against them.

Organic animal husbandry should be designed to satisfy the conditions for a sustainable animal husbandry. The work issue is the key to this design. Do we work with animals for money or to share their lives? If we knew why we worked with animals, we would know how to work and live with them, and the animals would know why and how to work and live with us. To make the difference, organic animal husbandry has to say why. Then, and only then, will animals, farmers and consumers together be able to say how.

References

Allen P, Kovach M (2000) The capitalist composition of organic: the potential of markets in fulfilling the promise of organic agriculture. Agric Human Val 17:221–232

Alroe HF, Vaarst M, Kristensen ES (2001) Does organic farming face distinctive livestock welfare issues? A conceptual analysis. J Agric Environ Ethics 14:275–299

Botreau R (2008) Evaluation multicritère du bien-être animal. Exemple des vaches laitières en ferme (Multicriterion evaluation of animal welfare. Example of on-farm dairy cows). Thèse AgroParisTech. Spécialité Productions animales

Budiansky S (1992) The covenant of the wild. Why animals chose domestication. Yale University Press

Caillé A (2009) Théorie anti-utilitariste de l'action. Fragments d'une sociologie générale. (Anti utilitarist theory of action. Fragments of a general sociology). La Découverte/MAUSS

Caillé A (1994) Don, intérêt et désintéressement (Gift, interest and disinterest). La Découverte

Chartier A, Porcher J (2009) Influence des conditions d'élevage en plein air sur le comportement des truies gestantes. Influence of extensive rearing conditions on the sow's behaviour) Journées de la Recherche Porcine, 41 http://www.journees-recherche-porcine.com/texte/2009/conduite/pbe03.pdf

Dejours C (2009) Travail vivant (living work). Tome 1 et 2. Editions Payot

Edelman PD, Mcfarland DC, Mironov VA, Matheny JG (2005) *In vitro*-cultured meat production. Tissue Eng 11(5/6):659–662

FAO (2006) Livestock long shadow. Environmental issues and options. http://www.fao.org/docrep/010/a0701e/a0701e00.HTM

Farm Animal Welfare Council (2009) Farm animal welfare in Great Britain: past, present and future. http://www.fawc.org.uk/pdf/ppf-report091012.pdf

Godbout JT, Caillé A, Winkler D (1998) The world of the gift. Mac Gill-Queen's University Press, University Press, Montreal

Goodland R (1997) Environmental sustainability in agriculture; diet matters. Ecol Econ 23:189–200

Haraway D (2003) The companion species manifesto: dogs, people and significant otherness. University of Chicago press

Hemsworth PH, Coleman GJ, Barnett JL, Borg S, Dowling S (2002) The effects of cognitive behavioral intervention on the attitude and behavior of stockpersons and the behavior and productivity of commercial dairy cows. J Anim Sci 80:68–78

Herman P (2008) Les nouveaux esclaves du capitalisme: agriculture intensive et régression sociale (New slaves of capitalism. Intensive agriculture and social regression). Editions Au Diable Vauvert

Holm J, Jokkala T (2008) The livestock industry and climate—EU makes bad worse. Published by the delegation of the Swedish Left Party in GUE/NGL—http://www.meatclimate.org/home

Hopkins PD, Dacey A (2008) Vegetarian meat: could technology save animals and satisfy meat eaters? J Agric Environ Ethics 21:579–596

Kaltoft P (1999) Values about nature in organic farming practice and knowledge. Sociologia Ruralis 39(1):39–53

Larrère R, Larrère C (2000) Animal rearing as a contract. J Agric Environ Ethics 12:51–58

Lauvie A (2007) Gérer les populations animales locales à petits effectifs: approche de la diversité des dispositifs mis en œuvre (Managing rare breeds: diversity of actor strategies). Thèse d'Agroparistech. Gestion des ressources génétiques animales. http://pastel.paristech.org/3464/01/th%C3%A8se_LauvieAnne2007.pdf. Accessed 15 June 2007

Lund V, Anthony R, Rocklinsberg H (2004) The ethical contract as a tool in organic animal husbandry. J Agric Environ Ethics 17:23–49

Marx K, Engels F (1975) L'idéologie allemande, Editions sociales

Mauss M (1990/1924) The gift (with an introduction by Mary Douglas). Routledge, London

Mepham B (2006) The ethical matrix as a decision-making tool, with specific reference to animal sentience. In: Turner J, D'Silva J (eds) Animals, ethics and trade: the challenge of animal sentience. Earthscan, London, pp 134–145

Mouret S (2012) *Élever et tuer des animaux*, Paris, Puf, Collection "Partage du savoir"

Osteen M (ed) (2002) The question of the gift: essays across disciplines. Routledge

Pelletier N, Tyedmers P (2010) Forecasting potential global environmental costs of livestock production 2000–2050. Proceedings of the National Academy of Sciences of the USA. 8–2010

Porcher J (2002a) Eleveurs et animaux, réinventer le lien (Animal farmers and animals, reinventing the bond). PUF

Porcher J (2002b) L'esprit du don, archaïsme ou modernité de l'élevage: éléments pour une réflexion sur la place des animaux d'élevage dans le lien social (The gift spirit, arcaism or modernity of animal farming? Fragments for a reflection about the rôle of animals in the social bond). *Revue du Mauss* n°20, Quelle autre mondialisation? Editions La Découverte/MAUSS, pp 245–262, online http://www.cairn.info

Porcher J (2003) La mort n'est pas notre métier (Death is not our job). Editions de l'Aube

Porcher J (2005) Le "bien-être animal" existe-t-il? (Does animal welfare exist?) Economie rurale n 285:88–94

Porcher J (2006) Well-being and suffering in livestock farming: living conditions at work for people and animals. Sociologie du travail 48(1):e56-e70 (August 2006)

Porcher J (2010a) Cochons d'or. L'industrie porcine en questions (Gold pigs. Questionning pig industry). Editions Quae

Porcher J (2010b) Le stade ultime des productions animales: la production de viandes in vitro (the ultimate stage of animal production: in vitro meat). La Revue Parlementaire. no. 1057. Europe: quelle PAC pour 2013? Oct/nov/dec 2010, pp 97–104

Porcher J (2011a) The relationship between workers and animals in the pork industry: a shared suffering (2011). J Agric Environ Ethics 24:3–17

Porcher J (2011b) Vivre avec les animaux, une utopie pour le 21ème siècle (living with animals, a 21st century utopy). Editions La Découverte/MAUSS

Porcher J, Daru E (2005) Concevoir des alternatives à l'organisation industrielle du travail. Un camion pour le transport et l'abattage des animaux (Designing alternatives to industrial organization of work. A truck for transport and slaughter of animals in farm). FaçSADe (INRA-SAD), n°2005–23 http://www.inra.fr/sciences_action_developpement/publications/facsade/concevoir_des_alternatives_a_l_organisation_industrielle_du_travail_en_elevage

Porcher J, Schmitt T (2012) Dairy cows: workers in the shadows? Soc Anim 20(1):39–60

Porcher J, Cousson-Gélie F, Dantzer R (2004) Affective components of the human-animal relationship in animal husbandry. Development and validation of a questionnaire. Psychol Reports 2004 95:275–290

Singer P (1975) Animal liberation. A new ethics for our treatment of animals. Random House, New York

Sorente I (2012) La viande in vitro, cuisine cellulaire. Le Monde Sciences et Techno. http://www.lemonde.fr/sciences/article/2012/06/21/la-viande-in-vitro-cuisine-cellulaire_1722680_1650684.html. Accessed 21 June 2012

Steinfeld H, Gerber P, Wassenaar T, Castel V, Rosales M, de Haan C (2006) Livestock's long shadow: environmental issues and options. FAO, Rome (Italy). Livestock, Environment and Development Initiative; FAO, Rome (Italy). Animal Production and Health Div., p 390

Straus E (2000) Du sens des sens (The meaning of senses). Editions Jérome Millon

Terenschenko M (2010) Amartya Sen, Martha Nussbaum et l'idée de justice (Amartya Sen, Martha Nussbaum and the idea of justice). Revue du MAUSS permanente. [http://www.journaldumauss.net/spip.php?article719]. Accessed 14 Oct 2010

Tuomisto HL, Teixeira de Mattos MJ (2011) Environmental impacts of cultured meat production. Environ Sci Technol 45(14):6117–6123

Wilkie RM (2010) Livestock/deadstock. Working with farm animals from birth to slaughter. Temple University Press, Philadelphia

Chapter 16
Food Quality and Possible Positive Health Effects of Organic Products

Denis Lairon and Machteld Huber

Abstract For decades, organic agriculture developed strategies to produce plant and animal foods with mandatory high standards based on certification at the production and processing levels. This coincided with the growing demand of consumers for accessible, environmentally-friendly, nutritional and safe foods. In this context, although limited and difficult to generalise due to the existence of many conflicting factors, comparative studies have been dedicated to the nutritional content and safety characteristics of organic vs. conventional foods. In this chapter, we review the main characteristics of organic foods in terms of their nutritional, safety and health aspects. The main findings of this review are: (i) a number of organic plant products tend to contain more dry matter, some minerals (Mg) and antioxidants (phenolics/flavonoids, salicylic acid); (ii) organic cow and chicken meats and cow's milk contain more omega-3 (n-3) polyunsaturated fatty acids; (iii) the vast majority (94–100 %) of organic food does not contain any residues of synthetic pesticides; (iv) organic vegetables contain significantly less nitrates; and (vi) organic cereals generally contain less protein but overall comparable mycotoxin levels are the same as conventional ones. Additionally, some health studies have highlighted benefits from organic dairy products for ectopic allergy in young children and some positive health indications in animals. Potential methods to evaluate the authenticity and quality of organic foods are discussed.

Overall, it appears that organic agricultural systems, just like pioneers in sustainable agriculture, have already proven to be capable of growing foods with high

D. Lairon (✉)
Univ d'Aix Marseille-INSERM-INRA, UMR1062 NORT Nutrition, obésité et risque thrombotique, 13385, Marseille, Cedex 5, France
e-mail: denis.lairon@univ-amu.fr

INSERM, UMR1062 NORT Nutrition, obésité et risque thrombotique, Cedex 5, 13385, Marseille, France

INRA, UMR1260 NORT Nutrition, obésité et risque thrombotique, Cedex 5, 13385, Marseille, France

M. Huber
Department of Healthcare & Nutrition, Louis Bolk Institute, Hoofdstraat 24, 3972 LA, Driebergen, The Netherlands
e-mail: m.huber@louisbolk.nl

S. Bellon, S. Penvern (eds.), *Organic Farming, Prototype for Sustainable Agricultures*,
DOI 10.1007/978-94-007-7927-3_16, © Springer Science+Business Media Dordrecht 2014

quality standards, but scientific evidence regarding the effects of organic foods on health is still lacking.

Keywords Sustainable agriculture · Organic agriculture · Human food · Nutrition · Food safety · Contaminants · Health

Definition

Phenolic acids, flavonoids and polyphenols: These elements are sub-groups of compounds naturally produced by plants as secondary metabolites and possessing anti-oxidant properties.
Pathogenic microorganisms: Among the wide range of microorganisms, some bacteria, fungi and viruses are pathogenic for other living organisms because they generate toxic molecules.
Contaminants: These substances are harmful compounds accumulated in foods, of either endogenous (e.g., nitrate) or exogenous origin (e.g., pesticide residues, heavy metals).
Immune responses: The immune system protects the organism from threatening contacts with the outer world via two sub-systems: an inborn system, which is innate and based on natural resistance, and an acquired system, which is geared to specific targets.
Rumenic and trans-vaccenic acids: Members of the large family of fatty acids, these acids belong to the conjugated linoleic acids or the trans fatty acids and are produced by the action of rumen microflora on some dietary unsaturated fatty acids.
Metabolomics: This new area of science refers to the measurement of the metabolome, i.e., the wide array of small-size metabolites present in living tissues or fluids (blood, urine); they are analyzed using nuclear magnetic resonance or chromatography coupled to mass spectrometry.
Robustness: An indication that an organism can function despite the presence of disturbances.
Resilience: A dynamic developmental process that has been operationalised as an individual's/organism's attainment of positive adaptation and competent functioning, despite having experienced (chronic) stress or detrimental circumstances, or following exposure to prolonged or severe trauma.
Proteomics: This new area of science refers to the measurement of the proteome, i.e., the large number of proteins present in living tissues or fluids (blood); they are analysed using 2-D chromatography and mass spectrometry.

16.1 Introduction

16.1.1 Food and Nutritional Challenge: Ecological vs. Industrialised Agriculture

Since the 1950s in industrialised countries, food production has evolved towards a mass production system based on intensive use of land and water, as well as external inputs such as machinery, fuel, seeds and breeds, chemical fertilisers, a large variety of chemicals (pesticides, fungicides, herbicides, antibiotics, regulators and hormones), long-distance transportation and considerably reduced human labour (De Schutter 2010). This has gone hand-in-hand with an enormous increase in food processing, including milling, leading to the fact that about 80 % of the foods consumed at this time are processed. The limitations and potential health hazards of such an intensive production system, including contamination of the food chain and water by persistent pesticide residues or nitrates, or reduced nutrient and flavour content, have been highlighted for decades (Lairon 2010). Nevertheless, it is only during the last decades that worldwide awareness has led to the search for sustainable food production and consumption systems (Gomiero et al. 2011). It should be stressed that approximately one billion humans are still under- or malnourished, emphasizing the fact that food insecurity is still an alarming and unsolved problem (FAO 2010). Although it mainly concerns third-world countries, it is becoming an increasing concern for certain population segments in industrialised countries as well. Only recently, the combined awareness of environmental protection and biodiversity, food safety and security and well-being has considerably raised institutional and public concern, as well as the demand for ecologically-grown animal and human foods (El-Hage Scialabba 2007; FAO 2010; Niggli et al. 2007; De Schutter 2011; SCAR 2011). Worldwide, emphasis is increasingly placed on the relationships between food, nutrition and health, especially for the prevention of increasing non-communicable diseases such as obesity, type-2 diabetes, cardiovascular/cerebro-vascular diseases, neurodegenerative diseases and cancers (WCRF 2007; WHO 2004). Awareness of health hazards linked to contamination by chemicals and pesticide residues is clearly rising. Since the 1950s, alternative methods of ecological agriculture have also been developed with the aim to tackle these growing concerns. In industrialised countries, they are generally referred to as ecological or organic methods of agriculture (EC 2007, 2008). The question has been raised about the nutritional and toxicological value of food produced under such methods of production, as well as their potential effects on animal and human health. During the last decades, several bibliographic reviews have been carried out and published in this field (AFSSA 2003; Bourn and Prescott 2002; Brandt and Mölgaard 2001; Finesilver et al. 1989; FSA 2009; Lairon et al. 1984, Lairon 2010; Magkos et al. 2006; Rembialkowska 2007; Winter and Davis 2006; Heaton 2001; Woëse et al. 1997; Worthington 1998; Huber et al. 2011). In most cases, these reviews have attempted to use data from published original studies, but a few of them only used selection criteria based on the quality of the work (AFSSA 2003; FSA 2009). The conclusions drawn from these two bibliographic surveys will be summarised and discussed below.

The potential advantages of ecologically-grown foodstuffs on health markers have also been addressed, but in a more limited number of animal and human studies (Dangour et al. 2010; Velimirov et al. 2010). These too are presented and discussed below.

The question of authenticity, i.e., identifying organic and conventional products, is an issue of interest today within the context of a growing market where consumer trust is highly valued. Promising new methods are being developed that go beyond traditional analytical methods that measure single markers and that have a potential for future food quality assessment.

The authors finally discuss how these data can support the concept that organic agriculture systems provide relevant alternatives for reaching optimal food quality and safety. Potential progress is also considered.

In this review, organic agriculture and organic food quality is defined according to the European Union regulations, CCE/2092/91 and CE/1804/99. Although many concepts still remain vague (Kahl et al. 2010), the Organic Food Quality and Health Research Association, FQH[1], takes great pains to clarify and define these notions.

16.2 Nutritional Value of Organic Foodstuffs

The two most recent and important collective and critical evaluations of the nutritional and sanitary quality of organic food were made by the French Food Safety Agency in France (AFSSA 2003; updated in Lairon 2010) and the Food Standard Agency in the United Kingdom (FSA 2009). Although such evaluations are limited by the insufficient number of studies published in this area, they provide worthwhile information about the effects of contrasted agricultural methods and the possible causal relationships that underlie them, as reviewed and discussed in this article. Box 16.1 contains the main findings of these two large and critical evaluations based on selected original studies but using different approaches. The detailed methodologies and data can be found in the original reports and are summarised in the footnote to Box 16.1. While the data obtained during these two independent evaluations are rather convergent, the major differences in the methodologies used could account for some of the discrepancies observed in the results.

Box 16.1 Comparative data between organic and conventional food items provided by reports published by AFSSA in France (2003) and the FSA in the UK (2009).

Nutrient	AFSSA Report 310 studies	FSA Report 162 studies
Dry matter	Org>	Org>(+9.8%)
Magnesium	Org>	Org>(+7%)
Iron	Org>	not diff

[1] www.organicfqhresearch.org

Nutrient	AFSSA Report 310 studies	FSA Report 162 studies
Zinc	not diff	Org>(+11.3%)
Ca, P, K, Cu, Mn	not diff	not diff
Vitamin C	Trend Org>	not diff
Phenolics/Flavonoids	Org>	Org>(+13.2/38.4%)
Polyunsaturated fatty acids (animal products)	Org>	Org>(+10%)
Nitrates (vegetables)	Org<(−50%)	(Org N<)*

The French Food Safety Agency report (AFSSA 2003) identified trends for differences based on scientific papers (310 original works published in journals, theses and conference proceedings) selected for their appropriate description of methods, sampling and statistics, with organic products from certified farms only and their sound interpretation of data. Evaluations were conducted separately for relevant plant or animal products for a number of parameters. All parts of the report were critically examined and finalised by the expert group.

The Food Standard Agency report (FSA 2009) used a systematic approach for selected original scientific papers (162) published in English in scientific journals, including field trials, farm surveys and food basket studies, and with clear origin of samples, description of laboratory analyses and appropriate statistics. The authors performed statistical tests for differences for a number of parameters based on all selected studies on all plants and animals analysed. Not diff: no difference found; Org>or Org<: higher or lower levels in organic products. * Nitrate data were not calculated for vegetables and are therefore not presented.

Amplitude of variation (±%) between methods is provided when available.

Dry matter content The data available generally refer to vegetables and fruit. A trend towards higher dry matter content in organic foodstuffs was observed, especially in vegetables as opposed to fruits.

Macronutrient content While the protein content of grains can be lower in organic ones, no differences were reported for eggs or milk. Some studies reported higher levels of some essential free amino acids in organic cereals that could, at least partly, compensate for lower protein content, but this could depend on the cultivar. For lipids, it was acknowledged that meat from organic cows and chickens, unlike pigs, contains less total fat.

Polyunsaturated fatty acids (PUFAs) The meat from organic cows, chickens, pigs and rabbits contains more recommended polyunsaturated fatty acids (Bellon et al. 2009), especially omega-3 fatty acids and, to a lesser degree, omega-6 fatty acids.

It is generally assumed that the level of polyunsaturated fatty acids is higher in organic cow's milk than in conventional cow's milk. Several studies have reported higher levels of omega-3 (n-3) fatty acids and CLA (conjugated linoleic acid), in addition to antioxidants such as vitamin E (Butler et al. 2008; Butler et al. 2011; O'Donnell et al. 2010). In 2009, the FSA (FSA 2009) reported 27.8 % more n-3 fatty acids in organic cow's milk, while a recent evaluation based on all 14 comparative studies published since 2003 reported an average of 67 % (21–116 %) more n-3 fatty acids in organic vs. conventional cow's milk (Aubert and Lairon, unpublished review), whereas omega-6 (n-6) polyunsaturated fatty acids had comparable levels overall. This results in a much higher and more favourable n-3/n-6 fatty acid ratio, as is widely recommended.

A positive ratio of n-3 fatty acid levels in milk is found with outdoor grazing, whereas increased fertilisation intensity (including organic fertiliser) and concentrate feed has a reverse effect. This applies to n-3 content in milk as well.

Mineral content No noticeable differences were found for most of the mineral elements studied. Both evaluation methods in the cited reviews indicated higher magnesium levels, but diverged regarding possible increases in zinc or iron levels in organic vegetables (Box 16.1). This was confirmed by another evaluation made by Rembialkowska (Rembialkowska 2007) based on numerous studies that reported higher levels for iron (+21 %) and magnesium (+29 %) in organic crops.

Vitamin and carotenoid content Overall, on the basis of the limited data available, no significant differences were found for vitamin levels in foods obtained by the two methods, despite a trend towards higher vitamin C content in some organic fruits and vegetables, especially potatoes. Limited data are available for fat-soluble vitamins only, and carotenoid levels may be higher or lower, depending on the study design. Nevertheless, the FSA reported significantly more (+53.6 %) beta-carotene content in organic plant foodstuffs (FSA 2009).

Antioxidant and anti-inflammatory phytomicronutrients For secondary metabolites such as antioxidant phenolic acids, flavonoids and polyphenols, a majority of studies showed higher levels in organic plant foods, including almost all of the frequently consumed fruits and vegetables (Box 16.1). A ten-year study on flavonoids in tomatoes provided particularly relevant observations that supported this concept (Mitchell et al. 2007). Numerous new studies have been published since 2009 on various fruits, vegetables, grains and other foodstuffs, and the vast majority (about 70 %) report that organic foodstuffs have higher phenolic acid and/or flavonoid levels. Two studies found higher levels of anti-inflammatory salicylic acid in organic foodstuffs (Baxter et al. 2001; Rossi et al. 2008). These generally higher levels in organic products can be explained by the fact that those plants, grown organically without the use of chemical pesticides, need to more effectively develop their own defence system against fungi and insects and, therefore, to produce more of these secondary metabolites that are involved in plant protection. A pre-condition is that crops are not too heavily fertilised, especially with nitrogen, even organically, since increased levels of nitrogen fertiliser are linked to reduced secondary metabolite levels in plants (Stamp 2003; Brandt et al. 2011).

Based on these consensual comparative data, summarised in Box 16.1, it can be observed that both of the production systems generate some differences in nutritional value overall. This phenomenon can be observed on the basis of controlled studies, whereas a large variability can occur within the two production systems, depending on the intensity level and the technical approaches used (Wang et al. 2008). Indeed, plant and farm animal composition depends on intrinsic (species, cultivar/breed) factors as well as on numerous extrinsic ones including soil and climatic conditions, environmental quality and stress, disease and pest pressure, agronomic and husbandry management. This explains why it is difficult to draw clear-cut conclusions from comparative studies performed under such varied conditions and in such limited number. Nevertheless, on the basis of recent evidence, differences resulting from the agricultural systems compared here have become more convincing for some nutritional parameters.

16.3 Safety Aspects of Organic Foodstuffs

We are mainly concerned here by contamination by bacteria, viruses, worms, mycotoxins and agro-chemicals. Not all aspects of this contamination have been comparatively studied as of this time, but we will summarise some relevant points based on some of the more reliable information available (AFSSA 2003; Lairon 2010).

Pathogenic Microorganisms It appears that the systematic use of aerobic composting in organic agriculture is an appropriate way to maximise the hygienic properties of the organic fertilisers used and, thus, to avoid contamination of organic plant foods by pathogenic microorganisms. No differences were found for dairy products.

Chemical contaminants Apart from the unauthorised use of synthetic chemicals in organic agriculture, the question has been repeatedly raised about low levels of contamination of organic foodstuffs by environmental pollution due to the systematic use of the same synthetic chemicals in industrialised agriculture. Several nationwide (AFSSA 2003; Bourn and Prescott 2002; DGAL/COOPAGRI/ESMISAB 2001; Ghidini et al. 2005; Poulsen and Andersen 2003; Tasiopoulou et al. 2007) and European surveys (EU-DG SANCO 2007) have clearly shown that only a very small percentage of organic food samples are minimally contaminated by chemical residues, whereas a large proportion (41 %) of conventional foodstuffs are contaminated, with about 5 % of the foods sampled above the legal maximum residue levels (MRL). In fact, these hundreds of molecules have a highly toxic capacity (including mutagenesis and carcinogenesis), and the long-term detrimental effects of the chronic low-dose ingestion of single or mixed residues are basically unknown but seriously questioned based on recent epidemiological observations (REACH 2006). Some studies have already confirmed that biological disorders may be induced by low-dose ingestion in some situations (Casals-Casas and Desvergne 2011; Lee et al. 2011; Merhi et al. 2010). This raises a great concern for consumers and scientists regarding possible major long-term health damage to living organisms, including human beings. Indeed, when comparing children who were fed either a conventional or an organic diet for 5 days, organo-phosphorus pesticide metabolite levels in the urine of those fed an organic diet were ten times less than those fed a conventional diet (Lu et al. 2008).

Some natural pesticides are used in organic agriculture. However, when they are studied, they are generally not quantifiable due to their low stability and, as a result, their reduced persistence in the environment.

Mycotoxins Mycotoxins are a large family of highly toxic molecules synthesized by fungi such as *Aspergillus*, *Penicillium* and *Fusarium* that develops on plants and that can then be transferred via the food chain from plants to animals and, subsequently, to humans. Contamination of foodstuffs, especially raw or processed cereals, is widespread at a low level. Overall, differences are not generally observable between organic or conventional production modes (AFSSA 2003; Lairon 2010). On the basis of the 20 comparative studies published since 2003 concerning cereal contamination, it appears that a majority of studies found lower mycotoxin levels in organic grains (Bernhoft et al. 2010; Klinglmayr et al. 2010), whereas few studies showed levels that were either higher or comparable to those found in conventional ones. Overall, this confirms that the preventive measures used in organic systems, despite the non-use of synthetic fungicides, generally appear to be capable of maintaining mycotoxin contamination at a low level (AFSSA 2003).

Nitrate Nitrate can accumulate in plants depending on various factors. While fruits do not accumulate nitrate, root, tuber and leafy vegetables readily store it if nitrogen levels are too high to be fully transformed into protein. Comparisons at the farm or retail market level generally revealed lower nitrate accumulation levels in most of the organic vegetable species studied (Box 16.1). On a yearly basis, organic vegetables generally contain at least 30–50% less nitrate than conventional ones (AFSSA 2003; FSA 2009; Lairon et al. 1982, 1985; Temperli et al. 1982) and these differences have been widely confirmed. These observational data have been fully supported by data obtained during controlled fertilisation trials under various conditions and with a large number of vegetable species.

Nitrates are generally a matter of concern for public health, especially for newborns and young children, pregnant women and elderly people, due to their conversion into nitrites, which are highly reactive molecules. More than 30 surveys have been carried out on the possible relationship between nitrate ingestion from water and human disease, and the vast majority found positive relationships with several types of cancer (stomach, bladder, colon, prostate, thyroid), as well as other diseases (e.g., diabetes), thus supporting current water regulations. In contrast, some other studies showed a hypotensive and cardioprotective effect of nitrate ingestion (Webb et al. 2008; Lundberg et al. 2011). It is not known whether high nitrate levels in vegetables can have detrimental effects comparable to those found in water, given the possible preventive interactions that could occur with other compounds present (e.g., vitamin C, polyphenols).

16.4 Possible Positive Health Effects of Organic Products

Consumers expect organic food to be healthier than conventionally produced food. This is a strong argument for buying organic foods (Chen 2007; Wier and Anderson 2003; Zakowska-Biemans 2008; Falguera et al. 2012). The expectation is either that products that have grown in a balanced ecosystem will benefit the personal health of the consumer, or 'just' that it is better to consume as little pesticide residue

Fig. 16.1 Scientific value of different study designs on the health effects of the consumption of organically- and conventionally-produced foods. (adapted from the GRADE Working Group 2004)

Intervention studies
Controlled studies in *humans*

Observational epidemiological studies
Prospective cohort studies
Retrospective cohort studies

Intervention studies
Controlled studies in *animals*

Supportive studies
Bioavailability studies
In vitro studies

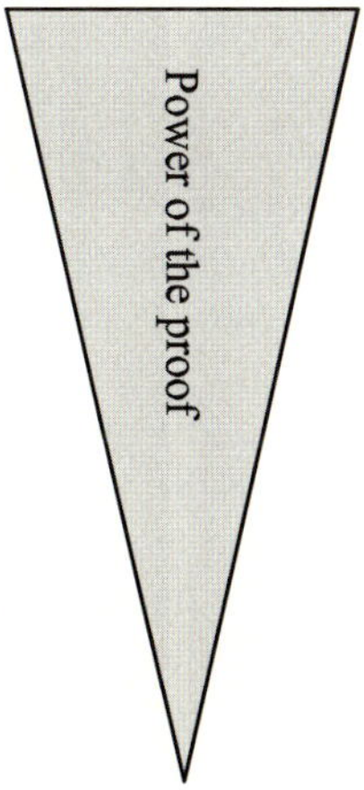

as possible. However, until now, these expectations have lacked sound scientific proof. Differences in the nutrient composition of products from different production systems are usually interpreted as having an impact on health. However, this should be done with great care. First, there is uncertainty about the 'bioavailability' of nutrients for the body after consumption. Second, the physiological response of the body cannot be predicted since it is not a question of a linear dose-response reaction to a single nutrient, but the result of a complex interaction with other nutrients in a food. The latest research in plant physiology suggests that there are up to 7,500–10,000 different compounds in a single 'nutritious' plant. Thus, feeding studies that explore effects after consumption are more effective. Such studies are complex and expensive, which is the reason why so few studies about the effects of organic food consumption on health have been done. Different types of research to investigate health effects from food consumption have different scientific value. Figure 16.1 gives an overview of study types.

Studies have been performed at each level. Most of the research was done with animals in intervention studies. Velimirov et al. (2010) recently presented a mini-review of feeding studies in animals. Their paper describes the choices that have to be made in designing these types of studies, e.g., choosing to make comparisons of whole feeds or just of a single product, and whether to compensate for clear differences between feeds in order to look for more subtle effects, or not. They report that research results until now have suggested positive influences from organic feeds on health, e.g., on weight, growth, immune responses and fertility performance in animals. The largest intervention study in animals to date (Huber et al. 2010) focused on two generations of chickens. It reported a lower body weight in the organically-fed group and a stronger immune reaction after an immune challenge, as well as quicker 'catch-up-growth' during recovery from a simulated illness. Studies like this are necessary to determine biomarkers of possible health effects from organic food since the availability of biomarkers is a pre-condition for future intervention studies in humans. Very few studies with humans are available. The largest prospective observational study so far is the KOALA Birth Cohort Study of life style effects on allergies and, at a later age, on obesity, among a group of 2700 children from birth until the present, and their mothers. Consumption of more than 90 % of organic dairy products by the children resulted in a 36 % decreased risk of eczema, compared to conventional dairy consumption, at two years of age (Kummeling et al. 2008). Analysis of mother's milk in this study showed a link between the proportion of intake of dairy products of organic origin and the levels of rumenic and trans-vaccenic acids, which are considered beneficial for health, in breast milk (Rist et al. 2007). Several trials that exposed humans to single foods from different production systems (wine, apples, tomato puree, carrots and apples), in addition to their regular diet, and that mainly measured antioxidant capacity, did not report significant differences (Briviba et al. 2007; Caris-Veyrat et al. 2004; Stracke et al. 2009). Grinder-Pedersen et al. (2003) exposed healthy volunteers to menus with several products from organic or conventional systems and found higher urinary excretion of the flavonoids, quercetin and kaempferol, and lower plasma antioxidant capacity after exposure to the organic menus. It should be noted that the amount of absorption and excretion of specific nutrients by the body is suggestive, but cannot be directly translated into effects on health since these depend on a complex interaction of nutrient substances with many other substances in the body. Only a few experimental *in vitro* studies have been published. Olsson et al. (2006) found significantly greater inhibition of cancer cell proliferation when exposing cell cultures to extracts from organic vs. conventional strawberries. Dangour et al. (2010), reporting for the FSA in the UK, recently concluded that there is no *evidence* of health effects based on the very few studies available. We agree with Dangour about this, yet we believe that *indications* for health effects exist since several animal studies and one human study (Velimirov 2010; Huber et al. 2011) showed an effect on weight, growth and immune responses, an indication of improved general health status (Huber et al. 2012). More research is needed to support this hypothesis.

16.5 The Question of Authenticity

Consumer trust is valuable in a growing organic product market. As a result, the question of authenticity is important and of relevance at this time. As described above, traditional analytical methods that measure single markers are useful but have a limited potential for future food quality assessment. In connection with the ongoing discovery of the large amounts of compounds in food products, new methods are being developed that aim at multiple markers and that go beyond traditional analytical methods. The 'selective fingerprinting profiling technique', using metabolomics and spectroscopic techniques combined with chemometrics, has shown its ability to discriminate between organically—and conventionally -produced products (Ruth et al. 2010). Another promising new approach involves 'complementary' image forming techniques that evaluate technically produced patterns of samples as a whole using human and computerised image analysis. The 'biocrystallisation' or 'copper chloride crystallisation method' and the 'Steigbild method' are presently being validated by a partnership between several institutes (Kahl et al. 2009) and have shown their potential to discriminate between organic and conventional products (Szulc et al. 2010; Zalecka et al. 2010). There is also the 'fluorescence excitation spectroscopy' method (Strube and Stolz 2010) that measures the capacity and duration of a product to store small amounts of imposed light. In addition to authentication, the methods described above may lead to new perspectives on food quality and food quality parameters (Kahl et al. 2009).

16.6 Perspectives and Research Requirements

16.6.1 Organic Agriculture as a Prototype for Sustainable Food Production

As briefly addressed in the introduction, we are still facing the challenge of food insecurity for approximately one billion people worldwide. This unfortunately but obviously underlines the fact that the widespread 'industrialised agriculture system' might not be able to solve this continuing problem. While organic production systems could slightly lower (by about 10–20 %) food production in industrialised rich countries, these alternative methods are reported to have the capacity to considerably increase (by about +80 %) food yields in poor and developing countries that are faced with the biggest food insecurity (El-Hage Scialabba 2007; De Schutter 2010). In view of climate change, increasing heat and drought and the expectation that the present number of six billion people will grow to approximately nine billion in 2050, the UN Special Rapporteur on the Right to Food called for a fundamental shift towards agro-ecological production methods in 2011 since they are expected to outperform conventional farming (De Schutter 2011).

At the same time, industrialised countries and, more recently, developing ones, are facing the growing epidemic of obesity and its health-related complications (Crombie et al. 2009; Popkin and Gordon-Larsen 2004. This is essentially due to dietary patterns characterised by high energy density and low/moderate nutrient density and over-feeding in terms of energy needs (Drewnowski et al. 2007; Darmon and Maillot 2010). As discussed above, organic agriculture can provide foods with somewhat higher dry matter, minerals (magnesium, iron, zinc), vitamin C and other antioxidants (carotenoids, phenolic acids/flavonoids) and n-3 fatty acids, thus increasing nutrient density overall, with a trend towards lower protein content in cereals. When combining these observations with yield performances of organic agriculture as reported above, it can be deduced that (i) a slightly lower production of foods but with higher nutrient density would be beneficial for people in industrialised countries, and (ii) a considerable increase in available staple foods with higher micronutrient content in developing countries would contribute to a much better coverage of essential nutrient needs and to securing their food self-sufficiency. This should be combined with a necessary minimal refining of foods in order to retain most of the fibre and micronutrients. It is interesting to note that while most traditional dietary patterns are plant food-based, organic food consumers in industrialised countries tend to shift towards more plant-based foods and less animal-based ones. This is indeed in line with worldwide dietary recommendations (WCRF 2007; WHO 2004).

The potential links between the type of production system and crop composition can be discussed in greater detail. Some data confirmed that organic farming systems work better in the context of natural interactions within their environmental constraints. This is illustrated by the generally higher levels of antioxidant secondary metabolites (carotenoids, phenolics), and the comparable/lower levels of mycotoxins (without chemical fungicides) in plants grown under organic agriculture management. It additionally suggests a better adaptation and, therefore, resistance to environmental stress.

The repeated observations of lower to much lower levels of nitrates in root and leafy vegetables also illustrate the different relationships that develop between soil and plants in organic agriculture systems. The non-use of chemical fertilisers (especially those containing readily available nitrogen) necessitates a stronger root system as well as an optimal soil biological activity (as in natural/forest soils) to ensure progressive and efficient release of nitrates from organic sources for root uptake. A likely result is that nitrate accumulation in plants is quite limited thanks to the efficient processing of plant nitrates for amino-acid synthesis as promoted by photosynthesis. It is also worth noting that the nitrate level in vegetables results from nitrogen availability for roots, temperature, light exposure, cultivar and species. Situations where big differences between nitrate accumulation systems in plants can be limited or eliminated are: (i) when geo-climatic conditions are too detrimental such as during winter or under greenhouse shading; and (ii) when large amounts of readily-mineralised organic fertilisers are used (such as guano). This points to the fact that while the optimisation of organic agriculture systems should be a goal, the upper limits of intensification must be defined to avoid considerable loss of benefits.

As mentioned above, low-input dairy farming and, in particular, low-input organic farming produces milk with more healthy fatty acids (Butler et al. 2008). The KOALA study showed the influence of such milk on the quality of mother's milk. It also showed that children who consumed this milk or who directly consumed organic dairy products had less allergies. This is an outstanding example of how sustainable agricultural practices could positively alter food composition and, as a result, improve human health.

Finally, the fact that the use of toxic chemicals, pesticides, fungicides and herbicides is banned in organic agriculture systems is clearly a gold standard in terms of the protection of the health of those who work on the land and of people who live in production areas, the protection of the biodiversity of living organisms, and consumer health. Evidence supporting the long-term detrimental effects on health upon professional exposure to such chemicals and the long-term effects of chronic exposure to low doses of pesticides has just begun to be documented after decades of intensive use of these substances due to the lack of sufficient research efforts.

Indeed, recent research supports the hypothesis that chronic ingestion of low doses of these substances through accumulation and long residence time in the body could have repercussions at the cellular and molecular levels, leading to metabolic imbalances, followed by syndromes and pathologies such as cancer, neuro-degenerative disorders, reproduction disorders, obesity, cardiovascular diseases and diabetes (Bailey et al. 2010; Elbaz et al. 2009; Elobeid et al. 2010; Farooq et al. 2010; Lee et al. 2011; Montgomery et al. 2008). This would most likely have a greater implication for people with a genetic predisposition or low resistance (children, the elderly or sick people), or during critical stages such as *in utero* foetal development through contamination of the mother.

Dangour et al. (2010) concluded that there is no evidence that organic products have an impact on health. As a marker for 'health effects', they used effects on defined diseases in humans ('negative health'). We hypothesize that it is more advantageous to study effects from the intake of organic food on the robustness and resilience ('positive health') of humans and animals that consume organic food. The organic agriculture system aims at increasing the robustness of plants and animals in order to avoid pesticides and antibiotics. Such robust organisms might represent good food for enhancing health, as a recent animal study suggested (Huber et al. 2010). However, to confirm this hypothesis, markers for robustness, resistance and resilience first need to be developed, e.g., better adaptation to physical or psychological challenges in a research setting. Publications have begun to appear on this subject (Huber et al. 2011).

To conclude, in contrast with the main trends in industrialised agriculture (breeding for yield, system simplification, etc.) and food processing (refining, use of additives, standardisation, etc.), the use and conservation of biodiversity in plant and animal production have long been important for the success of organic agriculture and are now increasingly recognised as key issues for attaining sustainable diets worldwide (De Schutter 2010; El-Hage Scialabba 2007; FAO 2010). A consensual definition of a sustainable diet was recently published (FAO 2010):

> Sustainable diets are those diets with low environmental impacts which contribute to food and nutrition security and to healthy life for present and future generations. Sustainable diets are protective and respectful of biodiversity and ecosystems, culturally acceptable, accessible, economically fair and affordable; nutritionally adequate, safe and healthy; while optimizing natural and human resources.

Organic farming systems are consistent with this concept of sustainable biodiversity-based food production systems and could largely improve the variety as well as the quality of food for human consumption.

16.6.2 Research Requirements to Improve Sustainable Food Security and Optimal Quality and Health

In view of the points raised and discussed above, we suggest that the following aspects urgently require research efforts to improve knowledge and allow for new improvements and developments of sustainable organic agriculture. Some of these points have already been suggested in the Vision, Strategic Research Agenda and Implementation Action Plan of the 'Organics' Technology Platform of IFOAM (Niggli et al. 2008; Schmid et al. 2009; Padel et al. 2010)

- Selecting or raising cultivars and breeds for optimal yield, taste and nutritional value (especially for limiting nutrients), suited for organic production systems.
- Testing methods for better agronomical efficiency and pest and disease resistances.
- Optimising 'careful' processing methods and limiting refining.
- Investigating the potential of exhaustive molecular fingerprinting (e.g., metabolomics, proteomics) or complementary image forming methods for authentication and quality determination (whole product assessment vs. targeted analytics).
- Performing comparative studies on the nutrient content of plant and animal foodstuffs with high quality standards for design, relevant information and methodologies.
- Discriminating the impact of organic agriculture production systems from that of low-input/extensive systems on the environment, farmers' health, animal welfare, nutritional status and the health of consumers.
- Focusing on the processing and preservation methods used and on distribution systems (local vs. remote), as well as on preparation methods (raw/cooked) in view of the nutritional quality of foods.
- Developing measurement systems and markers for robustness and resilience ('positive health'), and the enhancement of these, in humans and animals.
- Performing epidemiological surveys that link food intake to health and disease markers in various segments of the population.
- Performing adequate intervention studies to compare organic vs. conventional food-based diets to test causal relationships between dietary intake and 'positive health' levels.

References

AFSSA (2003) Report on evaluation of the nutritional and sanitary quality of organic foods (Evaluation nutritionnelle et sanitaire des aliments issus de l'agriculture biologique, in French), AFSSA, 164 pages. Accessible on line at www.anses.fr. Accessed 20 Jan 2011

Bailey HD, Armstrong BK, de Klerk NH, Fritschi L, Attia J, Scott RJ, Smibert E, Milne E, for the Aus-ALL Consortium (2010) Exposure to professional pest control treatments and the risk of childhood acute lymphoblastic leukemia. Int J Cancer Nov 15. [Epubahead of print]

Baxter GJ, Graham AB, Lawrence JR, Wiles D, Paterson JR (2001) Salicylic acid in soups prepared from organically and non-organically grown vegetables. Eur J Nutr 40:289–292

Bellon S, Prache S, Benoit M, Cabaret J (2009) Recherches en élevage biologique: enjeux, acquis et développements. Inra Prod Anim 22:271–284

Bernhoft A, Clasen PE, Kristoffersen AB, Torp M (2010) Less fusarium infestation and mycotoxin contamination in organic than in conventional cereals. Food Addit Contam Part A Chem Anal Control Expo Risk Assess. 27(6):842–52

Bourn D, Prescott J (2002) A comparison of the nutritional value, sensory qualities and food safety of organically and conventionally produced foods. Crit Rev Food Sci Nutr 42:1–34

Brandt K, Mölgaard JP (2001) Organic agriculture: does it enhance or reduce the nutritional value of plant foods?. J Sci Food Agric 81:924–931

Brandt K, Leifert C, Sanderson R, Seal CJ (2011) Agroecosystem management and nutritional quality of plant foods: the case of organic fruits and vegetables. Crit Rev Plant Sci 30(1):177–197

Briviba K, Stracke BA, Rüfer CE, Watzl B, Weibel FP, Bub A (2007) Effect of consumption of organically and conventionally produced apples on antioxidant capacity and DNA damage in humans. J Agric Food Chem 55:7716–7721

Butler G, Nielsen JH, Slots T et al (2008) Fatty acid and fat-soluble antioxidant concentrations in milk from high- and low-input conventional and organic systems: seasonal variation. J Sci Food Agric 88:1431–1441

Butler G, Stergiadis S, Seal C, Eyre M, Leifert C (2011) Fat composition of organic and conventional retail milk in North East England. J Dairy Sci 94:24–36

Caris-Veyrat C, Amiot MJ, Tyssandier V et al (2004) Influence of organic versus conventional agricultural practice on the antioxidant micro-constituent content of tomatoes and derived purees; consequences on antioxidant plasma status un humans. J Agric Food Chem 52:6503–6509

Casals-Casas C, Desvergne B (2011) Endocrine disruptors: from endocrine to metabolic disruption. Annu Rev Physiol 73:135–162

Chen MF (2007) Consumer attitudes and purchase intentions in relation to organic foods in Taiwan: moderating effects of food-related personality traits. Food Qual Prefer 18:1008–1021

Crombie IK, Irvine L, Elliott L, Wallace H (2009) Targets to tackle the obesity epidemic: a review of twelve developed countries. Public Health Nutr 12(3):406–413

Dangour A, Lock K, Hayter A et al. (2010) Nutrition-related health effects of organic foods: a systematic review. Am J Clin Nutr 92: 203–210

Darmon N, Maillot M (2010) In foods, energy is cheap where it is abundant and expensive where it is scarce: this is a fact, not an artifact. Am J Clin Nutr 91(4):1068–1069

De Schutter O (2010) The shift to low-carbon agriculture is becoming a matter of urgency. Statement by the UN Special Rapporteur on the right to food, World Food Day – 16 October 2010. Accessible at http://www2.ohchr.org/english/. Accessed 7 March 2011

De Schutter O (2011) via http://www.srfood.org/index.php/en/component/content/article/1-latest-news/1174-report-agroecology-and-the-right-to-food. Accessed 29 May 2011

DGAL/COOPAGRI/ESMISAB (2001) Evaluation de l'exposition des consommateurs et produits issus de l'agriculture biologique et de l'agriculture conventionnelle aux résidus de pesticides, métaux lourds, nitrates, nitrites et mycotoxines, Notre Alimentation, 37, I–VI

Drewnowski A, Monsivais P, Maillot M, Darmon N (2007) Low-energy-density diets are associated with higher diet quality and higher diet costs in French adults. J Am Diet Assoc 107(6):1028–1032

EC. Council Regulation (European Community) No. 834/2007 of June 2007 on organic production and labeling of organic products and repealing Regulation (EEC) No. 2092/91

EC. Commission Regulation (European Community) No. 889/2008 of Sept. 5, 2008 laying down detailed rules for the implementation of Council Regulation (EC) No. 834/2007 on organic production and labeling of organic products with regards to organic production, labeling and control

EU-DG SANCO (2007) Monitoring of pesticide residues in products of plant origin in European union, Norway, Iceland and Liechtenstein, 2005 Report. Accessible at http://ec.europa.eu/food/fvo/specialreports/pesticide_residues/report_2005_en.pdf. Accessed 17 Apr 2009

Elbaz A, Clavel J, Rathouz PJ, Moisan F, Galanaud JP, Delemotte B, Alpérovitch A, Tzourio C (2009) Professional exposure to pesticides and Parkinson disease. Ann Neurol 66(4):494–504

El-Hage Scialabba N (2007) Organic agriculture and food security, Food and Agriculture Organization of the United Nations, Report of the International conference on organic agriculture and food security, May 3–5, Roma, Italy, 22p. Accessible at www.fao.org. Accessed 11 Apr 2009

Elobeid MA, Padilla MA, Brock DW, Ruden DM, Allison DB (2010) Endocrine disruptors and obesity: an examination of selected persistent organic pollutants in the NHANES 1999–2002 data. Int J Environ Res Public Health 7(7):2988–3005

Falguera V, Aliguer N, Falguera M (2012) An interggrated approach to current trends in food consumption: moving toward functional and organic products? Food Control 26:274–281

FAO (2010) Final document. International scientific symposium biodiversity and sustainable diets united against hunger, 3–5 November 2010, FAO headquarters, Rome. Accessible at www.fao.org. Accessed 25 Nov 2010

Farooq U, Joshi M, Nookala V, Cheriyath P, Fischman D, Graber NJ, Stellman SD, Muscat J (2010) Self-reported exposure to pesticides in residential settings and risk of breast cancer: a case-control study. Environ Health 9:30

Finesilver T, Johns T, Hill SB (1989) Comparison of food quality of organically versus conventionally grown plant foods, a review, Ecological Agriculture Projects, Report. McGill University, Canada

FSA, Food Standards Agency (UK) (2009) Comparison of composition (nutrients and other substances) of organically and conventionally produced foodstuffs: a systematic review of available literature. 31 pages. Accessible at www.food.gov.uk. Accessed 1 Sept 2009

Ghidini S, Zanardi E, Battaglia A, Varisco G, Ferretti E, Campanini G, Chizzolini R (2005) Comparison of contaminant and residue levels in organic and conventional milk and meat products from Northern Italy. Food Add Contam 22:9–14

Gomiero T, Pimentel D, Paoletti MG (2011) Is there a need for a more sustainable agriculture? Crit Rev Plant Sci 30:6–23

GRADE-Grades of Recommendation Assessment, Development and Evaluation—Working Group, corresponding author A.D. Oxman (2004) Education and debate, grading quality of evidence and strength of recommendations. BMJ com 328:1490

Grinder-Pedersen L, Rasmussen SE, Bugel S, Jorgensen LV, Dragsted LO, Gundersen V, Sandstrom B (2003) Effect of diets based on foods from conventional versus organic production on intake and excretion of flavonoids and markers of antioxidative defense in humans. J Agric Food Chem 51:5671–5676

Huber M, Vijver van de LPL, Parmentier H, Savelkoul H, Coulier L, Wopereis S, Verheij E, Greef van der J, Nierop D, Hoogenboom R (2010) Effects of organically and conventionally produced feed on biomarkers of health in a chicken model. Br J Nutr 103:663–676

Huber M, Rembialkowska E, Srednicka D, Bügel S, van de Vijver LPL (2011) Organic food and impact on human health: assessing the status quo and prospects of research, NJAS-Wageningen. J Life Sci 58:103–105

Huber M, Bakker MH, Dijk W, Prins HAB, Wiegant FAC (2012) The challenge of evaluating health effects of organic food; operationalization of a dynamic concept of health. J Sci Food Agric 92(14): 2766–2773. doi: 10.1002/jsfa 5563

Heaton S (2001) Organic farming, food quality and human health. a review of the evidence. Soil Association, Bristol, p 86. (ISBN 0 905200802)

Kahl J, Burgt van der GJ, Kusche D et al. (2010) Organic food claims in Europe, better regulatory guidelines, improved testing methods, and additional research methods into product quality criteria are needed. Food Technology 03(10): 38–46

Kahl J, Busscher N, Doesburg P, Mergardt G, Huber M, Ploeger (2009) A First tests of standardized biocrystallization on milk and milk products. Eur Food Res Technol. 229:175–178

Klinglmayr C, Nöbauer K, Razzazi-Fazeli E, Cichna-Markl M (2010) Determination of deoxynivalenol in organic and conventional food and feed by sol-gel immunoaffinity chromatography and HPLC-UV detection. J Chromatogr B Analyt Technol Biomed Life Sci 878(2):187–193

Kummeling I, Thijs C, Huber M, Vijver van de LPL, Snijders B, Penders J, Stelma F, Ree van R, Brandt van den P, Dagnelie P (2008) Consumption of organicfoods and risk of atopicdisease-during the first 2 years of life in the Netherlands. Br J Nutr 99:598–605

Lairon D, Lafont H, Léonardi J, Hauton JC, Ribaud P (1982) Comparaison de l'intérêt nutritif de légumes produits par l'agriculture conventionnelle ou biologique. Sci Aliments 2(HS II):203–205

Lairon D, Termine E, Lafont H (1984) Valeur nutritionnelle comparée des légumes obtenus par les méthodes de l'agriculture biologique ou de l'agriculture conventionnelle. Cah Nutr Diet 6:331–339

Lairon D, Termine E, Gauthier S, Lafont H (1985) Teneurs en nitrates des productions maraîchères obtenues par des méthodes de l'agriculture biologique. Sci Aliments 5(HS V):337–343

Lairon D (2010) Nutritional quality and safety of organic food. A rev Agron Sustain Dev 30:33–41

Lee DH, Steffes M, Sjödin A., Jones R, Needham L, Jacobs D (2011) Low dose organochlorinepesticides and polychlorinated biphenyls predict obesity, dyslipidemia, and insulin resistance among people free of diabetes. PLoS One. 6(1):e15977

Lu C, Barr DB, Pearson MA, Waller LA (2008) Dietary intake and its contribution to longitudinal organophosphorus pesticide exposure in urban/suburban children. Environ Health Perspect 116(4):537–542

Lundberg JO, Carlström M, Larsen FJ, Weitzberg E (2011 Feb 15) Roles of dietary inorganic nitrate in cardiovascular health and disease. Cardiovasc Res 89(3):525–32

Magkos F, Arvaniti F, Zampelas A (2006) Organic food: buying more safety or just peace of mind? A critical review of the literature. Crit Rev Food Sci Nutr 46:23–56

Merhi M, Demur C, Racaud-Sultan C, Bertrand J, Canlet C, Estrada FB, Gamet-Payrastre L (2010) Gender-linked haematopoietic and metabolic disturbances induced by a pesticide mixture administered at low dose to mice. Toxicol 267(1–3):80–90

Mitchell AE, Hong YJ, Koh E, Barrett DM, Bryant DE, Denison RF, Kaffka S (2007) Ten-year comparison of the influence of organic and conventional crop management practices on the content of flavonoids in tomatoes. J Agric Food Chem 55:6154–6159

Montgomery MP, Kamel F, Saldana TM, Alavanja MC, Sandler DP (2008) Incident diabetes and pesticide exposure among licensed pesticide applicators: agricultural health study, 1993–2003. Am J Epidemiol 167(10):1235–1246

Niggli U, Leifert C, Aföldi C, Lück L, H Willer (2007) Proceedings of the 3rd International Congress of the European integrated project Quality Low Imput Food (QLIF), University of Hohenheim, March 20–23, 2007, Accessible at www.qlif.org. Accessed 17 Apr 2009

Niggli U, Slabe A, Schmid O, Halberg N, Schlüter M (2008) Vision for an organic food and farming research agenda to 2025. Published by IFOAM-EU Group, 44 pages Accessible at www.ifoam-eu.org. Accessed 17 Apr 2009

O'Donnell AM, Spatny KP, Vicini JL, Bauman DE (2010) Survey of the fatty acid composition of retail milk differing in label claims based on production management practices. J Dairy Sci 93(5):1918–1925

Olsson ME, Andersson CS, Oredsson S, Berglund RH, Gustavsson KE (2006) Antioxidant levels and inhibition of cancer cell proliferation in vitro by extracts from organically and conventionally cultivated strawberries. J Agric Food Chem 54:1248–1255

Padel S, Niggli U, Pearce B, Schlüter M, Schmid O, Cuoco E, Willer H, Huber M, Halberg N, Micheloni C (2010) Implementation action plan for organic food and farming.research.TP Organics. IFOAM- EU Group. Brussels. Accessible at www.ifoam-eu.org

Popkin BM, Gordon-Larsen P (2004) The nutrition transition: worldwide obesity dynamics and their determinants. Int J Obes Relat Metab Disord 28(Suppl 3):S2–9

Poulsen ME, Andersen JH (2003). Results from the monitoring of pesticide residues in fruit and vegetables on the Danish market, 2000–01. Food Addit Contam 20:742–757

REACH (2006) European Community Regulation EC 1907/2006 on chemicals and their safe use (Registration, Evaluation, Authorisation, and restriction of Chemical substances). Accessible at http://ec.europa.eu/environment/chemicals/reach. Accessed 5 Feb 2008

Rembialkowska E (2007) Quality of plant products from organic agriculture. J Sci Food Agric 87:2757–2762

Rist L, Mueller A, Barthel C, Snijders B, Jansen M, Simões-Wüst A, Huber M, Kummeling I, Mandach von I, Steinhart H, Thijs C (2007) Influence of organic diet on the amount of conjugated linoleic acids in breast milk of lactating women in the Netherlands. Br J Nutr 97:735–743

Rossi F, Godani F, Bertuzzi T, Trevisan M, Ferrari F, Gatti S (2008) Health-promoting substances and heavy metal content in tomatoes grown with different farming techniques. Eur J Nutr 47:266–272

Ruth van S, Alewijn M, RogersK et al (2010) Authentication of organic and conventional eggs by carotenoid profiling. Food Chemistry, online 2010, j.foodchem.2010.11.081

SCAR, Standing Committee on Agricultural Research of the European Committee: the 3rd SCAR Foresight Exercise 2011, Sustainable food consumption and production in a resource-constrained world

Schmid O, Padel S, Halberg N, Huber M, Darnhofer I, Micheloni C, Koopmans C, Bügel S, Stopes C, Willer H, Schlüter M, Cuoco E (2009) Strategic research agenda for organic food and farming. TP Organics, Brussels: IFOAM EU Group. Accessible at www.ifoam-eu.org.Soil Association (2001) Organic farming, food quality and human health. A review of the evidence. 87 p

Stamp N (2003) Out of the quagmire of plant defense hypotheses. Q Rev Biology 78(1):23–55

Stracke BA, Rüfer CE, Bub A, et al (2009) No effect of the farming system (organic/conventional) on the bioavailability of apple (Malusdomestica Bork., cultivar Golden Delicious) polyphenols in healthy men: a comparative study. Eur J Nutr 49(5):301–310

Strube J, Stolz P (2010) The Application of Fluorescence Excitation Spectroscopy of Whole Samples for Identification of the Culture System of Wheat and Carrots—Method, Validation, Results. Biol Agric Hort 27:59–80

Szulc M et al. corresponding author Kahl J (2010) Discrimination between organically and conventionally grown winter wheat farm pair samples using the copper chloride crystallisation method in combination with computerised image analysis. Comput Electron Agric 74:218–222

Tasiopoulou S, Chiodini AM, Vellere F, Visentin S (2007) Results of the monitoring program of pesticide residues in organic food of plant origin in Lombardy (Italy). J Environ Sci Health B 42:835–841

Temperli A, Kunsch V, Scharer H, Konrad P, Suter H, Ott P, Eichenberger M, Schmid O (1982) Einfluss zweier Anbauweisen auf den Nitratgehalt von Kopfsalat. Schweiz Landw Fo 21:167–196

Velimirov A, Huber M, Lauridsen C, Rembialkowska E, Seidel K, Bügel S (2010) Feedingtrials in organicfoodquality and healthresearch. J Sci Food Agric 90:175–182

Wang SY, Chen CT, Sciarappa W, Wang CY, Camp MJ (2008) Fruit quality, antioxidant capacity, and flavonoid content of organically and conventionally grown blueberries. J Agric Food Chem 56:5788–5794

Webb AJ, Patel N, Loukogeorgakis S, Okorie M, Aboud Z, Misra S, Rashid R, Miall P, Deanfield J, Benjamin N, MacAllister R, Hobbs AJ, Ahluwalia A (2008) Acute blood pressure lowering, vasoprotective, and antiplatelet properties of dietary nitrate via bioconversion to nitrite. Hypertension 51(3):784–790

WHO (2004) Food and health in Europe: a new basis for action. WHO Regional Publications, European series, N°96, 385 pages

Wier M, Anderson L (2003) Demand for organic foods-attitudes, values and purchasing behaviors. Newsletter from Danish Research Centre Farming, 2:1–3

Winter CK, Davis SF (2006) Organic foods. J Food Sci 71:R117–R124

Woëse K, Lange D, Boess C, Bögl KW (1997) A comparison of organically and conventionally grown foods—results of a review of the relevant literature. J Sci Food Agric 74:281–293

WCRF, World Cancer Research Fund (2007) Food, nutrition, physical activity, and the prevention of cancer: a global perspective. American Institute of Cancer, Washington DC, USA, 517 p. Accessible at www.wcrf.org. Accessed 17 Nov 2008

Worthington V (1998) Effect of agricultural methods on nutritional quality: a comparison of organic with conventional crops. Altern Ther 4(1): 58–69

Zakowska-Biemans S (2008) Consumers values and motives regarding organic food products in poland. Proceedings of the 2nd Scientific Conference of the International Society of Organic Agriculture Research ISOFAR, Modena, Italy, June 18–20, 2008, 2:506–509

Zalecka A, Kahl J, Doesburg P et al. (2010) Standardization of the steigbild method. Biol Agric Hort 27:41–57

Chapter 17
Advances, Issues and Challenges in Organic Lamb Meat Quality

Sophie Prache

Abstract Organic farming embodies extrinsic features that are valued by consumers, but studies on the intrinsic properties of organic products remain scarce. This paper first highlights the inherent difficulties in comparing the qualities of food products of organic systems to those of conventional systems. The paper then gives an overview of the current state of knowledge and the main issues and challenges regarding the different facets of organic lamb meat and carcass quality. Organic farming promotes pasture-feeding, which is favourable from a nutritional point of view, since meat from pasture-fed lambs has a more desirable fatty acid composition than meat from lambs fed concentrate diets. However, pasture-feeding may lead to a greater occurrence of off-flavour in meat and to a less desirable meat colour; these effects are the result of both the animal's diet and its increased age at slaughter. Pasture-feeding may also lead to a higher variability in sensory qualities because of a higher variability in animal age at slaughter. These sensory defects may be even greater in organic systems, which promote legumes within pastures, because these plant species have a prominent role in the ruminal synthesis of unpleasant smelling volatile compounds that are stored in the fat. One of the main challenges is to further experiment with management strategies to help minimise the occurrence of these sensory defects and control the variability in animals' performances. Additionally, the paper gives an insight into the advances in analytical methods for authenticating meat from low-input grassland-based systems.

Keywords Organic farming · Lamb · Meat · Pasture-feeding · Quality · Sensory · Nutritional · Fatty acids · Authentication

17.1 Introduction

Organic meat and milk products have 'added value' among consumers because of their perceived healthiness and environmental acceptability. However, the 'Organic' label mainly offers production process guarantees, and the intrinsic quality of

S. Prache (✉)
INRA, UR1213 Herbivores, Site de Theix, 63122, Saint-Genès-Champanelle, France
e-mail: sophie.prache@clermont.inra.fr

S. Bellon, S. Penvern (eds.), *Organic Farming, Prototype for Sustainable Agricultures*,
DOI 10.1007/978-94-007-7927-3_17,

organic-labelled products is therefore often questioned (Prache et al. 2011a). We therefore probably should anticipate a demand for product quality guarantees and for an onus to be able to authenticate organic products.

Organic farmers commit to a list of specifications that govern animal care, welfare and feeding, requiring them to provide herbivores with access to pastures during the grazing season and to organic raw feed material outside the grazing season. Moreover, they frequently raise rustic animal breeds (Leroux et al. 2009). All these factors influence the quality of the meat, although most of them are not exclusive to organic farming systems but can also be found in low-input or quality-based systems of production such as those with the labels, Protected Denomination of Origin (PDO), Protected Geographic Indication (PGI) or Label Rouge. It then follows that most of the considerations regarding organic products may be useful for low-input or quality-based systems of production for which organic systems may therefore be considered as a prototype. Moreover, it should be observed that there is a great diversity in production practices, in organic as well as in conventional production systems, which may limit the robustness and generalisation of any given comparison between products obtained from organically- or conventionally-reared herbivores. Most of the studies in this area have limited their comparisons to organic vs. very intensive production systems (Bellon et al. 2009) or have failed to provide sufficient robustness because of an insufficient description of the corresponding production practices. For example, the comparison of the quality of organic and conventionally-produced lambs performed by Angood et al. (2008) was based on lamb chops purchased from supermarket chains. Beyond its interest due to the large number of samples evaluated and to the fact that these samples directly corresponded to those potentially purchased by consumers, this design can be criticised by arguing that it is sensitive to confounding effects such as animal diet, animal breed, and sex and age at slaughter—all factors that were unknown in the study but that can greatly affect the sensory and nutritional quality of the lamb chops.

This paper highlights the main advances, issues and challenges in organic lamb meat and carcass quality. We attempted to provide an overview of the state of knowledge (and the gap in knowledge) of the different facets of meat and carcass quality, from the nutritional and sensory point of view to the regularity of food supply, the authentication of lamb meat from organic production systems and the carbon footprint of the meat produced.

17.2 Nutritional Quality

Organic farming consciously promotes pasture-feeding, which is favourable from a nutritional point of view since meat from pasture-fed lambs has been shown to have a nutritionally more desirable fatty acid (FA) composition than meat from lambs fed concentrate diets (Fisher et al. 2000; Aurousseau et al. 2004; Santé-Lhoutellier et al. 2008). Pasture-feeding is a production practice that actually has a positive impact on the nutritional quality of the meat because of higher contents of healthy compounds (such as alpha-linolenic acid and conjugated linoleic acids (CLA), and a lower n-6 polyunsaturated FA/n-3 polyunsaturated FA ratio). It is also responsible for a lower content of 16:0 FA, which is pro-atherogenic (Aurousseau et al. 2004). The composition of the saturated FA (SFA) and polyunsaturated FA (PUFA) from pasture-fed lambs actually meets the recommendations of the European Food Safety Authority (2010) and the French Nutrition Society (Legrand et al. 2001), i.e., a n-6/n-3 PUFA ratio lower than 5 and a limited level of 16:0. Similar results have been observed in bovine milk and meat lipids, with higher concentrations of n-3 PUFA and conjugated linoleic acid (CLA) in animals fed fresh grass compared to concentrates or maize silage, and corresponding beneficial changes in PUFA/SFA and n-6/n-3 PUFA ratios (Moloney et al. 2008). Differences in feeding regimens therefore largely explain why significantly higher amounts of PUFA, CLA, n-3 FA and antioxidant compounds are found in organic milk (Collomb et al. 2008; Butler et al. 2011). However, although organic farmers are committed to providing herbivores with access to pasture during the grazing season, pasture-feeding is not exclusive to organic systems and stall-feeding with concentrate-based diets is actually permitted by organic specifications outside of the grazing season. Stall-feeding is currently used and concerns a large part of the organic lambs produced in France.

17.3 Sensory Quality

17.3.1 Flavour/Odour

Flavour and odour are of overriding importance in the sensory characteristics of lamb meat since unacceptably flavoured meat could permanently deter a consumer. However, it should be observed that the sensory assessment of lamb largely varies among countries, depending on cultural backgrounds and sensory habits. This was clearly demonstrated in the landmark study by Sanudo et al. (1998) that compared the assessment of the flavour and odour of stall-fed lambs of the Spanish Merino and Rasa Aragonosa breeds with pasture-fed lambs of the British Welsh breed by both British and Spanish taste panels. When evaluating the same lamb chops, the British panel preferred the flavour of the British pasture-fed lambs, whereas the opposite was observed for the Spanish panel that preferred the flavour of Spanish stall-fed animals. Results from this study thus demonstrate that meat acceptability strongly depends on the consumer's previous knowledge of the product or consumption habits, as further confirmed for beef meat by Stassart and Jamar (2009).

Organic farming promotes pasture-feeding, and it has been shown that the intensity of the odour and flavour of the meat is greater for pasture-fed lambs than for lambs that are stall-fed with concentrate-based diets (Rousset-Akrim et al. 1997; Sanudo et al. 1998). This pastoral flavour is mainly due to the branched-chain fatty acids, 3-methylindole (skatole) and some products of linolenic acid oxidation and its derivatives (Priolo et al. 2001; Schreurs et al. 2008).

Moreover, lamb growth rate and age at slaughter can vary considerably in pasture-fed lambs (Prache and Thériez 1988), affecting the variability in odour and flavour of lamb meat as well. The study by Rousset-Akrim et al. (1997) using groups of lambs with contrasting ages at slaughter (means of 101 days vs. 217 days) demonstrated that age at slaughter and growth rate may strongly influence lamb meat flavour and odour. First, the 'sheep meat', 'cabbage' and 'confined animal' odours were slightly increased in pasture-fed lambs compared to stall-fed lambs when animals were slaughtered at a young age. Second, both 'sheep meat' and 'confined animal' flavours together with 'sheep meat', 'cabbage', 'roasty', 'rancid' and 'confined animal' odours were highly increased in pasture-fed lambs compared to stall-fed lambs when animals were slaughtered at a late age.

Therefore, since organic farming promotes pasture-feeding and the 'natural rhythm' of the animals, while limiting the incorporation of concentrate within animal diets, this production system may be more prone to the occurrence of strong pastoral odours and flavours of the lamb meat and to high variability in these sensory characteristics.

The odour and flavour of the meat may be even stronger when the pasture is rich in legume species such as white clover (*Trifolium repens*) or lucerne (*Medicago sativa*) due to their high degradable protein content and, therefore, their prominent role in the ruminal synthesis of skatole and indole (Schreurs et al. 2007a, b). This was recently confirmed by Prache et al. (2011a) with pasture-fed lambs reared either organically or conventionally, and whose average growth profiles were similar to avoid confusing the effects with lamb age or weight at slaughter. These authors observed that the intensity of abnormal odour of the fatty part of the chop was higher in lambs reared organically than in lambs reared conventionally. They explained this result by the higher proportion of white clover in organically-reared lambs' diets, linked to a higher proportion of white clover in the organic pasture. Recent research has actually demonstrated that meat from pasture-fed lambs has a stronger and less-preferred odour and flavour when the lamb has grazed a white clover-rich pasture compared to a grass-rich pasture (Schreurs et al. 2007a, b). Grazing white clover actually induces higher concentrations of skatole and indole in the fat than grazing ryegrass. These compounds, which are responsible for off-flavours and off-odours, are formed in the rumen from microbial deamination and decarboxylation of tryptophan, and their concentration in the fat is therefore increased when the forage proteins are highly soluble and rapidly degradable in the rumen (Schreurs et al. 2007a, b). Although there are convincing reasons to have high proportions of white clover in organic livestock systems to compensate for avoiding mineral fertilisers and to reduce reliance on purchased concentrate feed, the outcome of these studies demonstrates that this may increase the occurrence of off-odours in the meat end product.

17.3.2 Firmness of Subcutaneous Fat

Firmness of subcutaneous fat is an important sensory characteristic of lamb carcasses, and organically-reared pasture-fed lambs have been shown to have a greater occurrence of firmness defects in subcutaneous fat (Prache et al. 2011a). Again, this may be due to a higher proportion of white clover in the diet of organically-reared lambs. A recent study actually found a higher proportion of PUFA and a lower proportion of SFA, and, therefore, a higher PUFA-to-SFA ratio in the subcutaneous fat of lambs grazing a legume-rich pasture compared with lambs grazing a ryegrass pasture (Lourenço et al. 2007).

17.3.3 Meat Colour

Organic farming promotes pasture-feeding, and it has been shown that meat from pasture-fed animals is darker than meat from animals raised on concentrate-based diets (Priolo et al. 2001). The direct effect of the diet, i.e., when comparing pasture-fed vs. stall-fed animals with a similar growth pattern and slaughtered at a similar live weight, is mediated via a higher muscle myoglobin content (Renerre 1981; Hopkins and Nicholson 1999). Beyond the direct effect of the animal's diet, differences in meat colour between pasture-fed and stall-fed animals may also be due to the animal's age, meat ultimate pH, intramuscular fat content and physical activity. The animal's age at slaughter is generally higher in pasture-fed animals, which is known to decrease muscle lightness and to increase muscle redness due to higher myoglobin content (Renerre 1986). Meat ultimate pH is also generally higher in pasture-fed animals, which decreases muscle lightness (Renerre 1981). In fact, as the pH of meat increases above 5.6, its water-holding capacity increases and its light reflectivity decreases. As a result, the meat may be darker and have reduced microbiological stability (Sheath et al. 2001). Intramuscular fat content is also generally lower in pasture-fed animals, which is known to decrease muscle lightness (Hedrick et al. 1983). Physical activity may influence meat colour via an increased oxidative metabolism (Jurie et al. 2006). Moreover, it should be noted that there may be a great variability in pasture-feeding conditions and in animal growth rate at pasture, which may further enhance meat colour variability. Finally, since organic farming promotes pasture-feeding and the 'natural rhythm' of animals, and limits the incorporation of concentrate within animal diet, this production system may be more prone to the problems of high ultimate pH and darker meat colour, and to a high variability in these meat quality attributes and sensory characteristics.

Regarding the nature of the pasture, Prache et al. (2011a) found no difference in meat myoglobin and lipid oxidation between pasture-fed lambs reared either organically or conventionally. These outcomes are in line with results obtained by Petron et al. (2007) who found no difference in the meat colour or meat lipid oxidation in lambs that grazed either intensive ryegrass, legume-rich or botanically-diverse pastures. It should be noted that pasture-fed lambs exhibit lower thiobarbituric acid reactive substances (TBARS) levels (indicator for lipid peroxydation) than lambs fed concentrate-based diets, partly due to differences in antioxidant vitamin E levels of meat (Santé-Lhoutellier et al. 2008). Actually, the concentration of tocopherols (vitamin E) in dried grain is typically below 10 μg/g, whereas green pasture contains up to 300 μg tocopherols/g dry matter (Sheath et al. 2001).

17.4 Other Aspects of Lamb Meat Quality

Consumers, particularly those who purchase organic products, are increasingly environment-conscious, and studies urgently need to address the issue of the carbon footprint of meat produced under organic or conventional production systems (Prache et al. 2011b). The main factors that enable a low level of the use of non-renewable energy per kg of meat produced in sheep production systems are: (i) a large part of the animal nutrient requirements that are covered by forages and primarily by grazing; (ii) a high feed self-sufficiency; and (iii) limited use of mineral nitrogen (Benoit and Laignel 2010). The challenge in organic systems is therefore to combine high pasture utilisation to both lower the use of non-renewable energy per kg of meat produced and to preserve pastures as biodiversity reservoirs and carbon sinks, with high animal productivity in order to dilute the methane emission per ewe (Benoit and Laignel 2010; Prache et al. 2011b). The organic production process is further favoured because it avoids the use of synthetic fertilisers that are very non-renewable energy-consuming. All these factors contribute to reducing the carbon footprint of the meat produced, water pollution, the reliance on purchased concentrate and the sensitivity to input cost volatility (Benoit and Laignel 2010).

Beyond the intrinsic quality of organic food products, consumers and, therefore, commercial organisations demand a regular supply of organic food products all year round. This may be an issue for sheep production systems because the seasonality of the reproduction in this animal species and the prohibition of the use of hormonal treatments can strongly influence the supply pattern and make the supply of organic lamb meat throughout the year difficult. Commercial organisations actually observe a deficit in organic lamb meat supply around Easter and an excess in autumn. This issue is a key point and creates a bottleneck in organic lamb meat production, requiring additional research on the available methods for the natural control of reproduction in sheep and their interest for organic lamb production (Pellicer-Rubio et al. 2009).

The sanitary-health aspects of the meat also warrant further investigation. However, organic farming greatly limits the risk of having chemical residues within food products because the list of specifications forbids the use of synthetic pesticides and hormones and minimises the recourse to pharmaceuticals and veterinary drugs. Furthermore, organic farmers are committed to feeding animals with organic food, and recent studies have shown that residues from synthetic pesticides are rarely detected in organic vegetables. The level of contamination of organic vegetables by synthetic pesticides that is permitted in conventional farming is 2–6% vs. around 40% for conventional vegetables (Lairon 2009). As far as mycotoxins (toxins that may be transferred from feed to animal products) are concerned, the data available for vegetables do not make it possible to show large differences between organic and conventional farming, and there are no comparative data for organic livestock farming. Lastly, it is worth noting that organic farming professionals consider that genetically-modified organisms are not compatible with organic farming principles and that they therefore exclude them from the organic production process.

17.5 Authentication of Meat Products from Grassland-based Production Systems

The authentication of organic meat, i.e., meat derived from organic systems of production, is one of the challenges facing the sector. Certification of organic farming is currently done using on-farm controls by approved technicians. The last 15 years have been witness to major developments in the use of analytical methods to authenticate food products derived from particular production systems. However, it should be acknowledged that inherent difficulties exist for using such methods to authenticate organic food products because production practices are highly diverse in both organic and conventional agriculture.

Because organic farming specifications make it necessary to provide herbivores with access to pasture during the grazing season, and considering the nutritional advantages of the meat produced from pasture-fed animals, related research on pasture-feeding authentication is of interest. The current state of knowledge in this area will only be summarised in the present paper and full details may be found in Prache (2007, 2009).

The first approach developed to authenticate the meat produced from grassland-based production systems is based on the quantification of specific compounds in animal tissues that are directly transferred from the pasture to the end product or that are transformed or produced by rumen microorganisms under the effect of pasture-feeding (e.g., carotenoid pigments, volatile compounds such as terpenes and 2,3-octanedione, meat fatty acid composition and meat isotopic composition) (Prache 2007). The second approach is the use of fingerprints: differences in the meat and fat composition due to the nature of the diet actually induce differences in their optical properties and, therefore, in their spectral features, which can also be used for diet authentication (Prache 2007). As far as fingerprint approaches are concerned, promising breakthroughs have been made in the use of visible and near-infrared reflectance spectroscopy. As an example, Dian et al. (2008) investigated the spectral characterisation of perirenal fat using visible and near-infrared reflectance spectroscopy to discriminate between carcasses from two feeding practices (pasture-feeding vs. stall-feeding with a concentrate-based diet). Analysis of the optical information showed differences in perirenal fat resulting from the two different feeding practices, which led to the correct classification of 97.5% of fat samples from the pasture-fed lambs and 97.8% of the fat samples from the stall-fed lambs. The reliability of these methods for the discrimination of intermediate feeding conditions (such as concentrate supplementation at pasture, which is frequent even in organic lamb meat production) in interaction with animal characteristics such as the breed and the level of production is under active investigation.

The above methods could be further combined with the isotopic composition of meat for the authentication of products from animals that are pasture-fed on legume-rich grasslands. In fact, the N stable isotope composition of plants is modulated by the botanical family, i.e., with less ^{15}N enrichment of plant nitrogen compounds in leguminous plants that use nitrogen in the air as a nitrogen source (Schmidt et al. 2005; Prache et al. 2009). The stable isotope ratios of nitrogen ($^{15}N/^{14}N$) in the meat may therefore provide some information about the intensification level of the pastures used for grazing and be of some interest for identifying animal products derived from organic and low-input systems (Schmidt et al. 2005).

17.6 Ways of Limiting Identified Sensory Problems

The challenge for the pastoral production of meat is to guarantee average quality as far as it is economically possible, and to control the variability that is inherent in pasture-based production (Prache and Thériez 1988; Sheath et al. 2001). Although organic farming promotes the 'natural rhythm' of animals, the potential danger from a meat sensory quality perspective comes from slowing down the growth rate of lambs and, therefore, increasing their age at slaughter. Most of the quality traits that consumers seek are actually enhanced in the meat of young animals (Rousset-Akrim et al. 1997; Sheath et al. 2001).

The ways of promoting animal growth rate and reducing slaughter age and their variability are: (i) offering high allowances of good quality pastures with a minimum level of parasite contamination; (ii) supplementing lamb diets with carbohydrate-rich feeds within the limits of organic specifications (40% of the diet); and (iii) providing feed supplements that contain condensed tannins.

Although additional costs are associated with this practice that, as a result, should be well controlled, dietary supplementation with carbohydrate-rich feeds may make it possible to: (i) improve the lamb growth rate and lower the age at slaughter, (ii) increase the ratio of carbohydrate to protein and therefore minimise the flavour problems arising from skatole formation in the rumen, and (iii) minimise the problem of high ultimate pH and the subsequent darker colour and lower microbiological stability.

The dietary supplementation of pasture-fed lambs with forage containing condensed tannins (e.g., sainfoin) may also improve organic lamb meat quality, in addition to controlling parasitism in lambs in a 'natural' way. It may, in fact, improve lamb growth rate, lower the age at slaughter and protect ingested protein, thereby minimising the flavour problems arising from skatole biosynthesis in the rumen (Vasta and Luciano 2011).

Since sensory defects are of particular concern for lambs with slow growth rates that are slaughtered at a late age, one way of limiting sensory defects without undue economic cost is to sort animals on the basis of their growth rate, using specific management practices adapted to each lamb category. This may help to reduce the variability and to increase average lamb meat quality without having a negative impact on the production system's feed self-sufficiency and the corresponding production costs (Prache et al. 2011b). However, this management strategy increases the number of lamb batches and the corresponding labour load, and this should not be done at the expense of pasture management to ensure good quality pastures.

Finally, as underlined by Sheath et al. (2001) and Stassart and Jamar (2009), one challenge is also to persuade consumers that pastoral flavour is not a defect but is associated with a 'natural' pasture diet, and to take better advantage of the full flavour of grassland-based meat products. In this regard, it is worth noting that some sensory quality criteria that are considered as defects in the case of standard products may be well accepted by consumers if the product is differentiated, i.e., if it is the result of a quality-based production system such as those that bear the organic label.

17.7 Conclusion

Organic farming embodies extrinsic features that are valued by consumers, but studies on the intrinsic properties of organic products remain scarce. Having acknowledged the fact that production systems are diverse in organic as well as in conventional systems, the challenge is to get the measure of this variability and then ask what the factors are that may determine meat and carcass quality in organic farming to allow for robust conclusions.

Organic farming works according to specifications that guarantee pasture-feeding, which lead to nutritionally more desirable lamb meat fatty acid composition than that found in lambs that are stall-fed with concentrate-based diets. However, the meat from pasture-fed lambs is darker and has a higher odour and flavour intensity than that of lambs that are stall-fed with concentrate-based diets, which may be an issue for southern European consumers who generally prefer the second type of meat. Pasture-based production systems are also more prone to variability in lamb growth rate and age at slaughter and, therefore, in meat quality attributes than stall-fed production systems. Moreover, there is experimental evidence that organically-reared pasture-fed lambs present a greater risk of off-odours in chops and softer subcutaneous fat than conventionally-reared pasture-fed lambs. This risk is linked to a greater proportion of legumes in organically-grown pastures. However, since white clover is given as an explanation for these results, it is worth noting that low-input conventional systems may also be at a greater risk of occurrence of these sensory defects. Furthermore, since organic farming promotes the 'natural rhythm' of animals and limits the recourse to concentrate supplementation, it may be more prone to the occurrence of sensory defects in lamb meat and to a higher variability in sensory attributes. The challenge now is to further experiment with management strategies to help minimise the occurrence of these defects while taking advantage of the presence of legumes within pastures for animal nutrition and natural pasture fertilisation.

Regarding authentication issues, the combination of the isotopic composition of the meat together with the spectral features of the fat may present some interest for identifying animal products from organic and low-input pasture-based systems.

Much of this review has dealt with pasture-fed lambs because organic farming promotes pasture-feeding. However, stall-feeding with concentrate-based diets is actually permitted by organic specifications outside the grazing season. It is currently used and concerns a large part of the organic lambs produced in France. Further research is therefore needed to investigate sensory and nutritional meat quality in lambs that are stall-fed with concentrate-based diets and reared organically vs. conventionally.

Additional emphasis should also be placed on the carbon footprint of the meat produced, an increasing concern at this time, especially for consumers interested in organic products who are particularly environment-conscious. In addition to their primary function in food production, animal production systems now also face new challenges such as reducing their contribution to climate change and maintaining biodiversity, responding to societal demand in terms of ethics, quality and the safety of their food products, and contributing to the socio-economic sustainability of agriculture. Grassland-based systems such as organic systems are favoured because, beyond the nutritional advantages and typicality of their food products that are produced in a natural way, they contribute to maintaining and preserving pastures that play a key role on the environment by mitigating greenhouse gases through carbon sequestration and maintaining open landscapes and biodiversity (Bellon et al. 2009). Since organic specifications are committed to the pasture-feeding of herbivores, and since they forbid the use of synthetic fertilisers, pesticides and hormones, as well as minimising the recourse to pharmaceutical and veterinary drugs, organic systems may therefore be considered as a prototype for sustainable agro-ecological production systems.

References

Angood KM, Wood JD, Nute GR, Whittington FM, Hughes SI, Sheard PR (2008) A comparison of organic and conventionally-produced lamb purchased from three major UK supermarkets: price, eating quality and fatty acid composition. Meat Sci 78:176–184

Aurousseau B, Bauchart D, Calichon E, Micol D, Priolo A (2004) Effect of grass or concentrate feeding systems and rate of growth on triglyceride and phospholipid and their fatty acids in the M. *longissimus thoracis* of lambs. Meat Sci 66:531–541

Bellon S, Prache S, Benoit M, Cabaret J (2009) Recherches en élevage biologique: enjeux, acquis et développements. INRA Productions Animales 22(3):271–284

Benoit M, Laignel G (2010) Energy consumption in mixed crop-sheep farming systems: what factors of variation and how to decrease? Animal 4:1597–1605

Butler G, Stergiadis S, Seal C, Eyre M, Leifert C (2011) Fat composition of organic and conventional retail milk in northeast England. J Dairy Sci 94:24–36

Collomb M, Bisig W, Bütikofer U, Sieber R, Bregy M, Etter L (2008) Fatty acid composition of mountain milk from Switzerland: comparison of organic and integrated farming systems. Int Dairy J 18(suppl. 10–11):976–982

Dian PHM, Andueza D, Jestin M, Prado IN, Prache S (2008) Comparison of visible and near infrared reflectance spectroscopy to discriminate between pasture-fed and concentrate-fed lamb carcasses. Meat Sci 80:1157–1164

European Food Safety Authority (2010) Scientific opinion on dietary reference values for fats, including saturated fatty acids, polyunsaturated fatty acids, monounsaturated fatty acids, *trans* fatty acids, and cholesterol. EFSA J 8(3):1461, 107

Fisher AV, Enser M, Richardson RI, Wood JD, Nute GR, Kurt E, Sinclair LA, Wilkinson RG (2000) Fatty acid composition and eating quality of lamb types derived from four diverse breed x production systems. Meat Sci 55:141–147

Hedrick HB, Paterson JA, Matches AG, Thomas JD, Morrow RE, Stringer WC, Lipsey RJ (1983) Carcass and palatability characteristics of beef produced on pasture, corn silage and corn grain. J Anim Sci 57:791–801

Hopkins DL, Nicholson A (1999) Meat quality of wether lambs grazed on either saltbush (*Atriplex mummularia*) plus supplements or lucerne. Meat Sci 51:91–95

Jurie C, Ortigues-Marty I, Picard B, Micol D, Hocquette JF (2006) The separate effect of the nature of diet and grazing mobility on metabolic potential of muscles from Charolais steers. Livest Sci 104:182–192

Lairon D (2009) La qualité des produits de l'agriculture biologique. Innovations Agronomiques 4:218–287

Legrand P, Bourre JM, Descomps B, Durand G, Renaud S (2001) Lipides. Apports nutritionnels conseillés pour la population française. Editions Tec & Doc, Paris, pp 63–82

Leroux J, Fouchet M, Haegelin A (2009) Elevage bio: des cahiers des charges français à la réglementation européenne. INRA Productions Animales 22(3):151–160

Lourenço M, Van Ranst G, De Smet S, Raes K, Fievez V (2007) Effect of grazing pastures with different botanical composition by lambs on rumen fatty acid metabolism and fatty acid pattern of *longissimus* muscle and subcutaneous fat. Animal 1:537–545

Moloney AP, Fievez V, Martin B, Nute GR, Richardson RI (2008) Botanically diverse forage-based rations for cattle: implications for product composition, product quality and consumer health. In: Hopkins A, Gustafsson T, Bertilsson J, Dalin G, Nilsdotter-Linde N, Sporndly E (eds) Biodiversity and animal feed, future challenges for grassland production. Proceedings of the 22nd General Meeting of the European Grassland Federation. Uppsala (Sweden), 13, 361–374

Pellicer-Rubio MT, Ferchaud S, Freret S, Tournadre H, Fatet A, Boulot S, Pavie J, Leboeuf B, Bocquier F (2009) Les méthodes de maîtrise de la reproduction disponibles chez les mammifères d'élevage et leur intérêt en élevage biologique. INRA Productions Animales 22(3):255–270

Petron MJ, Raes K, Claeys E, Lourenço M, Fremaut D and De Smet S (2007) Effect of grazing pastures of different botanical composition on antioxidant enzyme activities and oxidative stability of lamb meat. Meat Sci 75:737–745

Prache S (2007) Developing a diet authentication system from the composition of milk and meat in sheep: a review. J Agric Sci 145(5):435–444

Prache S (2009) Diet authentication in sheep from the composition of animal tissues and products. Anais da 46ª Reunião Anual da Sociedade Brasileira de Zootecnia, 14–17 de julho 2009, Maringa, PR, Brasil, Revista Brasileira de Zootecnia, pp 362–370

Prache S, Thériez M (1988) Production d'agneaux à l'herbe. INRA Productions Animales 1:25–33

Prache S, Kondjoyan N, Delfosse O, Chauveau-Duriot B, Andueza D, Cornu A (2009) Discrimination of pasture-fed from lambs fed dehydrated alfalfa indoors using different compounds measured in the fat, meat and plasma. Animal 3(4):598–605

Prache S, Gatellier P, Thomas A, Picard B, Bauchart D (2011a) Comparison of meat and carcass quality in organically-reared and conventionally-reared pasture-fed lambs. Animal 5(12):2001–2009

Prache S, Benoit M, Tournadre H, Cabaret J, Laignel G, Ballet J, Thomas Y, Hoste H, Pellicer M, Andueza D, Hostiou N, Giraud JM, Sepchat B (2011b) Plateforme INRA de recherches en production ovine allaitante AB: de l'étude de verrous techniques à la conception de systèmes d'élevage innovants. Rencontres Recherches Ruminants 18:61–64

Priolo A, Micol D, Agabriel J (2001) Effects of grass feeding systems on ruminant meat colour and flavor. Anim Res 50:185–200

Renerre M (1981) La couleur de la viande et sa mesure. Viandes et Produits carnés 2:10–16

Renerre M (1986) Influence de facteurs biologiques et technologiques sur la couleur de la viande bovine. Bulletin Technique du CRZV de Theix, INRA 65: 41–45

Rousset-Akrim S, Young OA, Berdagué JL (1997) Diet and growth effects in panel assessment of sheepmeat odour and flavour. Meat Sci 45:169–181

Santé-Lhoutellier V, Engel E, Gatellier P (2008) Assessment of the influence of diet on lamb meat oxidation. Food Chem 109:573–579

Sanudo C, Nute GR, Campo MM, Maria G, Baker A, Sierra I, Enser M, Wood JD (1998) Assessment of commercial lamb meat quality by British and Spanish taste panels. Meat Sci 48:91–99

Schmidt O, Quilter JM, Bahar B, Moloney AP, Scrimgeour CM, Begley IS, Monahan FJ (2005) Inferring the origin and dietary history of beef from C, N and S stable isotope ration analysis. Food Chem 91:545–549

Schreurs NM, Marotti DM, Tavendale MH, Lane GA, Barry TN, Lopez-Villalobos N, Mcnabb WC (2007a) Concentration of indoles and other rumen metabolites in sheep after a meal of fresh white clover, perennial ryegrass or *Lotus corniculatus* and the appearance of indoles in the blood. J Sci Food Agric 87:1042–1051

Schreurs NM, Mcnabb WC, Tavendale MH, Lane GA, Barry TN, Cummings T, Fraser K, Lopez-Villalobos N, Ramirez-Restrepo CA (2007b) Skatole and indole concentration of fat from lambs that had grazed perennial ryegrass/white clover pasture or *Lotus corniculatus*. Anim Feed Sci Tech 138:254–271

Schreurs NM, Lane GA, Tavendale MH, Barry TN, Mcnabb WC (2008) Pastoral flavour in meat products from ruminants fed fresh forages and its amelioration by forage condensed tannins. Anim Feed Sci Tech 146:193–221

Sheath GW, Coulon JB, Young OA (2001) Grassland management and animal product quality. Proceedings of the XIX International Grassland Congress, 11–21 February 2001, Sao Pedro, Sao Paulo, Brazil, pp 1019–1025

Stassart PM, Jamar D (2009) Agriculture biologique et verrouillage des systèmes de connaissances: conventionalisation des filières agroalimentaires bio. Innovations Agronomiques 4:313–328

Vasta V, Luciano G (2011) The effects of dietary consumption of plants secondary compounds on small ruminants' products quality. Small Ruminant Res 101:150–159

Chapter 18
Is Organic Farming Unsustainable? Analysis of the Debate About the Conventionalisation of the Organic Label

Geneviève Teil

Abstract "Conventionalisation" hangs like the sword of Damocles over organic farming. Raised by certain agrobiologists, the threat has been examined by researchers who have attempted (with little success) to measure any shift in how organic standards are put into practice. This article takes up the issue, but instead of trying to make explicit what is meant by organic (a difficult task in light of the variety of interpretations) as earlier research has done, it attempts to enter into the controversy surrounding the development and sustainability of the movement. This investigation illustrates the opposition between two regimes of action, each based on very different visions of the acceptable use of the AB label: the first tends to reduce organics to a set of regulatory restrictions imposed by the label, while the other sees these restrictions as a reductive and insufficient framework. For this regime, organic production is more than just a set of restrictions. It is instead a "philosophy" or "spirit" guiding a broad examination of the production process and its result. Nevertheless, despite their disagreements, the two regimes are also in close and mutually-beneficial interaction. Two contradictory characteristics of organic production emerge, thus ensuring its sustainability: its capacity to spread through economic networks thanks to a more rigid framework, and its flexibility that enables it to continually redefine itself and adapt to new situations.

Keywords Sustainability · Conventionalisation · Label erosion · Certification · Objectivation · Organics farming · Labelling

18.1 Introduction

In all likelihood, the notion of sustainability owes its success to its ambiguity. Indeed, how do we know what is sustainable and what is not? Is there a threshold beyond which a condition or an approach may be considered sustainable? Moreover, examining the endurance of an object forces us to clarify its nature: what

G. Teil (✉)
INRA-AgroParisTech, UMR1048 SADAPT,
16 Rue Claude Bernard, 75005 Paris, France
e-mail: genevieve.teil@agroparistech.fr

S. Bellon, S. Penvern (eds.), *Organic Farming, Prototype for Sustainable Agricultures*,
DOI 10.1007/978-94-007-7927-3_18,

is organic production, which developments and adaptations can be considered tolerable and which unacceptable? How can we respond to questions such as these? Does the notion of sustainability even have a meaning?

Logic ought to incite us to quickly turn the page. Prudence stresses the fact that it is responsible for a great many discussions and activities, fills numerous pages and occupies a large number of people. We would surely already have grown weary of the notion if it were devoid of meaning. Following the counsel of prudence, this article then aims to examine how actors manage to deal with such a sensitive issue. Rather than looking at sustainability in general (a topic too vast for a short article), we focus on certain agrobiologists' recurrent fears that the organic label is being eroded and losing its capacity to differentiate between "conventional" and "organic" agricultural produce, and is thus impossible to sustain. In what way and why would organic production be unsustainable? What signs, evidence and facts arouse this fear?

We start by retracing the emergence of this concern, which arose among agrobiologists when the French organic label (AB[1]) was established to bring together a variety of approaches to better propagate agrobiology. Researchers heeded this alarm and investigated its reality without reaching a conclusion—at least the persistence of the controversy makes it seem so. We thus return to the actors and their debates on the matter in order to see how they manage to reconcile the two, somewhat contradictory elements of the adaptability and durability of organics. Examining these debates allows us to show the unexpected role this fear plays in sustaining organic production.

18.1.1 Fear of the Erosion of the AB Label

In 1981, French agrobiologists seeking to further promote organic production requested and obtained the authorities' recognition of farming that "does not use synthetic chemicals". In addition, organisations and private commercial brands that promoted the development of organic farming found themselves being offered the possibility of having their specifications and "endorsements" or interpretations of organics approved, thus lending a kind of official validation to their approach.

The first ratified specifications, those of *Nature & Progrès*, did not establish themselves as *the* organic specifications. In parallel to the approval of the 14 specifications and endorsements existing at the time, and in order to avoid the dispersal of the organic movement into a myriad of different specifications, a single French label was negotiated.

This led all those involved in the organic movement to seek a consensual definition based essentially[2] on the rejection of the technical innovation of "chemical" agronomy. Since the invention of synthetic ammonia by Fritz Haber in 1918, the chemical industry has developed a large number of solutions to the "problems" of

[1] AB stands for Agriculture Biologique, or organic agriculture.

[2] The 1991 European regulation (R CEE n° 2092/91) that provides the framework for certified European organic production also encourages biodiversity, pluriannual crop rotations and the association between agriculture and breeding.

plant growth, disease, pests and competition with weeds. The effectiveness of pesticides, fungicides and herbicides increases yields and matches the changing lifestyles of farmers, who also want to be able to take holidays, for example. However, this technical success is also transforming farming and depends on a reconstruction—ever more controllable and thus increasingly "artificial"—of biological processes, isolating and reducing them to a food production process with very clearly predefined characteristics.

The resulting regulations of the AB label[3] with which organic actors seek to promote alternative farming mainly consist in the prohibition of synthetic fertilisers and treatments. They monopolise the designation "organic" and have been a great success. Of the 14 pre-existing private specifications and charters for organic production, only two remain: *Nature & Progrès*[4] for organic farming ("in limbo" after the departure of many of its members, both producers and consumers), and *Déméter* for biodynamic production.

However, alarm is growing in the shadow of this success. Producers and those running organisations have pointed out the threat looming in the absence of the alternative socio-economic dimension in the label's regulations, as well as the recourse, voluntary or not, of certain producers to the usual marketing channels.

> We could say that the institutionalisation of organics has now marked its territory (or its objectives). The organic movement is currently in a phase of "market takeover": some people would like to restrict its development to a copy-&-paste of conventional food production and distribution methods. (Harrouch 2003)

In particular, they point to supermarket distribution, the pressure on productivity and the prices imposed on producers, which will, in their opinion, only lead to an erosion of organic standards. According to them, this is the same economic pressure that caused the recent transformation and excesses of conventional farming and against which they rebelled. The truncated interpretations of organics, limiting it to the label's purely technical regulatory requirements, are in the process of taking the movement in exactly the same direction.

18.1.2 The Contribution of Research to the Problem of the Label's Erosion

Researchers also find these developments disturbing. Two articles published in the same issue of the journal, *Sociologia Ruralis*, with very different reasoning but of-

[3] The AB label is combined with the CCREPAB F, the French specifications that are stricter than the European organic label, especially with regard to livestock farming.

[4] In order to be completely precise, we should say three, since the SIMPLES specifications that focus on gathering medicinal plants has also endured. In France, Nature & Progrès and Déméter are the two main organisations that defend organic production at this time. They offer private specifications independent of the public French AB label that became European after the abolition of European subsidiarity for organic certification. The two organisations bring together producers, distributors and consumers, while the FNAB (France's National Federation of Organic Farming) is exclusively a union of certified AB producers.

ten cited together (Buck et al. 1997; Tovey 1997), both come to the same conclusion about the impossible mission of the organic movement.

Buck et al. (1997) observe the rapprochement between conventional and organic farming in California. To explain it, they cite the market economy mechanism itself and the shortcomings of the barriers set up by organic stakeholders to this very profitable segment of the market. "Conventional" producers attracted by this organic production mode are converting and bringing with them a rationalisation of production by investing large amounts of capital. The comfortable margins they enjoy allow them to put pressure on prices and also lead to an increase in the value of land, which is gradually driving out the pioneers of the organic movement who generally do not have much capital to invest. Moreover, consumers are more interested in the qualities specific to organic produce (health, environment, taste, etc.) than questions of organisation and market access. They, therefore, put up little or no resistance to the restructuring of organic production and the abandoning of the alternative, sustainable socio-economic principles that comprised the foundations of the original organic movement.

In the second article, Tovey highlights an "institutional" mechanism: the Irish authorities are using European subsidies for agri-environmental measures to encourage the development of organic production on the sole basis of the agronomic principles of organic certification, ignoring any socio-economic goal. They are thus bringing about a growth in organic production that outflanks the initial, dissenting socio-economic vision, replacing it with a mere variant of classic agronomy that uses natural products instead of chemical treatments and fertilisers.

Far from being unanimously accepted, these articles sparked a lively debate about what is known as "conventionalisation", the name given to the implacable logic of the erosion of organic production and its attendant identified mechanisms of "appropriation", "commodification" and "bifurcation". In fact, the idea of a capitalist logic of capturing value evoked by Buck, Getz and Guthman is reversed in the name of capitalism's capacity to incorporate its own contradictions in order to continue its development and thus escape from its own deadlocks (Coombes and Campbell 1998). Guthman (2004) persists by evoking the flaws of organic labelling and its inability to impose an organic quality that is not only agronomic but socio-economic as well[5]. In early 2001, a special issue of *Sociologia Ruralis* (Michelsen 2001) was published, devoted to "conventionalisation". It broadened the debate which, until then, had been confined to economics, in order to set "conventionalisation" against a series of logics and competing social forces. In an attempt to move beyond disciplinary confrontation, it also provided the contribution of empirical studies that led to a series of case studies (Coombes and Campbell 1998; Dantsis et al. 2009) and "measurements" of conventionalisation (Hall and Mogyorody 2001; Lockie and Halpin 2005).

These case studies demonstrate that the erosion is not equally severe everywhere. In certain countries such as New Zealand, "genuine" organic production seems to be resisting alongside an industrialised and watered-down version. The authors of these studies therefore contest the universality of the erosive forces, which, until then, had been identified but not explained.

[5] Guthman's argument is very similar to the fear of "market takeover" mentioned above.

As for measuring conventionalising forces, this is not without its own problems. In order to appreciate the effect of the transformation of organic production, the authors compare the "values" of the new organic farmers, particularly those who convert, with those of the pioneers. However, in making this comparison, they consider the pioneers as fossilised dinosaurs, the unanimous (and particularly inflexible) bearers of a testimony of another era. As Best (2008) observes, following the publications of Darnhofer (2006) and Tovey (1997), we need a more precise definition of conventionalisation. At the moment, it is impossible to distinguish that which, in the changes observed, is part of an appropriation of organic production by conventional production, from what is specific to the development of the organic movement and its continual adaptation to new and changing situations.

In view of these difficulties, Darnhofer et al. (2010) suggest comparing practices with what the authors consider to be *the* definition of "organic". The article is thus based on the four principles listed by the International Federation of Organic Agriculture Movement (IFOAM undated) as forming the "roots" of organic farming, and follows the adjustment of practices to these principles by means of sets of practice indicators.

However, the bias of the choice of the IFOAM definition overlooks the historical dimension of any drafting of principles. It also completely disregards the diversity of interpretations of organics, particularly all those that do not recognise themselves in the IFOAM definition, or find these principles inadequate. The IFOAM, which is supposed to support the international development of organic production, is based on consensual criteria that exclude, in particular, alternative socio-economic dimensions. Prohibiting supermarket distribution is not, therefore, one of the criteria selected in the article, even though this is a crucial point for many of the organic militants, defenders or promoters involved, as shown in the interviews[6].

Another path explored by more "comprehensive" human sciences takes the diversity of actors' interpretations more seriously. Researchers following this path attempt to analyse conventionalisation not as a set of hidden forces imposed upon actors, but as the fruit of divergences of interpretations of organic production and its sustainability. The theory of conventions on which many of these studies are based (Murdoch and Miele 1999; Campbell and Liepins 2001; Rosin and Campbell 2009) transforms "conventionalisation" and the erosion of organics by the rationalised agricultural system into a play of oppositions between ideological tenets or value conventions. The interpretation of organic as a "value" turns it into the exact opposite of earlier studies (which attempted to list the criteria in order to enable its objective content to be grasped), into a simple, subjective construction. Consequently, the threat of the erosion of the label becomes nothing but performative autosuggestion[7].

[6] We should also mention that like "organic-formula", the article reduces organic farming to indicators of practices that alarmed adepts of organic farming have accused of being the very cause of the threat of the erosion of organics.

[7] The ensuing problem is well known. It is impossible to predict which of the "objective" or "subjective" forces will lead to a conventionalisation of a given label and, therefore, impossible to decide on the correct preventative measures.

Table 18.1 Breakdown of interviews by field

Field	*N*
Production	115
Sales and distribution	36
Central administration	12
Technical and research	14
Farming union	7
Quality certification	15
Media	10
Catering industry	3
Consumers	13
Phytosanitary industry	6
Total	231

Table 18.2 Breakdown of interviews by region

Region	*N*
Languedoc-Roussillon	80
Val de Loire	98
Paris	51
Other (Jura)	2
Total	231

However, the actors do not only live in a realm of ideas that dictate their laws to the plants, fields and annual turnover, regardless of their day-to-day experience of them. Rather than continuing along one of the two previous objective or subjective paths, we have chosen to come back to the actors and the problem they raise. Unlike earlier studies, we did not confine ourselves to the "values" or "ideologies" as subjective moral principles affixed to the actors' experience of the world. In accordance with the tradition of socio-technical studies (Latour 2007) or pragmatic studies (James 1996), we sought to define the threat sensed by some people through the actors' experience of it, the visible signs of it, the accusations they make against some of their colleagues and the responses made to these accusations, in order to understand the difference between "good" organics and the "bad" organics that threatens it. We thus carried out a pragmatic analysis that sees organics, like its objectivity, as the result of the collective—but not necessarily shared—experience.

The analysis presented in this article is partly based on a project carried out for the French Ministry of Ecology by a team of researchers—Barrey, Blanchemanche, Charpigny, Floux, Hennion and Teil—on the environmental quality of wine (Teil et al. 2011). The study devoted to organic production is based on over 70 interviews with organic or biodynamic vintners, certified (62) or not certified (10), in the Val de Loire, Jura and Languedoc-Roussillon regions, as well as people working for the authorities and technical centres in charge of organic production, organic distributors, consumers, journalists, certification bodies, etc. (see Tables 18.1 and 18.2). The detailed list of interviews is presented in the appendices. We supplemented this with seven interviews with representatives and staff from organisations that promote organic production and the FNAB, and four half-days of participant observation.

A significant corpus of textual data has been added to this corpus of interviews. It is composed of various articles, journals and studies, blogs and websites, research reports from public and private institutions, etc., related to environmental issues in viticulture as well as agriculture in general, since the qualifications are not limited to viticulture alone.

Our field study examined the example of viticulture and we questioned only vintners and wine grape producers. However, in most cases, the remarks made by these organic vintners are not exclusively confined to the field of wine-growing, and relate to organic production in general. Nevertheless, the application of actions correlated to their conclusions was, of course, carried out on the basis of their specific production and, therefore, their wines. Wine is usually regarded as one organic product among others and, occasionally, on the contrary, as a specific product. In the following section, the article reproduces the connection they make or do not make between wine and organic production as a whole. Lastly, just as we use agrobiology here in its widest, polysemous sense, in line with the diversity of actors' interpretations, the notion of organic wine is also a flexible notion here. Although the INAO[8] has not authorised the notion of "organic wine", keeping instead "wine made from organically-grown grapes", actors constantly use the term and various private charters even outline specifications.

18.2 The Contested Threat of the Label's Erosion

The AB label provoked debates and then rejections as soon as it was established in 1991. Initially, rejections were individual, and were then followed by a boycott by the *Nature & Progrès* organisation in 1995. Although the boycott had little support at the time, it has undergone a revival since 2000: the number of "boycotting" members (who apply exclusively for *Nature & Progrès* certification and refuse AB certification) has grown from a third to half the organisation's members in 2010. Would the truth of the threat gradually be spreading? However, in that case, why would the others resist? This section presents the experience that led those who denounce the AB label erosion to call a certain exploitation of the AB label into question, and the response they are given.

18.2.1 The Dispossession and Reappropriation of Organics

Detractors dissatisfied with the label[9] describe the "institutionalisation" of organic production. Some claim that the AB label "dispossesses" them of the vigilance they

[8] INAO: Institut National de la Qualité et de l'Origine, in charge of French policies regarding designations of origin.

[9] This is a criticism of the AB label made by militants from organisations, especially Nature & Progrès, and organic producers who are often also militants within organisations. To avoid the text

formally exercised over organic farming, how it is put into practice and its various interpretations and innovations—an impression that resurfaced in 2010 with the publication of the new regulations for organic production that abolish French subsidiarity[10].

> "...the text of European specifications is consensual and one that, for some, sounds the alarm regarding the loss of control of the founding principles of organic farming. [...&... It] is now being discussed out of the reach of organic producers and is slipping out of our hands. It could well lead to lower standards since it is based on a broad consensus, and this is even truer with the extension of Europe." (Interview with the Nature & Progrès organisation: 1&2[11])

Organisation members discuss the issues at stake for organic farming, as well as good and bad practices. They make adjustments to the concept and periodically revise the specifications in order to take new experiments, interesting or undesirable innovations, etc., into account. However, with the label, organic production is entrusted, on the one hand, to an independent monitoring body and, on the other, to the wisdom of those who consume organic produce, whose skill or expertise is distrusted by concerned producers. They think that consumer interest in organic farming is limited to banning "unnatural" chemical products and that the consumer shows little interest in specifications or the debates surrounding practices. Buyers "place their trust" in the label, i.e. they delegate the identification of organic produce to the label. With consumers withdrawing from the debate, the crucial discussion of what constitutes organic farming and how it is monitored or controlled therefore depends on the choices and judgment of staff or members of certifying bodies on a day-to-day basis[12].

The interviews show two recurrent examples that highlight the absence of monitoring: productivist agriculture that uses only natural treatments, or crops grown without synthetic products but which are "soilless". Both are seen as betraying the idea of organics while respecting the bare minimum of the label's specifications. These two examples are cited as the epitome of organic production's takeover by large operators, or market "capitalist logic". Since the label does not present a single economic criterion, control procedures are regarded as incapable of preventing the perversion of organics that they cause.

These interpretations of organic production are often qualified as "organic-formula" in that, like a formula, they are based on the pre-established list of the allowed organic techniques but lack the "spirit" of the organic movement. The adhesion to this sprit sustains a wider, contextualised analysis that, with the aim of developing

---

from becoming too cumbersome, we have not repeated this whenever possible.

[10] The subsidiarity rule allowed each EU member state to adopt specifications stricter than the common EU regulation; the new one (CE) N°834/2007 prohibits this "overruling" and forces all EU countries to adopt the same EU organic regulation.

[11] Figures indicate the page numbers of interviews, all transcribed in a standardised format.

[12] This designation includes everything that comes within the "sphere of control" with its monitoring commissions, advisory councils, etc., and which actors often call "monitoring" or "control" without further precision.

an alternative agriculture that is durable because it is protected from harmful market forces, reintroduces the socio-economic dimensions.

In anticipation of the foreseen slipping of standards, the two organisations, *Déméter* and *Nature & Progrès*, have revised their specifications, charters and guiding principles[13]. Agronomic restrictions have been tightened and the principles of sustainable farming—from the point of view of nature as much as from a human perspective—have been made explicit. Difficult (if not impossible) to translate or define with testable socio-economic criteria, these principles are written into the charters as farmers' commitments, monitored by an internal participatory guarantee system. The organisation's members themselves guarantee the vigilance of the interpretation, application and commitment to the organic or biodynamic producer or transformer by the other members of the organisation.

> Biodynamic production requires that one feels strongly connected to the essence of biodynamic methods, principles and aims. For this purpose, it is necessary to fathom natural processes through observation, thought processes and perception. An ever-deeper understanding of connections in nature can be acquired through continual effort. Working groups in various organisations, public events, journals and books are all significant sources of help and support.
> However, if someone should wish to use these standards only as is often done with laws, i.e., uniquely in relation to the formal aspect or by seeking loopholes because it is economically advantageous, this person would do better to practise a different type of farming. It is the mission of the Déméter France organisation, its representatives and consultants, to prevent this from happening. (Déméter 2004)

18.2.2 Using the Label to Spread the Organic Movement

For a great many supporters, organic farming is a state of mind, a different way to conceive of agricultural production that is constantly under discussion in the numerous collective decision-making bodies that are set up and form the "network" of the *Fédération Nationale de l'Agriculture Biologique* (FNAB).

However, not everybody rejects the label—far from it, in fact.

For some people, protecting the environment is very much a question of means and the banning of bad practices. For them, the main thing is that synthetic products are banned, making farming non-polluting once again, respectful of the environment and the consumer. In that case, the market should not be feared but rather used for an essential mission—to develop and generalise organic production. From this point of view, the label offers considerable advantages: it enables organic produce to become widespread, increases its credibility and does away with rival interpretations, bringing organic produce to wherever the market goes.

For those who support the AB label, the qualitative one-upmanship and charters to develop new socio-economic relations merely confine organic production to a

[13] The abolition of French specifications and their replacement by European specifications caused a similar reaction with the creation of the brand "Bio Cohérence" whose standards are stricter and attempt to incorporate socio-economic criteria to prevent the "hijacking" of the organic movement.

"niche" reserved for the elite, the only ones able to afford the products seen as very expensive. Supporters, on the contrary, hope to contribute to the spread of the organic movement, making it "commonplace". They scoff at their inefficient distribution channels that cause prices to rise, and later at the Parisian "*bobos*" ["bourgeois Bohemians"] who hijack the organic movement in order to turn it into a source of luxury product with the "obvious" complicity of the very exacting specifications of private organisations. They strongly reject these strategies of "confinement" and, on the contrary, use every resource offered by the market to disseminate their wines or products as widely as possible. For the vintners interviewed, it is crucial to market wines with prices comparable to the others so that they do not suffer from any discrimination.

The first producer we encountered during our survey had "rationalised" and "optimised" his equipment, become a wine merchant as well as a producer, and adapted the labelling of his wine at the request of his clients. One of his wines for New Zealand is called "bin", in reference to a renowned Australian wine merchant. The second producer gambled on economies of scale and set up a 220-hectare vineyard with AB certification and used biodynamic methods. Both producers supply their wines to supermarkets and sell and export at prices similar to those of other wines from the same *appellation* in order to facilitate the distribution of their products. They explained how, unlike the vast majority of organic producers, they have bowed to the demands of volume, standardisation and regularity made by supermarkets and large exporting wine merchants. In return, they benefit from the AB label's credibility abroad and this form of distribution relieves them of the burden of marketing their products themselves.

In the 1980s, these producers were the only ones (two out of seven producers with private certification before 1990 in our sample) to denounce the "irrational fear of the market" that they saw among their colleagues, and relations were stormy at times. However, they became less of a fringe element with the arrival of new producers with organic certification who are very concerned about the environment and who, like them, are convinced that rejecting synthetic products provides a radical solution to the problem of agricultural pollution. They are therefore less or only minimally involved in the discussion ranging from agronomy in the strictest sense to the economics of agricultural production and its marketing. Relatively unconcerned about the quirks of the market economy, they are more concerned with managing to ensure the economic health of their businesses—an essential weapon, according to them, for expanding the organic movement.

There are, therefore, two fairly contradictory emerging interpretations of organics. In one interpretation, organics is a subject in the process of development, moving towards the goal of sustainable farming for both the environment and its inhabitants, constantly reconsidering what it actually is and continually rethinking its ends and means. In the other interpretation, organics is an object sufficiently defined by a set of practices and prohibitions. The question of monitoring remains equally crucial for both camps as they cannot rely on consumers for any kind of vigilance. Unlike the gustative quality of wines or other products, the quality of organic production cannot be directly verified by the consumer. The seriousness of monitoring is one of the unavoidable issues when marketing and distributing organic produce.

Militants from organic organisations and partisans of the organic market do not have the same vision of how organic standards ought to be put into practice. For the first group, the label's criteria are only a minimum framework of interpretations whose "spirit" needs to be discussed. For the second group, on the contrary, the label's criteria define and guarantee organic production. The latter do not feel "dispossessed" of organics by the label, which "belongs" to no one. This does not necessarily mean that they are merely lukewarm supporters. Organic production is a production mode they support and, like the others, hope to develop. What's more, some of them are very active in the organic community. One of the two producers in our sample launched an organic wine competition to try and improve the gustative quality of the wines, which he believed to be a handicap to their development.

Together, they differ less in their acceptation or rejection of the market than in the modality of the elaboration of the organics they support. For partisans of the market, organic production is the circumscribed, defined and standardised quality of an object. During production, whatever does not fall within the norm makes up part of the "other" qualities of the product and, on the contrary, is part of the producer's choice. They are therefore opposed to those who, unlike them, see a global quality in organic production whose assessment must be continually reconsidered.

18.2.3 An Occasionally Stormy Coexistence

Relations between advocates of the two positions are somewhat conflicting. Each reflects an interpretation of organic ("niche-organic" versus "business-organic") that casts doubt on the sincerity or effectiveness of the other's commitment. To avoid using these pejorative designations, we will refer to the first group as "eco-alternative organic" to stress the broad and global nature of their interpretation, and to the second as "label supporters" who delegate responsibility for organic identity to the label.

The eco-alternative advocates in our sample fear that the market will end up watering down or hijacking the conception of agriculture that they promote, often at the cost of great abnegation. They feel that the development of organic production has been entrusted to consumers who are generally ill-informed and lacking in commitment, and producers who may be self-serving.

They challenge the authorities that regulate the markets and their reductive vision of organic production. They cite recent events and the abolition of French subsidiarity as an illustration that the authorities consider market development as a fluid circulation of homogenous organic products and, therefore, reduced to a single label. This invariably results in bringing organic production down to the lowest common denominator, lowering standards and ignoring the broader requirements of each of the original specifications.

The "label supporters" reproach the eco-alternative camp for the stricter demands of their private charters, the additional costs that these involve and their specialised distribution channels that risk turning organic goods into products reserved for the elite. Furthermore, this expensive organic production seems to them to open the way to an interpretation of "organic" as an increase in the quality of produce that justifies a higher premium, whereas an interpretation such as this can only be an obstacle in developing organic production, demanding that consumers agree to pay more for their organic produce. They therefore reiterate that, contrary to a widely held belief, organic production is not more costly than other production systems:

> …there is also the belief in consumers' minds that in any case, organic wine is more expensive. Someone decided that 30 years ago, that it's 30 % more expensive and it's stuck regardless of the production. And when you put two bottles, one organic and one not organic, next to each other and you look at the prices, generally it never works. It's not 30 % but someone decided that it was 30 % more expensive and so it's an obstacle (Organic vintner: 4)

The eco-alternative group admits that organic produce is often—but not always—slightly more expensive, but it should be reconsidered in a new way of perceiving consumption: less waste, greater attention paid to what is consumed, etc.—an attitude that easily compensates for the extra production and marketing costs. For them, "organic" means considering not only the "definition" of organic production, but also the human and economic consequences of our actions. This comprehensive reflection should lead to a fairer distribution of wealth. The eco-alternative advocates thus defend themselves against accusations of elitism by insisting upon the politico-social combat they are engaged in to encourage everyone to think further than "my tastes", "the price" and above all "*the* label", since this reduction of organics is an abdication of our duty to reflect, and for them, this is a high price to pay for the fluid distribution that enables organic production to reach the entire planet. In fact, with the label that "commodifies" organic production (as economists describe it), making an organic "thing" all the more successful because the label is considered "credible" and therefore not called into question, the opposite occurs. The credible label relieves consumers of their responsibility to think about the consequences of their actions and enables businesses to invent and implement organic solutions that are not organic in "spirit", such as "soilless" organic systems, so-called "intensive" organic systems and the transportation of converted soil to other places to supply organic greenhouses, for example.

18.3 Organic—A Global Quality or an Objectified Characteristic?

Is it possible to reconcile the "sworn enemies" or should we separate them?

> "When we set a rule, we create borderline cases, and this is a problem!" a member of the FNAB told us.

By "objectivising" organic production, the label opens the way to innovations that satisfy the label's criteria but breaks away from the collective procedure by which the global quality of organic production is drawn up and monitored. The assessment of these innovations is entrusted to customers who delegate the scrutiny of organic production to a label that makes only a very incomplete examination of the organic quality of the product and is blind to questions of transport, water, economic organisation, ethics, justice, etc.

The eco-alternative organic camp accuses the label certified by a third party of creating the possibility of its misappropriation since it allows for a "non-committed" use of organic production. Recourse to the label can suspend the critical interrogation of organic production—what it is, its good or bad interpretations, adjustments, etc.—to turn it into a quality determined by the criteria that designate it. The merit of the objectification of organic production is that it makes it a finite, autonomous quality, independent of place and the person applying it. To use Latour's (1987) term, organic quality has become an "immutable mobile", something that can be appropriated by anyone and "applied" to any supporting object as long as its manufacture respects the label's criteria and restrictions. Organic production "boxed up" by a label thus acquires an existence independent of those who conceived and created it. However, the label also makes another transformation possible. Whereas for eco-alternative organic advocates, "organic" designates a global quality that examines the organic product as a whole (its manufacture, transport, consumption and recycling, for example), it becomes, because of the label, a circumscribed quality that can be incorporated (with a varying degree of ease) into the product's other qualities, objectified or not, anticipated or imposed by the producer/manufacturer, the buyers, or market regulations.

In their opinion, the way in which organic standards are put into practice should be subject to an *ex post* integrative assessment. It should be based on the entire process, from manufacture and distribution to consumption, since these all have an impact on the "organic" quality of the product, and can no longer be seen as a set of criteria but, instead, as the ever ongoing development of an alternative to the "deadlock" of "conventional" farming.

The organic quality that emerges from this collective usage is not "something" autonomous or predefined. It is the result of multiple applications, each inseparable from the concrete situation of its use, and an interpretation of what organic production could and should be, all subject to a critical discussion led by other people who are also committed to developing the organic alternative. Whether this is to produce, find or consume these products, the eco-alternative advocates insist upon the need for each person to keep informed, gather judgments, inquire about the validity of these judgments, learn how to judge them and thus contribute to the collective task of critiquing and developing the global quality of organics. The sign or certification that establishes the acquisition of organic quality in the first instance constitutes a "reductive" vision, incomplete and at times even inappropriate in the latter. It loses its capacity to designate and is subject to discussion and judgment.

18.3.1 The Bifurcation of Organics?

The organics currently being developed cannot be monitored for compliance by a third party. Organisations (especially *Nature & Progrès*) opposed to the AB label since 1995, offer an alternative. They have invented and perfected a participatory guarantee system that, rather than delegating the task to independent third parties, submits all their members' organic production projects to a collective critique by the organisation's committed and vigilant members (May 2008; Darlong 2008; Fonseca et al. 2008; IFOAM 2007).

In the interviews, *Nature & Progrès* and *Déméter* were described as closed and sectarian groups and their internal monitoring procedures disparaged for their lack of "transparency", "independence" and "objective guarantee". Admittedly, the vigilance procedures regarding commitment are internal and organised by their members, the only people they recognise as competent for this mission. The lack of criteria to define *a priori* organic production is not a lack of transparency. Instead, it stems from the impossibility of definitively making "organic" explicit, taking every possible factor into consideration, i.e. depending on the region, the production, the farm and… the future. The lack of the evaluators' independence is, on the contrary, a guarantee of their competence, their sharing of ideas and the discussion of production projects. The proficiency of those evaluating organics as a global quality in the making is tied to and therefore inseparable from their participation in the drawing up of the quality itself. The notion of independence is meaningless when it involves evaluating the interpretation of an idea or a concept, as is an "objective guarantee", since both assume that the idea or concept has a defined and objectifiable existence[14].

Should, then, the AB label and its independent monitoring be replaced by an overall participatory review procedure? This is not an unusual situation. Consumers interested in a particular "quality" or object investigate, gather information and even engage in actual "surveys" to make their choices. They thus make use of and contribute to the collective critique that goes with the market circulation of these products. We are also aware of the limitations of this. A strong investment by a large number of buyers is required. Their acquisition of information and the circulation and confrontation of judgments limit the possibilities of developing and expanding these markets. On the contrary, certifications make it possible to take responsibility for this survey and to relieve consumers of the task in order to facilitate the profit sharing of new consumers and thus the expansion and growth of the markets. The modest size[15] of the organisations that have distanced themselves from the AB label is usually associated with the strict demands of their specifications. This point, however, does not concern consumers. It is hard to see the low membership cost for these organisations[16] as a significant limiting factor. Instead, we should prob-

[14] For a full discussion of the objectivity of engaged or disengaged monitoring, see Teil (2001).

[15] In France, Nature & Progrès has around 350 producer-members, whereas 20,000 producers have the AB label certification.

[16] For consumers, membership usually costs between 10 and 20 euros.

ably see their small size as a consequence of the implementation of strict vigilance, promoting small projects on a human scale and short networks where one can still gather information and where participatory monitoring can still take place and remain effective. However, when products travel all over the world, when farms sell hundreds of different products or very large volumes, members are no longer sufficient.

Are we now doomed to divergent organics as suggested in the idea of "bifurcation" (Coombes and Campbell 1998; Campbell and Liepins 2001), with short networks and participatory monitoring of all the actors, local markets and human scale, on the one hand, and, on the other, international markets, international labels delegated to independent monitoring bodies and organic produce of dubious identity?

18.3.2 Sustainable Organics: An Active and Framed Goal

The idea of bifurcation suggests separating into two distinct paths where "eco-alternative" organics remains out of the reach of and protected from "objectified" organics, thus avoiding conventionalisation.

First and foremost, we should not deceive ourselves. These two "organics" are not two different and homonymous versions of organic production. There is not "eco-alternative" organic production on one side and a different, "objectified" organic production on the other, any more than these two organics are the result of two different uses by producers engaged in organic reflection, on the one hand, and commercial profitability, on the other. They are two regimes of action that cause different modes of presence of organics to emerge, one in a form reduced to criteria and the other as an object constantly under construction. Each regime uses different instruments of proof or evaluation of the object's presence, but these are only judged to be incompatible insofar as the actors esteem that the ways in which they are applied are too different to continue to cohabit under the same name without mutual harm. This is the same question highlighted by the threat of erosion or conventionalisation raised by the eco-alternative camp. Would it be appropriate to separate the two regimes?

Eco-alternative organics would always experience the same difficulties in developing and extending its dense and informed circuits. For its part, objectified organics, like all standards, should be subject to constant revisions, additions and adjustments to adapt to the incessant changes and vagaries and to ensure its sustainability.

Seeking to protect eco-alternative organics by endowing it with more numerous, more objective criteria, as suggested by Conner (2004), Guthman (2004) and Darnhofer et al. (2010) (as *Nature & Progrès* and *Déméter* have been doing for a long time), would mean that the overflowing of the organic movement could be reframed by giving it new limits. However, this process still reproduces (like any guarantee of means or ends) the divergence between organics seen as a goal or as a predefined object. Therefore, it is not "the" solution but a stage of the process in which organics constantly rethinks and revises itself. Like any set of criteria or restrictions that only imperfectly and temporarily captures an object in progress such as organics, it always ends up flowing over because of the arrival of new people concerned. Once again, new explicitations and adjustments of organics must try to frame it[17].

This framing and overflowing is not due to approximations of criteria and restrictions; it is constitutive of the sustainability of organics, which holds together two different modalisations of the existence of the organics notion. Objectified organics derives its strength from the critical monitoring practised by eco-alternative organics that guides its slow content evolution and guarantee its credibility. Reciprocally, the latter increases its development capacities tenfold if it can benefit from the capacity to enlist objectified organics. Together, they form the two "pillars" that provide its attractiveness and resilience; without one another, they are nothing.

As the notion of organics is transformed and diversified, the signs of objectified organics must be readjusted to continue to benefit from the credibility bestowed by the critique. This is done by reinforcing charters and criteria—in the case of *Nature & Progrès* and *Déméter*, or by creating new specifications[18]. Their divergences, shown by a variety of charters, pluralise the notion of organics. Above all, they result in a growing detachment from the certification that limits the notion of organics. In fact, it is, above all, the certification that should also be adjusted and revised. It is the same process of exchange between a global vision of organics and its reduced interpretation that should, therefore, be regularly revised by all the members of the organic movement and not merely within its sub-groups to protect producers from sliding standards resulting from innovations brought about by the label and deemed unacceptable.

18.4 Conclusion

How can the development and sustainability of organic farming be helped? Are its development and sustainability now under threat, as many agrobiologists and researchers claim? Should we, like them, seek to guarantee the durability of organics by means of increasingly thorough knowledge of what it is? This idea supposes that organics is a finite object existing independently of those who contribute each day to make it exist. It is doomed to come up against historical evolutions that destroy definitions one after the other. Starting off from the opposite conception (that analyses organics as the result of the conception and implementation by the actors themselves), we have attempted to sketch the original solution they provide to the problems that may be caused by the diversification of a growing movement.

---

[17] We borrow the very appropriate image here used by Callon to describe the succession of movements of pausing and questioning that guarantee the durability of action groups (Akrich et al. 2010).

[18] This is the case in France with the new brand, Bio-Cohérence.

"*Traduttore, traditore*". As highlighted by the sociology of translation[19], this adage perfectly captures the difficulties faced by many movements that hope to expand. Translation, in other words, the enrolment of new adepts, always brings new interpretations of the goals or message and new applications of practices, and accompanies the expansion of a movement, the spreading of knowledge or techniques. Seen from the narrow framework of individual or stabilised interpretations, these new interpretations are just so many aberrations or blows to their "true" meaning, as well as adjustments that enable them to win over a wider public, to expand and to endure. In this way, the development of a new movement often appears to be forced to accept a "pluralisation" of its message through the new interpretations brought by new adepts, and this is its strength as well as its weakness.

The solution that seems to be emerging consists in holding together (and it is here that the difficulty lies) two different regimes.

The AB label represents a certification that introduces an objectification of the organic quality. Established to ensure the expansion of organics, it brings interpretations that are sometimes innovative, but whose evaluation escapes the most committed people involved, and relies upon the producers or consumers whose commitment to the development of enduring organic agriculture is not guaranteed. Organisations then react by establishing private brands and tightening specifications to correspond to a different conception of organics, not as criteria that define the scope of a quality but as a minimum framework to outline the development of organics as a goal or global quality that is not predefined, extending without *a priori* limits to every area of production, distribution and consumption. Through their internal critical vigilance, they bring about the continual adjustment and revision of organics in order to make it sustainable and durable. These organisations, however, add a major constraint of active participation to the discussion of organics.

Organics now appears as an object that combines strongly opposed and rival conceptions. Rather than distinguishing them or relinquishing one or another of the conceptions, we have defended the idea that it is important to maintain their interaction so that organics can benefit from the development capacities provided by the organic quality objectified in certification, and so that the innovations this generates can be validated by a critical discussion of organics as a global quality and not simply as the respect of *a priori* criteria.

Lastly, the fear of conventionalisation appears as one line of inquiry (by the eco-alternative organic camp)[20] to examine the healthy coexistence of the two regimes. As long as it does not conclude that it is necessary to split up into two incompatible "visions" and give them two different names—which is what has happened in other certification cases (Teil 2011)—it will continue to contribute to their mutual interaction.

---

[19] See Law and Williams (1982), Callon et al. (1983), and Latour (1984), and the anthology of Akrich, Callon, and Latour (2006) for a new edition of the founding texts.

[20] The "label supporters" camp raises another, symmetrical line of inquiry that highlights the confinement of organics to the internal, collective critique of organisations.

Appendix

Detail of the distribution of the survey interviews of vintners

Type of farming unit	Type of environmental quality	Number		
		V de L	LR	Jura
Cooperative	Without	1		
Cooperative	Industrial quality certification	1	2	
Cooperative	Integrated viticulture	1	3	
Cooperative	With part of the production certified as organic		2	
Farm	Without	8	1	
Farm	Industrial quality certification		4	
Farm	Integrated viticulture	7	5	
Farm	Non-certified integrated viticulture	5	5	
Farm	Certified AB	26	14	
Farm	Certified biodynamics	13	5	
Farm	Uncertified biodynamics	4		
Farm	"Natural wine" or "terroir wine"	2	4	2
	Total	68	45	2

Total number of survey interviews = 115
Although also organic, biodynamic producers do not appear under the certified count
V de L Val de Loire, *LR* Languedoc Roussillon

All samples

Activity		Nb			
Producers	All	115	Technique and research	Technical training	6
	AB organic certified	62		Organic technical training	3
	AB "committed"	10		Research	4
Retail and sales	Superstores	9		Agronomy teaching	1
	Wine trade	1	Farming syndicate	Farming syndicate	7
	Wine seller	8	Quality certification	Integrated viticulture	4
	Organic coop outlet	5		Industrial quality certification	1
	Organic associative outlet	1		Organic certification	1
	Franchised organic outlet	1		AOC certification	9
	Wine retailer with mixed (organic and non-organic) products	11	Catering	Organic catering	3
Administration	Ministry of Agriculture	6	Media	Wine critic	4
	Organic agency	2		Regional press	2
	AOC administration	4		Organic critic	1
Agro-chemical industry	Agro-chemical firm	5		Natural wine critic	2
	Agro-chemical product retailer	1		Economic press	1
	Total	*231*	Consumers	Consumers	13

References

Akrich M, Barthe Y, Muniesa F, Mustar P (eds) (2010) Débordements. Mélanges offerts à Michel Callon. Presses de l'Ecole des Mines, Paris

Akrich M, Callon M, Latour B (2006) Sociologie de la traduction: Textes fondateurs. Presses de l'Ecole des Mines, Paris

Best H (2008) Organic agriculture and the conventionalization hypothesis: a case study from West Germany. Agr Human Values 25(1):95–106

Buck D, Getz C, Guthman J (1997) From farm to table: the organic vegetable commodity chain of Northern California. Sociol Ruralis 37(1):3–20

Callon M, Courtial JP, Turner WA, Bauin S (1983) From translation to problematic networks: an introduction to co-word analysis. Soc Sci Inform 22:191–235

Campbell H, Liepins R (2001) Naming organics: understanding organic standards in New Zealand as a discursive field. Sociol Ruralis 41(1):22–39

Conner D (2004) Beyond organic: information provision for sustainable agriculture in a changing market. J Food Distrib Res 35(1):34–39

Coombes B, Campbell H (1998) Dependent reproduction of alternative modes of agriculture: organic farming in New Zealand. Sociol Ruralis 38(2):127–145

Dantsis T, Loumou A, Giourga C (2009) Organic agriculture's approach towards sustainability; its relationship with the Agro-Industrial Complex, a case study in Central Macedonia, Greece. J Agr Environ Eth 22(3):197–216

Darlong V (2008) Harmonizing jhum (shifting cultivation) with PGS organic standards in North-east India: key features and characteristics of jhum for process harmonization. Cultivating the future based on science. Volume 1: Organic Crop Production. Proceedings of the Second Scientific Conference of ISOFAR, held at the 16th IFOAM Organic World Conference in Co-operation with the International Federation of Organic Agriculture Movements (IFOAM) and the Consorzio ModenaBio in Modena, Italy, 18–20 June, 2008:718–721

Darnhofer I (2006) Organic farming between professionalisation and conventionalisation—the need for a more discerning view of farmer practices. The organic Congress, Odense

Darnhofer I, Lindenthal T, Bartel-Kratochvil R, Zollitsch W (2010) Conventionalisation of organic farming practices: from structural criteria towards an assessment based on organic principles. A review. Agron Sustain Dev 30(1):67–81

Déméter (2004) Cahier des Charges. Identification des produits

Fonseca MF de AC, Wilkinson J, Egelyng H, Mascarenhas GCC (2008) The institutionalization of Participatory Guarantee Systems (PGS) in Brazil: organic and fair trade initiatives. Cultivating the future based on science. Volume 2: Livestock, socio-economy and cross disciplinary research in organic agriculture. Proceedings of the Second Scientific Conference of ISOFAR, held at the 16th IFOAM Organic World Conference in Cooperation with the International Federation of Organic Agriculture Movements (IFOAM) and the Consorzio ModenaBio in Modena, Italy, 18–20 June, 2008:368–371

Guthman J (2004) The trouble with 'organic lite' in California: a rejoinder to the 'conventionalisation' debate. Sociol Ruralis 44(3):301–316

Hall A, Mogyorody V (2001) Organic farmers in Ontario: an examination of the conventionalization argument. Sociol Ruralis 41(4):399–322

Harrouch A (2003) Le rôle de Nature & Progrès dans l'histoire de la bio en France: témoignage d'une actrice engagée. Nature & Progrès 44(Novembre-décembre 2003)

IFOAM (2007) Systèmes de garantie participatifs: vision idéale

IFOAM (undated) Principles of organic agriculture—Préambule

James W (1996) Essays in radical empiricism. University of Nebraska Press & Longmans Greens and Co, New York. (1° édition (1912) ed)

Latour B (1984) Les Microbes: guerre et paix, suivi de Irréductions. A.-M. Métailié, Paris

Latour B (1987) Science in action, how to follow scientists and engineers through society. Open University Press (G.B.), Harvard University Press (U.S.), Harvard

Latour B (2007) Changer de société, refaire de la sociologie, Poche. La Découverte, Paris. (Original edition, en anglais, 2005, Oxford)

Law J, Williams RJ (1982) Putting facts together. Soc Stud Sci 12:535–558

Lockie S, Halpin D (2005) The 'Conventionalisation' thesis reconsidered: Structural and ideological transformation of Australian organic agriculture. Sociol Ruralis 45:284–307

May C (2008) Petit guide des SPG ou comment développer et faire fonctionner les Systèmes participatifs de garantie. IFOAM, Germany

Michelsen J (2001) Recent development and political acceptance of organic farming in Europe—introduction au numéro. Sociol Ruralis 41(1):3–20

Murdoch J, Miele M (1999) 'Back to Nature': changing 'Worlds of Production' in the food sector. Sociol Ruralis 39(4):465–483

Rosin C, Campbell H (2009) Beyond bifurcation: examining the conventions of organic agriculture in New Zealand. J Rural Stud 25(1):35–47

Teil G (2001) La production du jugement esthétique sur les vins par la critique vinicole. Revue de Sociologie du Travail 43(1):67–89

Teil G (2011) No such thing as terroir? Objectivities and the Regimes of Existence of Objects Science technology and Human Values 37(5):478–505

Teil G, Barrey S, Floux P, Hennion H (2011) Le vin et l'environnement: faire compter la différence. Presses de l'Ecole des Mines, Paris

Tovey H (1997) Food, environmentalism and rural sociology: on the organic farming movement in Ireland. Sociol Ruralis 37(1):21–37

Part III
Development Dynamics

Chapter 19
Framing the Social, Ecological and Economic Goods and Services Derived from Organic Agriculture in the Canadian Context

Derek H. Lynch, Jennifer Sumner and Ralph C. Martin

Abstract Consumer support for organic products continues to grow in Canada and the US. At the same time, the characteristics of organic agriculture and the wider social and political context in these countries have limited broader endorsement of organic and other forms of alternative agriculture, with the result that consumer understanding in North America of the 'value proposition' of organic agriculture is lagging in comparison with the rest of the world. The recent growth in targeted research funding for organic agriculture is providing much-needed documented evidence from Canada and the US, summarized in this document, of the broad social, ecological and economic goods and services (SEEGS) derived from organic agriculture. However, to further transform recognition of these benefits, the interrelated issues inherent in SEEGS will increasingly have to be tackled by multidisciplinary teams of researchers partnering with organic producers. In addition, we propose two approaches, one regional in scope, a pilot-scale watershed initiative to demonstrate the diverse benefits of organic agriculture, and more broadly, promotion and use of the concept of organic agriculture as a form of 'civil commons', as a meaningful framework and tangible concept to help promote sustainability and a shift in social consciousness to encourage broader support and endorsement of organic agriculture in North America as a prototype of sustainable agriculture.

Keywords Civic agriculture · SEEGS · Sustainability · Gender equality · Knowledge · Global warming potential (GWP) · Biodiversity

D. H. Lynch (✉)
Department of Plant and Animal Sciences, Dalhousie University,
Truro, Nova Scotia B2N 5E3, Canada
e-mail: derek.lynch@dal.ca

J. Sumner
Department of Leadership, Higher and Adult Education,
Ontario Institute for Studies in Education, University of Toronto,
Toronto, Ontario M5S 1V6, Canada
e-mail: jennifer.sumner@utoronto.ca

R. C. Martin
Department of Plant Agriculture, University of Guelph, Guelph,
Ontario N1G 2W1, Canada
e-mail: rcmartin@uoguelph.ca

S. Bellon, S. Penvern (eds.), *Organic Farming, Prototype for Sustainable Agricultures*, 347
DOI 10.1007/978-94-007-7927-3_19, © Springer Science+Business Media Dordrecht 2014

19.1 Introduction

Organic agriculture, it can be argued, is the first codified form of 'civic agriculture' (Lyson 2004), based on a set of principles that aims to go beyond solely a focus on production and to enhance the benefits of the social, ecological and economic goods and services (SEEGS) it provides to society. Such an approach, emphasizing the joint production of both agricultural commodities and SEEGS, has been a defining characteristic of organic agriculture since its inception (Lynch 2009; Drinkwater 2009; Gomiero et al. 2011; Lynch et al. 2012a). Although controversies continue as to the degree to which environmental benefits from organic agriculture are achieved (Kirchman and Bergström 2009) in some regions, primarily Europe, organic producers have received, for many years, direct government payments through various agri-environmental schemes (AES) for goods and services (protection of water, biodiversity, etc.) of benefit to society. In the UK alone, such AES schemes apply to almost 45 % of agricultural land with organic farmers routinely benefiting (Feehan et al. 2005; Zander et al. 2008; Fuentes-Montemayor et al. 2011).

Reganold et al. (2011) argue that since agriculture is a complex socio-ecological system, multiple sustainability concerns can only be addressed through a transformative, whole system redesign, rather than more incremental and solely technological approaches. In Canada, as is the case globally, national organic standards and organic producers also broadly adhere to elements of such a transformative approach to agriculture. In the Canadian national standards for organic agriculture (CGSB 2006), for example, five of the seven guiding principles refer to goals related to environmental and ecological benefits of organic farm design and management, including minimizing soil degradation and erosion, decreasing pollution, optimizing biological activity and 'health', maintaining biological diversity within the system, and recycling materials and resources whenever possible within the production system. The motivations of many Canadian organic producers also reflect these guiding principles. A recent survey of more than 600 Canadian organic farmers' production research interests found holistic management of their farming systems, including rotations, soil quality, ecological interactions and energy use, as their top-ranked research priorities (OACC 2008).

With respect to the social and political context, Canada is among those countries in the world with more than 80 % of the population now in urban centres, many of whom have no recent family history of involvement in any aspect of the agricultural industry. In addition, the majority of organic farms in Canada are located in the Canadian Prairies region, far from most of the rapidly growing urban centres. As a result of this dwindling direct social connection to agriculture and agricultural landscapes, instilling an appreciation of the social, ecological and economic goods and services derived from organic agriculture is a challenge. In addition, Canada is directly affected by the neoliberal political climate in the United States, which values economic returns over social and environmental benefits. In this vein, the neoliberal cult of the individual renders collective outcomes invisible or undesirable, especially if they create barriers to personal gain. Within this context, the need to support farming systems and landscapes that exhibit floral and wildlife diversity not just for ecological reasons alone, but also for aesthetic, recreational and tourism-related benefits, is less tangibly obvious to consumers and policy makers than is the case in Europe. For example, when North American consumers are polled regarding their reasons for purchasing organic products, their motivation spans a narrower range than that of consumers elsewhere. Most (78 %) purchase organic foods solely as a perceived healthier choice for themselves or their children, with only 11 and 2 %, respectively, citing environmental benefits or improved animal welfare as motivating factors influencing their purchasing choices. In contrast, environment and animal welfare concerns were respectively cited as key concerns by 19 and 12 % of European consumers, and 15 and 7 % of consumers globally (ACNielsen 2005).

From a political and policy perspective, there has been a lack of government support in Canada to promote an integrated or whole-system framework (Reganold et al. 2011) for addressing multiple sustainability concerns related to agriculture, in general, and to organic agriculture, in particular. The limited exceptions include a new farm 'multifunctionality' pilot program in the province of Quebec, which is providing financial support to producers for both the establishment and maintenance of important habitats such as hedgerows (MAPAQ 2011),as well as help to cover the costs of organic certification being adopted by some Canadian provinces. MacRae et al. (2007) have argued that expansion of organic farming could help solve broad multiple policy goals, including social and financial goals, for Canadian agriculture. The authors concluded that factors such as lower input costs and price premiums, more diversified production and marketing channels, resilience in the face of variable market conditions and even weather conditions often result in improved economic performance for organic farms across North America. With respect to social impacts, the increased demand for labour (which may be less limiting as fossil fuel prices rise and variable weather increases), and local goods and services, and possibly even a greater commitment to participation in civic institutions were linked to organic farms, although the data were considered less conclusive.

Overall, support for organic farming may result in a diverse range of goods and services. From an economic perspective, goods and services are understood as any tangible economic product and intangible economic activity that contributes directly or indirectly to the satisfaction of human wants (Pass et al. 1991). Costanza et al. (1997) added an environmental perspective to this basic understanding when they argued that "ecosystem goods (such as food) and services (such as waste assimilation) represent the benefits human populations derive, directly or indirectly, from ecosystem functions." We move beyond Costanza et al.'s (1997) conceptualization of ecosystem goods and services to propose social, ecological and economic goods and services (SEEGS), which represent the suite of benefits human populations may derive, directly or indirectly, from organic agriculture.

In the following section (Sect. 2), we review the current knowledge of SEEGS benefits derived from organic agriculture with particular reference to Canada and, secondarily, to North America. We then put forward two alternative approaches (Sect. 3)—a regional watershed approach and a civil commons approach—for framing SEEGS in the Canadian context. Finally, in the conclusion (Sect. 4) we summarize and point out promising recent developments in support of the promotion of organic agriculture as a prototype of sustainable agriculture.

19.2 Social, Ecological and Economic Goods and Services (SEEGS)

Organic agriculture can provide a range of goods and services that contribute to sustainability. These benefits can be divided into social, ecological and economic goods and services (SEEGS), which help to establish organic agriculture as a prototype for sustainable agriculture.

19.2.1 Social Goods and Services

Research in North America and beyond has found a range of social goods and services provided by organic agriculture that can contribute to a more sustainable form of agriculture: rural community sustainability, rural-urban linkages, gender equality, knowledge production and social learning opportunities, as well as animal welfare considerations.

Rural Community Sustainability

A study composed of semi-structured interviews with a representative sample of 41 organic farmers in southern Ontario, Canada, revealed that they made economic, social and environmental contributions to rural community sustainability (Sumner 2005). Economically, they positively affected both supply and demand. On the supply side, 56% of organic farmers made direct sales to local businesses, while 27% were involved in farm sales, and 21% ran community-supported agriculture programs (CSAs). On the demand side, 97% purchased farm supplies and household needs as locally as possible (see additional information on organic agriculture development in Canada in Text-Box).

Socially, 76% of organic farmers volunteered in their communities, and 70% were members of a local club or organization. Seventy-six percent of organic farmers also supported local cultural events and were politically active, with 76% involved in their local government and 61% participating in local roundtables and panels.

Environmentally, all of the farmers in the study followed local waste management guidelines. An overwhelming percentage—93%—made donations to an environmental group, and 79% supported local environmental initiatives. Eighty-eight percent sold their produce locally, and more than half of them defended the environment—69% spoke at public gatherings and 55% spoke to their politicians.

Many of these findings were corroborated by a second study, which found strong farm-community linkages among organic farmers (MacKinnon 2006). Organic farmers made economic contributions through local purchasing, job creation and viable farms and also made social contributions through a number of channels. They demonstrated strong involvement in education, networking and leadership, and they built social capital—the invisible social infrastructure thought to underlie a community's capacity for development.

Rural–Urban Linkages

A recent report found that organic agriculture can strengthen the relationships between rural and urban communities through both traditional arenas of interest and more recent ones (Sumner 2009). Within the traditional arenas, the report highlighted that organic agriculture could either begin or continue to strengthen relationships through farmers' markets and outreach provided by non-governmental organizations (NGOs) such as the Canadian Organic Growers and the Ecological Farmers Association of Ontario. Within more recent arenas, the report found that a number of new opportunities had arisen where organic agriculture could strengthen the relationships between rural and urban communities. These included a number of opportunities strongly associated with organic agriculture, such as 100-mile markets, which sell predominantly organic food from within a 100-mile radius, social movements with a preference for organic food (such as the Slow Food movement and the local food movement), and the growing phenomenon of urban agriculture, which follows organic principles.

Some Advances in Gender Equality

> Gender relationships are fundamental worldwide to the way farm work is organized, the way assets such as land, labor, seeds and machinery are managed, and to farm decision-making. Given this, the lack of adequate attention to gender issues within the organic and sustainable farming movements is worrying. The revolutionary potential of sustainable approaches to farming to reshape our food systems, and the way humans interact with those systems, will not be realized unless there is a concerted effort by committed sustainable farmers and consumers to work towards gender equality. Indeed, the question addressed by this paper can be turned on its head. As well as asking how participation in organic and sustainable farming can empower women, we can ask: How does the participation of women broaden and deepen the multiple goals of organic and sustainable farming? (Farnworth and Hutchings 2009)

In their study of organic agriculture for the International Federation of Organic Agriculture Movements (IFOAM), Farnworth and Hutchings (2009) contended that gender relations in organic agriculture did not differ substantially from conventional agriculture. In Canada, however, advances in gender equality are beginning to occur. Hall and Mogyorody (2001) found that a significant percentage of farmers working together in heterosexual couples (38%) reported that decisions were shared equally. This finding was corroborated by Maceachern (2008), who argued that the decision-making process on organic farms offered some opportunity for differences with conventional farms. In the US, Trauger et al. (2010) concluded that "in sustainable agriculture systems, the construction of masculinity and femininity, and their relationships to work roles and decision-making, are changing." The authors warn, however, that these changes are not total or transformative because "women still shoulder the burden of domestic work in addition to taking on more of the productive work of the farm."

In combination with their own primary research conducted among organic women farmers participating in the social economy in Ontario, Canada, Sumner and Llewelyn (2011) found that these and other studies in North America yielded a number of commonalities. They reported that, first, organic agriculture can offer an opportunity for addressing rural gender relations but, second, this opportunity has often been squandered. For example, Hall and Mogyorody (2001, p. 313) disclosed that even among dedicated organic farmers, gender relations were often "not on their radar screens." Third, glimmers of hope are nevertheless appearing as organic women farmers continue to increase in number and make their presence felt—Farnworth and Hutchings' (2009) report represents one example of this positive turn. Fourth, women's knowledge networks are vital to this process but, fifth, the spectre of conventionalization—the process by which "agribusiness is finding ways to industrialize organic production" (Buck et al. 1997) threatens the recently documented advances in gender equality. As McMahon (2005) warns, "the organic movement does not recognize that the conventionalization of organic agriculture, like earlier developments in non-organic agriculture, is itself a gendered process." And sixth, it is important to remember that alternatives to industrial agriculture (either conventional or organic) will not realize their full potential unless they focus on non-production issues such as gender. In North America, organic agriculture is beginning to pay attention to gender issues, but it has a long way to go before it can realize its full potential.

Knowledge Production and Social Learning Opportunities

In their study of multifunctional agriculture in the US, Jordan and Warner (2010) argue that to capitalize on opportunities and address problems, we need "new modes of perception, knowledge production, and decision-making." This, they maintain, will help to develop the policies and markets required to stimulate a diversified flow of goods and services from multifunctional landscapes. They call for the development of new management regimes for agriculture that will "support an ongoing process of inter-coordinated 'knowledge innovation' across social, technical, market, and policy sectors." In addition, the authors propose that advanced multifunctionality depends upon social learning, which they define as "participatory research by diverse stakeholders to manage specific agro-ecosystems," in order to coordinate and integrate different types of biological and practical knowledge to generate a multifunctional benefit stream.

Knowledge is central to organic agriculture—it is a knowledge-based system (Aeberhard and Rist 2009; El-HageScialabba 2007; Kummer et al. 2010; Morgan and Murdoch 2000; Warner 2007) that creates, shares and applies knowledge. Since knowledge has long been associated with power—whose knowledge counts and whose does not—organic farmers have come to understand that no university, government or corporation would teach them how to farm (Francis 2009).

> They have learned to teach themselves and each other, over decades, to compensate for any knowledge gaps. They read extensively, form learning associations, set up mentoring networks, hold kitchen table meetings, host farm tours, organize field trips and convene conferences. Organic farmers move seamlessly between the roles of teacher and learner, sharing knowledge, building knowledge networks and refining their knowledge through scientific experimentation. (Sumner 2008)

The breadth and depth of knowledge of organic farmers, and their commitment to social learning, has been recognized by the FAO in its position paper on organic agriculture and food security:

> Inexperience and lack of adequate extension and training for knowledge-intensive management systems and location-specific science require long-term investments in capacity building. With the objective of creating a critical mass and the necessity to strive in settings with limited opportunities, many organic communities have responded by establishing collective learning mechanisms and have become innovators or ecological entrepreneurs. (El-HageScialabba 2007)

In this way, organic farmers' modes of knowledge production and social learning position them to participate in a transformative redesign of agriculture and to help us learn our way out of unsustainable ways of producing food.

Animal Welfare

The rise of industrial agriculture and the accompanying proliferation of confined animal feeding operations (CAFOs) have raised questions in the minds of many people about the welfare of animals kept in these conditions. In the United States, for example, Jordan and Warner (2010) report "there are many concerns about animal agriculture for meat production". As a result, some are choosing to avoid meat altogether, rather than contribute to animal cruelty. Jordan and Warner (2010) suggest that by articulating an "integrated social vision, supportive of MFA," many of the concerns about animal agriculture could be addressed because of new grazing systems.

In Canada, a recent government market analysis study of socially conscious consumer trends (AAFC 2011) identified a growing niche for products offering animal welfare assurances. In response, a number of livestock sectors, government regulations and retailers have begun to address the animal welfare issue. In the US, in response to voter referendums, individual states have brought in statutes or regulations that deal primarily with livestock housing systems (AAFC 2011), and a ban on battery cages for poultry is scheduled to be implemented by some Canadian provinces. In the province of Ontario, a non-profit organization that aims to establish a local sustainable food system and certifies food as local and sustainable—Local Food Plus (LFP)—includes a requirement for improving animal welfare in its certification standards (Friedmann 2007).

The new Canadian Organic Standards (CGSB 2006) include animal welfare as one of seven general principles: "provide attentive care that promotes the health and meets the behavioural needs of livestock" (CGSB 2006) and corresponding specific management requirements regarding stocking density, housing, feeding and health management, handling and transport, etc. Furthermore, for sick or injured livestock, the Canadian standards require that all appropriate medications be used to restore them to health if methods acceptable to organic production fail. Recent organic sector industry efforts have been targeted to improve management practices on organic farms with respect to animal welfare and organic producer skills in this regard (Animal Welfare Task Force 2011). However, as noted above in Sect. 1, consumers in North America still lag behind their international counterparts in terms of associating animal welfare benefits with livestock management on organic farms. In this book, Porcher (Chap. 15) provides a more thorough review of the challenges related to more comprehensively providing for animal welfare while under a continuing paradigm focused on 'organic animal production'.

19.2.2 Ecological Goods and Services

Ecological goods and services from agriculture are tangible outcomes and benefits to society as influenced by farm and landscape management. These ecosystem services include:(1) *supporting* services such as biological pest control and soil processes that maintain soil structure and nutrient cycling; and (2) broader *regulating* services that extend beyond agriculture, such as maintenance of biodiversity, ground and surface water quality, and climate regulation, etc. (Drinkwater 2009). Agriculture is a key driver of environmental pressure and ecosystem degradation globally, through its impact on water use, loss of habitat, climate change, and pollution, and this is also true of North America (Mooney et al. 2005; Jordan and Warner 2010; Reganold et al. 2011; Lynch 2009; Lynch et al. 2012a). Against this backdrop, does organic agriculture live up to its claims of a reduced environmental and ecological footprint? In recent years, this has been the topic of a number of extensive reviews and meta-analyses drawing primarily upon European research data (Hole et al. 2005; Gomiero et al. 2008, 2011), and the reader can refer to these sources for a more detailed coverage of this complex topic. Lynch et al. (2012a) concluded, following a review of the published literature available to date, that organic farming system attributes as typically found within Canada (regarding cropping, floral, and habitat diversity, soil management and intensity of nutrient use, and energy and pesticide use) are sufficiently distinct as to impart potentially important environmental benefits (including maintenance of soil organic matter, improved soil health, reduced energy use and off-farm nutrient losses, as well as enhanced vegetative diversity and support for pollinators and wildlife (bird) diversity), but noted that much more research is needed to validate these results. The relatively rapid increase in research funding for organic agriculture in the past decade, including in North America (Francis and Van Wart 2009), is contributing to our ability to evaluate the ecological impact of these farming systems. However, as noted by Drinkwater (2009), most studies in organic agriculture in North America, while often impressively systems-based, continue to lack any reference to ecology or ecological processes.

Energy Use and Global Warming Potential

Research in Canada examined the energy use and efficiency and global warming potential (GWP) of organic production at the farm level (Lynch et al. 2011). Organic sectors assessed included field crops, beef, dairy, hogs, poultry, vegetables, fruit and greenhouse production. Field crops (grains, grain legumes, oilseeds and forages) represent one of the largest organic production sectors in Canada and the US. Even if limited to only North American studies ($n=7$), the strong consensus was that organic field cropping systems require less energy and improve energy efficiency both per hectare and per unit product compared to conventional field crop production. This was also found to be the case for organic dairy, beef and some vegetable sectors, but not for poultry and fruit. Data were relatively sparse for greenhouse, poultry and hog production, and generally insufficient for assessing GWP impact. In agreement with Gomiero et al. (2008), and Muller and Aubert (Chap. 13) in this book, the higher energy efficiency and climate change mitigation potential, respectively, found for organic systems can be attributed to a lack of input of synthetic N-fertilizers (which require a high E consumption for production and transport), low input of other mineral fertilizers (e.g., P, K) and much lower use of highly energy-consumptive foodstuffs (concentrates).

Soil Organic Carbon Storage

Soil organic carbon (SOC) storage, soil quality and soil health, as well as the nutrient loading impacts of farming systems are closely related to agronomic and livestock management practices of farming systems. In North America, organic farming systems are sometimes criticized for their continued reliance on mechanical tillage, especially within the context of gains in SOC over the past few decades, attributed to the adoption of no-till and minimum tillage practices (Smith et al. 1997). However, the few studies that have comprehensively assessed the *net* effects of organic management systems on SOC, including the added C return to soil from green manures, show no consistent trend of SOC depletion (Teasdale et al. 2007; Lynch et al. 2012a). Indeed, while changes in SOC between farming systems are often considered neutral (Lynch et al. 2011; Zentner et al. 2011a), the added return of C to soil in organic farming systems, mineralized and respired through microbial activity, is perhaps most important in its contribution to maintaining soil health and biodiversity (Lynch et al. 2012a). The extended rotations and more frequent use of perennial legume crops in organic systems have been shown to sustain SOC levels and promote soil biological activity, sometimes referred to as 'soil health', including microbial biomass and earthworms, as shown in Atlantic Canada by Nelson et al. (2009). Lynch (2009) and Gomiero et al. (2011) also concluded that organic systems are superior with respect to improving soil biophysical and biological properties.

Nutrient Loading

Intensity of nutrient use and farm-level loading of N and P, in particular, are negatively related to improving biodiversity and water quality outcomes (CCME 2002; UNEP 2010; Gomiero et al. 2011). In contrast to conventional production systems, the relatively low P inputs and farm surpluses of most organic cropping and livestock systems in North America are increasingly well documented (Martin et al. 2007; Roberts et al. 2008; Knight et al. 2010; Rick et al. 2011). Due to the reliance, in particular, on legume sources of nitrogen supplemented by organic amendments, organic farming systems similarly utilize N less intensively, with correspondingly low N surpluses per hectare and per farm (Roberts et al. 2008; Lynch 2009; Gomiero et al. 2011). Some North American studies have also documented lower residual soil nitrates following harvest in organic systems, and reduced nitrate leaching (Pimentel et al. 2005; Kramer et al. 2006; Lynch et al. 2012b). Few studies have directly assessed the comparative impact on water use and water quality (Gomiero et al. 2011).

Integration of Crop and Livestock Production

A substantial component of organic crop production is under arid and semi-arid conditions of the northern Great Plains of North America, with no direct connection to livestock production. As a result, livestock manure and compost sources are often in limited supply. Entz and Thiessen Martens (2009) report that only 12 % of organic fields in this region typically receive manure on a regular basis. Low nutrient loading, particularly of N and P, on organic crop and livestock farms in Canada is an important distinctive beneficial characteristic of these systems, which reduces off-farm impacts and enhances biodiversity (Lynch 2009; Lynch et al. 2012a). In the long term, critical deficiencies may have a negative impact on the primary ecological service of *provisioning* of food and fodder. As noted above, increasing data suggest low farm P imports, and potential P deficiencies may be as important a management consideration as the challenges of optimizing N use in some Canadian and US crop production and livestock systems. Thiessen Martens and Entz (2011) argue that greater integration of livestock on organic field-crop farms is critical to sustainability in order to address some of these fertility concerns, but could also provide both economic and agronomic benefits. For example, using livestock to graze novel and traditional green manures in rotation is an example of the integrated approach that could be promoted. Enhanced mixed crop livestock farming would in turn encourage biodiversity and allow for increased eco-agriculture tourism and the promotion of the animal welfare-focused approach of organic livestock management.

Biodiversity

Biodiversity broadly refers to the abundance, variety and variability of living organisms in a given environment, both within or between species. A growing body of literature, primarily European in origin (Hole et al. 2005; Gomiero et al. 2011) suggests that species abundance and richness across a wide range of taxa (including arable flora, birds, mammals and invertebrates) benefit from organic farm design and management. In North America, increased specialization in intensive arable cropping has resulted in landscapes characterized by low within-field and between-field variability, combined with reduction or elimination of field margins and other non-crop habitats.

As noted in reviews by Lynch (2009) and Lynch et al. (2012a) Canadian studies including those on birds (Freemark and Kirk 2001), vegetation (Boutin et al. 2008), arthropods (Boutin et al. 2009), moths (Boutin et al. 2011) and in the US on vegetation (Wortman et al. 2010) suggest that among taxa, plants show the most consistent and pronounced responses to organic farming systems. Bird species richness, abundance and frequency of occurrence were also enhanced by organic farming (Freemark and Kirk 2001). In contrast, Geigera et al. (2010) in Europe found that the effects of organic farming on farmland birds proved to be limited to simplified landscapes. Responses of other taxa were more variable and depend on interactions with habitat type and landscape complexity (Lynch et al. 2012a) as found in Europe as well (Gomiero et al. 2011). Moths species assemblages, for example, were found not to be influenced by farming systems, compared to the effects of landscape features such as hedgerows and field margins, in the study of Boutin et al. (2011) in Ontario, Canada. While these advances in our understanding of the relationship between farming system, landscape and biodiversity are encouraging, we are just beginning to understand the relationship between such biodiversity and specific ecosystem services such as pollination, pest control, SOC storage and water quality maintenance (Lynch et al. 2012a). Ultimately, linking the benefits of enhancing biodiversity through organic farming to these *provisioning* and *regulating* ecological services, plus *cultural* services such as aesthetic, recreational and tourism benefits, is an essential next step in deepening the appreciation of the multifunctional nature of organic agriculture. A more detailed discussion of the issue of biodiversity as affected by organic agriculture and landscape factors, and subsequent provision of ecological services such as conservation biocontrol, is provided by Simon et al. (Chap. 5).

19.2.3 Economic Goods and Services

Of the three goods and services (social, ecological and economic), economic goods and services are traditionally cited as indicators of agro-ecosystem value. Economic metrics are certainly necessary and are noted, although not emphasized, to complete the framing of SEEGS. The clearest indicator of economic success is profit.

Organic agriculture systems are usually more profitable than conventional farming systems, given a combination of yield changes, input cost reductions and price premiums (MacRae et al. 2007). In more extensive organic systems like those practiced in North America, input cost reductions are often sufficient to maintain margins.

Total yields rather than profit are frequently used as indicators of success. On the basis of a meta-study, De Ponti et al. (2012) showed that conventional crops yield 20 % more than organic crops at the crop level. They hypothesized that at crop rotation, farm and regional levels, conventional crops have an advantage of 20 % or more, given the greater availability of nutrients. MacRae et al. (2007) also acknowledged higher crop yields on conventional farms (20–40 % higher in Europe and about 20 % higher in North America). However, given the lower input costs and price premiums on organic farms, profits are usually higher than on conventional farms. If conventional input costs increase, particularly those of nitrogen and phosphorus fertilizer, organic farms may then be more economically resilient.

Marketing farm products as conventional commodities usually requires increasing volumes to increase gross income and/or to reduce costs per unit of production. An economic opportunity for profitable organic agriculture in Canada has been to provide high-quality products to meet consumer demand for organic products that are grown and processed in Canada. The Foodland Ontario and Canada Organic logos increased the likelihood of purchase (Campbell et al. 2010). Profitability is then a function of reassuring consumers about quality and a domestic source, as well as reducing input costs. Reassurance is provided through certification standards and labelling and/or by relationships, sometimes extended, between growers, processors and retailers. In the US, the estimated net income of $ 20,249 per farm per year averaged over 14,540 organic farms was higher than the average of all US farms (Bowman 2010).

Profits and evenness of income were higher for organic treatments than for high-input and reduced-input conventional treatments on the Canadian prairies (Zentner et al. 2011b). However, rotations including perennial forages, whether organic or conventional, were consistently less profitable and sometimes resulted in losses. There is ample evidence that forages improve soil quality, and that to maintain profits in the long term, either an adequate market for farm products or payment for the ecological service of including perennial forages is required. Organic farmers are well positioned to qualify and benefit from payments for ecological services, given their extensive record keeping and goals of feeding the soil.

19.3 Framing SEEGS in the Canadian Context

This section describes two distinct, but complementary, approaches for framing SEEGS in the Canadian context—a regional watershed approach and a civil commons approach. Both approaches recognize values that go beyond market parameters and offer frameworks for evaluating organic farming as a prototype for sustainable agriculture.

19.3.1 A Regional Watershed Approach

A regional watershed approach could enhance the benefits of SEEGS. The impact of organic agriculture is diluted when organic farms are "islands" among large areas of conventional farms. A landscape defined by the limits of a watershed provides otherwise unavailable opportunities to identify interactions among bio-physical factors such as soil erosion and water quality, as well as socio-economic factors such as human health, social well-being and income (Sachs et al. 2010). By concentrating organic expertise, materials and services required by organic farmers and by branding local organic products, opportunities for sales and tourism could increase. An organic watershed could be appealing for tourists seeking a clean vacation. Given that organic systems are certified along the entire value chain on the basis of processes, it is possible to distinguish an "organic" watershed, where all or even a majority of agriculture and food practitioners agree to become organically certified, from other watersheds with none or just a few certified practitioners. The interactions described by Sachs et al. (2010) can be contrasted in two different watersheds to holistically elucidate differences between sharing food production and ecosystem services or farming intensively and thus sparing ecosystems from agricultural production. To date, such contrasts tend to favour the sparing option, even though the results have been inferred from a farm scale (Green et al. 2005) rather than a watershed scale.

Gabriel et al. (2010) concluded from their spatial analysis that management beyond the farm scale is required to maximize conservation effects. They acknowledge the difficulty of persuading farmers within a landscape to collectively adopt practices. We suggest that incentives to become organically certified for famers within relatively small watersheds could be cost-effective as a pilot project and could be justified if a multidisciplinary team of researchers could build a dataset at the watershed scale. In the US, Farm Bill subsidies could be re-allocated to measure multiple sustainability indicators at the watershed scale (Reganold et al. 2011).

Although 80 % of Canadians reside in urban areas, and less than 10 % of the land is used for agriculture, many Canadians appreciate the value of preserving farmland and avoiding the externalities of fertilizer and pesticide pollution. Tangible demonstrations in specific organic watersheds are expected to reinforce these values.

19.3.2 A Civil Commons Approach

The concept of the civil commons describes any co-operative human construct that protects and/or enables universal access to life goods (McMurtry 2002). Not the same as public goods, life goods are means of life, which are whatever allows life to be preserved or extended on the three planes of being: thought, feeling and action. Life goods can be divided into three overlapping categories: social life goods such as education, healthcare and old-age security; environmental life goods such as clean water, a stable climate and healthy soil; and economic life goods such as safe working conditions, fair wages and unemployment protection. Examples of civil commons projects providing social life goods include public education systems, universal healthcare programs and pension plans. Civil commons projects providing environmental life goods include anti-dumping regulations, the Kyoto Protocol and bylaws banning the cosmetic use of pesticides. Civil commons projects providing economic life goods include labour regulations, minimum wage laws and unemployment insurance plans.

Many of the SEEGS derived from organic agriculture can be understood as public goods and services because they are available to everyone and do not exclude anyone. They would also qualify as life goods and services if they provide the means of life. Advances in gender equality, the creation of a knowledge commons, the minimization of soil degradation and erosion, decreased pollution, increased energy efficiency and a sustainable livelihood all highlight the connection between SEEGS and the means of life. Using this approach, organic agriculture could be considered as an umbrella form of the civil commons—a co-operative human construct that protects and/or enables universal access to a variety of life goods.

The civil commons, in turn, is connected to sustainability, which has been defined as a set of structures and processes that build the civil commons (Sumner 2005). According to this definition, sustainable agriculture would entail agriculture that built the civil commons. On this basis, we can argue that the more organic agriculture provides life goods and services, the more it can be understood as sustainable agriculture. The less organic agriculture provides life goods and services—by narrowing its scope to a limited set of production practices that encourage co-option by large corporate interests—the less it contributes to sustainability.

Understanding organic agriculture as a form of the civil commons may help to influence the recognition and acceptance of the suite of benefits it provides. Such an understanding would make a place for organic agriculture among other vital systems of life support, like universal healthcare, childcare and eldercare. This would raise the profile of the benefits derived from organic agriculture and also encourage people to better grasp the dangers of the modern enclosure of the commons. Just as common lands, which provided vital life goods such as homes, gardens, grazing and firewood to tens of thousands of farmers in the past, were fenced off during the Industrial Revolution, so too are the modern forms of the commons—public healthcare, public education, public pensions, public libraries—being threatened and defunded to make way for private forms of accumulation. Organic agriculture is not immune to enclosure. As it becomes more mainstream, the pressures of conventionalization—the process by which agribusiness industrializes organic agriculture—squeeze it toward simply being a profitable niche market in an increasingly monopolistic global corporate food system (Lynch et al. 2012a; Oelofse et al. 2011; Goldberger 2011; Guptill 2009; Guthman 2000,

2004; Hall and Mogyorody 2001), thus endangering the holistic set of life values on which it was founded, the suite of benefits it provides and its potential as a prototype for sustainable agriculture.

19.4 Conclusion

In summary, while consumer support for organic products continues to grow in Canada and the US, the characteristics of organic agriculture and the wider social and political context in which it operates have limited the greater endorsement of organic agriculture as of this time. As a whole, consumer understanding of the broad 'value proposition' of organic agriculture in North America is lagging in comparison to the rest of the world, although recent steady growth in the interest of consumers and producers for animal welfare, among other issues, provides an opportunity for the organic sector.

On the positive side, the growth in targeted research funding for organic agriculture in Canada and the US in recent years is providing much-needed documented evidence from North America, summarized in this document, of the broad SEEGS derived from organic agriculture. However, as also noted by Drinkwater (2009) and Reganold et al. (2011), we consider that to truly advance a paradigm shift in social consciousness and the relationship of an increasingly urban consumer base to a new 'transformative' agriculture, the interrelated issues inherent in SEEGS will increasingly have to be tackled by multidisciplinary teams of researchers partnering with organic producers.

Increasing sector organization through national stakeholder bodies such as the Organic Value Chain Round Table (OVCRT) formed in recent years in Canada, consisting of organic producers, retailers, processors, researchers, extension personnel and government representatives, provides a forum that is helping to strengthen policy support, research and targeted educational activities throughout the market and value chain. The OVCRT is one of a series of national roundtables within agriculture funded by the federal government. Interestingly, the OVCRT is the only roundtable that is truly cross-cutting and that transcends the typical commodities (grains and oilseeds, livestock, horticulture, etc.), which comprise all other roundtables. The inclusive and multi-stakeholder process of decision-making with the OVCRT can be considered as an important contribution to 'transforming' agriculture within the Canadian context.

To encourage this transformation, we have outlined the social, ecological and economic goods and services (SEEGS) derived from organic agriculture in this chapter and suggested two approaches for framing SEEGS in the Canadian context: one regional in scope, namely a pilot-scale watershed initiative to demonstrate the diverse benefits of organic agriculture, and more broadly, the concept of the 'civil commons' as a tangible concept that can link organic agriculture and sustainability. Both approaches offer a framework for evaluating organic farming as a prototype for sustainable agriculture.

Information on Organic Agriculture in Canada

- The most recent data (2009) show that there are 3,914 certified organic farms in Canada—around 1.7% of all farms—and that they occupy an estimated 695,463 ha of farmed land (not including maple and wild blueberry production).
- Retail sales of organic products in Canada total approximately US$ 2 billion (~2.5% of total retail sales), with an annual growth rate of 20%. This compares to an estimated US$ 24.8 billion (4% of total) of organic food and beverage sales in the US.
- Field crops, hay, vegetables, fruit and greenhouse products, livestock and maple syrup represent the majority of Canada's organic farm production, and its main export markets for organic food and beverage products are the US, the European Union and Japan.

References

AAFC (Agriculture and Agri-Food Canada) (2011) Socially conscious consumer trends: animal welfare. Market analysis report. International Markets Bureau. Agriculture and Agr-Food Canada, Ottawa, Ontario. AAFC No. 10749E. p 15

ACNielsen (2005) Functional food and organics: a global AC Neilsen online survey on consumer behaviour and attitudes. http://www.acnielsen.com/reports/documents/2005_cc_functional_organics.pdf. Accessed 10 Aug 2008

Aeberhard A, Rist S (2009) Transdisciplinary knowledge production in organic agriculture. Agrarforschung 16:414–419

Animal Welfare Task Force (2011) Factsheets and guidance documents. http://oacc.info/Extension/ext_welcome.asp. Accessed on 21 Dec 2011

Boutin C, Baril A, Martin PA (2008) Plant diversity in crop fields and woody hedgerows of organic and conventional farms in contrasting landscapes. Agric Ecosyst Environ 123:185–193

Boutin C, Martin PA, Baril A (2009) Arthropod diversity as affected by agricultural management (organic and conventional farming), plant species, and landscape context. Ecoscience 16:492–501

Boutin C, Baril A, McCabe SK, Martin PA, Guy M (2011) The value of woody hedgerows for moth diversity on organic and conventional farms. Environ Entomol 40:560–569

Bowman G (2010) USDA census shows profitability of organic farming. Rodale Institute. http://www.rodaleinstitute.org/20100319/nf_USDA-census-shows-profitability-of-organic-farming. Accessed 18 Dec 2011

Buck D, Getz C, Guthman J (1997) From farm to table: the organic vegetable commodity chain of Northern California. SociologiaRuralis 37:3–20

Campbell BL, Lesschaeve I, Bowen AJ, Onufrey SR, Moskowitz H (2010) Purchase drivers of Canadian consumers of local and organic produce. HortScience 45:1480–1488

CCME (2002) Linking water science to policy: effects of agricultural activities on water quality. Proceedings of a CCME sponsored workshop; 2002 Jan 31–Feb 1. Canadian Council of Ministers of the Environment, Quebec City, Canada

CGSB (2006) Organic production systems general principles and management standards. Canada General Standards Board/SCC (Standards Council of Canada) [Amended Oct. 2008] National Standard of Canada. CAN/CGSB-32.310–2006. Canadian General Standards Board, Gatineau, Quebec

Costanza R, d'Arge R, de Groot R, Farber S, Grasso M, Hannon B, Limburg K, Naeem S, O'Neill RV, Paruelo J, Raskin R, Sutton P, van den Belt M (1997) The value of the world's ecosystem services and natural capital. Nature 387:253–260

De Ponti T, Tijk B, van Ittersum MK (2012) The crop yield gap between organic and conventional agriculture. Agric Syst 108:1–9

Drinkwater LE (2009) Ecological knowledge: foundation for sustainable organic agriculture. In: Francis C (ed) Organic farming: the ecological system. American Society of Agronomy, Madison, pp 19–47

El-HageScialabba N (2007) Organic agriculture and food security. Food and Agriculture Organization of the United Nations. www.fao.org/organicag

Entz M, Thiessen Martens J (2009) Organic crop–livestock systems. In: Francis C (ed) Organic farming: the ecological system. ASA-CSSA-SSSA Book & Multimedia, Madison, pp 69–84

Farnworth C, Hutchings J (2009) Organic agriculture and womens' empowerment. IFOAM, Germany, p 86

Feehan J, Gillmor DA, Culleton N (2005) Effects of an agri-environment scheme on farmland biodiversity in Ireland. Agric Ecosyst Environ 107:275–286

Francis C (ed) (2009) Organic farming: the ecological system. American Society of Agronomy, Madison

Francis C, Van Wart J (2009) History of organic farming and certification. In: Francis C (ed) Organic farming: the ecological system. American Society of Agronomy, Madison, pp 3–18

Freemark KE, Kirk DA (2001) Birds on organic and conventional farms in Ontario: partitioning effects of habitat and practices on species composition and abundance. Biol Conserv 101:337–350

Friedmann H (2007) Scaling up: bringing public institutions and food service corporations into the project for a local sustainable food system in Ontario. Agric Hum Values 24:389–398

Fuentes-Montemayor E, Goulson D, Park KJ (2011) The effectiveness of agri-environment schemes for the conservation of farmland moths: assessing the importance of a landscape-scale management approach. J Appl Ecol 48:532–542

Gabriel D, Sait SM, Hodgson JA, Schmutz U, Kunin WE, Benton TG (2010) Scale matters: the impact of organic farming on biodiversity at different spatial scales. Ecol Lett 13:858–869. doi:10.1111/j.1461-0248.2010.01481.x

Geigera F, de Snooa GR, Berendsea F, Guerreroc I, Moralesc MB, Oñatec JJ, Eggersd S, Pärtd T, Bommarcod R, Bengtssond J, Clemente LW, Weissere WW, Olszewskif A, Ceryngierg P, Hawrog V, Inchaustih P, Fischeri C, Flohrei A, Thiesi C, Tscharntkei T (2010) Landscape composition influences farm management effects on farmland birds in winter: a pan-European approach. Agric Ecosyst Environ 139:571–577

Goldberger JR (2011) Conventionalization, civic engagement, and the sustainability of organic agriculture. J Rural Studies 27:288–296

Gomiero T, Paoletti MG, Pimentel D (2008) Energy and environmental issues in organic and conventional agriculture. Crit Rev Plant Sci 27:239–254

Gomiero T, Pimentel D, Paoletti MG (2011) Environmental impact of different agricultural management practices: conventional vs. organic agriculture. Crit Rev Plant Sci 30:95–124

Green RE, Cornell SJ, Jörn PW, Scharlemann JPW (2005) Farming and the fate of wild nature. Science 307:550–555

Guptill A (2009) Exploring the conventionalization of organic dairy: trends and counter-trends in upstate New York. Agric Hum Values 26:29–42

Guthman J (2000) Raising organic: an agro-ecological assessment of grower practices in California. Agric Hum Values 17:257–266

Guthman J (2004) Agrarian dreams: the paradox of organic farming in California. University of California Press, Berkeley

Hall A, Mogyorody V (2001) Organic farmers in Ontario: an examination of the conventionalization argument. SociologiaRuralis 41:399–422

Hole DG, Perkins AJ, Wilson JD, Alexander IH, Grice F, Evans AD (2005) Does organic farming benefit biodiversity? Biol Conserv 122:113–130

Jordan N, Warner KD (2010) Enhancing the multifunctionality of US agriculture. Bioscience 60:60–66

Kirchmann H, Bergström L (eds) (2009) Organic crop production—ambitions and limitations. Springer, New York

Knight JD, Buhler R, Leeson JY, Shirtliffe SJ (2010) Classification and fertility status of organically managed fields across Saskatchewan, Canada. Can J Soil Sci 90:667–678

Kramer SB, Reganold JP, Glover JD, Bohannan BJM, Mooney HA (2006) Reduced nitrate leaching and enhanced denitrifier activity and efficiency in organically fertilized soils. Proc Natl Acad Sci U S A 103:4522–4527

Kummer S, Aigelsperger L, Milestad R, Chowdhury A, Vogl CR (2010) Knowledge systems, innovations and social learning in organic farming—an overview. Proceedings of the 9th European IFSA Symposium, pp 664–669. www.ifsa-europe.org

Lynch D (2009) Environmental impacts of organic agriculture: A Canadian perspective. Can J Plant Sci 89:621–628

Lynch DH, MacRae RJ, Martin RC (2011) The carbon and global warming potential impacts of organic farming: does it have a significant role in an energy constrained world? Sustainability 3:322–362

Lynch DH, Halberg N, Bhatta G (2012a) Environmental impacts of organic agriculture in temperate regions. CAB Rev 7(10):1–17

Lynch DH, Sharifi M, Hammermeister A, Burton D (2012b) Nitrogen management in organic potato production. In: Zhongi H, Larkin RP, Honeycutt CW (eds) Sustainable potato production: global case studies. Springer Science + Business Media, New York, pp 209–231

Lyson TA (2004) Civic agriculture: reconnecting farm, food, and community. University Press of New England, Lebanon NH

Maceachern AB (2008) The work of women on organic farms: a gendered analysis. Unpublished Master's Thesis, University of Guelph, Guelph, Ontario, Canada

MacKinnon SL (2006) Identifying and differentiating farm-community linkages in organic farming in Ontario. Unpublished Master's Thesis, Department of Geography, University of Guelph

MacRae RJ, Frick B, Martin RC (2007) Economic and social impacts of organic production systems. Can J Plant Sci 87:1037–1044

MAPAQ (2011) Programme pilote d'appui a la multifunctionnalite de l'agriculture. Ministere Agriculture, Pecheries et Alimentation Quebec. http://www.mapaq.gouv.qc.ca/fr/Productions/md/Programmes/Pages/multifonctionnalite.aspx. Accessed 24 July 2011

Martin RC, Lynch DH, Frick B, VanStraaten P (2007) Phosphorous status on Canadian organic farms. J Sci Food Agric 87:2737–2740

McMahon M (2005) Engendering organic farming. Fem Econ 11:134–140

McMurtry J (2002) Value wars. Pluto Press, London

Mooney HA, Cropper A, Capistrano D, Carpenter SR, Chopra K, Dasgupta P et al (2005) Ecosystems and human well-being: synthesis, Millennium Ecosystem Assessment. Island Press, Washington, DC

Morgan K, Murdoch J (2000) Organic vs. conventional agriculture: knowledge, power and innovation in the food chain. Geoforum 31:159–173

Nelson KL, Lynch DH, Boiteau G (2009) Assessment of changes in soil health throughout organic potato rotation sequences. Agric Ecosyst Environ 131:220–228

OACC (2008) Final results of the first Canadian organic farmer survey of research needs. Organic Agriculture Centre of Canada, Nova Scotia Agricultural College, Truro NS

Oelofse M, Høgh-Jensen H, Abreu LS, Almeida GF, El-Araby A, Hui QY et al (2011) Organic farm conventionalisation and farmer practices in China, Brazil and Egypt. Agron Sustain Dev 31:689–698

Pass C, Lowes B, Davis L, Kronish SJ (1991) The HarperCollins dictionary of economics. Harper Perennial, New York

Pimentel D, Hepperly P, Hanson J, Douds D, Seidel R (2005) Environmental, energetic, and economic comparisons of organic and conventional farming systems. Bioscience 55:573–582

Reganold JP, Jackson-Smith D, Batie SS, Harwood RR, Kornegay JL, Bucks D, Flora CB, Hanson JC, Jury WA, Meyer D, Schumacher Jr A, Sehmsdorf H, Shennan C, Thrupp LA, Willis P (2011) Transforming US agriculture. Science 332:670–671

Rick TL, Jones CA, Engel RE, Miller PR (2011) Green manure and phosphate rock effects on phosphorus availability in a northern Great Plains dryland organic cropping system. Org Agric 1:81–90

Roberts CJ, Lynch DH, Voroney RP, Martin RC, Juurlink SD (2008) Nutrient budgets of Ontario organic dairy farms. Can J Soil Sci 88:107–114

Sachs J, Remans R, Smulder S, Winowiecki L, Andelman SJ, Cassman KG, Castle D, DeFries R, Denning G, Fanzo J, Jackson LE, Leemans R, Lehmann L, Milder JC, Naeem S, Nziguhe-baG, Palm CA, Pingali P, Reganold JP, Richter DD, Scherr SJ, Sircely J, Sullivan C, Tomich TP, Sanchez PA (2010) Monitoring the world's agriculture. Nature 466:558–560

Smith WN, Rochette P, Monreal C, Desjardins RL, Pattey E, Jaques A (1997) The rate of carbon change in agricultural soils in Canada at the landscape level. Can J Soil Sci 77:219–229

Sumner J (2005) Sustainability and the civil commons: rural communities in the age of globalization. University of Toronto Press, Toronto

Sumner J (2008) Protecting and promoting indigenous knowledge: environmental adult education and organic agriculture. Studies Educ Adults 40/2:207–223

Sumner J (2009) Building bridges: how organic agriculture can strengthen the relationships between rural and urban communities. Discussion paper commissioned by the Organic Agriculture Centre of Canada. www.organicagcentre.ca

Sumner J, Llewelyn S (2011) Organic solutions? Gender and organic farming in the age of industrial agriculture. Cap Nat Social 22:100–118

Teasdale JR, Coffman CB, Mangum RW (2007) Potential long-term benefits of no-tillage and organic cropping systems for grain production and soil improvement. Agron J 99:1297–1305

Thiessen Martens JR, Entz MH (2011) Integrating green manure and grazing systems: a review. Can J Plant Sci 91:811–824

Trauger A, Sachs C, Barbercheck M, Braiser K, Kiernan NE (2010) 'Our market is our community': women farmers and civic agriculture in Pennsylvania, U.S.A. Agric Hum Values 27:43–55

Warner K (2007) Agroecology in action: extending alternative agriculture through social networks. The MIT Press, Cambridge

Wortman SE, Lindquist JL, Haar MJ, Francis CA (2010) Increased weed diversity, density and above-ground biomass in long-term organic crop rotations. Renew Agric Food Syst 25:281–295

Zander K, Nieberg H, Offermann F (2008) Financial relevance of organic farming payments for Western and Eastern European organic farms. Renew Agric Food Syst 23:53–61

Zentner RP, Basnyat P, Brandt SA, Thomas AG, Ulrich D, Campbell CA et al (2011a) Effects of input management and crop diversity on non-renewable energy use efficiency of cropping systems in the Canadian Prairie. Eur J Agron 34:113–123

Zentner RP, Basnyat P, Brandt SA, Thomas AG, Ulrich D, Campbell CA, Nagy CN, Frick B, Lemke R, Malhi SS, Olfert OO, Fernandez MR (2011b) Effects of input management and crop diversity on economic returns and riskiness of cropping systems in the semi-arid Canadian Prairie. Renew Agric Food Syst 26:208–223

Chapter 20
Does the Seed Sector Offer Meet the Needs of Organic Cropping Diversity? Challenges for Organic Crop Varieties

Dominique Desclaux and Jean-Marie Nolot

Abstract This chapter aims to study the current balance between the offer and demand of the organic seed sector and how the breeding system can be adapted to the diversity of needs.

The offer is assessed by the evolution, over time, of the number and types of species and varieties registered in the French and European catalogues. This number has greatly increased, but some species remain or become orphans. There is an increasing number of varieties listed for the major crops, whereas the choice concerning organic seed varieties is severely limited. Moreover, the standardised varieties listed in the catalogue are currently in dispute because of the new diversification of cropping systems, outlets and social organisation. Do we need to change ideotypes, breeding methods, breeding criteria and evaluating methods? In other words, do we need to reform the seed system or to adapt it to this new diversity?

The diversity of organic seed variety requirements is represented by four models: Label, Brand, Autonomy and Empowerment. Each model requires a specific breeding scheme and relevant breeding actors. These different ways to look at new varieties and plant breeding must not exclude each other but must be considered as complementary and capable of renewing ways to implement plant improvements for agriculture. In such a context, there is a need for new references to evaluate and register varieties (new criteria, new protocols, changes in legislation, etc.).

Keywords Seed Variety · Catalogues · DUS · VCU · GxE interaction

D. Desclaux (✉)
INRA, UE398 Diascope, Domaine de Melgueil,
34130 Mauguio, France
e-mail: dominique.desclaux@supagro.inra.fr

J.-M. Nolot
INRA-Université de Toulouse-INPT, UMR1248 AGIR,
31326 Castanet-Tolosan, Cedex, BP 27, France
e-mail: jean-marie.nolot@toulouse.inra.fr

S. Bellon, S. Penvern (eds.), *Organic Farming, Prototype for Sustainable Agricultures*,
DOI 10.1007/978-94-007-7927-3_20,

20.1 Introduction

The common catalogues of varieties of agricultural plant and vegetable species list the varieties that can be marketed in Europe. Catalogues are based on the registration of plant varieties in EU countries after they have been technically examined and notified to the Commission. They are published in the Official Journal. Variety registration is a precondition for the certification of seeds.

To be listed, crop varieties must meet standards on Distinctness, Uniformity, Stability (DUS) and on Value for Cultivation and Use (VCU). This value is based on yield, resistance to harmful organisms, response to the environment and quality characteristics (EU catalogue 2013).

The new proposal for a regulation of the European parliament and of the council indicates that "The Member States shall adopt more detailed criteria for the VCU examination of (these) plant species as regards their yield, quality characteristics, resilience and suitability for low input production systems including organic production. Thus, given the specific characteristics required for organic farming, the methodology and requirements established for variety examination should take due account of the specific needs" (EU proposal 2013).

In France, an organic network is implemented for VCU testing since 2009. Is it enough to cover the increasing need for organic varieties in France or do we also need to take alternative criteria and breeding methods into account? The objective of this chapter is to use the French case to study the balance between organic seed variety offer and demand, and how the breeding system may be adapted to the needs.

20.2 The French Conventional and Organic Seed Variety Offer

20.2.1 Comparison Between the French and the European Catalogues

Some Orphan Species

Concerning agricultural species, the EU catalogue consists of 86 species (except potato), whereas the French catalogue proposes some 60 crop species.

The comparison between the current French catalogue and those published 25 years before (1987) shows the evolution of the number of species considered (Table 20.1). Some species such as field pea (*Pisum sativum L.*), smooth-stalked meadowgrass (*Poa pratensis L.*), crimson clover (*Trifolium incarnatum*), hybrids resulting from crosses with *Festulolium Asch*, California bluebell (*Phacelia tanacetifolia*), birdsfoot trefoil (*Lotus corniculatus*), sheep's fescue (*Festuca ovina L.*), alsike clover (*Trifolium hybridum*), rescue grass (*Bromus catharticus*), black oat/bristle (*Avena strigosa Schreb.*), hairy vetch *(Vicia Villosa)* and buckwheat (*Fagopyrum esculentum Moench)* were not listed in the 1987 catalogue and appeared later.

In 2013, across all species, French varieties represent less than 20 % of the European catalogue. However, this average percentage conceals a great diversity, from 67 % (rescue grass) to 0 % (rye or birdsfoot trefoil) (Table 20.1).

Species representing a low percentage undergo a continuous decrease of the number of varieties over time. Indeed, a comparison between the current French catalogue and the one published in 1987 illustrates this evolution (Fig. 20.1), especially for some cereals (oat, rye) and legumes (lupin, sainfoin) that are useful in organic cropping systems. There are no available varieties in the French catalogue for rye and birdsfoot trefoil at this time. The question therefore arises as to whether or not the varieties registered for these crops in the European catalogue are well adapted to French and to organic conditions.

A Wide Choice of Varieties for Major Crops

An impressive increase in the number of varieties can also be observed between the 1987 and 2013 catalogues for other crops such as swede rape, wheat, sunflower and maize (Fig. 20.2). At the European level, these major crops offer more than 50 % of the total number of varieties registered in the agricultural species catalogue. With 4,959 varieties, maize offers more than 25 % of the catalogue varieties!

20.2.2 Organic Seed Variety Offer

Organic agriculture regulations, in particular, European regulation EC 889/2008, prescribe the use of organically-produced seed.

At the European level and for many cultivated plants, however, organic seed is often not available (Döring et al. 2012).

At the French level, considering available organic seeds (http://semences-biologiques.org), the choice of varieties is extremely limited, even for the major crops (Fig. 20.3). Therefore, the number of derogations requested is increasing (Fig. 20.4) for all types of species.

Increasing Derogation Requests

The main cause (87 %) remains "variety not available in the national organic database" (Ministry of Agriculture 2011). Organic seed production is in fact insufficient. Concerning fodder seeds, derogation requests are numerous and represent half of the total requests. Several reasons are mentioned: the development of the organic market, the renewal of meadows, technical difficulties in producing enough seeds of the required quality. Among the other reasons for derogation requests is the

Table 20.1 Comparison between the numbers of varieties (NV) listed in the French and in the European catalogues for common crop species (classified according to the decreasing number of varieties in the EU catalogue)

Species	NV in the French catalogue (year 1987)	NV in the French catalogue (May 2013)	NV in the EU catalogue (May 2013)	Difference of NV between french and EU catalogues (%)
Maize—Zea mays L.	474	1184	4959	24
Wheat—Triticum aestivum	205	303	1979	15
Sugar beet—Beta vulgaris L	244	340	1435	24
Sunflower—Helianthus annuus	46	230	1403	16
Hordeum vulgare (two and six rows)	183	216	1303	17
Swede rape—Brassica napus L.	14	247	1185	21
Perennial ryegrass—Lolium perenne L.	71	253	1086	23
Durum wheat—Triticum durum Desf	34	63	527	12
Italian ryegrass—lolium multiflorum—Ssp. Alternativum + non alt	35	95	422	23
Field pea—Pisum sativum L.	–	55	415	13
Lucerne–Medicago sativa	38	58	380	15
Oat, Red and small naked oat - Avena sativa L., nuda	42	33	357	9
Soya bean—Glycine max (L.)	40	50	352	14
Red fescue—Festuca rubra L.	44	65	344	19
Rice—Oryza sativa L.	7	17	333	5
Hybrids resulting … —×Triticosecale Wittm	5	61	290	21
Tall fescue—Festuca arundinacea…	34	89	275	32
Sorghum—Sorghum bicolor …	30	82	216	38
Red clover—Trifolium pratense …	29	31	215	14
Smooth-stalked meadowgrass—Poa pratensis L.	–	13	201	6
Flax, Linseed—Linum usitatissimum	16	64	182	35
Rye—Secale cereale L.	10	0	159	0
White mustard—Sinapis alba L.	8	19	158	12
Cocksfoot—Dactylis glomerata …	17	53	148	36
White clover—Trifolium repens …	22	29	139	21
Field bean—Vicia faba L. (Partim)	23	25	139	18
Common vetch—Vicia sativa L.	12	23	137	17

Table 20.1 (continued)

Species	NV in the French catalogue (year 1987)	NV in the French catalogue (May 2013)	NV in the EU catalogue (May 2013)	Difference of NV between french and EU catalogues (%)
Timothy—Phleum pratense …	11	8	126	6
Fodder beet—Beta vulgaris L.	35	34	120	28
Hybrid ryegrass—Lolium × boucheanum …	8	44	104	42
Fodder radish—Raphanus sativus …	5	7	97	7
Meadow fescue—Festuca pratensis	8	11	87	13
Hemp—Cannabis sativa …	7	12	52	23
Turnip rape—Brassica rapa L. …	5	3	37	8
Hard fescue—Festuca trachyphylla …	8	11	36	31
Crimson clover—Trifolium incarnatum	–	1	36	3
Hybrids of Festulolium Asch. …	–	3	35	9
Fodder kale—Brassica oleracea …	11	2	31	6
California bluebell—Phacelia tanacetifolia …	–	1	30	3
Birdsfoot trefoil—Lotus corniculatus	–	0	30	0
White lupin—Lupinus albus L.	8	6	23	26
Brown mustard—Brassica juncea …	1	4	22	18
Sainfoin—Onobrychis viciifolia …	4	2	22	9
Sudan grass—Sorghum sudanense.	11	3	21	14
Sheep's fescue—Festuca ovina L.	–	2	20	10
Alsike clover—Trifolium hybridum …	–	1	18	6
Rough-stalked meadowgrass-Poa trivialis L.	25	3	14	21
Rescue grass—Bromus catharticus …	–	6	9	67
Black oat/Bristle—Avena strigosa Schreb	–	3	8	38
Alaska brome-grass—Bromus sitchensis …	8	3	6	50
Small timothy—Phleum nodosum L.	–	2	5	40
Hairy vetch—Vicia Villosa	–	2	–	–
Buckwheat-Fagopyrum esculentum Moench	–	2	–	–
Fodder kale brassica oleracea	21	4	–	–
Phleum bertolonii	6	–	–	
Sum (s), Mean (m)	1838 (s)	3870 (s)	19728 (s)	20% (m)

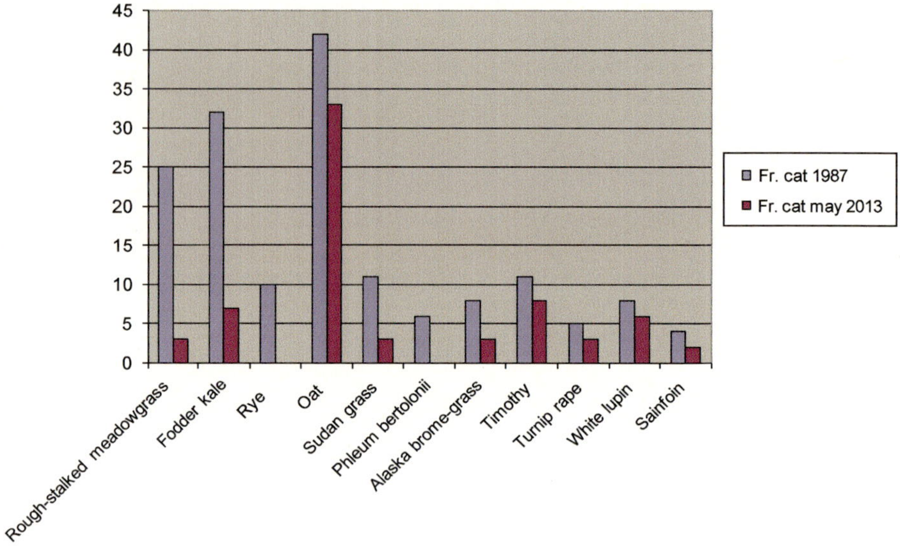

Fig. 20.1 Decrease in the number of varieties registered in the French catalogue in 2013 compared to the French catalogue in 1987

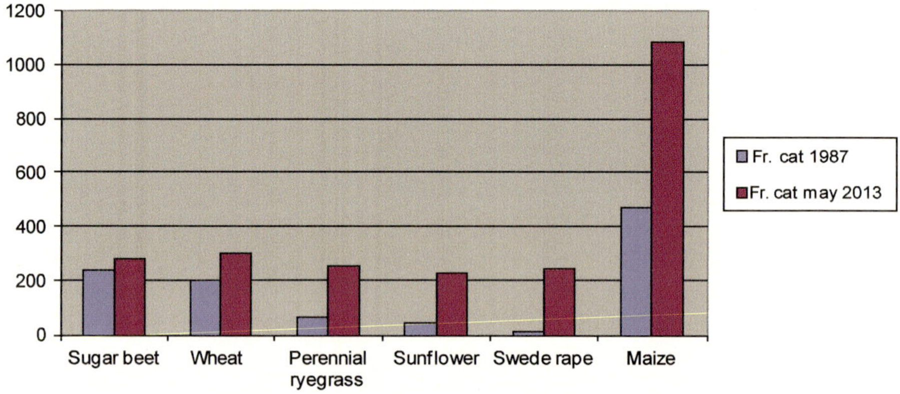

Fig. 20.2 Increase in the number of varieties registered in the French catalogue in 2013 compared to the French catalogue in 1987

fact that varieties are often not adapted to organic agriculture (Ministry of Agriculture 2011). The possibility of derogation decreases on a regular basis and no more derogations are granted for maize species. The reason given for this is that "there is a sufficient number of available organic seeds varieties". A list of species "exempt from derogation" is available online (http://www.semences-biologiques.org/pages/actu.php#actu7).

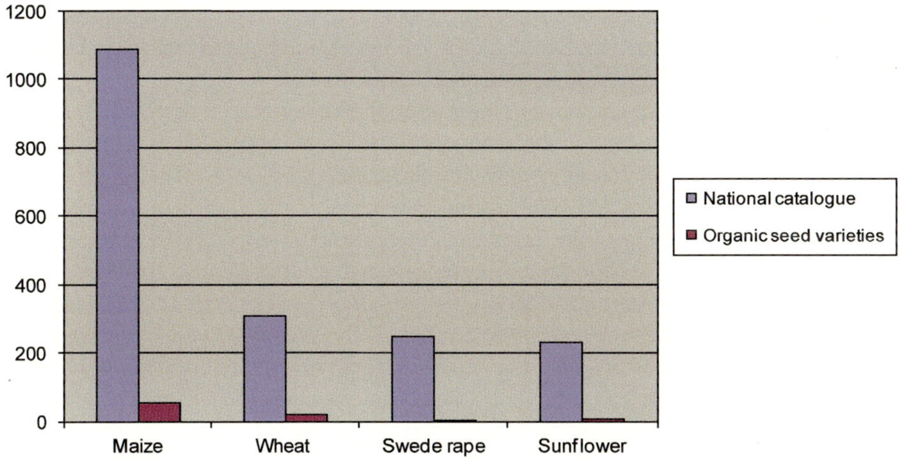

Fig. 20.3 Comparison between the number of varieties registered in the French catalogue (database consulted in May 2013) and the number of organic seed varieties available on the French market (official database: http://www.semences-biologiques.org/)

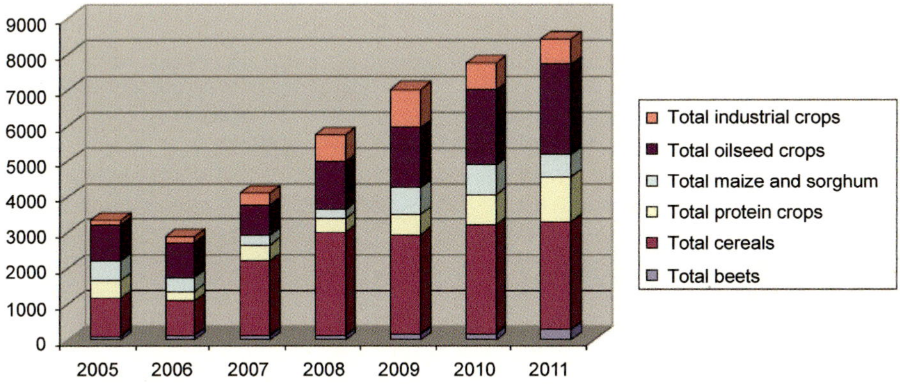

Fig. 20.4 Evolution of the agreed derogation request number (source: http://www.semences-biologiques.org/pages/Rapport-semences-bio-2011-final.pdf—Table 20.3, page 11)

A Standardised Type of Varieties: The Causes

For economic reasons, breeding firms target the profitable market, and the main objective of breeding programmes is to develop varieties that can be widely distributed and adapted to the dominant market (Bonneuil and Thomas 2013). Moreover, most breeding firms are subsidiaries of multinational companies. Among them, the four biggest multinationals (Monsanto, Dupont/Pioneer, Syngenta and Limagrain/Vilmorin) represent more than 25 % of the seed market. The concentration of the seed sector does not help the diversity of ideotypes (Desclaux et al. 2010).

Breeders look for a standard ideotype. The aim is to avoid interactions between genotype and environment and to adapt the environment (E) to the varieties by standardising E. This standardisation consists of eliminating all limiting factors present in the biophysical environment through input supply. The varieties created are therefore very similar within a species. The genetic uniformity is accompanied by a spatial uniformity, with some French regions growing the same varieties (Goffaux et al. 2011).

Because of the U of DUS (distinctness, uniformity and stability), the varieties listed in the catalogue are pure lines for autogamous species, hybrids for allogamous species, and clones for species with vegetative reproduction. To create such varieties, the classic plant improvement system, described as a centralised, sequential, linear process (Sperling et al. 2001), is generally used by private plant breeders. A breeding programme consists of five main stages: (1) identifying the objectives; (2) creating variability; (3) selection; (4) evaluation; and (5) dissemination. This approach contributes to a standardised variety that under regulations for registration imposes uniformity, and that under the dominant market conditions imposes specific yield and technological quality (Desclaux et al. 2008).

20.2.3 Organic Seed Varieties Offer: The Synthesis

Three main points must be highlighted concerning the French organic seed variety offer:

- the lack of varieties for some species in the French catalogue
- the large number of varieties for the major crops but, in fact, a very poor choice for organic farmers forced to use organic seed varieties.
- the low within-species variety diversity

This standardisation and lack of choice are currently in dispute because of the increased diversification of organic cropping systems, outlets and social organisation. The offer (varieties listed in the catalogue) and the demand are becoming increasingly divergent. For each new demand, the challenge is to determine whether or not the current seed and breeding systems must be completely renewed or only adapted to the new diversity.

20.3 A Great Diversification of the Organic Varieties Demand

20.3.1 A Request for a Wide Diversity of Species

The Interest in Crop Rotation or for Specific Adaptation

Organic farming systems are usually based on a wide diversity of species integrated into a crop rotation. Crop rotation is a key strategy that farmers use to maintain soil fertility, soil organic matter levels and soil structure, to control pests, diseases and weeds, and to ensure that enough nutrients are available for different crops each year (Baldwin 2006). Legumes are generally the core of the rotation because of the level of nitrogen they leave in the soil. A diversity of legumes is needed to fulfil the different demands according to the ecological zone of production. For example, birdsfoot trefoil *(Lotus corniculatus)* may be used as an alternative to alfalfa in poor soils. However, without any variety available in the French catalogue, this species is no longer cultivated in France.

In the same way, others species are required. This could include species more adapted to specific conditions, such as buckwheat (*Fagopyrum esculentum*), a short season crop, well adapted to low-fertility or acidic soils, provided that the soil is well drained. Too much fertiliser, especially nitrogen, will reduce yields. This species was cultivated in France and in Europe, but cultivation sharply declined in the 20th century due to the use of nitrogen fertiliser, to which maize and wheat respond strongly. Only two buckwheat varieties are listed in the French catalogue and none in the EU catalogue.

These "niche" species are not economically profitable for plant breeders. In addition to these, sorghum, winter pea, oat and lupine also belong to the orphan species class. The lack of breeding and, therefore, the lack of varieties lead farmers to progressively eliminate these species from their cropping system. The need for a diversification of crop rotations requires breeding work on several species, implying a considerable effort on the improvement of several legumes as well.

Interest of New Species that Satisfy Specific Criteria

Rye (*Secale cereale*) is a grass grown extensively as a grain and as a forage crop. It is a member of the wheat tribe (Triticeae). Concerning the quality of protein, rye flour is high in gliadin but low in glutenin. It therefore has a lower gluten content than wheat flour and contains a higher proportion of soluble fibre. These nutritional interests are currently very important because of gluten allergy, but no varieties of rye are listed in the 2013 French catalogue.

In the same way, some species capable of playing an ecological role (remediation of polluted soils, refuges for pests, impact on soil microbial communities, water improvement, reduction of gas emissions, etc.), and some targets requiring a long breeding period (global change, multifactorial modifications: less water, more CO_2, more T, etc.), are not compatible with the economic and temporal stakes of private breeders.

20.3.2 A Demand for Different Types of Varieties Based on Organic Farming Diversity

On the basis of the work of Sylvander et al. (2006), Desclaux et al. (2008) proposed to represent organic farming diversity using four models defined by two axes, one socio-economic (individual logics vs. collective governance), and the other agro-ecological

Table 20.2 Relevant varieties and traits in the four models represented in Fig. 20.5

Model	Illustrations	Relevant varieties and traits
Label	Conventional or organic agriculture strictly respecting specifications or guidelines	Registered DUS, VCU varieties. Traits: genetic progress, yield, standard quality
Brand	Integrated sectors (segmented market)	Varieties imposed by specifications Specific traits
Autonomy	Farmers-bakers	Local varieties, flagship varieties, evolving populations; Traits: consumer satisfaction
Empowerment	Participatory plant breeding	Multiple and multi-functional varieties; Traits: Social, ethical, economic progress

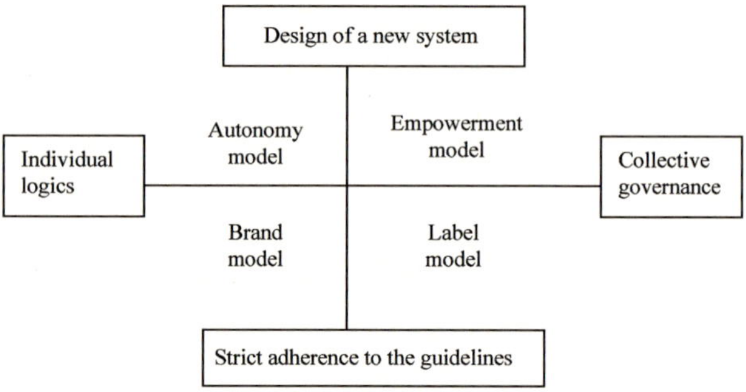

Fig. 20.5 Four models of agriculture and environment. (according to Sylvander et al. (2006) and Desclaux et al. (2008))

(reductionist vs. systemic approaches). The latter axis distinguishes a reductionist or factorial approach from a holistic approach (Lammerts van Bueren et al. 2003).

This representation links the notion of spatial and temporal scale (from local to worldwide, short or long food supply chains), to status (public, associative, private) and to the sharing of tasks and competences (specialising vs. delegating) (Table 20.2). The challenge is to analyse whether or not the diversity of the objectives inherent in these models can give impetus to the concept of plant breeding for organic agriculture.

The Organic "Label" Model Satisfied by Registered Varieties

In response to new concerns about health and environment, and embodied in national or European public policies (environmental cross-compliance, reinforcement of the second pillar of the CAP, etc.), some farmers have developed organic or sustainable agriculture systems. They strictly respect organic specifications mentioned in the official IFOAM guidelines and are interested in acquiring the organic label. Their system can be referred to as an "agriculture of substitution", consisting of replacing conventional with organic inputs. The registered varieties are of interest in this "label model" because they correspond to the targeted markets: standardised organic products for long-chain markets.

The first breeding stage (setting of objectives) can sometimes be collectively defined with stakeholders, and breeding activity is then defined by professional teams that may be private, cooperative or public. This method is currently being explored by SZD in Austria (Löschenberger et al. 2008), Ojo Seeds in the Netherlands, ACW and Peter Kuntz in Switzerland and, more recently, by INRA DGAP (bread wheat) in France (Rolland 2006). Since 2003, new wheat cultivars in Austria have been specifically registered for organic farming. The selection is reoriented for farmers and users, and new criteria specifically oriented to organic farming are taken into account, including height, lodging resistance, weed competitiveness and nitrogen use efficiency. The example of weed competitiveness for cereals is one example of specific needs in organic farming (Fontaine et al. 2008, 2010).

The Organic "Brand" Model Requires "Varieties Reserved for Industrial Uses"

Some farmers are linked by contract to collectors or industrialists to produce specific varieties. A well-known example in France is provided by the Limagrain company that breeds special wheat varieties adapted to producing a particular type of bread (Pain Jacquet). These varieties may be registered on a special list ("varieties reserved for industrial uses", known as VUIR in France) or not registered at all. However, they can only be grown by farmers under contract. The objective for the company is no longer to ensure the wide spread of a varietal innovation but to control and target the dissemination (Stage 5; Sperling et al. 2001) of a specific final product by imposing a variety, its guidelines and the exclusive return of the harvest. The evaluation stage of the breeding scheme (Stage 4) may thus be narrow or even circumvented. The variety may not even have to be registered in the official catalogue because seeds are only distributed within specified limits, e.g., within an integrated value chain or a club. The purchase of the harvest at a guaranteed price is one of the main reasons that the package – comprising variety, crop management and biophysical area – is accepted by farmers. The selection stage (Stage 3) is either conducted generically by choosing from the genetic diversity, or is considerably simplified by introducing the gene of technological interest into a variety to obtain, for example, a waxy maize or oleic sunflower. The logic of this model, referred to as a "brand" model, can be extended to include the privatisation of genetic resources and their economic valorisation via the integration of an entire sector (by firms involved both in plant breeding and agro-industrial sectors). The objective identification stage (Stage 1) creates opportunities for the combination or the emergence of value chains and specific market niches. It is as if the stages proposed by Sperling et al. (2001) were inverted (from Stage 5 to Stage 1) by the desire to first control the dissemination stage. However, if this contractual flexibility offers an opportunity to escape from classical seed regulations, it has surprisingly been reintroduced as an official rule by the creation of a special list dedicated to "varieties reserved for industrial uses" in the catalogue (Anvar 2008).

The Organic "Autonomy" Model Targets "Patrimonial Varieties"

Some farmers, referred to by sociologists as 'whole chain farmers', ensure not only the production but also the processing of their harvest. They look for a genetic resource with patrimonial and identity traits, capable of becoming a "flagship" variety at a low cost (a symbol of a social movement) or a "sentry" variety (considered by the Slow Food movement as a shield against uniform industrialised products). The variety of interest here is a designated phenotype, referred to as "local population" or "old variety". The objective is an extremely localised individual adaptation – at the scale of one farm or even of one field. We therefore refer to this model as the "autonomy" model. "Farmers must have an enormous range of varieties at their disposal that are as adaptable as possible, i.e., are accessible to different types of evolution and, therefore, neither very uniform nor very stable" (Kastler 2006). Irrespective of the biology of the species (self or open pollinated), the means range from the cultivation of populations under natural selection to mild pressure of mass selection by dynamic management (Goldringer et al. 2007). Evaluation and diffusion are no longer based on the classical criteria that define genetic progress (e.g., yield or technological quality) but on consumer satisfaction instead.

The Organic "Empowerment" Model Needs "Multifunctional" Varieties

The objective in this case is to obtain diverse varieties that correspond to diverse functions. At the same time, the search continues for multifunctional genotypes as a contribution to: (i) enhancement of the landscape (e.g., through colour); (ii) health (through nutrients, etc.); or (iii) a balanced agro-ecological system (capacity for mycorrhisation, competition, remediation of polluted soils, etc.).

The aim is to reconcile the design of a new system for plant breeding and collective action. This model gives more equal weight to agro-ecological interactions (environmental aspects of sustainability) and socio-economic interactions (between actors). The organisation of the emergent system of complex interactions may be facilitated by a participatory approach.

"Participatory plant breeding" (PPB) was originally developed in the southern countries. In Europe today, PPB concerns local projects for the creation of varieties adapted to environments in which organic and low-input agriculture is practiced (Desclaux and Hedont 2006). PPB is described as an approach involving all the actors of a given sector, not only in drawing up breeding objectives, but also in managing the breeding process and the creation of varieties (Gallais 2006). It aims to respond to systemic issues and demands for which classic breeding appears to be unsuited (Almekinders and Hardon 2006; Cecarelli et al. 2001; Witcombe et al. 2003).

The reason this empowerment model is of considerable heuristic interest in that it deeply modifies the stages: each stage becomes a function that will tend to exacerbate and reveal GxE interactions in both the agro-ecological and socio-economic dimension of the environment.

These models differ in their objectives, their variety requirements, their breeding schemes and their breeding actors. Concerning the type of varieties required, two of these models are under regulations (Label and Brand models), whereas the two others (Autonomy and Empowerment) are not allowed at this time.

20.4 The Need to Renew the References for Organic Agriculture

20.4.1 An Evolution of Regulations: The Case of DUS and VCU

The concept of DUS was based on the need of uniformity to facilitate the characterisation of varieties and to fight against frauds. This concept may be useful in the case of long-chain seed markets to provide a guarantee to farmers. However, in the case of short-seed chains based on confidence, the concept of DUS is questionable.

It is true that within-variety heterogeneity and variety mixtures have a number of agronomic advantages, including disease control and better adaptation to uncontrolled variability of the climate-soil environment (Pope de V et al. 2007; Wolfe 1997). Heterogeneity is also an economic necessity when it enables the diversification or differentiation of final products and markets, particularly in the case of organic farming and of products that valorise particular specifications (as in the Brand model) or a particular local territory labelled with a Geographical Indication (as in the autonomy model).

The challenge is thus to design new plant improvement systems and to attempt to change the legislation governing variety registration since there is an increasing gap between the uniformity inherited from the productivist model and the requirements of sustainable agricultures, in particular, of organic agriculture in which heterogeneity is a key factor for management, and the further development of organic agro-ecosystems.

Concerning VCU, organic systems impose a change in priority. The yield remains an important target, but provided that the quality is acceptable. This should include not only process quality but nutritional quality and other traits as well that are particularly important for organic farming.

20.4.2 A Change of Paradigms

In addition to the classical types of varieties present in the catalogue, a wide diversity of varieties is required to meet new models of agriculture. Indeed, other models of agriculture are emerging based on a wide diversity of farming systems (low input, organic, agroecology, agroforestry, integrated pest management, etc.), and of farmers' and consumers' requirements (Table 20.3).

A different breeding logic is required for sustainable agriculture. A limited margin of manoeuver does not allow crop management to compensate for the limiting factors of the biophysical environment, and the wide range of uses prevents the emergence of

Table 20.3 Change of Paradigms. (taken from Bardsley 2003)

Registered crop varieties	What is also needed
Conventional breeding for conventional farming	Alternative breeding for innovative farming (organic, agroecology, low input, etc.)
Seed = economic good	Seed = cultural good
Wide diffusion	Targeted diffusion
Quality: adapted to technological process	Quality: adapted to human nutritional needs
Uniformity: pure lines, hybrids or clones	Heterogeneity: populations, mixtures, etc.
Breeding from ex situ germplasm	Breeding from in situ and dynamic management of genetic resources
Innovations: from biotechnologies	Innovations: on the basis of farmers' know-how
Farmers: users	Farmers: breeders or co-breeders

a uniform, universal logic for variety. The aim is no longer to adapt the environment to the variety but just the reverse, to try to adapt the variety to each environment.

This change is potentially disruptive to established plant breeding. The effort required partially explains the difficulty in rethinking breeding systems for new environments.

20.5 Conclusion

The diversity of farming systems and of valorisation leads to a diversity of requirements for variety innovation. The offer of organic seed varieties is not consistent with this diversity of demands.

- The lack of varieties for some species in the French catalogue and the increasing orphan demands require the emergence of new actors in plant breeding.
- The obligation to use organic seed varieties and the very limited list cannot adequately fulfil the demand for a wide diversity of variety types.

Innovation depends on the integration of knowledge derived from the world of uses, the knowledge of farmers and of processors, the desire of consumers and, as always, the experience of breeders (McMeekin et al. 2002, cited by Bonneuil and Thomas 2009).

Depending on the breeding actors and on their objectives, target varieties, selection criteria and breeding schemes differ. These different ways to consider varieties and plant breeding must not exclude each other but must be considered as complementary and capable of renewing ways to implement plant improvements for agriculture. The involvement of a wide diversity of actors in plant breeding will help to integrate the concept of phenotype into a whole that takes systemic realities and environmental heterogeneity into account.

Because the phenomenon of concentration of breeding companies is increasing, questions arise as to the ability of this sector to propose variety innovations that correspond to the demands of society.

References

Almekinders C, Hardon J (eds) (2006) Bringing farmers back into breeding. Experiences with participatory plant breeding and challenges for institutionalisation. Agromisa Special 5. Agromisa, Wageningen, p 125

Anvar L (2008) Semences et droit—L'emprise d'un modèle économique dominant sur une réglementation sectorielle—Thèse pour le doctorat en droit, Université Paris I, p 470

Baldwin KR (2006) Crop Rotations on Organic Farms- Published by North Carolina cooperative extension service

Bardsley D. (2003) Risk alleviation via in situ agrobiodiversity conservation: drawing from experiences in Switzerland, Turkey and Nepal. Agriculture, Ecosystems & Environment 99:149-157

Bonneuil C, Thomas F (2009) Gènes, pouvoirs et profits- Recherche publique et régimes de production des savoirs de Mendel aux OGM- 619 pp- Editions Quae

Bonneuil C, Thomas F (2013) Semences: une histoire politique. ed. Charles Leopold Meyer. ISBN 978-2-84377-165-1, pp 213

Cecarelli S, Grando S, Bailey E, Amri A, El Felah M, Nassif F, Rezgui S, Yahyaoui A (2001) Farmer participation in Barley breednig in Syria, Marocco and Tunisia. Euphytica 122:521–536

Desclaux D, Hedont M (eds) (2006) Proceedings of ECO-PB Workshop: "Participatory plant breeding: relevance for organic agriculture?" Ed; ITAB, p 112

Desclaux D, Nolot JM, Chiffoleau Y, Gozé E, Leclerc C (2008) Changes in the concept of genotype × environment interactions to fit agriculture diversification and decentralized participatory plant breeding: pluridisciplinary point of view. Euphytica 163:533–546

Desclaux D, Nolot JM, Triboulet P, Lorentz B, Chiffoleau Y (2010) A necessary complementarity of actors for variety innovation. ISDA

Döring TF, Bocci R, Hitchings R, Howlett S, Lammerts van Bueren ET, Pautasso M, Raaijmakers M, Rey F, Stubsgaard A, Weinhappel M (2012) The organic seed regulations framework in Europe—current status and recommendations for future development; Organic Agriculture December 2(3–4):173–183

EU catalogue (2013) http://ec.europa.eu/food/plant/propagation/catalogues/database/public/index.cfm?event=SearchForm & cat=A

EU proposal (2013) Proposal for a regulation of the european parliament and of the council "On the production and making available on the market of plant reproductive material (plant reproductive material law)" Brussels, 6.5.2013COM (2013) 262 final 2013/0137 (COD)

Fontaine L, Rolland B et al (2008) Contribution to organic breeding programmes of wheat variety testing in organic farming in France. In: Cultivating the future based on science. Volume 1: Organic Crop Production. Proceedings of the Second Scientific Conference of the International Society of Organic Agriculture Research in Modena, Italy, 18–20 June 2008

Fontaine L, Rolland B et al (2010) Value for cultivation and use testing under organic conditions: experience in winter wheat in France. Conference poster. In Breeding for resilience: a strategy for organic and low-input farming systems? EUCARPIA 2nd Conference of the Organic and Low-Input Agriculture Section, France, p 1–3

Gallais A (2006) Preface. In: Lançon J, Floquet A, Weltzien E (eds) Partenaires pour construire des projets de sélection participative. Ed CIRAD, p 207

Goffaux R, Goldringer I, Bonneuil C, Montalent P, Bonnin I (2011) Quels indicateurs pour suivre la diversité génétique des plantes cultivées? Le cas du blé tendre cultivé en France depuis un siècle. Rapport FRB, Série Expertise et synthèse, p 44

Goldringer I, Galic N, Rousselle Y, Demeulenaere E, Bonneuil C, Payement J, Berthellot JF, Chesneau V, Mercier F, Ferté H, Pommart A, Zaharia H (2007) On-farm dynamic management of wheat populations/varieties in organic agriculture: a way to valorize GxE interactions. In: Proc. of Eucarpia symposium.7–9 nov. Wageningen

Kastler G (2006) Les semences paysannes. Les Dossiers de l'environnement de l'INRA, Quelles variétés et semences pour des agricultures paysannes et durables? 30:53–56

Lammerts van Bueren ET, StruiK PC, Tiemens-Hulscher M, Jacobsen E (2003) The concept of intrinsic value and integrity of plants in organic plant breeding and propagation. Crop Sci 43:1922–1929

Löschenberger F, Fleck A, Grausgruber H, Hetzendorfer H, Hof G, Lafferty J, Marn M, Neumayer A, Pfaffinger G, Birschitzky J (2008) Breeding for organic agriculture: the example of winter wheat in Austria, Euphytica 163:469–480

Ministry of Agriculture (2011) Rapport de synthèse- année 2011, France. Fonctionnement de la base de données sur les variétés disponibles en semences issues du mode de production biologique Synthèse annuelle des dérogations accordées http://www.semences-biologiques.org/pages/Rapport-semences-bio-2011-final.pdf

Pope de Vallavieille C, Belhaj Fraj M, Mille B, Meynard JM (2007) Associations de variétés de blé pour stabiliser le rendement et la qualité de la récolte. In Rencontres du Cirad- Journée d'agronomie- 30 août 2007. ed CIRAD

Rolland B (2006) Les Dossiers de l'environnement de l'INRA. Quelles variétés et semences pour des agricultures paysannes et durables? 30:53–56

Sperling L, Ashby JA, Smith ME, Weltzien E, McGuire S (2001) A framework for analyzing participatory plant breeding approaches and results. Euphytica 122:439–450

Sylvander B, Bellon S, Benoit M (2006) Facing the organic reality: the diversity of development models and their consequences on research policies. Paper presented at Joint Organic Congress, Odense, pp 30–31

Witcombe JR, Joshi A, Goyal SN (2003) Participatory plant breeding in maize: a case from Gujurat, India. Euphytica 130:413–422

Wolfe MS (1997) Variety mixtures: concept and value. In: Wolfe MS (ed) Variety mixtures in theory and practice. European Union Variety and Species Mixtures working group of COST Action 817. Online at: http://www.scri.sari.ac.uk/TiPP/Mix/Booklet/default.htm

Chapter 21
Seeds for Organic Agriculture: Development of Participatory Plant Breeding and Farmers' Networks in France

Véronique Chable, Julie Dawson, Riccardo Bocci and Isabelle Goldringer

Abstract The lack of seeds and varieties suited to organic agriculture has been a problem for a long time. Conventional breeding strategies do not fit the needs of organic agriculture, which requires specific adaptation to the environment. Moreover, several current breeding methods do not respect the principles of organic agriculture. To overcome these limitations, organic farmers and their organisations initiated participatory plant breeding (PPB) programmes, together with researchers.

In France, this process became part of a movement of re-appropriation of breeding practices by farmers that aims at re-establishing their autonomy. These practices promote an agriculture linked to the "terroir" and culture with strong social and ethical values and, therefore, share many of the needs in terms of varieties with organic farmers. They often supply the same local markets. In 2003, the "French Farm Seed Network" (Réseau Semences Paysannes) was created in the aim of supporting PPB initiatives by facilitating collaboration with researchers and authorities concerning seed regulations.

Most often, farmers organise plant breeding activities by means of collective structures (cooperatives, associations, professional organisations, etc.). Some 30 species are concerned, including arable and vegetable crops. Researchers participate in projects from the outset to the evaluation of the process (e.g., evolution of the

V. Chable (✉)
INRA, UR980 SAD-Paysage, 65 rue de Saint Brieuc, 35000 Rennes, France
e-mail: chable@rennes.inra.fr

J. Dawson
Department of Horticulture, University of Wisconsin-Madison,
1575 Linden Drive Madison, WI 53706, USA
e-mail: dawson@hort.wisc.edu

J. Dawson · I. Goldringer
INRA-CNRS-INAPG-Université de Paris 11, UMR0320 Génétique Végétale,
91190 Gif-sur-Yvette, France
e-mail: isabelle.goldringer@moulon.inra.fr

R. Bocci
AIAB, Florence, Italy
e-mail: r.bocci@aiab.it

S. Bellon, S. Penvern (eds.), *Organic Farming, Prototype for Sustainable Agricultures,*
DOI 10.1007/978-94-007-7927-3_21, © Springer Science+Business Media Dordrecht 2014

biodiversity). In this chapter, we analyse why PPB is particularly well adapted to breeding requirements for organic farming in general and, more specifically, to the situation in France. The consequences in terms of seed regulation in Europe are then discussed, opening up prospects for PPB and its contribution to breeding for organic varieties.

Keywords Participatory plant breeding · Seeds · Population · Landraces · Organic agriculture · Network · Biodiversity · Selection

21.1 Introduction

Seeds have become a commercial good. Looking back on the history of plant breeding, maize hybrids can be seen as a change in approaches to breeding and the beginning of the concept of modern varieties.

> When hybrid maize was invented and presented to US farmers in the first decades of the 20th century, it was based on two new operations, one biological and the other socio-economic. First, strange manipulations (forced inbreeding and controlled hybridization) produced biological products that never before existed in nature. Second, farmers gave up their time-honoured practice of saving their own varieties of seeds in favour of the annual purchase of hybrid maize seed (Duvick 2001).

Together with the pure line varieties that were simultaneously developed in self-fertilising species, F1 hybrids perfectly fit the homogeneity and stability criteria that defined modern varieties, and could be mass-produced by larger and larger seed companies.

Both aspects are in contrast with the foundations of the organic agriculture movement, which includes a preference for local markets and the protection of farm autonomy (Paull 2006). The four major principles of organic agriculture (Health, Ecology, Fairness and Care) defined by the International Federation of Organic Agriculture Movements (IFOAM 2005a) clearly establish the concomitant biological/agronomic and socio-economic aspects of organic agriculture. In keeping with these principles, varieties for organic agriculture could include plant populations that adapt to their environment because their intrinsic heterogeneity maintains their ability to develop positive interactions within the crop and to evolve (Enjalbert et al. 2011).

> Positive interactions such as early crop vigour for nutrient uptake, weed competition and disease resistance are needed. Incorporation of all characteristics into the crop can be helped by diversification within the crop, allowing complementation and compensation among plants (Wolfe et al. 2008).

The recognition that specific seeds and varieties were needed for organic agriculture progressively arose at the end of the last century. Several factors combined to make the situation increasingly difficult for organic farmers: (i) varieties selected for conventional agriculture were increasingly less well adapted to organic systems; (ii) breeding methods using biotechnologies[1] became more prevalent; (iii) European

[1] Biotechnologies in plant breeding: all techniques that encompass the natural species barriers and reproductive processes, incompatible with organic principles (Lammerts van Bueren and Struik 2004).

regulations (EU Regulation 1452/2003) forced certified organic farmers to use organically produced seeds; and (iv) the organic seed market was not economically attractive for seed companies. In addition to the seed market, several encouraging elements have helped the actors of organic agriculture to face this situation: (i) several professional/farmer organisations have decided to engage in plant breeding and seed production for organic agriculture (Bocci and Chable 2008); (ii) different scientific trends (such as research that focuses on the benefits of within-field diversity to increase agricultural sustainability (Ostergard et al. 2009), PPB (Ceccarelli et al. 2007), evolutionary plant breeding (Suneson 1956; Murphy et al. 2005), and on-farm dynamic management of crop populations (Thomas et al. 2011, 2012) have led to similar conclusions on the interest of growing and breeding genetically heterogeneous varieties; and (iii) a part of civil society is increasingly looking for safe food and for quality, and is now conscious of the endangered environment[2]. These changes within society are not all directly linked to seed provision, but will contribute to the development of alternative research strategies in breeding (Valenquoren and Baret 2009), new market organisation (Adams and Salois 2010; Sahu 2011) and seed regulation propositions (Bocci et al. 2011; Chable et al. 2012).

Current seed regulations were established during the last century and conform to the dominant concept of cultivated varieties, which include the criteria of DUS (Distinction, Uniformity, Stability) and VCU (Value for Cultivation and Use). During this period, plant breeding activities in public research centres and private firms led to the development of varieties that fit the needs of industrialised agriculture and the need for standardised products for the international market (Bonneuil 2008). Currently, no seeds can be sold or exchanged for agricultural use if the varieties are not registered in the official catalogue of varieties. The catalogue system is not adapted to evolving and heterogeneous varieties bred by farmers (Serpolay et al. 2011a).

PPB organisation is in agreement with four major principles of organic agriculture (Health, Ecology, Fairness and Care) through: (i) the use of breeding processes that respect the biological characteristics and the integrity of the species (Lammerts and Struik 2004, 2005); (ii) the enhancement of local adaptation, which sustains the ecological system; (iii) the promotion of small-scale seed markets where trust between operators must be the first rule; and (iv) participatory research for healthy seed production and adapted crops, and the development of cultivated diversity for future generations (Döring et al. 2012). At the European level, a project (Farm Seed Opportunities) has developed a proposal for regulation scenarios that takes the development of on-farm breeding activities into account.

In this paper, we illustrate how participatory plant breeding (PPB) may fulfil the needs of organic agriculture, and how organic French farmers and their organisations have met with researchers to build joint experiments of on-farm plant breeding. We emphasize how farmers took the initiatives and how new farmers' associations emerged to organise collective means, to involve society at large, to address legal questions and to ensure the continued existence of on-farm breeding. PPB is mainly an organisation of plant breeding that was particularly applied in France over the

[2] www.tporganics.eu/.

last ten years. We describe how and why several "seed associations" were created and how they are federated within a national organisation, the "French Farm Seed Network" (Réseau Semences Paysannes)[3].

21.2 Definitions of Participatory Plant Breeding

21.2.1 Participatory Plant Breeding in the World

Vernooy (2003) broadly defined Participatory Plant Breeding (PPB) by approaches that involve close collaboration between researchers and farmers and, potentially, other stakeholders, to bring about genetic improvements within crops. Participatory approaches have emerged in the past two decades and have been described by a number of authors from all continents (Almekinders and Elings 2001; Ceccarelli et al. 2000, 2003; Machado and Fernandes 2001; Sperling et al. 2001; Vernooy 2003; Weltzien et al. 2008; Witcombe et al. 2001). These examples include several different crops, and are primarily focused on projects involving resource-poor farmers in developing countries who cannot afford to modify their environment through additional inputs. Conventional plant breeding has been more beneficial to farmers in high-potential environments, and PPB is seen by some scientists as a way to overcome the limitations of conventional breeding by offering farmers the possibility to choose and to create the varieties that better suit their needs and conditions (Ceccarelli and Grando 2007). Moreover, the joint selection process is an exchange of knowledge and information at the same time (Bishaw and Turner 2008).

The organisation of PPB varies according to the degree of involvement of the farmers, which may include the orientation of selection activities (Witcombe et al. 2005), the evaluation of plants during the breeding process (Morris and Bellon 2004), or may be expanded to include the entire breeding process. Several other authors report that farmers prefer to conduct their own selection from the beginning because they are thus able to combine more characteristics within a single variety (Almekinders and Hardon 2006).

21.2.2 Plant Breeding for Organic Agriculture

Organic Agriculture (OA) in developed countries has several points in common with the small-scale agriculture of the marginal environments of developing ones: a highly variable environment without chemical inputs, a combination of characteristics required for one crop, and a wide diversity of crops, of uses and of markets (Desclaux et al. 2008). Heterogeneous environments make it difficult to apply

[3] www.semencespaysanne.org.

consistent selection pressure because it is often hard to identify a single or several superior genotypes across all sets of conditions (Dawson et al. 2008).

Moreover, the conservation and enhancement of biodiversity are mentioned in the IFOAM standards and in all national guidelines as essential pillars for organic agriculture. OA is recognised as having a beneficial impact on biodiversity (Bengtsson et al. 2005; Wyss and Pfiffner 2008) but it also requires biodiversity to function. OA is also characterised by its choice of plant breeding and multiplication methods, which must respect the natural characteristics of the species (Lammerts van Bueren et al. 2002) and the integrity of the organisms (Lammerts van Bueren et al. 2003; Lammerts van Bueren and Struik 2005). IFOAM has proposed compatible methods in draft standards (IFOAM 2005b).

21.2.3 Participatory Plant Breeding in Europe

In Europe, when the EU Regulation 1452/2003 requiring the use of organic seed for planting went into force in 2004, organic seed professionals were not ready to fulfil the demand. Thus, organic farmers and their organisations started to meet with researchers and to build PPB projects. For example, in Germany, knowing that organic farming requires cultivars that are specifically adapted to this low-input cropping system, organic farmers and scientists joined together in a participatory breeding approach to develop region-specific genotypes of spring faba bean for organic conditions (Ghaouti et al. 2008). In the Netherlands, PPB was also the most efficient strategy to address the needs of organic onion producers because commercial onion breeders select varieties solely for conventional farming (Lammerts van Bueren et al. 2005). Seed companies consider the organic market too small to justify a specific programme. Breeders only give priority to storage and bulb quality traits, but organic farmers also need varieties that perform well in the field. Breeders give low priority to field selection, e.g., to make mechanical weed control easier, farmers prefer plants with a more erect growth habits. Moreover, the cytoplasmic male sterility system used to produce these hybrids does not comply with organic principles (Osman et al. 2008).

In Portugal, PPB was initiated with two objectives not specifically linked to OA: the conservation of white maize populations for traditional bread-making, and the maintenance of evolutionary processes in the farmers' fields, which could be valuable for continuing adaptation to future environmental conditions (Patto et al. 2008).

21.2.4 Participatory Plant Breeding in France

PPB simultaneously began in 2001 in three areas in France within three groups of farmers. Analysis of the functioning of each pioneer group showed a distinct organisational strategy for each of them, with different roles for researchers and farmers.

- In Brittany (western France), a regional organic umbrella organisation (IBB, Inter Bio Bretagne) and researchers from INRA (the French National Institute for Agricultural Research) initiated a participatory plant breeding programme for organic cabbage and cauliflower (Chable et al. 2008a). The aim was to include all concerned actors (farmers, processors, traders, advisors, researchers, etc.) in defining the objectives and the means to reach them. On the basis of the evaluation of genetic resources for seed production, the experimental agrobiological station of IBB (PAIS) was the meeting point for all of the project's partners. In the beginning, the farmers found technical and scientific information at the PAIS and were then able to share their experiences from plant selection to seed production, even though they were conducting the breeding steps on their own farms.

- In the Mediterranean region in southern France, organic farmers needed durum wheat varieties, with the aim to produce grain with sufficient protein content and vitreousness for pasta processing (Desclaux 2005). Pasta manufacturers were involved in the project. Although the programme was carried out in close collaboration with INRA scientists and organic farmers, the choice of the breeding populations and methods was led by researchers.
- Maize and sunflower farmers in south-western France had several objectives for breeding, including quality, rusticity and adaptation to dry conditions. The project was led by a local farmers' organisation (AgroBio Périgord), and locally funded by the region, with research support from an independent breeder. Farmers and the breeder have worked in parallel from the same populations to evaluate two breeding strategies for organic agriculture: PPB on-farm and the creation of composite varieties for farmers who are not ready to breed their own varieties.

In France, at the same time as officially certified organic production, a movement of re-appropriation of the farmers' breeding practices is emerging with the aim to regain autonomy for the farmers (referred to in France as "agriculture paysanne", or "peasant agriculture"). A group of farmers-bakers includes both organic farmers and "peasant" farmers[4]. Because their breads are made using traditional baking methods, because they do not use chemical pesticides and fertilisers and because they use locally adapted varieties/populations, modern certified short straw varieties are not appropriate. Thus, they develop their own bread wheat varieties based on historical resources including landraces (19th century) or old varieties (1900–1950)[5] and using different selection strategies (mass selection, designing mixtures, crossing, etc.). Their selection aims to manage and increase the intra-varietal diversity, whereas modern wheat breeding only aims at the homogeneity of the varieties.

More than 30 species are currently concerned, mainly arable crops and vegetables, and nearly all parts of the country are involved. Only the central plain region

[4] These farmers promote an agriculture linked to their region and culture with strong social and ethical values, and thus share many of the same needs in terms of seeds with organic farmers, often supplying the same local markets. The official certification is sometimes too expensive for small-scale farmers who prefer participatory certification (for example, Nature & Progrès, a private label in France).

[5] Based on anecdotal evidence from consumers, they feel that these varieties induce less gluten intolerance (Mercier 2008).

with very large farms with arable culture is not concerned. In 2009, several other species were introduced, especially forage and fibre cultures. For the farmers, the objectives can be considered at several levels: the crops themselves, the cropping systems, the development of organic farming and the recognition of social and ethical issues. One point common to all the objectives is increasing diversity.

- At the crop level:
  Improving the quality of the product and the adaptation of plants to organic conditions were the two main aims at the beginning of each experiment. Many farmers have direct contact with consumers, which provides direct indications for selection. With the continuous adaptation of the crops to their environment, the farmers perpetuate the history of cultivated plants, and the development of specific adaptation leads to varietal diversification across farms and regions.
- At the cropping system level:
  On-farm breeding also allows adaptation to new agronomic practices. Considering their agronomic bottleneck, farmers try to improve their production techniques at the same time they adapt the plants, i.e., living mulches, date of sowing, plant associations, etc.
- Development of OA and all forms of low-input agriculture:
  Innovation for products, techniques and species is becoming more and more important for the future of this type of agriculture. Farmers are aware that they need to deal with the challenges of climate change and economic difficulties. In order to buffer future variations at both levels, the farmers do not only consider their current market and farming system but often enlarge the number of species they choose to grow.
- Social and ethical issues:
  At the beginning, all the farmers' groups mentioned the importance of autonomy for seed supplies. With PPB, they also have the control of the breeding methods and choose to avoid all biotechnologies. Moreover, they often report that their interest in their work has increased: breeding is a pleasure, a passionate adventure with the plants.

21.3 The Organisation of PBB for Organic Agriculture in France

An analysis of different PPB approaches led Sperling et al. (2001) to distinguish 'Formal-led PPB' and 'Farmer-led PPB', with the former being developed primarily by public-sector professional plant breeders, and the latter developed through farmers' associations and NGOs, with the involvement of plant breeders at the request of the farmers. The French case is a 'Farmer-led PPB', which brings together public-sector scientists from INRA and several networks of farmers. The researchers take a supporting role and the farmers remain responsible for the process.

Five main steps have been described for the establishment of a participatory plant breeding programme for organic farming (Chable and Berthellot 2006):

1. Constitution of the group and creation of exchange space: farmers, researchers and other actors can collect specific funds to work together and to define the means of common action;
2. Definition of priorities in terms of crops and research of the genetic resources, with priority given to native and locally adapted varieties;
3. Discovery, adaptation and selection in the farmers' fields, under the cropping conditions defined by the group;
4. Seed production and distribution in collective organisations;
5. Exchange of experiences and genetic resources through meetings—formal and informal, regional, national or international, farmers and professionals in the organic sector (accompanied by researchers and gardeners and advisors as well).

PPB is generally a collective action. The origin or nature of the initial group may strongly vary and many are formed within existing agricultural organisations (private or public, cooperatives for market, trade unions, etc.). The coordinator of the group, who is generally employed by the organisation, has the fundamental role in the beginning of cooperating with researchers. For the researchers, it is not possible to work with many species all over the country without the help of these coordinators. All of the participants together determine the main objective of the group, the genetic resources that are needed, the means to collect them and the organisation of the trials.

The key step of the PPB project is the choice of genetic resources that fit the diversified objectives, that have not been genetically modified and that will be appropriate for local agriculture. The farmers involved in these projects seek to avoid numerous aspects of modern breeding and are therefore looking for varieties that have not been modified by biotechnology. Landraces or historical varieties that have maintained their intrinsic variability for adaptation and qualities are the primary genetic resources used in these projects. This investigation into the history of the species and the search for seed samples in genebanks, research centres, or from private breeders is usually the role of the researcher. Sample collection has to address the farmer's needs but it is also important for the first "discovery trial" to explore the variability of the species and its performance in terms of yield and quality in the area considered.

After the first collective trials (on a farm or within a collective structure), the farmers make their choices and begin observation, breeding and seed production on their own farms. Some experienced farmers may become autonomous for an innovative breeding project, but remain in their group for many reasons: the need for seed exchanges, the use of collective materials, and to deal with complex legal issues (e.g., only small quantities of seeds are exchanged since seed cannot be sold unless it is registered in the official catalogue). The action nevertheless remains collective and the initial structure remains the meeting point for all of the actors involved (farmers, traders, advisors, researchers, etc.). The group members actively exchange scientific and technical information and share experiences on different aspects of the breeding process.

The two last points of the organisation of the PPB of the French farmer groups concern seed production and exchanges. In vegetable production, one farmer is not able to produce seeds for all his crops. At this step, the initial group may create a new association devoted to the PPB actions and biodiversity conservation. In addition, being part of a formal organisation allows the farmers to propose research activities and to submit projects to regional grant programmes in order to finance shared equipment (mainly for seed harvesting, cleaning, storage and quality assessment). In addition to these economic and technical reasons, the creation of farmers' networks has been essential to undertake negotiations with policymakers to carry out farm seed production within a legal framework.

In fact, one result of this PPB process (concerning its organisational component) is the creation of seed associations in which farmers are autonomous in terms of their plant breeding activities. We will illustrate this type of result with two examples:

- Wheat in France, often associated with other cereals and arable crops: Table 21.1 shows the numerous seed associations created since 2000 to promote selection, conservation and breeding of ancient varieties, landraces or farmers' creations. These associations may have between less than a hundred to several hundred members and as many varieties. Next year, three more associations will complete this list. They are at the creation phase in two more regions of France: Lorraine and Normandie.
- Vegetables in the Pays de Loire region in the case of a group of farmers, Bio Loire Ocean: the colour plate 21 illustrates the PPB activity of this group that concerns some 100 farmers, 8 species, 40 trials and 200 varieties since 2005.

The development of researcher/farmer partnerships evolves by means of participatory research that helps farmers to overcome bottlenecks (agronomic, breeding methods, quality assessment, developments in seed laws, etc.). This research may be organised on-farm with farmers from one or several associations and the French Farmer Seed Network (Réseau Semences Paysannes). A European project (SOLIBAM[6], Strategies for Organic and Low-Input Breeding and Management) involves farmers from several of the associations described and illustrates their activities.

21.4 Origin and role of farmers' networks and the evolution of seed regulations at the EU level

An umbrella organisation was created to federate the first initiatives and the seed associations that have arisen where PPB initiatives have taken place.

In France, before the development of PPB actions, several pioneer farmers, mainly using organic practices, had initiated the conservation of and/or selection from traditional or historic varieties for many species. In 2003, national or-

[6] www.solibam.eu.

Table 21.1 Seed associations involved in wheat PPB in France: region, name of the association, date of creation, type of members and main goals. See colour plate 21

Region of France	Associations	Date of creation, actors, main goals, other species
Brittany— western France	Triptolème	2006—farmers-bakers, gardeners and consumers Collection, conservation, breeding, innovation in agronomic practices, baking, training and education Many other species of arable crops
Poitou-Charentes	Cultivons la Bio Diversité en Poitou-Charentes (CBD)	2009—farmers-bakers, gardeners and consumers Collection, conservation, breeding, innovation of agronomic practices Maize, forage and vegetables
South-West	Centre d'Etudes et de Terre d'Accueil des Blés (CETAB)	2005—farmers-bakers, gardeners and consumers Collection, conservation, breeding, baking Some other species of arable crops
Midi Pyrénées-Tarn	Association Pétanielle (formerly- "Semeurs et semeuses de biodiversité des champs et des jardins")	2010 (2008)—Farmers-bakers, gardeners and consumers Collection, conservation, breeding, baking
Midi Pyrénées-Hautes Pyrénées	Terre en Vie	2000—Farmers-bakers and consumers Collection, conservation, breeding, several types of transformation (bread, pasta, bulgur, etc.) Local markets
South-East	ARDEAR Rhône Alpes	2004—farmers-bakers, gardeners and consumers Collection, conservation training, breeding Other species such as maize
Mediterranean region	Syndicat de Promotion Touselle	2005—farmers-bakers and consumers Collection, conservation, breeding, baking Several cereals, forage and other Mediterranean species
South-East	Parc Naturel Régional du Queyras	2003—Farmers, bakers and consumers Collection, conservation, breeding, several types of transformation (bread, pasta, bulgur, etc.) Local markets with traditional products.
South-East	Producteurs bio des Alpes de Haute Provence	2004—farmers, millers, bakers, consumers Collection, conservation, baking, local markets They created the "blé meunier d'Apt" network

Table 21.1 (continued)

Region of France	Associations	Date of creation, actors, main goals, other species
Bourgogne	Graines de Noé	2010—Farmers, millers, bakers and gardeners Collection, conservation, breeding Other species: cereals and vegetable
Alsace	Kerna ùn Sohma (Grain and Seed)	2011—farmers, farmers-bakers, gardeners, artisans, consumers Collection, conservation, breeding, sharing of knowledge and material Other species: vegetables, grapes

ganic organisations and several agricultural trade unions (Fédération Nationale de l'Agriculture Biologique, Confédération Paysanne, Nature & Progrès, Fédération Nationale d'Agriculture Biologique des Régions de France, Mouvement de Culture Bio-Dynamique, Bio d'Aquitaine, Groupement de Développement de l'Agriculture Biologique Midi-Pyrénées, Syndicat des Semences et Plants Bios du Languedoc-Roussillon) organised a meeting in Auzeville, France. There, they met other farmers involved in similar initiatives and shared their first experiences with researchers who had initiated PPB actions or were involved in the management of genetic resources and small-scale breeding companies. This event consolidated the farmers' determination to take charge of the future of their varieties and seeds. The farmers also considered the starting point for a new form of collaboration between researchers and farmers in France. The French Farmer Seed Network was created after this meeting and brings together several local networks, ensuring a link between farmers and authorities to stimulate the necessary adaptation of French registration laws (Bocci and Chable 2008). In addition to the formal seed "market", the network calls for ongoing seed exchanges between farmers, gardeners and the other actors involved in biodiversity conservation. The ITPGRFA (International Treaty of Plant Genetic Resources for Food and Agriculture) recognised the past, current and future contribution of the farmer in creating and renewing cultivated biodiversity (Andersen and Tone 2008)

In Italy and Spain, similar groups of farmers who refuse to adopt industrial agricultural practices and instead practice organic agriculture have existed since the beginning of this century. In recent years, they have become more organised into networks. In addition to the Réseau Semences Paysannes in France, we also have the Red de Semillas in Spain and the Rete Semi Rurali in Italy. Their members are also farmers, consumers, scientists, and all types of stakeholders concerned by the landraces, working together in order to reconsider the scientific, technical and legal aspects of seed production (Bocci and Chable 2008). Starting from their initial activities—preserving farm seeds (France) or safeguarding ancient varieties (Italy and Spain)—the networks' aims and activities have broadened. The development of varietal innovation produced by the farmers themselves has been particularly pronounced in France.

In 1998, the European Commission mentioned for the first time that it is essential to ensure the conservation of genetic resources and therefore necessary to introduce a legal basis for that purpose. They agreed to allow the *in situ* use of varieties threatened with genetic erosion with the intention of conserving them, within the framework of legislation concerning the seed trade. The last step was in June 2008 with EU Directive 2008/62/CE, "providing for certain derogations for acceptance of agricultural landraces and varieties which are naturally adapted to the local and regional conditions and threatened by genetic erosion and for marketing of seed and seed potatoes of those landraces and varieties". Nevertheless, no legal status has yet been created for new population varieties and for on-farm PPB activities.

A European project (2007–2010), Farm Seed Opportunities (FSO), was initiated mainly to respond to the needs of European policy makers to implement a review of the seed regulation framework. It also aimed at evaluating and analysing the role of farmers in the conservation of varieties and in the breeding of new varieties. FSO has been managed by a consortium of public and private scientific institutions, farmers' seed networks and organic farmers' associations in six European countries. French, Italian and Spanish farmers' networks have been involved in the project.

In addition to the implementation of the new directive, the FSO has taken the PPB experiences of several groups of organic farmers into account, with the goal of proposing regulation scenarios that will recognise the existence of the innovative activity of farm selection (Chable et al. 2008b). This project has added its reflection to those of developing countries in order to draw up regulation scenarios. Farmers' systems of seed supply and crop development are by far the most important source of seed in most farming systems worldwide. Most agricultural land in the world is still sown with seed that is informally produced by farmers (Almekinders and Louwaars 2002). Farm Seed Opportunities has also included a survey of existing experiences, of the expectations of stakeholders and of the limitations of the current laws in Europe. Moreover, field experiments were organised to collect data about three kinds of farmer varieties (landraces, mass selection within landraces and evolutionary mixtures) that exist in Europe (Serpolay 2011a). The goal was to describe how the farmers' dynamic management and the natural pressures of selection determine the evolution of these varieties. It has been shown that the farmers' varieties cultivated in traditional farming systems should not be considered as separate entities, but rather as an open genetic system. The functioning of the landrace system has been compared with the metapopulation concept established for wild species (Goldringer et al. 2001). The conservation of genetic diversity at the metapopulation level depends on the existence of contrasting environments at the subpopulation level (populations on different farms). The development of local adaptation leads to some loss of diversity within subpopulations, whereas a high level of genetic diversity is maintained across subpopulations. Seed exchanges among farmers allow for incorporation and renewal of the genetic diversity within the subpopulations (Enjalbert et al. 2011). In the case of maize landraces in traditional farming systems in Mexico, "farmers exchange, pool or replace seeds for several reasons, including seed loss due to poor harvests or insect damage in storage. A principal reason, however, is the belief that the same seed should not be planted in the same field over successive seasons because its yield will decline." This concept of a "tired" variety and the need to renew it through exchange has been reported for others crops and regions (Louette and Smale 2000; Pressoir and Berthaud 2004). French organic or "peasant" farmers involved in on-farm breeding also feel the necessity of seed exchanges to maintain diversity, adaptation and vigour and to guarantee the sustainability of their activities. It has been shown that the genetic diversity found in a seed exchange network was much greater than the diversity conserved in a genebank for a given variety "name", and that farmers' management practices and seed exchanges combined with natural selection have contributed to shape the variability of this metapopulation (Demeulenaere et al. 2008; Thomas et al. 2011; Thomas et al. 2012). In these cases, however, the seed exchange community involved farmers who share the same aims (practices, qualities, markets) at the regional or national level (Demeulenaere and Bonneuil 2012). Organic agriculture still represents only 2.4% of the cultivated area in France, and organic farmers involved in PPB represent a small part of them. Even if some farmers involved in organic agriculture without certification or low-input agriculture increased this number, there is a need to collaborate over fairly large geographic distances.

21.5 Perspectives

Farmers who are involved in PPB actions are motivated by both the breeding activity itself and by the potential gain (at the agronomic, social and market levels) represented by the improved varieties and the "free" seed. Even if breeding and seed production mean an extra quantity of work for them, most of them have succeeded in introducing breeding and seed multiplication within their production schedule. The rapidity of adaptation of the plant population has often surprised them, i.e., the precocity of the morphological evolution (Serpolay et al. 2011b; Dawson et al. 2012a; Dawson et al. 2012b) has given them confidence.

All the groups or associations agree that the next step is the creation of "seed community banks" (Chable et al. 2011) where the conservation could be collectively organised, where collective material and machines could be found to thresh and clean the seed and assess seed quality, and where know-how could be exchanged. The question now is how to find the means for these organisations, and the involvement of members of society (consumers) is very important to help farmers to support the safeguard of the diversity and the creation of locally adapted varieties.

The legal context has evolved over the last ten years. In addition to the UE project (Farm Seed Opportunities), the Food Chain Evaluation Consortium[7], in its final report presented in March 2009 in Brussels, has confirmed a perspective of evolution:

> At the moment the commercial breeding systems strongly influence the interpretation of both DUS and VCU characteristics and the testing systems. With environmental issues becoming more important there is a stronger case now for a wider interpretation of both DUS and VCU for varieties that are going to be used for organic and low input agriculture" (FCEC 2009).

[7] http://ec.europa.eu/food/plant/propagation/evaluation/index_en.htm.

> Moreover, "the two different systems of the large commercial breeding companies and the smaller market or regional breeders and producers could run side by side because they are targeting completely different markets.

Meanwhile, at the end of Farm Seed Opportunities, it was recognised that the farmer seed associations may represent a framework in which a new legal area will be built.

From a scientific point of view, the varieties bred on the farm encourage researchers to renew their genetic approach. We are no longer considering individual plants or genotypes, but plant populations (Wolfe 2000; Thomas et al. 2012). The rapid evolution of these plant populations, as well the notion of "tired" varieties, deal with genetic and epigenetic mechanisms that control plant plasticity in response to environmental stimuli. Statistical and/or quantitative population genetics methods, combined with epigenetic approaches, will be developed to be applied to these breeding populations submitted to various selection pressures or stimuli (e.g., agro-climatic conditions, crop management systems and selection practices).

21.6 Conclusion

Even if farmers involved in PPB are not numerous, they are very actively collaborating (Enjalbert et al. 2011; Dawson et al. 2011) with scientists, other stakeholders and consumers, and they significantly contribute to change the plant breeding landscape for organic agriculture. Regional political bodies recognise their actions and the innovative potential for increasing the sustainability of agricultural systems. They are also supported by the ITAB (French Technical Institute for Organic Agriculture) at the national level through the publication and diffusion of research results to many groups involved in organic agriculture.

Finally, it appears that the PPB and, more generally, on-farm breeding (with or without the help of researchers) reinforces the coherence of organic agriculture by increasing biodiversity, improving the technical performance of varieties on organic farms, supporting the evolution of farming systems and increasing fairness in the seed supply system. Thus, on-farm plant breeding helps organic farmers to increase crop and farm performances. The continued existence of on-farm breeding is ensured by the organisation of seed associations and "community seed banks" with infrastructures for storing seeds of varieties used by farmers and with the aim of facilitating the exchange and conservation of these varieties.

References

Adams DC, Salois MJ (2010) Local versus organic: a turn in consumer preferences and willingness-to-pay. Renew Agr Food Syst 25(4):331–341

Almekinders C, Hardon J (2006) Bringing farmers back into breeding. Experiences with participatory plant breeding and challenges for institutionalisation. Agromisa Special 5, p 135

Almekinders CJM, Louwaars NP (2002) The importance of the farmers' seed systems in a functional national seed sector. J New Seeds 4(1/2):15–33

Almekinders CJM, Elings A (2001) Collaboration of farmers and breeders: Participatory crop improvement in perspective. Euphytica 122(3) 425–438

Andersen R, Tone W (2008) The farmers' rights project—background study 7: success stories from the realization of farmers' rights related to plant genetic resources for food and agriculture. FNI Report 4/2008. Lysaker, FNI, 2008, p 72

Bengtsson J, Ahnstrom J, Weibull A (2005) The effects of organic agriculture on biodiversity and abundance: a metaanalysis. J Appl Ecol 42:261–269

Bishaw Z, Turner M (2008) Linking participatory plant breeding to the seed supply system. Euphytica 163:31–44

Bocci R, Chable V, Kastler G, Louwaars N (2011) Farm seed opportunities, recommendations for on farm conservation in Europe. In: Nigel Maxted, Ehsan Dulloo, Brian V. Ford-Lloyd, Jose M. Iriondo, M. A. A. Pinheiro de Carvalho (eds) Agrobiodiversity conservation: securing the diversity of crop wild relatives and landraces CABI Publishing, Oxford, p 400

Bocci R, Chable V (2008) Les Réseaux Semences Paysannes en Europe et l'évolution des lois sur les semences—The Peasant Seeds Network in Europe and the evolution of seed laws. Cahiers d'études et de recherche francophones/Agricultures 17:216–221

Bonneuil C (2008) Producing identity, industrializing purity. Elements for a cultural history of genetics. In: Müller-Wille S, Rheinberger H-J (eds) A cultural history of heredity IV. Preprint 343, Berlin, pp 81–110

Ceccarelli S, Grando S (2007) Decentralized-participatory plant breeding: an example of demand-driven research. Euphytica 155(3) 349–360

Ceccarelli S, Grando S, Singh M, Michael M, Shikho A, Al-Issa M, Al-Saleh A, Kaleonjy G, Al-Ghanem SM, Al-Hasan AL, Dalla H, Basha S, Basha T (2003) A methodological study on participatory barley breeding II. Response to selection. Euphytica 133(2) 185–200

Ceccarelli S, Grando S, Baum M (2007) Participatory plant breeding in water-limited environments. Exp Agric 43(4)(October):411–435

Ceccarelli S, Grando S, Tutwiler R, Baha J, Martini AM, Salahieh H, Goodchild A, Michael M (2000) A methodological study on participatory barley breeding—I. Selection phase. Euphytica 111(2) 91–104

Chable V, Berthellot J-F (2006) La sélection participative en France: Présentation des expériences en cours pour les agricultures biologiques et paysannes. In: Gasselin P, Clément O (eds) Quelles variétés et semences pour des agricultures paysannes durables? Dossiers de l'environnement de l'INRA 30:129–138

Chable V, Conseil M, Serpolay E, Le Lagadec F (2008a) Organic varieties for cauliflowers and cabbages in Brittany: from genetic resources to participatory plant breeding. Euphytica 164:521–529

Chable V, Goldringer I, Hazard L, Serpolay E, Bocci R, Cardarelli A, Micheloni C, Roviglioni R, Lammerts Van BuerenET, Osman A, Soriano JJ, González JM, López P, Sánchez H, Valero T, Zaharia H, Kastler G, Levillain T, Van der Burg WJ, Groot SPC, Pimbert M, Thommen A, Schärer HJ, Van den Berge P, Almekinders C, Pino S, Bressan M, Padovan S, Kik C, Louwaars NP (2008b) "Farm Seed Opportunities", a European program: what are the future prospects for organic varieties and seeds? 16th IFOAM Organic World Congress, Modena, Italy, June 16–20, 2008

Chable V, Kastler G, Kochko de P (2011) Seed in transition: the trends of participatory plant breeding and "farmer breeding" in France. Heuksalim native seed institute. 17th IFOAM Organic World Congress, 26th Septembre–4th October 2011 Seoul, Republic of Korea

Chable V, Louwaars N, Hubbard K, Baker B, Bocci R (2012) Plant breeding, variety release and seed commercialisation: laws and policies of concern to the organic sector. In: Lammerts van Bueren ET, Myers JR (eds) Organic crop breeding. Wiley-Blackwell, Hoboken. ISBN:978-0-470-95858–2

Dawson JC, Murphy KM, Jones SS (2008) Decentralized selection and participatory approaches in plant breeding for low-input systems. Euphytica 160(2) 143–154

Dawson JC, Rivière P, Berthellot JF, Mercier F, de Kochko P, Galic N, Pin S, Serpolay E, Thomas M, Giuliano S, Goldringer I (2011) Collaborative plant breeding for organic agricultural systems in developed countries. Sustainability 3(8):1206–1223. doi:10.3390/su3081206

Dawson JC, Serpolay E, Giuliano S, Schermann N, Galic N, Berthellot JF, Chesneau V, Ferté H, Mercier F, Osman A, Pino S, Goldringer I (2012a) Phenotypic diversity and evolution of farmer varieties of bread wheat on organic farms in Europe. Genet Resour Crop Evol 60:145–163

Dawson JC, Serpolay E, Giuliano S, Schermann N, Galic N, Chable V, Goldringer I (2012b) Multi-trait evolution of farmer varieties of bread wheat after cultivation in contrasting organic farming systems in Europe. Genetica 140(1–3):1–17

Demeulenaere É, Bonneuil C (2012) Des semences en partage. Construction sociale et identitaire d'un collectif 'paysan' autour de pratiques semencières alternatives. Technol Cult 57(2011/2):202–221

Demeulenaere E, Bonneuil C, Balfourier F, Basson A, Berthellot JF, Chesneau V, Ferté H, Galic N, Kastler G, Koenig J, Mercier F, Payement J, Pommart A, Ronot B, Rousselle Y, Supiot N, Zaharia H, Goldringer I (2008) Étude des complémentarités entre gestion dynamique à la ferme et gestion statique en collection: cas de la variété de blé Rouge de Bordeaux. Les Actes du BRG, 7:117–138

Desclaux D (2005) Participatory plant breeding methods for organic cereals—proceeding of the COST-SUSVAR/ECO-PB workshop on organic plant breeding strategies and the uses of molecular markers—17–19 janvier 2005, Driebergen, The Netherlands

Desclaux D, Nolot JM, Chiffoleau Y, Goze E, Leclerc C (2008) to fit agriculture diversification and decentralized Changes in the concept of genotype × environment interactions participatory plant breeding: pluridisciplinary point of view. Euphytica 163(3):533–546

Döring TF, Pautasso M, Finckh MR, Wolfe MS (2012) Concepts of plant health—reviewing and challenging the foundations of plant protection. Plant Pathol 61:1–15

Duvick DN (2001) Biotechnology in the 1930s: the development of hybrid maize. Nature Rev 2:69–74

Enjalbert J, Dawson JC, Paillard S, Rhoné B, Rousselle Y, Thomas M, Goldringer I (2011) Dynamic management of crop diversity: from an experimental approach to on-farm conservation. CR Biol 334(5–6):458–468

FCEC (2009) Evaluation of the community acquis on the marketing of seed and plant propagating material (S & PM). Food chain evaluation consortium final report, European Commission Directorate General for Health and Consumers. Retrieved at http://ec.europa.eu/food/plant/propagation/evaluation/s_pm_evaluation_finalreport_en.pdf, on December 2010

Ghaouti L, Vogt-Kaute W, Link W (2008) Development of locally-adapted faba bean cultivars for organic conditions in Germany through a participatory breeding approach. Euphytica 162(2): 257–268

Goldringer I, Enjalbert J, David JL, Paillard S, Pham JL, Brabant P (2001) Dynamic management of genetic resources: a 13 years experiment on wheat. In: Cooper HD, Spillane C, Hodgkin T (eds) Broadening the genetic bases of crop production. CAB International, Oxfordshire, pp 245–259

IFOAM (2005a) The principles of organic agriculture. http://www.ifoam.org/about_ifoam/principles/index.html

IFOAM (2005b) Plant breeding draft standards. http://www.ifoam.org

Lammerts van Bueren ET, Soest LJM van, Groot LC de, Boukema IW, Osman AM (2005) Broadening the genetic base for better-adapted varieties for organic farming systems: participatory characterisation, evaluation and selection of onion accessions for new base populations. Euphytica 146(1/2):125–132

Lammerts van Bueren ET, Struik PC, Jacobsen E (2002) Ecological concepts in organic farming and their consequences for an organic crop ideotype. Netherlands J Agric Sci 50(1):1–26

Lammerts van Bueren ET, Struik PC (2005) Integrity and rights of plants: Ethical notions in organic plant breeding and propagation. J Agric Environ Ethics 18(5):479–493

Lammerts van Bueren ET, Struik PC, Tiemens-Hulscher M, Jacobsen E (2003) Concepts of intrinsic value and integrity of plants in organic plant breeding and propagation. Crop Sci 43(6):1922–1929

Lammerts van Bueren ET, Struik PC (2004) The consequences of the concept of naturalness for organic plant breeding and propagation. Neth J Agric Sci 52(1):85–95

Louette D, Smale M (2000) Farmers' seed selection practices and traditional maize varieties in Cuzalapa, Mexico. Euphytica 113:25–41

Machado AT, Fernandes M (2001) Participatory maize breeding for low nitrogen tolerance. Euphytica 122:567–573

Mercier F (2008) Petite histoire des blés. In: Zaharia H. Voyage autour des blés paysans. Réseau Semences Paysannes, Arts Graphiques Langlade, France, pp 23–31

Morris M, Bellon MR (2004) Participatory plant breeding research: opportunities and challenges for the international crop improvement system. Euphytica 136(1):21–35

Murphy K, Lammer D, Lyon S, Carter B, Jones SS (2005) Breeding for organic and low-input farming systems: an evolutionary-participatory breeding method for inbred cereal grains. Renew Agr Food Syst 20(1):48–55

Osman AM, Almekinders CJM, Struik PC, Lammerts van Bueren ET (2008) Can conventional breeding programmes provide onion varieties that are suitable for organic farming in the Netherlands? In: Lammerts van Bueren ETL, ØOstergaård H, Goldringer I, Scholten O (eds) Euphytica 163(3):511–522

Ostergard H, Finckh MR, Fontaine L, Goldringer I, Hoad SP, Kristensen K, van Bueren ETL, Mascher F, Munk L, Wolfe MS (2009) Time for a shift in crop production: embracing complexity through diversity at all levels. J Agr Food Inf 89(9):1439–1445

Patto MCV, Moreira PM, Almeida N, Satovic Z, Pego S (2008) Genetic diversity evolution through participatory maize breeding in Portugal. Euphytica 161(1/2):283–291

Paull J (2006) The farm as organism: the foundational idea of organic agriculture. Elementals. Journal of Bio-Dynamics Tasmania 83:14–18. http://orgprints.org/10138/1/10138.pdf. Accessed 1 May 2012

Pressoir G, Berthaud J (2004) Population structure and strong divergent selection shape phenotypic diversification in maize landraces. Heredity 92:95–101

Sahu SK (2011) Localized food systems: the way toward sustainable livelihoods and ecological security—a review. J Anim Plant Sci 21(2 Suppl.):388–395

Serpolay E, Dawson JC, Chable V, Lammerts Van Bueren E, Osman A, Pino S, Silveri D, Goldringer I (2011a) Diversity of different farmer and modern wheat varieties cultivated in contrasting organic farming conditions in Western Europe and implications for European seed and variety legislation. Org Agr 1(3):127–145. doi 10.1007/s13165–011-0011–6

Serpolay E, Schermann N, Dawson JC, Lammerts van Bueren ET, Goldringer I, Chable V (2011b) Phenotypic changes in different spinach varieties grown and selected under organic conditions. Sustainability 3(9):1616–1636. doi:10.3390/su3091616

Sperling L, Ashby JA, Smith ME, Weltzien E, McGuire S (2001) A framework for analyzing participatory plant breeding approaches and results. Euphytica 122(3):439–450

Suneson CA (1956) An evolutionary plant breeding method. Agron J 48:188–191

Thomas M, Dawson JC, Goldringer I, Bonneuil C (2011) Seed exchanges, a key to analyze crop diversity dynamics in farmer-led on-farm conservation. Genet Resour Crop Ev 58:321–338

Thomas M, Demeulenaere E, Dawson JC, Khan A-R, Galic N, Jouanne-Pin S, Remoue C, Bonneuil C, Goldringer I (2012) On-farm dynamic management of genetic diversity: the impact of seed diffusions and seed saving practices on a population-variety of bread wheat. Evolutionary applications article first published online: 4 APR 2012. doi:10.1111/j.1752–4571.2012.00257.x

Vanloqueren G, Baret PV (2009) How agricultural research systems shape a technological regime that develops genetic engineering but locks out agroecological innovations. Res Policy 38:971–983

Vernooy R (2003) Seeds that give: participatory plant breeding. IDRC, Canada. www.idrc.ca/seeds

Weltzien E, Brocke KVom, Toure A, Rattunde F, Chantereau J (2008). Review and outlook for participatory plant breeding in West Africa. In: Bazile D, Weltzien E (eds) Cah Agric 17(2)165–171

Witcombe JR, Joshi KD, Rana RB, Virk DS (2001) Increasing genetic diversity by participatory varietal selection in high potential production systems in Nepal and India. Euphytica 122(3): 575–588

Witcombe JR, Joshi KD, Gyawali S, Musa AM, Johansen C, Virk DS, Stapit BR et al (2005) Participatory Plant Breeding is better described as highly client-oriented plant breeding. I. Four indications of client-orientation in plant breeding. Exp Agr 41 299–319

Wolfe MS, Baresel JP, Desclaux D, Goldringer I, Hoad S, Kovacs G, Loeschenberger F, Miedaner T, Østergård H, van Bueren ETL (2008) Developments in breeding cereals for organic agriculture. Euphytica 163(3) 323–346

Wolfe MS (2000) Crop strength through diversity. Nature 406:681–682

Wyss E, Pfiffner L (2008) Biodiversity in organic horticulture—an indicator for sustainability and a tool for pest management. In: Prange RK, Bishop SD (eds) Acta Horticulturae 767:75–80

Chapter 22
Considerations for Enabling the Ecological Redesign of Organic and Conventional Agriculture: A Social Ecology and Psychosocial Perspective

Stuart B. Hill

Abstract The main aim of this 'critical position' chapter is to provide a foundation and framework for developing effective local, contextual, collaborative, integrated planning and action for achieving sustainability within our food systems. Deep (eco-design/redesign-based) organics is distinguished from shallow (substitution-based) organics by the originator of these terms. A discussion is provided of the redesign implications for a transition from conventional agriculture and shallow organics to the more sustainable deep organics, using social ecology and associated 'testing questions' relating to personal, social, ecological and general aspects, as a framework for implementation. In addition to documenting the historical origins of these concepts and arguing for their relevance to achieving sustainability within food systems, emphasis is placed on the need to understand and address the psychological and psychosocial roots of the unsustainability challenges to modern societies. Failure to do this is considered as a main reason for the limited progress that has been made in addressing most current problems. Also, proactive, ecodesign-based preventative approaches to problems are advocated over reactive, curative approaches.

An Efficiency-Substitution-Redesign (ESR) progression in relation to change is illustrated using pest control to characterise the differences between the stages involved.

It is acknowledged that the changes being advocated here will require a fundamental shift in the paradigms underlying current dominant thinking and action. An innovative, proactive approach to enabling such a shift and facilitate meaningful change, based on 'lying', is provided.

Keywords Ecological redesign · Deep organic agriculture/farming/food systems · Transformation/change · Social ecology · Psychology · Psychosocial · Sustainability · Lying

---

S. B. Hill (✉)
School of Education, University of Western Sydney (Kingswood Campus),
Locked Bag 1797, NSW 2751, Penrith, Australia
e-mail: s.hill@uws.edu.au

S. Bellon, S. Penvern (eds.), *Organic Farming, Prototype for Sustainable Agricultures,* DOI 10.1007/978-94-007-7927-3_22, © Springer Science+Business Media Dordrecht 2014

22.1 Introduction

This chapter provides a critical reflection on key factors (many widely neglected) to consider when endeavouring to make agricultural systems, particularly organic ones, more sustainable. It is informed primarily by many years of experience using an *Efficiency-Substitution-Redesign* (ESR) framework for working on progressive change with farmers and policy makers.

The broad *higher goals* that have been emphasized throughout this work include enabling equitable access to nourishment, well-being, ecological sustainability, the conservation of biodiversity, wildlife habitats, resources and rural communities, and amelioration of changes to the global climate. Whereas such goals are compatible with sustainability, the more common goals of growing consumption, profit, power and privilege are unsustainable. In relation to such *higher goals*, most organic farms are already outperforming most conventional farms, particularly those that are heavily dependent on synthetic inputs (Mäder et al. 2002; Lynch et al. 2011). More importantly, however, the potential of organic farming to make further gains is considerable. This is particularly because much progress can be expected when:

- more funds are made available for 'deep organic' farming education, research, development and support; and
- current 'shallow organic' farms are redesigned, and new ones designed ecologically to avoid problems, function optimally and achieve the kinds of higher goals listed above.

In contrast to the 'bifurcation' argument (Coombes and Campbell 1998) that distinguishes between the mostly older, smaller, artisanal, true-to-organic-principles farms and the mostly newer, larger, industrialised organic farms—and that often advocates a return to the former, with their original organic values—I am advocating a need to improve those values, as well as the associated designs and management systems related to them. I am also advocating that these developments be applied to all farm and food systems, including the pre-industrialised organic farms (Rosin and Campbell 2008).

My 'deep organics' was developed as a much broader concept than 'agroecology' (*the science of applying ecology concepts and principles to the design and management of sustainable food systems*; Altieri 1987; Gliessman 1997; Wezel et al. 2009), as it applies to whole food and fibre systems, and considers all associated underlying aspects, including those characterised as philosophical, ethical, psychological, psychosocial socio-cultural and 'spiritual' (which for me, rather than being a 'religious' concept, is concerned with the important extensive 'unknown' aspects of all phenomena).

For over 30 years I have been arguing for a progression from *shallow (substitution-based) organics* towards *deep (eco-design/redesign-based [permanent]) organics* (Hill 1976, 1984a, 1985, 1998, 2006)[1]. This was conceived independently

[1] See also: White (1991).

from, and in parallel with, Arne Naess' (1973) distinctions between 'Shallow and Deep Ecology Movements'. More recently the outstanding organic horticulturist, Eliot Coleman (1999, 2004, 2009), has written about his version of 'deep organics'[2]. It was also conceived in parallel with the development of permaculture (discussed later), which also emphasizes eco-design (Mollison and Holmgren 1978).

My 'shallow organics' has been subsequently written about as 'input substitution' organics (Rosset and Altieri 1997), and reframed as the sociological 'second order' change process of 'conventionalisation' (the industrialisation of organics by agribusiness; Buck et al. 1997; Darnhofer et al. 2010; Lockie and Halpin 2005), resulting in 'organic lite' (Guthman 2004a, b). The need for 'redesign' approaches has been extensively advocated, although—as with the concept of sustainability—the meanings implied cover a wide spectrum of agendas (e.g., Williams and Gascoigne 2003).

The shallow to deep progression was originally conceived as part of a larger transition from conventional agricultural systems (highly simplified, controlled and synthetic-input dependent), through efficiency (optimal use of inputs with minimal waste), to substitution (of more benign, primarily non-synthetic inputs), to ecologically redesigned/designed systems (the upper visual in Fig. 22.1). These latter systems are ecologically redesigned/designed to minimise problems and dependence on purchased inputs, increase resilience, and enable self-maintenance, self-regulation, sustainability, and ability to provide the needed ecosystem services and support for achieving the well-being of all.

This framework is often referred to as the ESR (Efficiency-Substitution-Redesign) progression (Hill and MacRae 1995; Ikerd 2009; Lamine et al. 2014, Chap. 23). It has been, and can be, applied to diverse systems, beyond the pest management systems for which it was originally conceived. It has been effectively applied to the design of experiments (Bellon et al. 2010), research strategies and public policy (MacRae and TFPC 1999 McCullum et al. 2005; Sylvander et al. 2006). MacRae et al. (2010) have effectively used the ESR framework to propose numerous ways in which energy efficiency and greenhouse gas mitigation might be further improved on organic farms in Canada. Gliessman (1997) added a fourth level to the progression to address the need for associated social transformations and, subsequently, with colleagues, applied this model to 'conversion to sustainable agriculture' in general (Gliessman and Rosemeyer 2010). However, from my original and current perspective, institutional and sociocultural transformations must be considered at every stage in the ESR progression, and this progression applies to the forms that these sociological initiatives may take; so, adding it on as a fourth stage could be a barrier to considering transformation processes in the most progressive way.

To be able to conceptualise the progression in agroecosystem design and management from 'conventional' to 'redesign', as applied to pest management, the visuals shown in Fig. 22.1 were developed. The upper visual highlights the differences between the *efficiency, substitution* and *redesign* stages, and emphasizes the paradoxical point that the more effective that any efficiency and substitution initiatives

[2] This is an extension of Coleman's earlier 'plant-positive' approach to production; Coleman and Ridgeway (1983).

are, the more they are likely to protect and perpetuate the design and management characteristics of the systems that are the root causes of the problems. This is why, to be useful, these former strategies must serve as stepping stones to the improved design and management of the systems involved.

The aim of the lower image in Fig. 22.1 is to help visualise a progression across the following six overlapping stages, from conventional pest control to the redesign approach, and to make clear the degree to which the *redesign* (Integrated Agroecosystem Design and Management) stage differs from the still dominant *conventional* stage, with its dependence on pesticide interventions (along a spectrum from inefficient to efficient use):

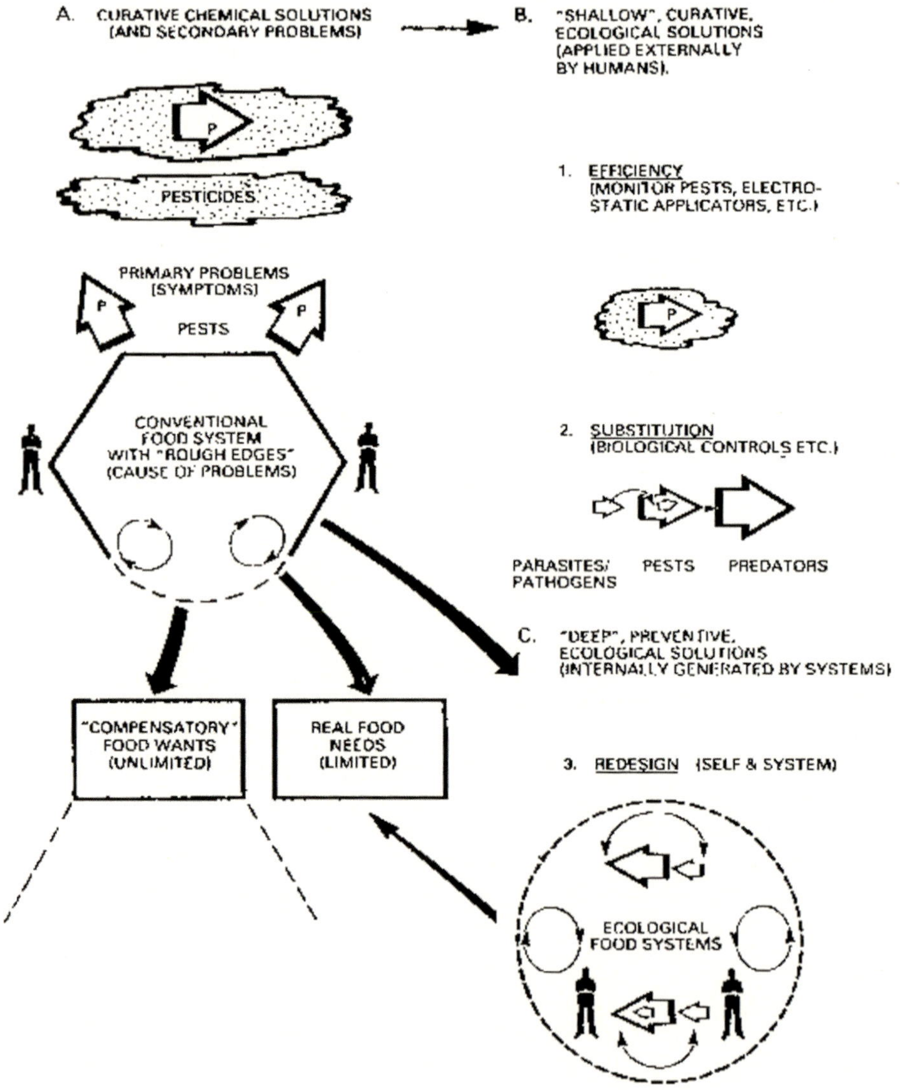

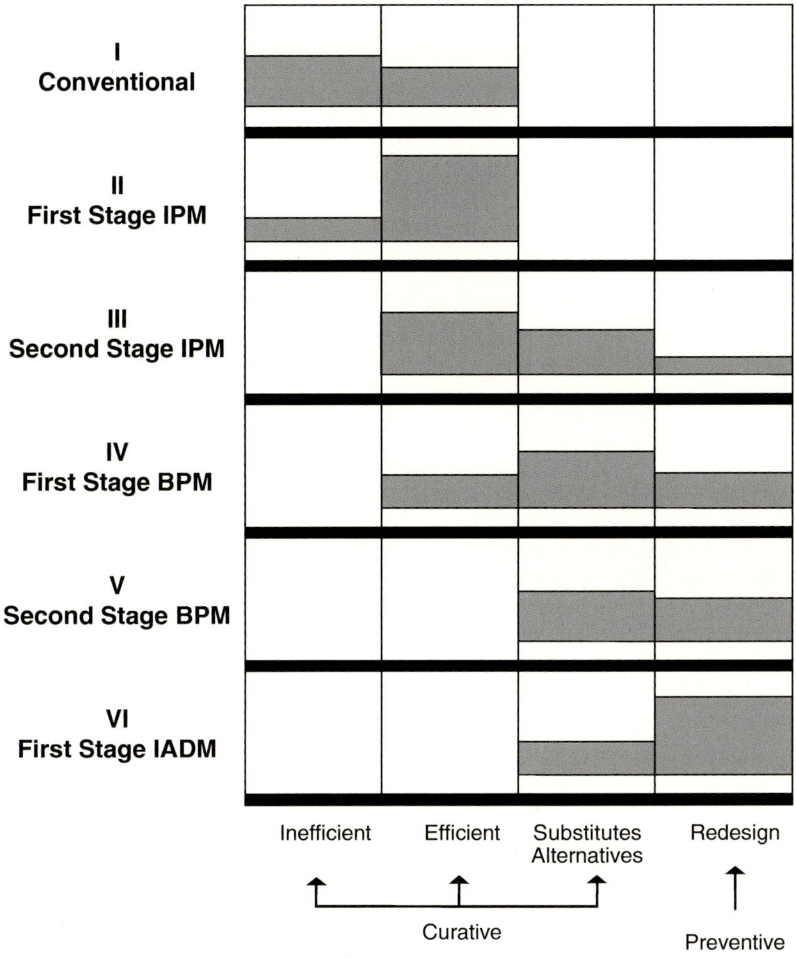

Fig. 22.1 Conceptual model for progressive implementation of improved pest control within agroecosystems (upper visual from Hill (1984b), and lower visual from Hill et al. 1999)

I: *Conventional Pest Control*: prophylactic, calendar-based and reactive, often inefficient use of broad-spectrum pesticide interventions;
II: *Integrated Pesticide Management* (First Stage Integrated Pest Management): with most of the inefficient use of pesticides eliminated;
III: *Genuine Integrated Pest Management* (Second Stage IPM of Prokopy 1994): efficient pesticide use is integrated with various, more benign substitutes such as biological controls, and with some ecosystem redesign such as groundcover management.
IV: *First Stage Biointensive Pest Management*: biological controls dominate the approach, together with compatible pesticides used only as a last resort, and supportive redesign strategies;

V: *Second Stage Biointensive Pest Management*: with biologicals being used as curative interventions instead of pesticides, together with ecosystem redesign (Bugg 1992);
VI: *Proactive Integrated Agroecosystem Design and Management* (IADM; similar to Hill's 1984b, 2004a redesign approach). In this stage the aim is to proactively solve all pest problems through advanced ecosystem design/redesign and management, with the least disruptive curative interventions being used only in emergencies. Such redesign might include the use of several cultivars with complementary functions (resistant, trap and insectary plantings), the design and management of groundcovers and mulches, insectary islands, strips, field borders and adjacent areas (to trap or repel overwintering pests and to support natural controls), the careful balancing of plant nutrients, including trace elements, and the appropriate timing of all operations to favour the crops and natural controls and not the pests (Altieri et al. 1983; Bugg 1992; Prokopy 1994; Hill 2004a; Ratnadass et al. 2011). This kind of system-transformation thinking is advocated for all other parts of the food system and beyond.

Apart from the lack of funding for research and development[3], and lack of appropriate educational programs, extension and other support services (Warner 2007), the other main institutional barriers to advancing this needed progression have been and continue to be:

- the globalised nature of the industrialised world's food system, with its high levels of concentration, simplification (dependence on fewer and fewer cultivars and strains), and demand for cosmetic perfection of the produce;
- the lack of understanding and support for farmers and rural communities by increasingly urbanised populations, and governments' associated *cheap food policies*, which favour urban over rural communities; and
- government regulations and subsidies related to purchased inputs and control processes (MacRae 2011).

In addition to these largely institutional barriers to genuine progress are the underlying psychological and behavioural barriers of feelings of disempowerment and associated lack of awareness and confused (often compensatory) visions and values (Hill 1991; Sattmann-Frese and Hill 2008). Because most proposals for improvement neglect these underlying roots of so many of society's problems, progress has generally been disappointing and predictably limited.

Despite the basic design similarity between most current organic and conventional farms, one of the earliest (in the 1970s) comprehensive comparisons found that in the Midwestern (Corn Belt) states of the USA, organic farms used two-fifths the amount of energy of conventional ones, with comparable levels of produce and profits (Lockeretz 1978; Lockeretz et al. 1984). Such considerable savings of energy by organic farmers for their host nations, together with most of the other

[3] Until very recently, for nearly 100 years, virtually all research funding had been channeled to conventional agriculture.

benefits of organic approaches (Rodale Institute 2011), still remain largely unrecognised and unrewarded by governments. Although such savings indicate what has been achieved in the past, it is likely that current (non-deep) organic farms that are more dependent on inputs and that are less efficiently managed, would be unlikely to achieve such savings. More importantly, however, deep organic (eco-designed/redesigned) farms may achieve much greater energy savings and result in many other positive benefits.

Because of the potential for improved agroecosystem design and management, the focus in the rest of this chapter will be on how the benefits of 'deep organics' might be achieved, rather than on comparisons involving the largely maldesigned and unsustainable existing systems.

The primary prerequisites for the sound design of managed ecosystems are a profound and comprehensive understanding of their components and the relationships between them, and of the ecological processes that occur within natural and managed ecosystems. Ideally, such relevant competence needs to be gained through experience, experimentation and formal studies in ecology, natural resources management, agriculture and psychology (all in the broadest possible sense). Other areas that require some competence include engineering (particularly with respect to the effective selection, design, modification, repair, maintenance and operation of machinery), economics, business management, marketing (I have found this to be a particular weakness of many organic farmers), communications, politics and sociology.

Below, I first argue for the relevance of considering psychology and personal development when working with progressive change within food systems. I then propose a comprehensive framework for 'testing' all proposals for change, briefly describe ecological foundations for the design and management of sustainable systems, apply these concepts to pest control and prevention, examine key psychological factors that influence progress and, finally, describe an innovative approach to taking a first step (committing to achievable sustainable goals) in implementing progressive change.

22.2 Relevance of Psychology and Personal Development

It is important to understand what might enable or, conversely, inhibit the creation and implementation of high quality designs.

Humans have evolved psychosocially throughout our species' history; from a time when child rearing practices commonly included infanticide, through periods of enslavement and abandonment, to the socialisation of children (through manipulative education and training—still the dominant situation in most of the industrialised world)—to the beginnings of enabling next generations to realise *their* potential, rather than the agendas of the previous generation for them (deMause 2002)[4]. All of the stages prior to this last one lead to variable

[4] See also: Hill (2004b).

levels of *reactive living* (from the outside-in), recognisable as diverse expressions of compliance and rebellion. This is in contrast to the *enabling* stage[5], which supports *proactive living* (from the inside-out). Because 'reactive living' requires compromise and is limited and distractive, the predictable general outcomes are disempowerment, loss of awareness, and distortion of one's vision and values, as well as the emergence of unjustifiable fears and anger, and a wide range of compensatory behaviours, including attraction to external symbols of power (to compensate for internal feelings of powerlessness), an unreasonable need to control, growing unsatisfiable consumption, and a mixture of irrational over-reaction to minor disruptions and a tendency to postpone needed action. The characteristics of 'proactive living' are essentially the opposite of this, and because of this, they are likely to provide a much better foundation for effective eco-design/redesign and responsible system management.

With this awareness, it is easy to understand the driving forces for many of the dominant aspects of conventional agriculture. These include, in particular, conventional agriculture's large-scale, simplification, fragmentation, focus on control, and its associated low resilience and vulnerability in the face of change, its unsustainability, and the associated common denial of consequences, particularly of those that, for the detection of impacts, require an understanding of complex interrelationships among multiple factors and large time and space scales. Sadly, most large-scale, substitution-based organic farms share many of the negative characteristics of conventional farms. In contrast to this, support for 'proactive living' has the potential to enable the design and management of food systems with opposite characteristics, including complex multi-story polycultures and other designs with high resilience, as well as qualities of self-regulation, self-maintenance, diverse and fulfilling work experiences, and ecological sustainability (Hill 1991, 2011b).

Of similar importance to effective eco-design/redesign and sustainable system maintenance is having a high level of relational competence, an area that deserves much greater attention in human development. Neglect of this has played a major role in the development of most common personal, social and environmental problems, which invariably are, to a large extent, relational problems (Josselson 1996).

Suffice it to say here that if we want to have genuinely sustainable and well-being-enabling managed ecosystems, we will have to pay much more attention to how we raise and educate our children, to social attitudes towards worth and status, and to the design of institutional structures and processes so that they can be supportive of, and not barriers to, the achievement of higher values and goals. Without such awareness and change, food systems will predictably continue to be unsustainable, and the bulk of the population will remain in denial of this being the case. It should be noted that denial is often the first stage in a change process, and also one of the main survival strategies that humans use to deal with oppression and physical and psychological 'hurt'. Consequently, I would not be surprised if some readers may already be responding in this way to some of the ideas being presented

[5] deMause (2002) prefers the term helping.

here. A down side of this is that denial is frequently the main barrier that must be overcome when striving to achieve significant progress. Laing (1971) described this process of 'protective denial' as a *double hypnosis* (in which we are hypnotised twice: firstly, into accepting pseudoreality as reality and, secondly, into believing that we have not been hypnotised). This is why denial is such a formidable barrier to have to deal with when trying to implement progressive change (and possibly in engaging positively with some of the ideas in this chapter).

22.3 Goals, Testing Questions, and Strategy Options for Enabling Progressive Change

To achieve improvements in food systems, in addition to developing the kind of psychological awareness described above, we will also need to have clear goals and *testing questions* (Savory and Butterfield 1999) to help keep us on course. To be adequately comprehensive, such questions need to cover the personal, social, environmental and general aspects of any situation. My provisional list of such *testing questions* is provided in Box 22.1.

Box 22.1. Testing questions for 'challenging' all understandings, ideas and initiatives when designing and implementing 'progressive' change proposals (modified from Hill 2005a). Comparable criteria in Box 22.1 are underlined, in bold or in italics. For example, mutual relationships are implied between personal, social and natural capital, and between personal, cultural and ecosystem development; all of these must be taken into account in all eco-design/redesign initiatives.

To what extent, and in what ways, do the initiatives (policies, programmes, plans, regulations, decisions, initiatives, etc.) support the sustainable well-being of all, in relation to the following four main areas?

Personal considerations

1. *Building and maintaining personal capital*—**personal sustainability**: empowerment, awareness, creative visioning, values and worldview clarification, acquisition of essential literacies and competencies, responsibility, well-being and health maintenance, vitality and *spontaneity*?
2. Home and ecosystem maintenance: caring, loving, responsible, mutualistic, *negentropic (capital building) relationships* with diverse others (valuing equity & social justice), other species, place and planet ['negentropic' is the opposite of 'entropic': breaking down]?
3. Lifelong learning: positive total life-cycle *personal development* and 'progressive' change?

Socio-political considerations

4. *Building and maintaining social capital*—**cultural [including economic] sustainability**: trust, accessible, collaborative, responsible, creative, celebrational, *life-promoting community and political structures and processes*?
5. Inter-cultural and interpersonal capital: the valuing of 'functional' high *cultural diversity*, mutually beneficial relationships and the required competencies?
6. Co-evolutionary change: positive *cultural development* and evolutionary change that benefits all involved?

Environmental considerations

7. *Building and maintaining natural capital*: effective *ecosystem functioning*, **resilience** and **ecological sustainability**?
8. 'Functional' high *biodiversity*, and prioritised use and conservation of resources?
9. Positive *ecosystem development* and co-evolutionary change?

General considerations

10. **Proactive** (vs. reactive) **design/redesign** (vs. just efficiency & substitution) and **small meaningful collaborative initiatives** that one can guarantee to carry through to completion (vs. heroic, Olympic-scale, exclusive, high-risk ones), and **public celebration** at each stage—to facilitate the spread of initiatives—thereby making well-being and environmental caring 'contagious'?
11. A focus on key opportunities and **windows for change** (contextually unique change 'moments' & places)?
12. **Effective evaluation and monitoring**: (broad, long-term, as well as specific & short-term) by identifying and using **integrator indicators** and **testing questions**, and by being attentive to all **feedback and outcomes** (& redesigning future actions & initiatives accordingly)?

Reference to such a comprehensive list—in contrast to just considering yield and profit—makes it clear that humans have hardly scratched the surface of achieving our potential for responsible decision making and for enabling progressive developments within farms and food systems (and, indeed, within all other aspects of one's life). Paradoxically, because of this, I consider that humans have reason to be hopeful about the future.

Taking such a proactive, multifaceted, integrated, whole-system approach is central to my work as a social ecologist (Hill 2011a). I define social ecology as *the study and practice of personal, social [including all economic, political and other institutional considerations], and ecological sustainability and change, based on the critical application and integration of ecological, humanistic, relational, community and 'spiritual' values to enable the sustained wellbeing of all.* I consider that this (Australian) version of social ecology provides the most comprehensive foundation and framework for the ideas that I am advocating here (Hill 2004b), and it is the only one to explicitly include reference to the 'personal', without which all problem creation and solving tend to be regarded as the responsibility of others, as has so often happened with the use of the 'triple bottom line' (the economy, society and the environment), because it neglects 'personal' considerations.

When working with change, in addition to reflecting comprehensively on one's goals and regarding all outcomes as valuable sources of feedback, it is important to also consider the general strategy options that are available, and then the details of what/which, where, when and how we might go about the process of eco-design/re-design and implementation. Important—particularly political—strategy options that should be considered, and that will require coordination, include the full spectrum of supports, rewards and penalties (Hill 1999), and of the various economic and policy instruments (Hill 2006; Hill and MacRae 1995), plus the extensive strategic use of the media and of the increasing number of global computer networks. A recent analysis of policy options that need to be considered by bureaucrats and politicians in industrialised countries has been provided by MacRae (2011), although it does not consider the deeper needs for fundamental change that are being advocated here.

To address the many fundamental problems that the dominant economic approaches create for food systems, Georgescu-Roegen (1971) has argued that humans need to develop two economic systems, which he called *factory economics* (for technology-based systems) and *farm economics* (for living systems). He made the point that whereas the speed with which products such as cloth can be woven (by machines) has increased over time, the rate at which seeds planted by early agriculturists germinate and grow has not changed, and that it is the subjection of such natural processes to machine-based systems of thinking that are predictably causing so many of the problems that the managers of natural systems face. However, from my perspective, because people work in 'factories', it makes more sense to consider ways to enable a transition from factory to farm economics for all systems, including technology-based ones, and, more importantly, to start to enable the more challenging transition from an 'economic bottom line' (and system of values) to a 'life and well-being-enabling bottom line' that all institutional structures and processes (including, but not privileging, economics) are redesigned to be in the service of. McMurty (2002) refers to this as the 'life economy' and has discussed the philosophical underpinnings of elevating 'life capital' above all other forms of capital.

For many, the decision to preferentially purchase organically-grown produce is a significant step towards this needed change in values. When such consumers are asked why they are willing to pay more for organic produce, most reply that they consider that it is better for the health of both themselves and the environment (Aertsens et al. 2011). This may be interpreted as reflecting a decision to value the maintenance and well-being of living systems over cheapness, superficial appearance and—in the broader context—the dominant emphasis in most industrialised societies on growth in consumption, productivity, yield and profit.

22.4 Ecological Foundations and Models for the Design and Management of Sustainable Systems

Underlying all eco-design/redesign and management initiatives, there must be a comprehensive understanding of ecology, including the ecological foundational concepts presented in Box 22.2 (Hill 2005b):

Foundational Concepts for a comprehensive understanding of ecology:

- limiting factors and their substitutes;
- microhabitats, niches and territoriality;
- time and space specificity;
- numbers, biomass, energy flow and the specifics of resource partitioning and budgeting;
- guilds, roles and keystone species;
- system maintenance and service functions;
- resilience and ecosystem resistance;
- succession, developmental and intergenerational change;
- feedback loops, co-evolutionary processes, altruism and group selection;
- edge effect and boundary phenomena;
- functional diversity, system stability and homeostasis;
- specialists and generalists (eurytypic and stenotypic expressions of life-style), and r and K strategists;
- entropy and negentropy;
- specific indicators and integrator indicators;
- synergy and mutualism;
- catalysis and amplification;
- non-linearity, cyclic and threshold relationships;
- integrated web-like relationships;
- homeostatic, self-regulative and regenerative processes;
- adaptation, addiction, allergy and system degeneration (the result of adaptive processes, over time, becoming maladaptations); and
- hierarchical and systems phenomena at every level.

Information about these and other related phenomena may be found in most major ecology textbooks: Andrewartha HG, Birch LC (1984) *The Ecological Web*. University of Chicago Press, Chicago, IL; Begon M, Harper JL, Townsend CR (1996) *Ecology: Individuals, Populations and Communities*, 3rd edn. Blackwell Scientific, Cambridge, MA; Krebs CJ (2009) *Ecology: The Experimental Analysis of Distribution and Abundance*, 6th edn. Benjamin Cummings, San Francisco, CA; Odum EP, Barrett GW (2004) *Funda-*

mentals of Ecology. Brooks Cole, Belmont, CA; Smith TM, Smith RL (2009) *Elements of Ecology*, 7th edn. Pearson International Edition. Pearson Benjamin Cummings, San Francisco, CA); and also in the various dictionaries of ecology: Allaby M (ed) (1998) *A Dictionary of Ecology*, 2nd edn. Oxford University Press, Oxford; Lincoln R, Boxshall G, Clark P (1998) *A Dictionary of Ecology, Evolution and Systematics*, 2nd edn. Cambridge University Press, New York.

Such a long list of ecological concepts is provided to illustrate the potential—through their application—for enabling the improved design and management of farms to achieve greater sustainability, and it should be noted that this is an indicative rather than a comprehensive list. To date, few of these understandings have been taken into account in the design and management of current farms, although there is a growing literature on the need to learn from natural systems, referred to as 'mimicry', 'biomimicry' and 'ecomimicry' by some (Ewel 1999; Lefroy et al. 1999; Doré et al. 2011; Farley et al. 2011; Malézieux 2011), ecology (Thomas and Kevan 1993; Altieri 1987; Gliessman 1997), and traditional ecological knowledge (Martin et al. 2010). These approaches are part of the much needed strategy of substituting knowledge and skills for purchased, imported and disruptive inputs, most of which must be endlessly reapplied.

In the area of agroecosystem eco-design/redesign, the most significant contribution, from my perspective, has been P. A. Yeomans' development in Australia of the *Keyline* system for sustainable landscape design and management (Yeomans 1958, 1971, 1978)[6]. It provides a model of both the kind of thinking that is required and of what the application of such thinking can achieve. Predictably, some of Yeomans' approaches inspired and were integrated into *Permaculture* by its Australian creators (Holmgren 2002; Mollison 1988; Mollison and Holmgren 1978)[7], who, with their followers, have been one of the main global promoters of integrated eco-design approaches for food and energy systems. Sadly, rather than Yeomans' pioneering work[8] being used as a basis for ongoing design research and development, most agroecosystem research continues to be conducted within the simplified designs and management systems that comprise the roots of so many of conventional agriculture's problems, including its unsustainability.

Another Australian initiative that is worthy of note is Landcare (Campbell 1994; Hill 1999; Toyne and Farley 1989; Youl et al. 2006), which has now spread to over

[6] See also: Hill (2005c, 2006); Mulligan and Hill (2001); Yeomans (2005).

[7] See also: Hill (2005c, 2011b). A lecture based on the ideas presented in this chapter is available at: http://www.youtube.com/watch?v=mzY1eZLwOdk.

[8] The failure to appreciate the significance of Yeomans' design work in Australia is apparent from the recent rejection of an application to preserve his farm as a Heritage site where design research could have been continued, and its likely conversion into a housing estate (http://nrdcaa.blogspot.com/).

a dozen other countries (Catacutan et al. 2009), and has enabled many conventional producers to take important steps towards conserving resources and biodiversity, and becoming more sustainable (e.g., Fenton 2010). As with organic systems of production, examples of Landcare extend from 'shallow' to 'deep' expressions.

Examples of other ecologically-based agroecosystem design initiatives from around the world include:

- Voisin and Lecomte's (1962) rotational grazing;
- Jeavons' (1974) integration of Biodynamics (Koepf 1989) and French Intensive Raised Beds (culture maraîchère [vegetable production]; Weathers 1909);
- Jackson's (2002) Natural Systems Agriculture (originally developed in 1977);
- various initiatives of the New Alchemy Institute (e.g., Quinney 1984; Todd and Todd 1984);
- Fukuoka's (1985) Natural Farming;
- Savory and Parson's (1980) Holistic Resource Management[9];
- Andrews' (2008) Natural Sequence Farming; and
- Altieri's (1987)[10] synthesis of key ideas in agroecology.

Additional design ideas may be gained by consulting the parallel habitat eco-design literature[11]. Interestingly, the greatest developments in ecomimicry have been made by architects and non-agriculturists[12].

22.5 Ecological Design and Management for Pest Prevention

Many of my eco-design/redesign efforts have been focused on expanding and promoting cultural approaches to pest prevention and control (Hill 1984b, 2004b; Hill et al. 1999). The strategies involved are exemplary of the kind of design/redesign approach that I am advocating. In this case, they focus on identifying possible initiatives at every stage in the crop and livestock production process that might support the crop or livestock and (i) make it less attractive and (ii) more resistant to its pests, and (iii) make the habitat more supportive of the natural controls of the pest, and (iv) less supportive of the pests. For crops, this requires a consideration of the effects on these four aims of each of the following activities:

- cultivar and companion plant selection;
- site selection and preparation;

[9] See also: Savory and Butterfield (1999); Butterfield et al. (2006).

[10] See also: Gliessman (1997), Warner (2007), Wezel et al. (2009).

[11] Outstanding examples include: Aberley (1994); Alexander (2002); Alexander et al. (1977); Glickson and Mumford (1971); Papanek (1995); Soleri (2006); Todd (2006); Van der Ryn and Cowan (2007).

[12] Yeang (2013); see also: Marshall (2009).

- planting design and procedures;
- site maintenance (including cultivation, use of mulches, soil amendments, fertilisation, irrigation and drainage); and
- harvesting and distribution.

Such complex integrated approaches to pest prevention, which once established have the potential of providing ongoing control, contrast with the naive reactive/curative linear strategy of applying pesticides, which can only ever provide temporary control and which have numerous disadvantages, most of which are neglected by current decision-makers and practitioners in conventional agriculture. These disbenefits include the inevitable selection for pesticide resistance, harmful side effects on beneficial organisms within both the agroecosystem and other ecosystems, and on those involved in the production of the food and its consumption, as well as the perpetuation of systems of production that have low resilience, high vulnerability, and that are heavily resource dependent and inherently unsustainable (Pimentel 2005).

There is an extensive literature on the various components of *alternative* approaches to pest management. However, dealing comprehensively with whole systems is still at an early stage of development (Hill et al. 1999; Deguine and Penvern Chap. 06). Lamine's use of my ESR framework (Fig. 22.1), together with a sociological study of farmers' trajectories (Lamine and Bellon 2009; Lamine 2011; Lamine et al. 2010), along with Constance's work (2010) have provided important contributions to this area.

22.6 Psychological Roots of Unsustainable and Sustainable Initiatives

Sadly, most proposals and initiatives for progressive change that receive funding fail to meet their goals, and some create more problems than they solve. A better understanding of the influence of psychological factors can, I believe, provide some explanations of this. Thus, if the *change agents* involved in sustainability initiatives are 'living reactively', and are feeling anxious about doing something significant, they will be more likely to propose large, impressive (yet, invariably unachievable) goals. These will usually need more resources than are available, and longer time frames for implementation than can be supported. Most disturbing is that many such projects are often abandoned after the initial stage, which commonly just involves 'describing and measuring the problem' that is being addressed. Because the problem rarely needs such extra description, one cannot help but wonder if such studies are really (invariably subconscious) fear-based postponement and avoidance strategies. I suspect, however, that most projects are abandoned because the structural and procedural changes required to effectively address the causes of most problems would result in those in power losing their positions and privileges. Because of these reactive tendencies, rather than dreaming up such 'Olympic-scale' projects, it may help to focus on what I call 'small meaningful initiatives that one can guarantee to carry through to completion', and then doing all one can to enable others to learn from and copy such initiatives[13].

Paradoxically, it is likely that individuals who are 'living proactively'—who are more empowered, aware, clearer about their values, and with more realistic visions of better futures—are more likely to be willing to be 'anonymous' in the process of implementing change, and this, I believe, can help improve the probability of achieving genuine progress. My explanation for this is that such anonymity is most likely to be experienced if one is working in a large collaborative team (with rotating 'servant leadership'; Greenleaf 2002), over a long time-frame, and using low-powered, subtle and indirect, integrated, multi-faceted interventions (which may also be based largely on local, solar and renewable resources, be as self-regulating and self-maintaining as possible, repairable, and open to progressive improvement). These are essentially the qualities that are most likely to result in sustainable progressive change, i.e., the opposite of the more common single, simple, high-powered, expensive, imported, product-based interventions, which although providing short-term relief, invariably result in dependence and create further problems. Politically, this may be understood as progressing from democracy (with its naive dualism and inherent power struggles[14]) to co-operacy (collaborative pluralism in the service of well-being for all; Hunter et al. 1997).

22.7 Lying to Proactively Enable Prioritised, Achievable Progressive Change

As a facilitator, instead of using the usual 'visioning' exercises, I now ask individuals who want to achieve progressive change in a particular area to 'boldly lie' about changes that they have already successfully implemented in a particular area (knowing that everything they say is *fiction*). For example, when working with farmers, I might ask them to boldly lie about how they made their whole farm completely ecologically sustainable, and to describe in detail the initiative that they are most proud of achieving. This is an elegant strategy—based on psychological principles, explained below—that enables the participants to access the 'proactively living' (unwounded) part of their being. This part is able to think about and say what they (as well-functioning social beings) would really want to do. This is in contrast to what commonly happened when I asked them to visualise what they would like to achieve, when a subconscious battle would ensue between their 'reactive-living' (wounded) selves and their 'proactive-living self. What they then tended to say was something that they felt would be acceptable in the present company. In some cases, their aim was to not appear too radical; in others, it was to impress. Both cases are

[13] The best example and model I am familiar with for achieving this is from the health field: the Peckham Experiment in the UK (Williamson and Pearse (1980); Stallibrass (1989).

[14] For its effects on the Canadian food system, see: Koc et al. (2008).

not what, deep down, they really wanted to do and, as such, not things that they could make a commitment to achieving. Paradoxically, the opposite is the case of those initiatives that have their origin as lies! I have discussed some of the psychological bases for this approach elsewhere (Hill 1991)[15]. I encourage you, the reader, to try this for yourself, and to share your experiences with as many others as you can. You may be surprised by the results.

In workshops in which I use this 'lying' approach, we work in pairs, and after sharing our 'lies', we proceed to identify related initiatives that we have actually already implemented, or at least thought about or planned—this is to connect these with our lie(s)—then we identify possible barriers and needed resources, and ways to get around the former and gain access to the latter. Finally, we draw up a plan for implementation, emphasizing do-able actions, and all who are present serve as witnesses to one another's commitments. We also brainstorm ways to publicly celebrate the outcomes of our actions so that others may learn from our experiences. At subsequent meetings, we share our experiences, address any problems, celebrate outcomes, and plan further action.

My current 'lies' are that this facilitation technique has now been extensively adopted throughout the world, and that it has played a major role in enabling our species to take the next steps in our psychosocial evolution (deMause 2002), with significant benefits for the well-being of all.

22.8 Concluding Remarks

The complexity of the ideas and long lists of things to take into account in this chapter, while being possibly challenging for some, are intended: firstly, as a critique of the predominantly naive and fragmented thinking that still dominates most understanding, planning and action; secondly, to indicate the vast, largely untapped, potential for improvement; and thirdly, as an indicator of the urgent need to embrace the kinds of paradigm shifts—discussed here—that will be required for humans to be able to develop the competencies, technologies and institutional supports required for the improved eco-design/redesign and management of complex systems (to achieve sustainability).

For me, the two most important messages in this chapter are the need to include personal (psychological) and cultural (psychosocial) development considerations and initiatives in all efforts to enable progressive change, and the importance of emphasizing a deep, eco-design/redesign approach in the ongoing development of food and farming systems, and in the associated policies and regulations, research and educational programmes, extension and other supports, technology development and marketing strategies. I consider that only by doing these things will humans be able to make genuine, significant progress towards the kinds of goals mentioned in

[15] A lecture based on the ideas presented in this chapter is available at: (www.youtube.com/watch?v=mzY1eZLwOdk), and my research in this area is ongoing.

the introduction. Such progressive change should also acknowledge and build on the substantial contributions of organic farming to sustainability and well-being that have already been achieved.

I am hopeful and confident that the frameworks and initiatives discussed in this chapter, and the sharing and ongoing development of them, will enable our species to actually achieve genuine progressive change towards sustainability and well-being for all.

References

Aberley D (ed) (1994) Futures by design: the practice of ecological planning. New Society Publishers, Gabriola Island

Aertsens J, Mondelaers K, Verbeke W, Buysse J, Van Huylenbroeck G (2011) The influence of subjective and objective knowledge on attitude, motivations and consumption of organic food. Brit Food J 113(11):1353–1378

Alexander C (2002) The nature of order: an essay on the art of building and the nature of the universe (Book one: the phenomenon of life). Oxford University Press, New York

Alexander C, Ishikawa S, Silverstein M (1977) A pattern language. Oxford University Press, New York

Altieri M (1987, 2nd edn. 1995) Agroecology: the science of sustainable agriculture. Westview Press, Boulder

Altieri MA, Martin PB, Lewis WJ (1983) A quest for ecologically based pest management systems. Environ Manag 7(1):91–100

Andrews P (2008) Beyond the brink: Peter Andrews' radical vision for a sustainable Australian landscape. ABC Books, Ultimo

Bellon S, Desclaux D, Le Pichon V (2010) Innovation and research in organic farming: a multi-level approach to facilitate cooperation among stakeholders. In: WS1.8—Knowledge systems innovations and social learning in organic farming, 9th European IFSA Symposium, 4–7 July 2010, Vienna, Austria, pp 703–717

Buck D, Getz C, Guthman J (1997) From farm to table: the organic vegetable commodity chain in Northern California. Sociol Rural 37(1):3–20

Bugg RD (1992) Using cover crops to manage arthropods on truck farms. HortSci 27(7):741–745

Butterfield J, Bingham S, Savory A (2006) Holistic management handbook: healthy planet, healthy profits. Island Press, Washington. See also: www.holisticmanagement.org

Campbell A (1994) Landcare: communities shaping the land and the future. Allen and Unwin, St Leonards

Catacutan D, Neely C, Johnson M, Poussard H, Youl R (2009) Landcare: local action—global progress. World Agroforestry Centre, Nairobi

Coleman E (1999) Discussion session 1(The effects of plant health and soil health on susceptibility to pests). In: Stoner KA (ed) Alternatives to insecticides for managing vegetable insects (Proceedings of a Farmer/Scientist Conference, Dec 6–7 1998, New Haven, CT). Natural Resource, Agriculture and Engineering Service, Cooperative Extension, Ithaca, NY, p 7

Coleman E (2004) Can organics save the family farm? The Rake Mag [2002–2008, after which it was renamed Secrets of the City], September, Minneapolis, MN. http://archives.secretsofthecity.com/magazine/reporting/features/can-organics-save-family-farm. Accessed 5 Feb 2005

Coleman E (2009) The Winter Harvest handbook: year round vegetable production using deep organic techniques and unheated greenhouses. Chelsea Green Publishing, White River Junction

Coleman EW, Ridgeway RL (1983) Role of stress tolerance in integrated pest management. In: Knorr D (ed) Sustainable food systems. AVI Publishing, Westport, pp 124–142

Constance DH (2010) Sustainable agriculture in the United States: a critical examination of a contested process. Sustainability 2(1):48–72

Coombes B, Campbell J (1998) Dependent reproduction of alternative modes of agriculture: organic farming in New Zealand. Sociol Rural 38(2):127–145

Darnhofer I, Lindenthal T, Bartel-Kratochvil R, Zollitsch W (2010) Conventionalisation of organic farming practices: from structural criteria towards an assessment based on organic principles: a review. Agron Sustain Dev 30:67–81

deMause L (2002) The emotional life of nations. Other Press, New York. www.psychohistory.com

Doré T, Makowsi D, Malézieux E, Munier-Jolain N, Tschamitchian M, Tittonell P (2011) Facing up to the paradigm of ecological intensification in agronomy: revisiting methods, concepts and knowledge. Eur J Agron 34:197–210

Ewel JJ (1999) Natural systems as models for the design of sustainable systems of land use. Agrofor Syst 45:1–21

Farley J, Schmitt FA, Alvez J, de Freitas Jr NR (2011) How nature can transform agriculture. Solutions 2:64–73

Fenton JD (2010) The untrained environmentalist. Allen and Unwin, St Leonards

Fukuoka M (1985) The natural way of farming: the theory and practice of green philosophy. Japan Publications, Tokyo

Georgescu-Roegen N (1971) The entropy law and the economic process. Harvard University, Cambridge

Glickson A, Mumford L (1971) The ecological basis of planning. Martinus Nijhoff, The Hague, Netherlands

Gliessman SR (1997) Agroecology: ecological processes in sustainable agriculture (2nd edn. 2007, Agroecology: the ecology of sustainable food systems). CRC Press, Taylor & Francis, New York

Gliessman SR, Rosemeyer M (eds) (2010) The conversion to sustainable agriculture: principles, processes, and practices. CRC Press, Taylor & Francis, New York

Greenleaf RK (2002) Servant leadership: a journey into the nature of legitimate power and greatness, 25th Anniversary Edition. Paulist Press, Mahwah

Guthman J (2004a) The trouble with 'organic lite' in California: a rejoinder to the 'conventionalisation' debate. Sociol Rural 44(3):301–316

Guthman J (2004b) Agrarian Dreams? The paradox of organic farming in California. University of California Press, Berkeley

Hill SB (1976) Conditions for a permanent agriculture. Maine Org Farming Gard 3(5): 8–9 [Reprinted in the Soil Assoc Q Rev 3(2):1–4, 1977]

Hill SB (1984a) Deep organics. Nat Farming [USA] Fall: 3, 10

Hill SB (1984b) Ecological pest control: confronting the causes. Int J Biosoc Res 6:1–3

Hill SB (1985) Redesigning the food system for sustainability. Altern 12(3/4):32–36

Hill SB (1991) Ecological and psychological pre-requisites for the establishment of sustainable prairie agricultural communities. In: Martin J (ed) Alternative futures for prairie agricultural communities. Faculty of Extension, University of Alberta, Edmonton, pp 197–229

Hill SB (1998) Redesigning agroecosystems for environmental sustainability: a deep systems approach. Syst Res Behav Sci 15:391–402

Hill SB (1999) Landcare: a multi-stakeholder approach to agricultural sustainability in Australia. In: Dragun AK, Tisdell C (eds) Sustainable agriculture and environment: globalisation and the impact of trade liberalisation. Edward Elgar, Cheltenham, pp 125–134

Hill SB (2004a) Redesigning pest management: a social ecology approach. In: Clements D, Shrestha A (eds) New dimensions in agroecology. Haworth Press, Binghamton, pp 491–510

Hill SB (2004b) Camping to change the world. Austr J Outdoor Educ 8(1):3–10

Hill SB (2005a) Social ecology as a framework for understanding and working with social capital and sustainability within rural communities. In: Dale A, Onyx J (eds) A Dynamic balance: social capital and sustainable community development. University of British Columbia Press, Vancouver, pp 48–68

Hill SB (2005b) Critically acknowledging, celebrating and enabling further development in 'Deep Organics' contributions to sustainability and wellbeing: a Social Ecology perspective. Keynote to IFOAM 2005 Conference [Adelaide, SA; 20–23 Sept] web; pp 12. www.ifoam.org

Hill SB (2005c) Sustainable living through permaculture: a social ecology perspective (Keynote to the 8th Australian Permaculture Convergence, Melbourne, 8–15 April; available at: http://organicdesign.co.nz/Sustainable_Living_through_Permaculture:A_Social_Ecology_Perspective

Hill SB (2006) Enabling redesign for deep industrial ecology and personal values transformation: a social ecology perspective. In: Green K, Randles S (eds) Industrial Ecology and Spaces of Innovation. Edward Elgar, London, Chapter 12, pp 255–271

Hill SB (2011a) Social ecology: an Australian perspective. In: Wright D, Camden-Pratt C, Hill S (eds) Social ecology: applying ecological understanding to our lives and our planet. Hawthorn Press, Stroud, UK. Powerpoint presentations on applied social ecology are available at: www.stuartbhill.com and www.scribd.com/doc/55937783

Hill SB (2011b) Afterword: four key features of permaculture (applicable to 'everything'); and an opportunity for the future (also applicable to 'everything'). In: Dawborn K, Smith C (eds) Permaculture pioneers: stories from the new frontier. Melliodora Publishing, Hepburn, VIC

Hill SB, MacRae R (1995) Conceptual frameworks for the transition from conventional to sustainable agriculture. J Sustain Agric 7:81–87

Hill SB, Vincent C, Chouinard G (1999) Evolving ecosystems approaches to fruit insect pest management. Agric Ecosyst Environ 73:107–110

Holmgren D (2002) Permaculture: principles & pathways beyond sustainability. Holmgren Design Services, Hepburn

Hunter D, Bailey A, Taylor B (1997) Co-operacy: a new way of being at work. Tandem Press, Birkenhead

Ikerd J (2009) Fundamentals of environmental responsibility: ensuring a sustainable environment for global food production. Institute of food technologists annual meeting, Anaheim, CA, June 9, 2009. http://web.missouri.edu/~ikerdj/papers/California%20IFT%20Environmental%20Responsibiity.htm. Accessed 10 Jan 2010

Jackson W (2002) Natural systems agriculture: a truly radical alternative. Agric Ecosyst Envir 88:111–117

Jeavons J (1974) How to grow more vegetables than you ever thought possible on less land than you can imagine. Ecology action of the Midpeninsula, 2225 El Camino Real, Palo Alto, CA 94306 [6th edn., 2002; Ten Speed Press, Berkeley, CA], 240p

Josselson R (1996) The space between us: exploring the dimensions of human relationships. Sage, Thousand Oaks

Koc M, MacRae R, Desjardins E, Roberts W (2008) Getting civil about food: the interactions between civil society and the State to advance sustainable food systems in Canada. J Hunger Environ Nutr 3(2):122–144

Koepf HH (1989) The biodynamic farm: agriculture in the service of the earth and humanity. Anthroposophic Press, Hudson

Laing RD (1971) The politics of the family. Penguin, New York

Lamine C (2011) Transition pathways towards a robust ecologization of agriculture and the need for system redesign: cases from organic farming and IPM. J Rural Stud 27(2):209–219

Lamine C, Bellon S (2009) Conversion to organic farming: a multidimensional research object at the crossroads of agricultural and social sciences. A review. Agron Sustain Dev 29:97–112. doi:10.1051/agro:2008007

Lamine C, Barbier M, Blanc J, Buurma J, Haynes I, Lehota J, Maraccini E, Noe E , Paratte R, Szabo Z, Wierzbicka A (2010) Reducing the dependence on pesticides: a matter of transitions within the whole agri-food system, in WS4.4 – Transitions towards sustainable agriculture: from farmers to agro-food systems, 9th European IFSA Symposium, 4–7 July 2010, Vienna (Austria), pp 1943–1954

Lamine C, Navarrete M, Cardona A (2014) Transitions towards organic farming at the farm and at the local scales: the role of innovative production and organisational modes and networks. In: Bellon S, Penvern S (eds) Organic farming, prototype for sustainable agricultures. Springer, Chap. 23, pp 423–438

Lefroy EC, Hobbs RJ, O'Connor MH, Pate JS (1999) Agriculture as a mimic of natural ecosystems. Current plant science and biotechnology in agriculture. Vol. 37. Kluwer Academic Publishing, Dordrecht

Lockeretz WR (1978) Economic and resource comparison of field crop production on organic farms and farms using conventional fertilization and pest control methods in Midwestern United States. In: Besson JM, Vogtmann H (eds) Towards a sustainable agriculture. Proc. IFOAM Int. Conf., Sissach, Switzerland, 1977. Verlag Wirz Aarau, Switzerland, pp 157–169

Lockeretz W, Shearer G, Kohl DH (1984) Organic farming in the Corn Belt. Sci 211:540–547

Lockie S, Halpin D (2005) The conventionalisation thesis reconsidered: structural and ideological transformation of Australian organic agriculture. Sociol Rural 45(4):284–307

Lynch DH, MacRae R, Martin RC (2011) The carbon and global warming potential impacts of organic farming: does it have a significant role in an energy constrained world? Sustain 3:322–362

MacRae R (2011) A joined-up food policy for Canada. J Hunger Environ Nutr 6:1–33

MacRae R, Toronto Food Policy Council (1999) Not just what, but how: creating agricultural sustainability and food security by changing Canada's agricultural policy making process. Agric Hum Values 16:187–201

MacRae RJ, Lynch D, Martin RC (2010) Improving the energy efficiency and GHG mitigation potential in Canadian organic farming systems. J Sustain Agric 34(5):549–580

Mäder P, Fließbach A, Dubois D, Gunst L, Fried P, Niggli U (2002) Soil fertility and biodiversity in organic farming. Science 296(5573):1694–1697

Malézieux E (2011) Designing cropping systems from nature. Agronomy for sustainable development. http://www.springerlink.com/content/x810823737671q46/fulltext.pdf. Accessed 3 Jan 2012

Marshall A (2009) Wild design: ecofriendly innovations inspired by nature. North Atlantic Books, Berkeley

Martin JF, Roy ED, Diemont SAW, Ferguson BG (2010) Traditional ecological knowledge (TEK): ideas, inspiration, and designs for ecological engineering. Ecol Eng 36:839–849

McCullum C, Desjardins E, Kraak VI, Ladipo P, Costello H (2005) Evidence-based strategies to build community food security. J Am Diet Assoc 105:278–283

McMurty J (2002) Value wars: the global market versus the life economy. Pluto Press, London

Mollison B (1988) Permaculture: a designer's manual. Tagari Books, Tyalgun

Mollison B, Holmgren D (1978) Permaculture one: a perennial agriculture for human settlements. Transworld (Corgi, Bantam), Melbourne

Mulligan M, Hill SB (2001) Ecological pioneers: a social history of Australian ecological thought and action. Cambridge University Press, Melbourne (Ch. 8)

Naess A (1973) The shallow and the deep, long-range ecology movement. Inquiry 16:95–100

Papanek V (1995) The green imperative: ecology and ethics in design and architecture. Thames and Hudson, London

Pimentel D (2005) Environmental and economic costs of the application of pesticides primarily in the United States. Environ Dev Sustain 7:229–252

Prokopy RJ (1994) Integration in orchard pest and habitat management: a review. Agric Ecosyst Environ 50(1):1–10

Quinney J (1984) Designing sustainable farms. Mother Earth News 88(Jul/Aug):54–65

Ratnadass A, Fernandes P, Alevino J, Habib R (2011) Plant species diversity for sustainable management of crop pests and diseases in agroecosystems: a review. Agron Syst Dev 31:31. doi:10.1007/s13593-011-0022-4

Rodale Institute (2011) The farming systems trial: celebrating 30 years. Rodale Institute, Kutztown

Rosin C, Campbell H (2008) Beyond bifurcation: examining the conventions of organic agriculture in New Zealand. J Rural Stud 25(1):35–47
Rosset P, Altieri MA (1997) Agroecology versus input substitution: a fundamental contradiction of sustainable agriculture. Soc Nat Resour 10(3):283–295
Sattmann-Frese W, Hill SB (2008) Learning for sustainability: psychology of ecological transformation. Lulu, Morrisville. www.lulu.com
Savory A, Butterfield J (1999) Holistic management: a new framework for decision making. Island Press, Washington
Savory A, Parsons SD (1980) The Savory grazing method. Rangelands 2(6):234–237
Stallibrass A (1989) Being me and also us: lessons from the Peckham experiment. Scottish Academic Press, Edinburgh. See also: www.thephf.org
Soleri P (2006) Arcology: the city in the image of man, 4th edn. Cosanti Press, Mayer
Sylvander B, Bellon S, Benoit M (2006) Facing the organic reality: the diversity of development models and their consequences on research policies. OF and European Rural Development; Joint Organic Congress, Odense (DK), 30–31
Thomas VG, Kevan PG (1993) Basic principles of agroecology and sustainable agriculture. J Agric Environ Ethics 6(1):1–19
Todd NJ (2006) A safe and sustainable world: the promise of ecological design. Island Press, Washington
Todd NJ, Todd J (1984) Bioshelters, ocean arks, city farming: ecology as the basis of design. Sierra Club, San Francisco
Toyne P, Farley R (1989) A national land management program. Aust J Soil Water Conserv 2(2):6–9
Van der Ryn S, Cowan S (2007) Ecological design, 2nd edn. Island Press, Washington
Voisin A, Lecomte A (1962) Rotational grazing. Crosby lockwood, London
Warner KD (2007) Agroecology in action: extending alternative agriculture through social networks. MIT Press, Boston
Weathers J (1909) French market-gardening: including practical details of "Intensive Cultivation" for English growers. J Murray, London
Wezel A, Bellon S, Dore T, Francis C, Vallod D, David C (2009) Agroecology as a science, a movement and a practice: a review Agron Sustain Dev 29:503–515
White J (1991) Deep organics: avoiding inputs by design and management [based on viewing a video of a talk by SB Hill on 'Deep Organics' to the Northeast Organic Farming Association in 1984]. Maine Org Farmer Gard Jan/Feb: 12–13
Williams J, Gascoigne H (2003) Redesign of plant production systems for Australian landscapes. Proc. 11th Australian Agronomy Conference, Geelong, VIC. http://www.regional.org.au/au/asa/2003/i/4/williams.htm. Accessed 5 Feb 2010
Williamson GS, Pearse IH (1980) Science, synthesis and sanity. Scottish Academic Press, Edinburgh
Yeang K (2013) Ecomimicry: ecological design by imitating ecosystems. Routledge, New York, 256p
Yeomans PA (1958) The challenge of landscape: the development and practices of keyline. Keyline Press, Sydney
Yeomans PA (1971) The city forest: the keyline plan for the human environment revolution. Keyline Press, Sydney
Yeomans PA (1978) Water for every farm using the keyline plan. Murray Books, Ultimo
Yeomans AJ (2005) Priority one: together we can beat global warming. Keyline Press, Gold Coast City, QLD. See also: www.yeomansplow.com.au
Youl R, Marriott S, Nabben T (2006) Landcare in Australia: founded on local action (first published 1999). SILC and Rob Youl Consulting, Melbourne

Chapter 23
Transitions Towards Organic Farming at the Farm and at the Local Scales: The Role of Innovative Production and Organisational Modes and Networks

Claire Lamine, Mireille Navarrete and Aurélie Cardona

Abstract In a context of growing environmental constraints and economic uncertainties, how is it possible to facilitate transitions towards more ecological forms of agriculture? In this chapter, changes in practices from conventional agriculture towards organic farming (OF) are investigated by combining sociological and agronomical studies of farmers' trajectories conducted in the fruit and vegetable sectors. We specifically explore the potential of combinations of systems, both at the level of production and in terms of marketing outlets. We analyse the processes of adoption of alternative crop protection strategies using the Efficiency-Substitution-Redesign grid developed by biological and agricultural scientists. The combination of diversified systems of production (including organic and IPM) and marketing channels (including short and long food supply chains) might provide promising transition pathways for organic farmers. We also examine the conditions that enable such transitions, involving learning processes, collective and territorial dynamics and the ability of the networks to overcome the classical frontier between organic and conventional agricultures. Our three French case studies, which cover a wide range of marketing networks and diversification levels, show that a robust ecologisation of agricultural practices requires the redesign of both technical agricultural systems, as well as the larger interactions within agri-food systems and non-agricultural networks.

Keywords Transition · Fruits and vegetables · Alternative techniques · Marketing channels · Networks

---

C. Lamine (✉) · M. Navarrete · A. Cardona
INRA, UR0767 Écodéveloppement, 228 route de l'Aérodrome, Domaine Saint Paul, Site Agroparc, CS 40509, 84914 Avignon Cedex 9, France
e-mail: Claire.Lamine@avignon.inra.fr

S. Bellon, S. Penvern (eds.), *Organic Farming, Prototype for Sustainable Agricultures*,
DOI 10.1007/978-94-007-7927-3_23,

23.1 Introduction

At the European level, the banning of some chemical products planned in the recent directive on pesticide use,[1] as well as the perspective of a large transition of European agriculture towards Integrated Pest Management (IPM) by 2014 provides a supportive context for finding new production and organisational models. In France, in 2007, the government announced an ambitious goal for the growth of organic agriculture (from 2% in 2007 to 20% of the agricultural area by 2020) and for the reduction in pesticide use in conventional agriculture (a 50% reduction by 2018, “if possible”). All this invites policy-makers, scientists, professionals and the civil society to consider Organic Farming (OF) and other forms of ecological agricultures such as IPM as key models for achieving these objectives. We will focus here on OF, which is the most institutionalised of the alternative models. We will also consider the relevance of IPM practices for three main reasons: first, they can provide a transition stage to OF; second, OF and IPM can coexist on some farms; and third, both OF and IPM can contribute to a larger ecologisation of agricultural practices and provide insights into the conditions of such an ecologisation.

The issues linked to organic agriculture have generated a boom in research and publications in various scientific disciplines, including the social sciences (Lamine and Bellon 2009). Many sociological studies have emphasized the driving social and relational factors and motivations behind transitions towards organic farming. These have been based on both quantitative analyses of farmers’ motivations and attitudes, and on qualitative approaches to farmers’ practices and conceptions (Lockie and Halpin 2005; Best 2008). Most of these studies of motivations and of conversion-decision processes support the distinction between market-oriented (pragmatic) farmers and value-oriented (organic-committed) ones, where the former group is supposed to return to conventional farming if price premiums diminish (Fairweather 1999; Darnhofer et al. 2005).

A major technical challenge facing these new organic farmers is to be able to control pests and diseases without drastically lowering yields and commercial crop quality (Zehnder et al. 2007). In the agricultural sciences, studies assessing alternative control strategies tend to focus on particular pests (e.g., Chellemi 2002) and weeds (Anderson 2007), rather than on the effects of alternative management systems on the cropping systems (Navarrete et al. 2010) and on soil health (Doran and Safley 1997; Mateille et al. 2008). This may reflect the difficulty of conducting experiments within such complex systems and the scarcity of systemic models for predicting interactions between techniques and biological compartments. Other studies compare yields and economic performances of conventional and alternative systems (including OF), but the inherent diversity in the systems compared lowers the robustness of the comparisons (Sautereau et al. 2010).

[1] This Framework Directive on the Sustainable Use of Pesticides was adopted by the European Parliament on 13 January 2009 to replace Directive 91/414/EEC.

With a few exceptions (Guthman 2000), most studies in both the social and the agricultural sciences minimise the importance of transitional and long-term aspects within trajectories, and rarely approach conversion to OF as a longer process than its legal duration (Lamine and Bellon 2009). The impact of food-chain interdependencies on agricultural changes (or non changes) has received relatively little attention, an exception being the studies of short food-supply chains such as local markets and box schemes, which strive to minimise the stages between production and consumption (Renting et al. 2003). However, the growing literature about the 'conventionalisation' of the organic sector (Buck et al. 1997) and about lock-in and path-dependency effects within the agricultural sector and the food chains (Cowan and Gunby 1996; Possas et al. 1996; Vanloqueren and Baret 2008; Lamine et al. 2010) tackle these issues, although generally at the 'macro' scale. Consequently, there is a need to study the relationships between production and marketing in more detail, taking the long-term and broader scale (local to global) effects of such transformations into account.

This paper examines the organisational aspects of production and marketing, as well as the potential of collective and territorial initiatives for facilitating transitions towards organic fruit and vegetable production. Our approach, based on a transversal analysis of the results of three case studies carried out in recent research projects, will combine an agronomical grid of analysis and a sociological perspective on transitions and social interactions at the farm and local scales.

Data from these three recent case studies were analysed from both an agronomical and a sociological point of view. We analysed how the diversification of products and the combination of organic, IPM or conventional production and of various short and long food supply chains at the farm level might be able to provide transition pathways for organic farmers. We also investigated the potential of learning processes, of collective and territorial dynamics, and of networks in helping farmers to overcome the traditional frontiers between organic and conventional agricultures and cultures.

23.2 Methods and Data

Although we analysed several types of farming systems, we only focused here on the fruit and vegetable sectors, which are both highly pesticide-dependent (in France, 9.9% of pesticide use is for only 1.6% of the cultivated area; Butault et al. 2010), and the sectors most affected by the recently increasing rate of conversion to organic agriculture (Agence Bio 2010).

Pesticide dependence in these sectors is already under threat because of recent regulatory changes (e.g., banning of certain pesticides such as methyl bromide, which was commonly used against various soil pests), and because of reduced economic viability of pesticide production due to the increasing costs of development and registration (including alternative products). Thus, because of the declining range of registered pesticides, the transition to OF or IPM is quite challenging.

Societal expectations are particularly high in the fruit and vegetable sectors, with growing consumer concerns about the health effects of pesticide residues, access to local produce, and the development and maintenance of peri-urban agriculture. Although limiting pesticide use in the field lowers the risk of chemical residues in food, it may also reduce cosmetic quality, which is a major challenge, particularly for distant markets. It is therefore crucial to simultaneously analyse the transition processes in production and marketing networks.

We will describe and analyse these transition processes at:

1. the farm scale, focusing on crop protection strategies and their consequences for marketing,
2. and collective and local scales, by identifying how local networks may contribute (or not) to farm transition processes.

The demonstration is based on three French case studies (see text box) that differ in their characteristics (production systems, marketing channels, etc.) and the specific topics they focus on: technical and marketing changes and trajectories at the farm scale (case studies 1 and 2); and collective dynamics and interaction with the civil society (case studies 2 and 3). In these three regions, we conducted several series of surveys that combine sociological and agronomical studies at the farm level, with a total number of qualitative interviews of about 100 farms and 60 institutional, food supply chain and civil society actors.

Their comparison will allow us to explore and discuss the conditions for sustainable transitions to OF both at the farm and the local scale.

Short Description of the Three Case Studies

1. Provence and Roussillon (south-eastern France) are traditional market garden areas with many sheltered crops originally aimed at supplying urban areas with fresh fruits and vegetables and are largely oriented today towards export outlets. Most of the farms that converted to organic farming are involved in short marketing channels, but the main issue is to encourage conventional specialised farms involved in long channels to adopt more eco-friendly practices (Navarrete 2009).
2. Southern Ardèche (also in south-eastern France) is a diversified region in terms of natural conditions and production systems, where the geographical isolation of the area has led to a lack of competitiveness and to a critical economic crisis in the fruit sector. However, there are interesting ongoing and potential transitions at this time towards organic farming for small as well as larger farms that were formerly specialised in products that are in decline today (Lamine and Cambien 2011).

3. The Paris area is a much larger region specialised in arable crops with variable and locally intense pressures of the local civil society to reduce the impacts of chemical inputs. In this context, some producers recently started fruit and vegetable production as a means to diversify farms primarily devoted to arable crops. This activity is most often based on short food chains and organic (or sometimes IPM) farming and is generally localised in peri-urban areas (Cardona and Lamine 2010).

23.3 Transitions at the Farm Scale

At the farm scale, we first present how technical changes occur over time and then highlight how combinations of systems may favour transitions towards organic methods at the farm level.

23.3.1 Technical Changes and Consequences at the Farm Scale

In this study, we used both the Efficiency-Substitution-Redesign (ESR) grid (Hill and MacRae 1995) and the sociological approaches of farmers' trajectories (Lamine 2011) to describe the effects of the adoption of IPM or organic practices. This made it possible to tackle the issue of transitions from a combined agronomical and sociological point of view. The ESR grid was proposed as a framework that makes a distinction between three steps for the changes towards more sustainable practices:

- Efficiency (minor changes to existing operations that help to make them more efficient),
- Substitution (replacing one measure by an alternative one)
- Redesign (reorganising production systems according to ecological principles).

The E, S and R stages can be considered as three successive steps in the ecologisation of farming systems that may overlap within a given trajectory (Hill et al. 1999).

As regards changes in crop protection, they can be arranged in the ESR grid, depending on to what degree they modify farming systems (Table 23.1).

Efficiency strategies consist in reducing the pesticide dosage while minimising losses in terms of yield and visual quality without a drastic change in the crop protection strategy. It mainly refers to improving the conditions of pesticide application (e.g., local use of pesticides, improving pesticide application tools). The room for manoeuvre available at this time is rather slim for vegetable production. Since chemical application has long been restricted by the high cost of chemical products and by laws that regulate healthy foods, market gardeners are used to optimising the application of chemical products. As indicated by experts in the French Ecophyto

Table 23.1 The diversity of crop protection strategies in market gardening and their relative position in the Efficiency-Substitution-Redesign (ESR) grid

Crop protection strategy	Type of ecologised agriculture	Position in the ESR grid	Degree of farm specialisation/ diversification
Improving the efficiency and effectiveness of application of chemical pesticides to reduce their negative effects.	IPM	Efficiency	Specialised farms (3-5 species per farm)
Using biological antagonists. Using punctual alternative techniques (organic biocides, soil heat treatment, resistant cultivar). Using long-lasting alternative techniques (solarisation, green manure).	IPM, OF	Substitution	
Building crop rotations and cropping patterns to optimise pests and disease control.	(IPM), OF	Redesign	Diversified farms (up to 50 species or cultivars)

Solid line: most frequent relationship; Dotted line: least frequent relationship

report[2] (2009), increasing efficiency does not appear to be sufficient to significantly limit pesticide use.

Substitution strategies consist, for example, in introducing a resistant cultivar, replacing a fertiliser with an organic biocide to reduce soil-borne diseases, or using biological control. Since these strategies are punctual, they do not lead to deep changes in the system. Biological control is already used in routine procedures under shelter. Farmers plan regular inundative releases of biological control agents rather than trying to maintain them from one crop to the other or favouring their natural establishment in specific hedges as part of field margin management. That is why, at the moment, the application of biological organisms refers to the same paradigm as pesticide application, whereas more complex management strategies are proposed by scientists (see, for example, Zehnder et al. (2007), for arthropod control). Therefore, apart from the necessary adaptation in know-how to adjust the technical itineraries at the plot level, their introduction does not create major changes at the farm level *per se*. Similarly, the use of a resistant cultivar when it

[2] A collective scientific body of experts whose aim is to assess what would be the consequences of a 50 % reduction in pesticide use at the national production level.

is used in the same crop rotations and intended for the same marketing outlets as conventional ones also falls into the category of *substitution*.

Redesign strategies consist in promoting ecological principles within a systemic vision to control pests: introducing diversified rotation; designing soil management to enhance natural biological control (Altieri 1999; Jackson et al. 2007) and to promote healthier soil (Mateille et al. 2008). They usually lead to considerable adaptation at the farm level because they imply major changes in technical strategies, investments, tools, or the work force required. As an example, in south-eastern France nematode diseases are becoming increasingly severe under shelter. A sustainable control should rely on a redesign strategy based on large crop rotations and the cropping of non-host botanical families as cash crops or green manure in order to increase natural and cropped biodiversity, as suggested by scientists involved in agroecology (Altieri 1999; Jackson et al. 2007).

Moreover, in our study in the Provence and Roussillon regions, we pointed out that the same technique may refer either to the *substitution* or *redesign* level, depending on the farming systems in which it is introduced. Such is the case for solarisation and green manure. Solarisation consists in trapping solar radiation with a transparent plastic film laid on the top of the soil. When the technique is applied during the two hottest months, the increase in soil temperature is sufficient to kill some soil-borne pathogens (Stapleton 2000). However, it is in competition with commercial summer crops (tomato, cucumber, etc.) and leads to major cropping system adaptation. We showed that in some cases, the introduction of solarisation led to a *substitution* of summer crops by shorter ones (e.g., melon, potato) in order to free plots in July and August (Navarrete et al. 2006). Changes on the farm were minor because the marketing networks and the organisation of labour and equipment were not altered. For other farms, introducing solarisation led to the progressive suppression of summer crops within a few years. They specialised in winter crops (e.g., lettuce), and permanent labour was replaced by seasonal labour for lettuce cropping in winter. An unintended consequence of the latter case was that an alternative technique, assumed to reduce soil-borne pathogens, was finally linked to an increase in pathological and economic risks due to crop specialisation.

The previous analysis clearly focused on the description of final states of changes and showed that several degrees of change exist, even at the *redesign* level, ranging from "simple" adaptation in crop rotation to global farm organisation redesign. It also highlighted the necessity of understanding how transition occurs over time.

In order to analyse this ecologisation of practices on the long-term, from conventional to organic practices, we conducted an interdisciplinary study among market gardeners and fruit growers in south-eastern France in 2005–2006 (Lamine et al. 2008; Navarrete 2009). These pathways appeared to be varied.

Some were very gradual, when farmers progressively changed their conventional systems by adopting IPM strategies, experimenting with alternative techniques and, finally, converting to OF, with the whole process lasting for 5–20 years. In that case, they followed the three steps of the ESR grid, even if some steps took more time than others, and even if these steps overlapped. The efficiency phase took place while still in conventional agriculture, increasing farmers' attention to disease observation and the use of thresholds for pesticide application. The substitution phase often started while farmers were still in conventional agriculture and reached a complete elimination of chemical inputs when they converted to organic. The redesign phase started once farmers gained more experience in organic farming. Few farmers were able to quickly change their farming systems by adopting OF principles during the legal conversion time of 3 years, and jumping from the efficiency to redesign step. However, most growers whose trajectories appeared as more direct had practices that consisted in the substitution of conventional inputs with biological ones.

23.3.2 A Simultaneous Change in Marketing Channels and Production Systems

In addition to the technical practices, this interdisciplinary study also showed that the degree of specialisation in market gardening (assessed by the number of different species and varieties of vegetables) and marketing channels would also be adjusted in parallel to changes in technical practices, leading to a progressive redesign of both the farming and marketing system on the long-term.

Farmers who gradually and extensively redesigned their production system also considerably diversified their crops, whereas they tended to shift their marketing channels towards shorter ones. Indeed, in the case of local channels, marketing requirements in terms of product diversity go hand-in-hand with the enhancement of planned biodiversity and natural regulations, whereas visual quality criteria are less limiting (Navarrete 2009). On the other hand, farmers who converted more directly to OF and substituted organic inputs for chemical ones were also often more specialised and generally remained in the long marketing channels (see Table 23.1).

However, many farmers actually combined different types of marketing channels. Some who were in the process of an extensive redesign of their farming and marketing systems maintained long marketing channels for several years, while others built more lasting complementarities between different types of products (mainly fruits, wine grapes and vegetables) and outlets, thus combining short food chains (box schemes, open air markets, on-farm sales, collective farmers' shops) and longer food chains (public organic food procurement, organic wholesalers, local fruit cooperatives).

In addition to the progressive redesign of the farming and marketing system that involves a diversification of products as described above, a sociological study about organic or converting-to-organic farmers' trajectories carried out in 2009 in the south of the Ardèche department in France also revealed two other types of transitions.

The first type of transition involves conventional farmers who diversify towards organic farming in a context of crisis for their main products (fruits and wine). In their case, a partial transition to OF[3] makes it possible to balance economic risks, as well as to experiment with innovative technical management without endangering the sustainability of the entire farm. In fact, we observed that these farmers not only converted some plots and crops to OF, but also changed some of their techniques in the plots and crops that were still under conventional farming (e.g., they increased their use of biological control on their trees). In some cases we can expect this combination of OF and conventional (actually, IPM) modes of production to be temporary, a step on the way to a 100 % organic farm, but in others it might also be considered as being more sustainable in economic terms, especially when the local outlets are not ready to absorb a high increase of organic production.

The other type of transition corresponds to some farmers who would, in opposition to the very first type of transition described above (progressive redesign of the farming and marketing system and diversification of products), progressively change from such highly diversified systems towards more specialised ones, often linked to an extension of their surface areas and the development of mechanisation. A few years after the legal conversion period, this partial specialisation of the farm is generally a way to simplify cropping patterns and labour organisation, which are tricky on highly diversified market garden farms. This phenomenon was probably also involved in the farms previously analysed from an agronomic perspective, whose farmers specialised their crops after the introduction of green manure and solarisation. This progressive specialisation goes hand-in-hand with a combination of longer and shorter food chains.

As a conclusion, this study identified three types of farm transition pathways linking production and marketing strategies, the first one corresponding to the Redesign paradigm associated with the high diversification of crops:

1. Diversification of both production and marketing outlets (high added value food chains) within an already organic system; farmers adapt their system to its actual potentialities (surface areas, material, work load); they tend to specialise in high value short food chains.
2. Diversification of conventional farmers towards organic horticulture and towards new outlets; farmers diversify towards organic farming in a context of crisis in the fruit and wine sectors and of increasing opportunities in OF; they combine short and long food chains as well as organic and conventional (indeed, IPM) farming.
3. Rationalisation of production (surface areas, mechanisation, partial specialisation) and marketing (combination of short and long chains) within an already organic system; farmers rationalise their system for a better viability; they combine short and long food chains.

This typology confirms the idea that the ESR grid should not be understood within a normative and teleological perspective (the "good" goal in the long run being

[3] In Europe, according to organic farming legislation, transition to OF does not imply the conversion of the entire farm area, providing that organic and conventional crops and products can be clearly distinguished (RCE 834/2007).

the redesign step, generally linked to the high diversity of products), and that it encompasses not only technical practices, but marketing strategies and organisational issues as well because of their close interrelationships. Therefore, the results of our different studies allow us to suggest that possible combinations of products (fruits and vegetables, for example), marketing outlets (short and long chains) and sometimes even modes of production (OF and IPM) should be taken into account. Moreover they also lead us think that the redesign step might be given a larger definition: from certain perspectives, a process of re-specialisation (the third type of transition described above) might appear as a form of redesign, even though this might seem paradoxical at first glance.

23.4 Transitions at the Collective and Local Scales

The reconfigurations and interrelationships between production and marketing strategies previously analysed at the farm level also involve collectives of farmers or, in some cases, non-agricultural networks and the civil society. We will now analyse how professional networks (other farmers in the same area, advisory services, marketing networks) and non-agricultural networks (including local authorities and civil society) might facilitate or impede farmers' transitions.

23.4.1 Professional Networks

Many studies have pointed out the effects of collective dynamics and networks on learning and technical change: in professional groups and networks, whether local or not, farmers can support one another on technical aspects, moral and economic considerations (Darré 1994; Norton et al. 1999; Warner 2007). In the case of organic as well as IPM production, belonging to a professional group helped farmers to build a collective identity in a professional world that is (or was) largely sceptical about chemical input reduction or suppression. In fact, turning to low-input practices often hurts the bases of a farmer's professional identity and the image of professional excellence, since crops might host higher levels of pests and lead to lower yields (Lamine 2011). However, these organic and IPM farmers often said that they were also re-discovering the core identity of their profession and that they liked the experimental and technical sides of such changes. Belonging to these groups helped them to overcome the apprehension linked to profound changes in their practices. This also indicates that risk aversion is as much collective as individual – contrary to prevailing economic interpretations.

Another major influence is of course linked to advisory systems. Generally speaking, advisory systems appear to be market-led, which means that farmers are clients that the various extension services do not want to lose. Therefore, advisors become rather risk-adverse and reluctant to promote alternative strategies that could lead to lower yields. However, some actors who are considered to provide advice designed to promote their "sale and purchase" business interests (input suppliers, farmers' cooperatives and trade partners) have begun to have more positive attitudes towards low-input practices (Blanc et al. 2010). For example, in our studies, farmers who were in the process of converting to organic farming or considering it (and were currently adopting biological control strategies), reported that even "classical" input suppliers progressively enlarged the range of biological products they proposed. However, as previously mentioned, this favours substitution strategies more than redesign ones. Providing advice about alternative systems is particularly challenging because most of the techniques involved generally have long-term benefits and often only partial effects. As a consequence, advisors have to move from a prescriptive activity to a constructive one by accompanying farmers in a global reflexion on system redesign. Since the economic sustainability of the new farming systems is sometimes uncertain, especially on the short-term, their activity may become more risky. Agronomical models and systemic studies that make it possible to evaluate the long-term consequences of alternative models should be developed in the coming years as a way to foster system redesign (e.g., Navarrete et al. 2010).

As regards marketing networks, our study in southern Ardèche has also shown that the interrelationships between production and marketing strategies often concern not only the farm scale, but also farmers' collectives as well. The combinations of products and outlets that we described at the farm level in the previous section are facilitated by the insertion of farmers into various professional networks: organic farmer networks and informal local farmers' groups that gather and market products for wholesalers or that also supply schools, in addition to sharing equipment and technical advice. As an example, the development of collective farmers' shops, which offer a variety of farm products and share marketing time between several farmers, makes it possible to find a balance between (1) the necessary diversification as an agronomical lever, (2) the difficulty for one particular farm to supply short channels alone, and (3) the attempt to rationalise production systems as indicated above.

These transitions also partly rely on institutions that are not specifically devoted to organic farming, but that can support *de facto* the development of organic products and offer already available infrastructures. An example would be the case of a local fruit cooperative that decided to enlarge its range of organic products in the context of a sectorial crisis.

23.4.2 Interactions with Non-agricultural Networks and Civil Society at the Local Scale

As our second case study has shown, the transitions towards OF involve not only producers, but other local food chain actors and non-agricultural actors and networks as well, mainly local authorities and civil society. Many short food chains are linked to non-agricultural organisations (such as AMAP[4] or consumer cooperatives), and organic farmers are partly supported by localised public policies aimed at developing organic farming and/or food for local schools. In this regard, farmers' transitions have to be considered beyond the agricultural sphere.

Our third study in the Paris area was precisely centred on these possible roles of civil society and public policies in farmers' transitions to OF as well as IPM. Based on a study of farmers' transition processes, as well as on an ethnographic analysis of local projects linked to agricultural and food issues, we analysed how non-agricultural actors within the civil society (mainly local environmental and/or consumer organisations) may more or less directly influence the evolution of agricultural practices (Cardona and Lamine 2010). We identified three modes of action of such organisations:

- They may urge public authorities to deal with environmental issues such as landscape (preservation of agricultural land from urbanisation), water quality and providing food for public schools;
- They may initiate their own projects and either develop marketing outlets that facilitate farm transition towards OF such as box schemes or, in some cases, purchase agricultural land in order to facilitate the installation of organic farmers;
- They may also directly take responsibility for the implementation (and related advice) of agri-environmental measures (mainly for the reduction of inputs in conventional agriculture and its transition towards IPM) or the supply of food to schools (food hubs or platforms that centralise local organic or low input products) in the place of public institutions; in that case public actions are delegated to civil society organisations.

Concerning the effect of such actions, we observed that the first mode of action tends to bring about a favourable context for agricultural practice changes and transition pathways towards OF. With the second mode of action, the requirements of civil society organisations can be strongly expressed and can lead to the redesign of the farming systems or at least part of them. With the last mode of action, changes in farming practices seem to be limited to efficiency or substitution. In this case, civil society organisations have to reach a compromise with various actors and have to be less demanding in terms of farming system redesign.

We have highlighted here the importance of professional networks – some of them quite informal – and the role of local public policies and non-agricultural networks (civil society) for facilitating transitions towards organic farming as well as other forms of ecological agriculture. This should promote social studies on these networks and local scale initiatives in order to analyse their effects and limits.

[4] AMAP (Associations for the Maintenance of "Peasant" Agriculture) are comparable to Community Supported Agriculture (CSA) initiatives.

23.5 Conclusion

Based on the ESR grid and on a sociological study of farmers' trajectories, we demonstrated in this paper how the combination of different products, different marketing outlets and even, in some cases, of different production modes can facilitate transitions towards organic farming. This goes against several classical oppositions between specialisation and diversification, between long and short food supply chains and, of course, between organic and conventional farming. It also shows that what some researchers and organic farming advocates consider as the 'ideal' of OF is a much narrower view than what many farmers strive for.

At the farm scale, our analysis has emphasized the accurateness of the ESR grid for analysing current changes in practices and comparing several production systems. Efficiency and substitution are often a first step that leads farmers to initiate their reflection on their practices and possible changes, thus facilitating possible future redesign. The analysis also confirmed that there is often overlapping between the three levels of the grid (e.g., Hill et al. 1999): farmers maintain, for example, substitution practices in an overall redesigned farming system. Besides, our results show that the ESR grid should also take the possible complementarities between several systems of production into account. For example, the combination of OF and IPM on the same farm, which is allowed in Europe, might make it possible to balance economic risks (as could the combination of short and long marketing channels), as well as to experiment with innovative technical management without endangering the sustainability of the entire farm. It may also facilitate learning processes. The positive effects of such combinations, especially in economic terms, would require further research.

Finally, on the basis of our interdisciplinary comparisons, we can highlight a different and complementary definition and use of the ESR grid in our two disciplines. For the agronomist, this grid is useful for describing successive steps and for classifying the types of technical changes at the farm level. For the sociologist, it instead describes transition processes, whereas the three paradigms of the ESR grid define the overall objective of the farmer rather than the current state of his system. The combination of our agronomical and sociological points of view allowed us to demonstrate that few technical elements can be altered at the farm scale without rethinking other compartments of the farm system. We focused on the interactions between production choices and marketing issues here, but changes in labour organisation are also concerned (know-how, availability throughout the year, organisation at the farm level) and should also be investigated.

The sociological approach also showed that transitions had to be considered beyond the farm scale, at the collective and/or local scales. The local environment may favour or impede the technical evolutions at the farm scale, depending on whether it creates favourable conditions for technical change or not. The main triggers appeared to be the professional networks and the availability and organisation of advice, local policies and the civil society. Therefore, the notion of transition, whether approached through the ESR grid or through a sociological study of farmers' trajectories, has to be expanded from the farm level to the agrifood system level. In this context, our analysis is also a contribution to transition theories. Indeed, while studies of agricultural changes are often at the micro-level of plots or farms, and studies of transitions at the macro-level of the technological regimes (Geels and Schot 2007), it is necessary to insist on the necessity of considering the "meso" levels of the socio-technical agrifood systems, defined as encompassing not only agricultural production, transformation and distribution, but the diverse agricultural and food institutions, the advisory sector, the upstream industries, and the role of regulations and civil society as well. A growing number of authors have recently suggested applying these transition theories to the organic sector as a whole (Sautereau and Bellon 2010) or have focused on alternative food systems (Brunori and Rossi 2010), but without any clear reference to the actual agricultural models farmers are likely to apply. Our suggestion is that this socio-technical system should be dealt with at the scale of small regions, where interdependencies between the different components of the system can be studied and lead to the production of reliable scientific data, and where these interdependencies can be discussed and possibly redesigned. This is why we are exploring the notion of territorial agrifood systems in our current research projects (Lamine et al. 2012).[5] Besides, while these transition theories suggest the idea of a stable configuration both before and after the actual transition process, we argue instead that farms (and agrifood systems) often remain in a dynamic situation where farmers' experiences, social and economic contexts and opportunities continue to reconfigure farm activities on the long-term.

An important point we want to conclude with, whether at the local or national (or larger) scale, is the necessity to consider the different forms of ecological agricultures together. Whereas OF and IPM are usually considered as separate paradigms, more for historical and sociological reasons than for technical ones, our research clearly demonstrates that they are closely related. Thus, the transitions towards IPM or OF have many points in common in terms of learning processes, changes in conceptions, relationships to others and graduality (Lamine 2011). Indeed, as developed in this paper, IPM might appear as a step in a transition towards OF or as an element of a mixed system that combines IPM and OF. Finally, after a period where organic farming was disregarded by conventional farmers and technical advisers, it is now considered as a source of inspiration and alternative methods for all professionals in search of more sustainable farming systems (Cardona and Lamine 2010).

References

Agence Bio (2010) Exploitations et surfaces bio par production végétale et par region. http://www.agencebio.org/upload/Ch6_ChiffresCles2010.pdf

Altieri M (1999) The ecological role of biodiversity in agroecosystems. Agric Ecosyst Environ 74:19–31

[5] POPSY (2008–2012) and Dynrurabio research projects (2011–2014; http://www.inra.fr/comite_agriculture_biologique/media/les_recherches/programme/anr/systerra_2010_fiche_dynrurabio).

Anderson RL (2007) Managing weeds with a dualistic approach of prevention and control. A review. Agron Sustain Dev 27:13–18
Best H (2008) Organic agriculture and the conventionalization hypothesis: a case study from West Germany. Agric Hum Values 25(1):95–106
Blanc, J, Noe E, Burma J, Burnett M-L, Compagnone C, Marraccini E (2010) The role and attitudes of agricultural advisors in implementing sustainable pest management in European agriculture—a cross national case study in NL, FR; UK and DK. IFSA congress, Vienna, July 2010
Brunori, G, Rossi A (2010) Drivers of transformation in the agro-food system. GAS as co-production of Alternative Food Networks. IFSA congress, Vienna, July 2010
Buck, D, Getz C, Guthman J (1997) From farm to table: the organic vegetable commodity chain of northern California. Sociologia Ruralis 37(1):3–20
Butault JP, Dedryver CA, Gary C, Guichard L, Jacquet F, Meynard JM, Nicot P, Pitrat M, Reau R, Sauphanor B, Savini I, Volay T (2010) Ecophyto R & D. Quelles voies pour réduire l'usage des pesticides? synthèse du rapport d'étude, INRA Editeur (France) (90 p)
Cardona A, Lamine C (2010) Mobilisation et animation territoriale autour des réductions d'intrants: stratégies d'intéressement des agriculteurs dans trois territoires franciliens", oral communication at SFER, La réduction des pesticides: enjeux, modalités et conséquences, 11–12 March 2010, ENS Lettres et Sciences Humaines in Lyon
Chellemi DO (2002) Nonchemical management of soilborne pests in fresh market vegetable production systems. Phytopathology 92:1367–1372
Cowan R, Gunby P (1996) Sprayed to death: path dependence, lock-in and pest control. Econ J 106:521–543
Darnhofer I, Schneeberger W, Freyer B (2005) Converting or not converting to organic farming in Austria: farmer types and their rationale. Agric Hum Values 22:39–52
Darré J-P (1994) Pairs et experts dans l'agriculture. Dialogues et production de connaissance pour l'action. Erès, Paris
Doran JW, Safley M (1997) Defining and assessing soil health and sustainable productivity. In: Pankhurst C, Doube B, Gupta V (eds) Biological indicators of soil health. CAB International, Wallingford, Oxon, pp 1–28
Ecophyto R&D (2009) Vers des systèmes de culture économes en produits phytosanitaires, tome V: analyse comparative de different systems en culture légumière (63 p + annex)
Fairweather J (1999) Understanding how farmers choose between organic and conventional production: results from New Zealand and policy implications. Agric Hum Values 16:51–63
Geels FW, Schot J (2007) Typology of sociotechnical transition pathways. Res Policy 36(3):399–417
Guthman J (2000) An agro-ecological assessment of grower practices in California. Agric Hum Values 17:257–266
Hill SB, MacRae RJ (1995) Conceptual frameworks for the transition from conventional to sustainable agriculture. J Sustain Agric 7(1):81–87
Hill SB, Vincent C, Chouinard G (1999) Evolving ecosystems approaches to fruit insect pest management. Agric Ecosyst Environ 73:107–110
Jackson LE, Pascual U, Hodgkin T (2007) Utilizing and conserving agrobiodiversity in agricultural landscapes. Agric Ecosyst Environ 121:196–210
Lamine C (2011) Transition pathways towards a robust ecologization of agriculture and the need for system redesign. Cases from organic farming and IPM. J Rural Stud 27:209–219
Lamine C, Bellon S (2009) Conversion to organics, a multidimensional subject at the crossroads of agricultural and social sciences. A review. Agron Sustain Dev 29:97–112
Lamine C, Cambien L (2011) Les transitions vers l'agriculture biologique: une approche à l'échelle d'un système agri-alimentaire territorial, colloque Ecologisation, Avignon, mars 2011
Lamine C, Meynard J-M, Perrot N, Bellon S (2008) Analyse des formes de transition vers des agricultures plus écologiques: les cas de l'Agriculture Biologique et de la Protection intégrée. Innovations Agronomiques 4:499–511
Lamine C, Meynard J-M, Bui S, Messéan A (2010) Réductions d'intrants: des changements techniques, et après? Effets de verrouillage et voies d'évolution à l'échelle du système agri-alimentaire. Innovations Agronomiques 8:121–134

Lamine C, Renting H, Rossi A, Wiskerke H, Brunori G (2012) Agri-food systems and territorial development: innovations, new dynamics and changing governance mechanisms. In: Darnhofer I, Gibbon D, Dedieu B (eds) The farming systems approaches into the 21st century: the new dynamics. Springer, New York, pp 229–256

Lockie S, Halpin D (2005) The conventionalisation thesis reconsidered: structural and ideological transformation of Australian organic agriculture. Sociologia Ruralis 45(4):284–307

Mateille T, Cadet P, Fargette M (2008) Control and management of plant-parasitic nematode communities in a soil conservation approach. In: Cianco A, Mukerji KG (eds) Integrated management and biocontrol of vegetable and grain crops nematodes (Integrated management of plant pests and diseases, vol 2). Springer, Dordrecht, pp 79–97

Navarrete M (2009) How do farming systems cope with marketing channel requirements in organic horticulture? The case of market-gardening in south-eastern France. J Sustain Agric 33:552–565

Navarrete M, Le Bail M, Papy F, Bressoud F, Tordjman S (2006) Combining leeway on farm and supply basin scales to promote technical innovations in lettuce production. Agron Sustain Dev 26:77–87

Navarrete M, Tchamitchian M, Aissa-Madani C, Collange B, Taussig C (2010) Elaborating innovative solutions with experts using a multicriteria evaluation tool. The case of soilborne disease control in market-gardening cropping systems. International symposium on Innovation and Sustainable Development in Agriculture and food, Montpellier, June 28 to July 1, 10 p. http://hal.archives-ouvertes.fr/hal-00512273/fr

Norton GW, Rajotte EG, Gapud V (1999) Participatory research in integrated pest management: lessons from the IPM. Agric Hum Values 16:431–439

Possas ML, Salles-Filho S, Maria da Silveira J (1996) An evolutionary approach to technological innovation in agriculture: some preliminary remarks. Res Policy 25:933–945

Renting H, Marsden T, Banks J (2003) Understanding alternative food networks: exploring the role of short food supply chains in rural development. Environ Plan A 35:393–411

Sautereau N, Bellon S (2010) Stimulate transitions towards sustainable agri-food systems: the contribution of conversion to organic food and farming to analyze dynamics and governance, SISA Congress, Leylistad

Sautereau N, Geniaux G, Bellon S, Petitgenet M, Lepoutre J (2010) Quantity versus quality and profit versus values? Do these inherent tensions inevitably play in organic farming? International symposium on Innovation and Sustainable Development in Agriculture and Food, 28 June to 1 July 2010, Montpellier, France, 15 pp

Stapleton JJ (2000) Soil solarisation in various agricultural production systems. Crop Prot 1:837–841

Vanloqueren G, Baret PV (2008) Why are ecological, low-input, multi-resistant wheat cultivars slow to develop commercially? A Belgian agricultural 'lock-in' case study. Ecol Econ 66:436–446

Warner K (2007) The quality of sustainability: agroecological partnerships and the geographic branding of California winegrapes. J Rural Stud 23:142–155

Zehnder G, Gurr GM, Kuhne S, Wade MR, Wratten SD, Wyss E (2007) Arthropod pest management in organic crops. Annu Rev Entomol 52:57–80

Chapter 24
Contributing to a Transition to Sustainability of Agri-Food Systems: Potentials and Pitfalls for Organic Farming

Ika Darnhofer

Abstract Organic farming is built on systemic principles that include environmental as well as social and political aims. Indeed, the aims are to produce wholesome food in an environmentally-friendly way, as well as to contribute to social justice, e.g., by preserving the family farm and rural communities. However, as organic farming has grown out of its niche, a broad range of practices have emerged. The current diversity is partly a result of internal processes and partly a result of its involvement with the dominant agri-food system. In fact, as a result of its involvement with the agri-food system, organic farming attempts to modify it by resisting its reductionist logic. As such, organic farming could be conceived of as being in a co-evolutionary dance with the dominant agri-food system: being changed by it, as well as contributing to its transition to sustainability. It is argued here that if organic farming is to serve as a prototype for sustainable agriculture, it will not only have to show that it can produce high-quality food in an environmentally-friendly way, but also demonstrate its ability to work with and induce a transformation of the rest of the food chain (including food handling, marketing and consumption). To achieve this, it will have to reintegrate and better articulate issues related to economic sustainability and social justice, possibly through alliances with other alternative food systems.

Keywords Conventionalisation · Evolving practices · Adaptability · Systemic approach · Social equity · Public health · Economic viability · Alternative practices · Co-evolution · Co-construction

24.1 Introduction

Organic farming is usually understood as a systemic approach to agriculture and food production. It is associated with ecological aspects such as nutrient cycling, social aspects such as the preservation of family farms and providing nutritious foods to consumers, as well as economic aspects that ensure the viability of the farm

I. Darnhofer (✉)
Department of Economic and Social Sciences, University of Natural Resources and Life Sciences, Feistmantelstr. 4, 1180 Vienna, Austria
e-mail: ika.darnhofer@boku.ac.at

S. Bellon, S. Penvern (eds.), *Organic Farming, Prototype for Sustainable Agricultures*, DOI 10.1007/978-94-007-7927-3_24, as a cornerstone of rural areas (Woodward et al. 1996). These aims are reflected in the four principles of organic agriculture formulated by the International Federation of Organic Agriculture Movements (IFOAM): health, ecology, fairness and care (Luttikholt 2007). Organic farming would thus seem to be well positioned to be a prototype for sustainable agriculture, the proposition that is the underlying rationale of this book.

To contribute to this proposition, I clarify its various terms. Firstly, I would like to define the term 'organic farming' by pointing out that it seems helpful to differentiate between organic farming *principles* as an ideal, and the diversity of organic farming *practices*. This diversity has been highlighted within the framework of the conventionalisation debate, which I will briefly review. Secondly, it seems helpful to clarify how comprehensively the term 'sustainability' is meant: does it refer primarily to the environmental impact on-farm, or does it have a wider meaning all along the agro-food chain? The specific issues linked to sustainability have evolved over time, and organic farming—at least in its present form—might not be able to address them all equally. Finally, I clarify the meaning of 'prototype' by assessing whether this refers to a set of standardised practices, or whether it includes a dynamic component that allows organic farming to co-evolve with its context. I propose that for organic farming to contribute to a transition to sustainability (i.e., to transform the dominant agro-food regime), it might have to be reflexive and harness its internal diversity while building alliances with other alternative movements.

24.2 Conventionalisation and the Diversity of On-Farm Practices

Organic farming has made substantial contributions to various aspects of environmental sustainability, especially regarding on-farm production methods. While these contributions are acknowledged, it has also been pointed out that on-farm practices are very diverse, so that these potential environmental benefits are not necessarily achieved in each and every case. The observed heterogeneity of on-farm practices has been characterised as ranging from 'input substitution' to 'system re-design' (Lamine and Bellon 2009; Hill 2014 Chap. 22). Thus, while at one end of the spectrum, organic farmers redesign their whole farm over time in an ongoing effort to implement a holistic understanding of organic farming, on the other end, farmers implement only limited changes, in effect substituting prohibited for allowed inputs. As a result, on some organic farms, production practices differ only marginally from conventional practices. These observations have spurred a scientific debate around the 'conventionalisation' of organic farming (Guthman 2004; Lockie and Halpin 2005; Rosin and Campbell 2009).

The conventionalisation debate has highlighted the diversity within organic farming, pointing out that organic practices are not necessarily associated with the principles underlying organic farming. Instead, there is a common core, which is defined by the legal standards. Beyond this, a broad range of factors—characteristics of individual farms and farmers, institutional support, agricultural policy, markets—influence to what extent the principles of organic farming are implemented. The debate has especially identified the involvement of agribusiness in organic production, and of large retailers in marketing organic foods as a main driver of the conventionalisation process (Guthman 2004). Indeed, in many instances, 'conventionalised' practices are most pronounced on large farms that tend to be specialised and to focus on long food chains (De Wit and Verhoog 2007; Stassart and Jamar 2008). These farms are the most highly exposed to the demands of multiple retailers that require large batches of standardised, uniform goods (Guthman 2004).

A more contentious aspect of the conventionalisation debate is whether conventionalisation inevitably affects all organic farmers, i.e., whether the structural conditions set by agro-industrialisation processes systemically undermine the ability of even the most committed producers to practice an alternative form of farming, i.e., remain true to the principles of organic farming (Rosin and Campbell 2009). This proposition is based on market dynamics that lead to lower prices for organic products, thus pressuring all organic farmers to reduce production costs, e.g., by disregarding practices that serve to implement organic principles but that are not mandated by legal standards (Stassart and Jamar 2008; Pratt 2009). However, despite these economic constraints, small-scale organic producers persist in most regions, as do production practices that are in line with a comprehensive understanding of the principles of organic farming. Conventionalisation, while present to some extent, may not systematically undermine the ability of organic farmers to implement a comprehensive approach to organic farming (Rosin and Campbell 2009).

The conventionalisation debate has thus shown that the development of organic farming over the last 20 years is more complex than the initial dualism, 'true to principles' vs. 'conventionalised', suggested (Lockie and Halpin 2005). Such a dualism—as well as others, e.g., large vs. small producers, producing for the domestic market vs. for export, focusing on short vs. long chains—may thus not be useful since it tends to cloud the diversity within organic farming, as well as how it changes over time. Indeed, organic farming may not be a matter of 'either/or' but rather a matter of 'both/and'. It is not that a farm is either 'small, involved in direct marketing and true to principles' or 'large, oriented towards export and follows an input substitution approach'. In practice, many organic farms are involved in a multitude of evolving practices and several marketing chains. While the pioneers might have had fairly unitary organic farming practices, these practices have become diverse as they have spread. Not only has diversity increased, but this diversity remains fluid and dynamic as a result of ongoing contestation and negotiation processes. In other words, organic farming practices are not a homogeneous and static set, but a diversity that evolves in time and in space. The changes, driven by endogenous processes within the organic movements, are a response to opportunities and constraints in the context, and emerge from the interplay of various actors along the food chain.

While this diversity and fluidity might be problematic to those actors who have a stake in a specific definition of organic farming, it might well be a strength of organic farming: fuelling its adaptability and thus contributing to its durability. Indeed, taking a systemic and dynamic perspective, if organic farming is to be a prototype for sustainable agriculture, it needs to be able to ride the dynamics of change, taking advantage of new opportunities as they arise. However, this is not meant to imply that organic farming should embrace an ‘anything goes’ approach. While exploring new approaches is likely to help identify promising avenues, it needs to be balanced with a reflexivity that makes it possible to constantly reassess boundaries between practices that are desirable and those that are no longer justifiable (Woodward et al. 1996).

These negotiations need be context-specific, i.e., to respond to concerns that are voiced at a specific time and a specific place. To illustrate the type of dynamics involved, I briefly review the broad range of current sustainability issues linked to farming and food towards which organic farming might be expected to contribute. I group them along the three dimensions of sustainability, while highlighting their systemic implications.

24.3 A Systemic Approach to Sustainability

A systemic approach emphasizes the interdependences between elements and between sub-systems. The interdependencies within the agro-food system highlight that changes in one sub-system (e.g., on-farm practices) are both constrained and enabled by the dynamics in other sub-systems (e.g., processing, marketing and consumption practices). Thus, if organic farming is to be a prototype of sustainable agriculture, it might need to show that it can address practices at the various sub-system levels along the agro-food chain, as well as address the various dimensions of sustainability.

24.3.1 Environmental Protection

The environmental impact of intensive, industrial agricultural practices has received much attention since the 1980s, and has been at the centre of agricultural policies in Europe, leading to the agri-environmental programme and compulsory environmental standards on farms. However, this approach to environmental sustainability privileges a particular definition of sustainability, one that is measured by specific criteria at the level of individual farms, e.g., the reduction of nitrate leaching from fertilisers and the reduction in the use of harmful pesticides and herbicides. As a result, most of the assessments of the environmental impact of organic farming stop at the farm gate (e.g., Stolze et al. 2000; Nemecek et al. 2011), thus sidelining more systemic definitions of agro-food sustainability. In particular, the environmental cost of industrial food handling, processing and marketing is rarely taken into account. However, if organic food is as transport-intensive, undergoes the similar high-energy processing, and relies on the same retailing systems as conventional food (Yakovleva and Flynn 2009), then the difference in its ecological footprint is likely to be minimal. As the public concern shifts from the environmental impact of on-farm practices to climate change (e.g., energy use and CO_2 emissions), issues such as 'food miles' become more prominent (Pretty et al. 2005; Coley et al. 2009). This has led to doubts about the relative environmental benefits of organic foods if they too are flown around the globe, and to highlighting the fact that organic is no longer synonymous with regional food (Lobley et al. 2013).

The focus on farm-level practices has also tended to obscure the fact that the room for manoeuver that a farmer has to implement environmentally-friendly production practices is limited by the requirements of the transport and processing industry, as well as the specifications of large retail chains (Green and Foster 2005). This is problematic since these quality specifications may or may not be compatible with production methods that contribute to environmental protection.

In the context of environmental impact of agro-food practices, the role of household routines has also been highlighted (Shove and Walker 2010). Consumers of organic foods are not necessarily inclined to reduce the frequency of their shopping trips, to renounce globally sourced commodities such as coffee or cocoa, or to accept the restrictions imposed by the seasonality of regional fruit and vegetable production. However, all these elements contribute to the environmental impact of the weekly food basket (Pretty et al. 2005).

If organic farming aims to be a prototype for environmental sustainability, focusing only on farm-level practices may be too limited. The notion of sustainability in its deeper meaning requires thinking in systemic terms (Green et al. 2003; Thompson and Scoones 2009), i.e., requires taking the whole agro-food chain into consideration. In other words, enhancing the sustainability of organic farming requires an integrated perspective on the social, economic and organisational structures involved from an organic farm to the consumer's fork.

24.3.2 Social Equity

At its outset, social aspects were an integral part of the values of organic farming (Woodward et al. 1996). However, in order to expand, organic farming has heavily relied on market-based mechanisms, thus implicitly accepting the neoliberal logic of consumer choice and individual responsibility (Guthman 2008; Allen 2010). Since markets serve those who are able and willing to consume, they do not inherently encourage social equity or democratic participation (DeLind and Bingen 2008). As a result, issues linked to social sustainability such as production and trade relations based on democratic and participatory control, transparency and accountability, rights of labourers, fair and stable pricing, or the preservation of farming livelihoods and local knowledge (DFTA 2008; Brown and Getz 2008; Padel et al. 2010) have been sidelined. Indeed, values such as support for small farms and community cohesion are now often associated with local foods rather than with organic foods (Winter 2003). Similarly, consumers seeking direct contact with producers may turn to conventional as well as to organic farmers (Adams and Salois 2010).

Organic farming practices at this time have no privileged association with the concept of 'civic agriculture', with its emphasis on an agro-food system that is embedded in the community and that addresses the needs of local growers, consumers and rural communities (DeLind 2002; Lyson 2005). In other words, while these ideals are still present within organic farming, they are not a constituent part of organic farming practices (Woodward et al. 1996, Goldberger 2011).

In addition to issues linked to social justice, social sustainability can also be understood as including the contribution of agro-food systems to public health and human well-being. Indeed, in the consumer's motivation to purchase organic foods, issues related to human health play a key role (Wier et al. 2008). Organic foods may thus contribute to human health, e.g., through lower pesticide residue levels (Lairon and Huber 2014 Chap. 16). However, health is also related to diet composition. It is well known that the excessive consumption of energy-dense nutrient-poor foods—which tend to make up the wide-spread high fat, high sugar, high salt diets—while being pleasurable, is unhealthy (Kearney 2010). Such foods—often associated with refined and processed foods—have been linked to the obesity pandemic, heart disease and type 2 diabetes (McMichael et al. 2007). However, despite the fact that organic standards limit the use of certain additives, processing aids and methods, there is a growing range of processed food, convenience food and food supplements made from organic ingredients. This may be seen as an indication that there is no privileged relationship between organic food and a balanced, nutritious diet.

These emerging trends do not imply that organic farming actors make no efforts to include social and ethical practices in organic farming (e.g., Schäfer et al. 2010; Padel et al. 2010). However, currently organic farming is not inherently construed as contributing to social sustainability. As a result, organic farming practices do not necessarily address issues such as fair market relations, diets that contribute to human health, democratic participation in decision-making or civic responsibilities.

24.3.3 Economic Viability

Studies of economic aspects of organic farming are often focused on comparing the profitability (or gross margins) of conventional and organic farms. Comparisons are usually made at the whole farm level to take account of the fact that organic farms tend to have different crop rotations to ensure soil fertility, and that there may be crop-livestock interactions, i.e., through closing nutrient cycles and reducing the purchase of off-farm inputs such as animal feed. Many studies have shown that organic farms can be more profitable than conventional farms, usually due to a combination of lower production costs and price premiums (Nemes 2009). However, some authors point out that organic farms in Europe tend to be dependent on direct payments (Offerman et al. 2009). Also, it is unclear if economic profitability can be maintained once farm-gate prices for organic products drop (Smith and Marsden 2004).

However, most of these economic assessments focus on short-term efficiency and not on long-term sustainability. Indeed, if economics is to contribute to sustainability, it will have to shift from its ingrained short-termism to a long-term perspective. Such a longer-term perspective will need to take the concept of uncertainty seriously (Stirling 2010). At the farm level, uncertainty implies that profitability considerations need to be balanced with issues such as adaptability, flexibility and resilience since they are likely to be key attributes to ensure the long-term survival of a farm (López-Ridaura et al. 2005; Darnhofer et al. 2010).

To contribute to sustainability, economics will also have to reflect the true cost of resources (Pretty et al. 2005). This implies, among other things, the need to internalise currently externalised environmental and social costs (e.g., those related to greenhouse gas emissions or to the global sourcing of phosphate and soy-based animal feed). Indeed, whereas neoclassical economics aims at efficiently satisfying human needs and wants, sustainability aims at justice within and between generations, as well as in the domain of human-nature relationships (Fullbrook 2004). In other words, the tools currently used by economists seem to only partially capture the sustainability of farming. To adequately assess the economic viability of farming within a comprehensive understanding of economic sustainability, a farm should no longer be conceptualised as a profit-maximising firm but, instead, as a 'political economic organisation' (Söderbaum 2008). As such, an organic farm is a product of its history, has social responsibility, farmers are guided by their ideological orientation and make choices based on multiple criteria. Since economic assessments of organic farming are usually based on neoclassical economics, they may well underestimate the economic viability of organic farming.

24.3.4 Diversity

As this very brief overview has shown, there is a range of sustainability issues that are not necessarily addressed by all organic farming practices. However, this is less a failure of organic farming principles, but is instead linked to the diversity of organic farming practices. Indeed, organic farming does not preclude sustainable practices, nor does it impose these practices. It is thus dependent on the actors' choices and on the structural context that may encourage or discourage specific practices that have the potential to address specific sustainability issues.

At the same time as recognising the diversity of organic practices along the agro-food chain, it is helpful to recognise that the various sustainability issues are also being addressed by 'conventional' farming. Indeed, a whole range of alternative practices (e.g., permaculture, agro-ecology, local food, farmers' markets, traditional specialties, fair trade, slow food, community-supported agriculture) are just as likely to be based on organic as on conventional production methods. Through the growth of diversity in organic practices and, simultaneously, the growth of a diverse set of alternative practices, they have come to propose innovative approaches to a range of sustainability issues. What they all have in common is that they are niches developed in response to concerns regarding intensive, conventional agricultural practices that constitute the dominant regime.

24.4 A Transition to Sustainability

24.4.1 A Co-Evolutionary Perspective

What the various studies on the development and practices of organic farming have shown is that while the legal definition of organic standards has promoted its spread, it has not led to a transformation of the whole agro-food chain. However, for organic farming to be a prototype of sustainable agriculture, it not only needs to show that it can effectively address a range of sustainability concerns, but it also needs to show that it can successfully work with the dominant agro-food regime. Indeed, if organic farming remains a niche (one amongst a number of 'alternative' niches) without successfully working with and transforming the dominant regime, it may have only a limited contribution to sustainability.

Transition studies build on a co-evolutionary understanding of change (Kemp et al. 2007). In this perspective, agricultural practices, as well as supermarket and consumer habits are part of an ongoing co-adaptive dynamic, the outcome of which is neither linear nor uniform. The co-adaptive dynamics are fuelled by actors' evolving projects, by actors reacting to changes in their environment, as well as by actors anticipating changes in the context and thus preemptively changing their strategies (Vasileadou and Safarzyńska 2010).

A co-evolutionary understanding also points out that the ability to directly influence the trajectory of organic farming is limited. Thus, the past development of organic farming should not be seen as the direct outcome of strategic decisions by actors within organic farming (Kjeldsen and Ingemann 2009) but, instead, the result of the co-evolution of strategies and actions by organic actors and a range of exogenous factors. Indeed, organic farming had been a niche for well over 30 years before it received broader public recognition and policy support. In much of Europe, the 'breakthrough' of organic farming was not so much the result of one-sided activities by organic actors. Instead, a 'window of opportunity' was created through a constellation of broader societal trends (e.g., policy makers looking for ways to reduce over-production, neoliberalism becoming the dominant perspective, thus pushing for new agro-environmental policies, NGOs raising consumer awareness of health risks related to chemical use in agriculture, etc.). This broader context allowed various actors (e.g., stakeholders in organic farming, policy makers) to build coalitions and design measures that allowed organic farming to spread.

Similarly, the further development of organic farming will not be the direct result of actions taken by organic farmers or organic associations since the efficiency of these actions is context-dependent. How organic farming develops will be the result of the dynamic interaction between the actors and the context in which organic farming is embedded, i.e., farmers' associations, processors, retailers, policy makers, etc. Markets and user practices are co-constructed, changing as new options and new practices arise. This co-construction takes time and involves interactions between producers, retailers and consumers (Shove and Walker 2010; Spaargaren et al. 2012). For example, current, busier lifestyles are making it harder to meet nutritional requirements, so that some consumers are attracted to functional foods (Kearney 2010). It is unclear to what extent unhealthy diets will be addressed through 'more of the same' (e.g., highly processed convenience foods supplemented with functional foods), and to what extent organic and alternative agro-food systems will be successful in their efforts to reshape diets, consumption patterns and lifestyles to be more sustainable. Whereas individuals and groups will become involved with and further shape alternative practices, a system-wide transformation will depend on the spread of changes in the expectations, beliefs and perceptions of consumers, and the strategies of retailers and policy makers.

Various current trends might well converge to create a 'window of opportunity' for organic farming. Indeed, political awareness of a need to transform the agro-food system towards a more comprehensive understanding of sustainability is increasing (e.g., Heinberg and Bomford 2009; SDC 2011; Freibauer et al. 2011). The combined crisis of nutrition, energy, peak oil and climate change emphasize the need for a transition towards sustainability by shedding light on the 'externalities' produced by the industrial agro-food system (Pretty et al. 2005). Research on food regimes (Campbell 2009) has also shown that there might be an emergent regime that is building on various forms of localism promoted by movements such as Slow Food, Via Campesina, and Community Supported Agriculture. There is thus the distinct possibility of a decentralised and ecologically-grounded agro-food system in which the production methods and principles of organic farming could play a central role.

24.4.2 Obstacles and Potentials for a Systemic Transformation Towards Sustainability

To initiate a transition, organic farming needs to identify social innovations that make it possible to overcome the contradictions between the market-driven logic that focuses on input efficiency, high labour productivity and economies of scale, and the ability to enact organic farming's principles and values (e.g., equity, transparency, solidarity and reciprocity) on a broader scale. This search will obviously be constrained by powerful interests (e.g., some large retailers, transnational corporations, national and international bodies) that have high stakes in the current status quo (Jaffee and Howard 2010).

This dominant regime also includes consumer practices, e.g., shopping habits, food preparation practices and dietary choices. A systemic approach to sustainability crucially implies that a growing number of consumers be willing and able to shift their consumption practices (Green and Foster 2005; Shove and Walker 2010). It may include a transition from the currently dominant consumer expectations (convenience, choice, low cost) towards accepting less choice and cooking healthy food sourced regionally within a fair economic system. It may also include issues such as accepting the seasonality of fruits and vegetables, selecting food with less packaging, reducing the number of shopping trips with a car, and becoming involved in some form of local food system. Since the currently dominant practices are deeply embedded in modern lifestyles, only a systemic approach will bring about substantial steps towards a transition to sustainability.

Gillespie (2010) has pointed out that the biggest challenge in achieving a sustainable food system is not defining what it is, but understanding the collective action needed to implement it. While organic farming has done much to define sustainable production practices and to establish their feasibility, its ability to participate in collective action and social learning processes is now challenged. Indeed, for organic farming to be able to effectively adapt the dominant food regime, it will have to harness its internal diversity to propose a coherent answer to challenges such as fair trade relations and healthier diets. To do so, organic farming movements will have to overcome their internal inertia and resistance (Woodward et al. 1996). Currently, within organic farmer associations, there is little consensus on the way forward, e.g., on whether standards of animal welfare should be raised or whether social standards should be mandatory (Milestad et al. 2008). Similarly, little effort is made to find ways to build synergies with other alternative agro-food movements.

The ability of organic farming to enact its potential for being a prototype for a sustainable agro-food system may be contingent on its ability to harness the creative potential inherent in its internal diversity, as well as its ability to build alliances with other alternative agro-food chains. By seeking such alliances, the organic agro-food system may offer opportunities for inclusion, innovation and participation, thus—ironically—getting closer to its own original principles. It could become a discursive space, a space for reflection, communication and experimentation, with alternative social structures, structures that are oriented towards meeting people's needs rather than subjecting them to the market logic (Guthman 2008; Clarke et al. 2008; Niggli et al. 2008).

The question then concerns what social innovations are needed to build sustainable agri-food networks (Lamine et al. 2014 Chap. 23). The challenge for organic farming actors is to recognise the potential and to be able to mobilise a sufficiently large share of their constituency to experiment and commit themselves. This may allow for new development pathways to unfold, taking organic farming into a transition where it embodies a more comprehensive understanding of sustainability. Such synergies may make it possible to push for new policies, policies developed on the basis of inputs from both the agriculture and health sectors. Coherent intersectorial policies that systemically address food, health and the environment could enable the implementation of development policies that support agriculture and farm families, as well as the environment and healthy diets.

Within a co-evolutionary framework, organic farming is involved in a dance where each step redefines what is thinkable and what is possible, where each step challenges what has been achieved so far and extends boundaries. What these next steps are and how they may be coordinated depends on the ability of the organic movements to energise their constituency and build synergies with other movements that seek to promote a more sustainable future. Since future dynamics are a result of emergent co-evolutionary processes between the various actors and networks, they cannot be pre-determined, pre-ordained or imposed.

24.5 Outlook: Transition to Sustainability?

As research on societal transition has shown, any transition to sustainability will imply a substantial degree of socio-cultural change as well as technological change (Elzen and Wieczorek 2005). It is thus unlikely that organic farming will be a prototype for a sustainable food production system if the focus is exclusively on the technical aspects of production (i.e., agronomy and livestock production methods) without requiring more systemic approaches on the farm, the inclusion of social justice issues, more energy-efficient transformation and distribution systems, as well as sustainable consumption patterns. To achieve a transition, there needs to be an emphasis on the co-evolution of technical and societal change. Such a transition will involve the development of new (social) technologies embedded in new economic, social, institutional and cultural relations.

To achieve such a transition, barriers embedded in the current system, as set by production methods, regulations, user practices, cultural values, patterns of behaviour, infrastructure requirements, investment needs, technological lock-ins, power relations, etc., need to be overcome. This is not a trivial endeavour. However, the agro-food system is subject to co-evolutionary processes, subject to ongoing changes. The diversity of organic farming practices testifies to learning, networking, visioning and experimenting processes. These experiments take place both within organic farming and in the interaction between organic farming and other alternative agro-food practices. Within these experiments, various attempts at reintegrating a broader meaning of sustainability are playing out. They include issues such as environmental sustainability, democratic control over economic life, cultural diversity, food access and security, energy use, as well as nutrition and public health.

These social innovations are kernels that, given appropriate opportunities, can engender a transition process. Such a transition is a long-term process and is inherently complex (Elzen and Wieczorek 2005; Spaargaren et al. 2012). Thus, the question is whether organic farming will be able to fruitfully engage in the co-evolutionary process, i.e., by transporting values of deep sustainability to strengthen internal coherence, integrating social justice, building strategic alliances with other alternative food movements and working with consumers to develop healthier diets. The need for a transition towards sustainable agro-food systems is increasingly recognised. The challenge for organic farming is to recognise and build the alliances necessary to take advantage of the 'window of opportunity', thereby growing out of its niche and engendering a broad transition.

References

Adams D, Salois M (2010) Local versus organic: a turn in consumer preferences and willingness-to-pay. Renew Agric Food Syst 25:331–341

Allen P (2010) Realizing justice in local food systems. Cambridge J Reg, Econ Soc 3:295–308

Brown S, Getz C (2008) Towards domestic fair trade? Farm labor, food localism, and the 'family scale' farm. GeoJournal 73:11–22

Clarke N, Cloke P, Barnett C, Malpass A (2008) The spaces and ethics of organic food. J Rural Stud 24:219–230

Campbell H (2009) Breaking new ground in food regime theory: corporate environmentalism, ecological feedbacks and the 'food from somewhere' regime? Agric Hum Values 26:309–319

Coley D, Howard M, Winter M (2009) Local food, food miles and carbon emissions: A comparison of farm shop and mass distribution approaches. Food Policy 34:150–155

Darnhofer I, Bellon S, Dedieu B, Milestad R (2010) Adaptiveness to enhance the sustainability of farming systems. A review. Agron Sustain Dev 30:545–555

DeLind L (2002) Place, work and civic agriculture: common fields for cultivation. Agric Hum Values 19:217–224

DeLind L, Bingen J (2008) Place and civic culture: re-thinking the context for local agriculture. J Agr Environ Ethics 21:127–151

De Wit J, Verhoog H (2007) Organic values and the conventionalization of organic agriculture. NJAS-Wageningen J Life Sci 54:449–462

DFTA (2008) Principles for health, justice and sustainability. Domestic Fair Trade Association. http://www.dftassociation.org/principles. Accessed 31 Oct 2010

Elzen B, Wieczorek A (2005) Transitions towards sustainability through system innovation. Technol Forecast Soc 72:651–661

Freibauer A, Mathijs E, Brunori G, Damianova Z, Faroult E, Girona I Gomis J, O'Brian L Treyer S (2011) Sustainable food consumption and production in a resource-constrained world. 3rd SCAR Foresight Exercise. Standing Committee on Agricultural Research (SCAR)

Fullbrook E (ed) (2004). A guide to what's wrong with economics. Anthem Press, London

Gillespie G (2010) 2009 AFHVS presidential address: The steering question: challenges to achieving food system sustainability. Agric Hum Values 27:3–12

Goldberger J (2011) Conventionalization, civic engagement and the sustainability of organic agriculture. J Rural Stud 27: 288–296

Green K, Harvey M, McMeekin A (2003) Transformations in food consumption and production systems. J Environ Policy Plan 5:145–163

Green K, Foster C (2005) Give peas a chance: Transformations in food consumption and production systems. Technol Forecast Soc 72:663–679

Guthman J (2004) Agrarian dreams. The paradox of organic farming in California. Univ. of California Press, Berkeley

Guthman J (2008) Neoliberalism and the making of food politics in California. Geoforum 39:1171–1183

Heinberg R, Bomford M (2009) The food and farming transition: Toward a post-carbon food system. Sepastopol: Post Carbon Institute. www.postcarbon.org/food. Accessed 29 Oct 2011

Hill SB (2014) Enabling the ecological redesign of organic and conventional agriculture: a social ecology and psychosocial perspective. In: Bellon S, Penvern S (eds) Organic farming, prototype for sustainable agricultures

Jaffee D, Howard P (2010) Corporate co-optation of organic and fair trade standards. Agric Hum Values 27:387–399

Kearney J (2010) Food consumption trends and drivers. Philos T Roy Soc B 365:2793–2807

Kemp R, Loorbach D. Rotmans J (2007) Transition management as a model for managing processes of co-evolution towards sustainable development. Int J Sus Dev World 14:78–91

Kjeldsen C, Ingemann JH (2009) From the social to the economic and beyond? A relational approach to the historical development of Danish organic food networks. Sociol Ruralis 49:151–171

Lairon D, Huber M (2014) Food quality and possible positive health effects of organic products. In: Bellon S, Penvern S (eds) Organic farming, prototype for sustainable agricultures. Springer, Chap. 16, pp 295–312

Lamine C, Bellon S (2009) Conversion to organic farming: a multidimensional research object at the crossroads of agricultural and social sciences. A review. Agron Sustain Dev 29:97–112

Lamine C, Navarrete M, Cardona A (2014) Transitions towards organic farming at the farm and at the local scales: the role of innovative production and organisational modes and networks. In: Bellon S, Penvern S (eds) Organic farming, prototype for sustainable agricultures. Springer, Chap. 23, pp 423–438

Lobley M, Butler A, Winter M (2013) Local organic food for local people? Organic marketing strategies in England and Wales. Regional Studies in press 47:216–228
Lockie S, Halpin D (2005) The 'conventionalisation' thesis reconsidered: Structural and ideological transformation of Australian organic agriculture. Sociol Ruralis 45:284–307
López-Ridaura S, van Keulen H, van Ittersum M, Leffelaar P (2005) Multiscale methodological framework to derive criteria and indicators for sustainability evaluation of peasant natural resource management systems. Environ. Dev Sustain 7:51–69
Luttikholt L (2007) Principles of organic agriculture as formulated by the International Federation of Organic Agriculture Movements. NJAS-Wageningen J Life Sci 54:347–360
Lyson T (2005) Civic agriculture and community problem solving. Cult Agric 27:92–98
McMichael A, Powles J, Butler C, Uauy R (2007) Food, livestock production, energy, climate change and health. Lancet 370:1253–63
Milestad R, Wivstad M, Lund V, Geber U (2008) Goals and standards in Swedish organic farming: trading off between desirables. Int J Agric Resour, Gov Ecol 7:23–39
Nemecek T, Dubois D, Huguenin-Elie O, Gaillard G (2011) Life cycle assessment of Swiss farming systems: I. Integrated and organic farming. Agric Syst 104:217–232
Nemes N (2009) Comparative analysis of organic and non-organic farming systems: A critical assessment of farm profitability. Rome: FAO ftp://ftp.fao.org/docrep/fao/011/ak355e/ak355e00.pdf. Accessed 29 Oct 2011
Niggli U, Slabe A, Schmid O, Halberg N, Schluter M (2008) Vision for an organic food and farming research agenda to 2025. Organic knowledge for the future. Brussels:IFOAM EU Group. http://www.tporganics.eu/upload/TPOrganics_VisionResearchAgenda.pdf. Accessed 29 Oct 2011
Offermann F, Nieberg H, Zander K (2009) Dependency of organic farms on direct payments in selected EU member states: today and tomorrow. Food Policy 34:273–279
Padel S, Zander K, Gössinger K (2010) 'Regional production' and 'Fairness' in organic farming: Evidence from a CORE Organic project. Proceedings of the 9th IFSA Symposium, held 4-7 July 2010, Vienna (Austria), pp 1793–1802
Pratt J (2009) Incorporation and resistance: Analytical issues in the conventionalization debate and alternative food chains. J Agrar Chang 9:155–174
Pretty J, Ball A, Lang T, Morison J (2005) Farm costs and food miles: an assessment of the full cost of the UK weekly food basket. Food Policy 30:1–19
Rosin C, Campbell H (2009) Beyond bifurcation: Examining the conventions of organic agriculture in New Zealand. J Rural Stud 25:35–47
Schäfer M, Kröger M, Wirz A (2010) Mit „alten" Werten die Zukunft gestalten. Eckpunkte für ein faires Miteinander entlang der Bio-Wertschöpfungskette [Corner stones for a fair cooperation along the organic value chain]. Broschure. Technical University Berlin, Berlin. http://orgprints.org/17275/1/17275-Bio-Fair-Broschuere_2010.pdf. Accessed 6 Jan 2011
SDC (2011) Looking back, looking forward. Sustainability and UK food policy 2000-2011. Sustainable Development Commission, London
Shove E, Walker G (2010) Governing transitions in the sustainability of everyday life. Res Policy 39:471–476
Smith E, Marsden T (2004) Exploring the 'limits to growth' in UK organics: Beyond the statistical image. J Rural Stud 20:345–357
Söderbaum P (2008) Understanding sustainability economics. Towards pluralism in economics. Earthscan, London
Spaargaren G, Oosteveer P, Loeber A (2012) Sustainability transitions in food consumption, retail and production. In: Spaargaren G, Oosterveer P, Loeber A (eds) Food practices in transition. Changing food consumption, retail and production in the age of reflexive modernity. Routledge, New York, pp 1–31
Stassart P, Jamar D (2008) Steak up to the horns! The conventionalisation of organic stock farming: knowledge lock-in in the agrifood chain. GeoJournal 73:31–44
Stirling A (2010) Keep it complex. Nature 468:1029–1031
Stolze M, Piorr A, Häring A, Dabbert S (2000) The environmental impacts of organic farming in Europe. Organic Farming in Europe: Economics and Policy, Vol 6. University of Hohenheim, Stuttgart

Thompson J, Scoones I (2009) Addressing the dynamics in agri-food systems: an emerging agenda for social science research. Environ Sci Policy 12:386–397

Vasileadou E, Safarzyńska K (2010) Transitions: taking complexity seriously. Futures 42:1176–1186

Wier M, O'Doherty Jensen K, Mørch Andersen L, Millock K (2008) The character of demand in mature organic food markets: Great Britain and Denmark compared. Food Policy 33:406–421

Winter M (2003) Embeddedness, the new food economy and defensive localism. J Rural Stud 19:23–32

Woodward L, Fleming D, Vogtmann H (1996) Health, sustainability, the global economy-the organic dilemma. Reflections on the past, outlook for the future. Paper of a presentation given at the 11th International IFOAM conference, held in Copenhagen, Denmark

Yakovleva N, Flynn A (2009) Organic production: the adoption of a niche strategy by the mainstream food system. Int J Innov Sustain Dev 4:43–60

Chapter 25
Multi-Scale Integrated Assessment of Regional Conversion to Organic Farming (OF)

Santiago Lopez-Ridaura, Sylvestre Delmotte, Christophe Le Page, Laure Le Quéré, Gaël Goulevant, Philippe Chauvelon, Alain Sandoz and Jean-Claude Mouret

Abstract Is the regional conversion to Organic Farming (OF) possible? What could be the consequences at different scales in terms of social and economic development as well as nature conservation? Taking the heterogeneity of farms and farming systems in the region into account, are there farmers more prone to conversion and others that face greater obstacles?

The objective of this paper is to shed light on some of these question by presenting the results of a scenario assessment carried out with regard to the extension of OF in the Camargue region in southern France. The application of different modelling approaches with great potential for the multi-scale and multi-criteria evaluation of the extension of OF is presented: bio-economic models, agent-based models and land-use/cover change models. According to our results, the most probable conversion in the near future in the Camargue would take place in fields with low salt pressure that belong to livestock breeders and diversified cereal producers. However, the regional conversion to OF is plausible since the region could maintain its economic productivity while decreasing potential harmful impacts on the environment. Finally, the possible conversion trajectories suggest that certain farmers (specialised in rice production) might need greater assistance to ensure such conversion to OF since their economic performance would be hampered during that period.

The application of these three approaches to explore the same scenario in one region revealed their complementarity for tackling the complex issue of regional conversion to OF from different angles.

S. Lopez-Ridaura (✉) · S. Delmotte · L. Le Quéré · G. Goulevant · J.-C. Mouret
INRA-SupAgro-CIRAD, UMR0951 Innovation, 34000, Montpellier, France
e-mail: ridaura@supagro.inra.fr

C. Le Page
CIRAD, UPR Green, 34000, Montpellier, France

P. Chauvelon · A. Sandoz
Tour du Valat, Research Centre for the Conservation of Mediterranean Wetlands, 13200, Arles, France

S. Bellon, S. Penvern (eds.), *Organic Farming, Prototype for Sustainable Agricultures*,
DOI 10.1007/978-94-007-7927-3_25,

Keywords Scenario assessment · Trade-offs · Trajectories · Rice · Bio-economic models · Agent-based models · Land-use change analysis · Camargue region

25.1 Introduction

Organic farming (OF) has reached the status of a viable option for more sustainable agriculture among farmers, policy makers and consumers. In 2009, it was practiced on over 3.5 million hectares in the EU25, representing 5.1 % of the agricultural area. In some countries such as Austria, Switzerland, Sweden, Estonia and the Czech Republic, OF occupied more than 10 % of the agricultural area in 2009, and in other countries such as Spain, Greece and Portugal, the surface area allotted to OF has tripled in the last ten years.[1]

This rapid growth of OF suggests new horizons for the future evolution of agriculture in general and of organic agriculture in particular. At the farm scale, the conventionalisation of OF has been widely discussed by Darnhofer et al. (2010) with, on the one hand, an increase in the size of organic farms and their specialisation and, on the other, higher intensification of agricultural practices (e.g., greater use of concentrate and disease treatments for animals and the intensive use of organic fertilisers on arable land).

The effects of this extension of OF creates new challenges for agricultural research where new questions are arising as to the effect of total or partial conversion to OF of a given region (Acs et al. 2007).

For example, is the regional conversion to OF possible? What would its consequences be at different scales in terms of social and economic development as well as nature conservation? Taking the heterogeneity of farms and farming systems in a region into account, are there farmers more prone to conversion and others that face greater obstacles?

To answer such questions, a prospective analysis is needed to assess different future scenarios on the extension of OF. Such assessments must be able to take the different objectives (e.g., economic, environmental, social) of stakeholders related to agriculture in a given region into account by means of indicators relevant to the scales at which they operate (e.g., field, farm, watershed, region).

Such a multi-scale integrated assessment of scenarios is the core of this chapter. To illustrate the assessment of scenarios related to the extension of OF, different approaches based on modelling tools have been developed and applied to the Camargue region in southern France. The objectives of this chapter are: (i) to present the results of the assessments of different scenarios on the extension of OF in the Camargue; and (ii) to identify the potential complementarities of the different modelling approaches for a multi-scale, integrated assessment of scenarios.

In the following section, we briefly describe three modelling approaches used in the prospective analysis of agricultural change (bio-economic models, agent-based

[1] http://epp.eurostat.ec.europa.eu/portal/page/portal/agriculture/data/database. Accessed in April 2010.

models and land-use/cover change models) and the Camargue case study. Section 3 shows the results of the application of these approaches to assess the extension of OF in the Camargue, and Section 4 discusses some of the main findings of the scenario analysis and the complementarities of the three methods to make a prospective multi-scale, integrated assessment of agricultural change.

25.2 Three Approaches, One Case Study

25.2.1 Approaches for Prospective, Multi-Scale and Integrated Assessment of Agricultural Systems

In relation to agricultural systems and land use, we identified three approaches commonly used for scenario analysis (Delmotte 2011; Delmotte et al. 2013b): (i) land-use change modelling (LUC); (ii) bio-economic modelling (BEM); and (iii) agent-based modelling (ABM). These three approaches are briefly described in the next paragraphs.

The objective of LUC models is to describe the actual land use and to provide insight into the possible changes of land-use patterns that would occur in the near future following either some biophysical or demographic (Veldkamp and Fresco 1996), or economic and structural changes (Verburg et al. 2004). LUC approaches cover a wide range of methods but most of them are "descriptive models that aim at simulating the functioning of the land use system and the spatially explicit simulation of near future land use patterns" (Verburg et al. 2004).

LUC models are based on the identification of drivers that are correlated with the observed past or current land use. These driving factors can be socio-economic aspects such as demography, commodity demand and infrastructure development (e.g., a new road or the presence of a market), or biophysical aspects such as type of soil and climate. This makes it possible to identify the most probable spots of changes, i.e., locations where changes are more likely to happen. LUC has been widely applied in different case studies and different models exist in the literature for deforestation (Pontius et al. 2001), urban extension (White and Engelen 2000) and agriculture (Verburg et al. 2002). In CLUE (Conversion of Land Use and its Effect[2]), correlations are used to evaluate what would be the probable change in land use following, for example, a change in commodity demand, policy instruments and infrastructure development (de Koning et al. 1999).

Bio-economic models (BEM) aim at identifying optimum allocation of agricultural activities (e.g., cropping, livestock), in space and time, which maximise or minimise an objective (see Jansen and van Ittersum (2007) for a review on BEM). In integrated assessment, optimum systems are obtained by including several criteria and indicators, using a Multiple Goal Linear Programming model (MGLP) (van Keulen 1990). In MGLP, one criterion is defined as the objective function

[2] http://www.cluemodel.nl/.

and the other criterion is set as the constraint. This optimisation has been done for objectives defined at different scales, most commonly at the farm (Janssen and van Ittersum 2007) and regional scales (Laborte et al. 2007).

In BEM, cropping and livestock activities are quantitatively described at the scale of the field or livestock unit in terms of their contribution to indicators at higher levels of aggregation (e.g., the farm, the region). Such technical coefficients that describe agricultural activities include the inputs needed (e.g., capital, labour, fertilisers, pesticides, fuel) and the desired as well as undesired outputs (e.g., grain yield, N lost). These technical coefficients are the building blocks in the model and the data used as a basis for optimisation. Systems (e.g., farms or regions) are described by their resources (e.g., area of lands and their soil types, water for irrigation, labour available), which are set as constraints for optimisation. Different types of scenarios can be evaluated such as the impacts of a change of context (change of prices or subsidy levels) and the impacts of the introduction of an alternative system (e.g., organic farming) (Lopez Ridaura 2005).

Agent-based models (ABM) represent systems as agents in interaction, with a social structure and that use resources in an environment. Agents perceive, self-represent and act in their environment by making decisions and interacting with other agents. Each agent has its own tendencies and objectives (Bousquet and Le Page 2004; Ferber 2006). ABM is an approach originally developed in computer sciences to study the dynamics of complex systems and to reproduce phenomena that emerge from the addition and interactions of individual behaviours. ABM can be based on multiple formalisms for representing decision-making by the agents. In the case of human agents, decision-rules are often defined with thresholds and if-then rules.

These individual centred approaches are increasingly used to represent nature-society interactions (Ligtenberg et al. 2004; Monticino et al. 2007), particularly in the domain of natural resource management (NRM) (Mathevet et al. 2003; Bousquet and Le Page 2004). Specific platforms such as Cormas® (Bousquet et al. 1998) or NetLogo (Wilensky 1999) make it possible to create simulations where agents interact with one or more resources. In the context of NRM systems, ABM often integrates a spatial representation of the land to represent the decision-making process of individuals deciding on the use of these spatial units (Bousquet and Le Page 2004). This makes it possible to study the interaction between the resources and the agents' decisions in a dynamic manner and to calculate the impact of these decisions at different aggregation levels.

25.2.2 The Camargue Case Study

In Camargue, a deltaic region in the south of France, OF has been presented as a potential way for reducing the externalities of current agricultural practices. Agriculture plays a crucial role in the economic, ecological and social equilibrium of the region. The region has been recognised as a Biosphere Reserve (Man and Biosphere

Program of UNESCO) since 1977, and hosts a natural regional park, a national reserve and many other associative and private protected areas.

Farming systems based on the production of cereals and on livestock breeding represent 37,000 ha in the Camargue. The main crops of the region are rice, durum wheat and, in smaller quantities, sunflower, maize, oil seed rape and sorghum. Irrigated rice is the main cropping activity with about 20,000 ha devoted to it each year. Irrigated rice can grow in the four main soil types of the region: deep soils are sandy or loamy-clay soils since sandy soils are less favourable for rice cultivation due to the difficulty of maintaining water in highly draining soils. Shallow soils are clay loamy or salty and hydromorphic soils. Both require a high frequency of rice cultivation (a minimum of one year out of three in salty and hydromorphic soils to desalinate and allow the production of rainfed crops (Mailly et al. 2013).

Cropping systems play a crucial role in the water dynamics of this deltaic region. Most of the land is at sea level and salinisation is a natural process due to the negative water balance between rain and evapo-transpiration. Irrigation of rice then plays a role in desalinating the soils. Irrigation water that enters through pumping from the Rhone River also has a key function in maintaining the level of water and salt concentration of the central lagoon of the Camargue, the Vaccarès Lagoon (Colour plate 25 Fig. 1), which is the temporary habitat of several migrating bird species.

However, conventional rice production uses large quantities of pesticides, mainly herbicides. These herbicides are dispersed throughout the environment and, given the high diversity and interest of the local fauna and flora, ecologists have long called for a reduction in the use of pesticides (Comoretto et al. 2008). The extension of OF in the Camargue would certainly imply a decrease in the area of irrigated rice due to the difficulty of managing weeds in these systems (Mailly et al. 2013). In fact, one crop of organic rice on a single field has to be separated by at least five other crops (i.e., by five years), whereas in conventional cropping systems, it is possible to grow rice on a continuous basis by using herbicides (Lopez Ridaura et al. 2010; Mouret 2010).

For the evaluation of scenarios on the extension of OF in the Camargue, a wide range of cropping activities were defined and quantitatively described. The definition of the agricultural activities is based on the concepts developed by Hengsdijk et al. (1999) and corresponds to the combination of a biophysical environment and a crop or animal activity under specific management. Sixteen hundred agricultural activities are possible taking these features into account. However, due to the lack of references concerning OF systems in the region, only one intensity level was described for organic crops. All together, we obtained 1,200 possible agricultural activities. Each activity was described quantitatively by the calculation of different technical coefficients such as the labour demand, inputs used or the cost of production. Inputs for the different activities (fertilisers, pesticides, seeds, machinery, labour) were calculated based on several technical reports from the region, more than 20 reports from students who interviewed farmers and a series of interviews with key farmers to complete the data (LeQuere 2010). Rice yield was estimated based on a database analysis containing more than 350 fields surveyed in different years (Delmotte et al. 2011). Yield for other crops was estimated together with experts from local technical institutions and from average yields reported for the region by the grain millers. For the calculation of economic indicators, average prices of inputs declared by input suppliers and crop prices declared by the cooperative from 2009 and 2010 were used. Current subsidy levels were used and obtained from farmers' interviews.

To up-scale the assessment from field to farm level, a farm typology was built (Delmotte 2011). In the Camargue, about 180 farmers depend on crop production for their economic viability, according to the rice producers' union. Using several databases, we developed a farm typology based on farm size, the proportion of rice in their cropping system, the presence of livestock activities, and their orientation in terms of conventional or organic management. This typology resulted in nine farm types: (1) specialised large-size rice producers (farm area greater than 265 ha, more than 80 % of area cultivated with rice) (16 % of the total cultivated area of the territory); (2) specialised middle-size rice producers (farm area less than 265 ha, more than 80 % planted with rice) (20 % of the total cultivated area of the territory); (3) large-size rice producers (farm area greater than 267 ha, between 60 and 80 % cultivated with rice) (9 % of the total cultivated area of the region); (4) middle-size rice producers (farm area less than 267 ha, between 60 and 80 % cultivated with rice) (12 % of the total cultivated area of the region); (5) partially organic rice producers (same land use as middle-size rice producer but an average of 20 % of land under in organic) (18 % of the total cultivated area of the region); (6) livestock breeders (around 35 % rice, 35 % forage crops and 30 % other crops) (5 % of the total cultivated area of the region); (7) organic livestock breeders (same land use as livestock breeders but with partial or total area in organic) (9 % of the total cultivated area of the region); (8) diversified crop producer (more than 50 % of durum wheat and other crops, an average of 35 % rice) (9 % of the total cultivated area of the region); (9) organic diversified crop producer (same land use as diversified crop producer but with partial or total area in organic) (2 % of the total cultivated area of the region). Due to the existence of geo-referenced datasets with the farm perimeters in the region, a spatial distribution of different farm types in the region and sub-regions was possible.

25.3 Scenario Assessment on the Extension of Organic Agriculture in the Camargue

25.3.1 Probable Spots of Change to OF in the Camargue (LUC)

Land-use change models may help to identify the most probable fields and farms to be converted to OF based on past trends of conversion in the Camargue region. Instead of using external drivers like in most of the LUC analyses, an agronomic point of view was adopted where the cropping system characteristics where taken into account (i.e., soil suitability and crop rotations).

A shared vision among agronomists working in the Camargue is that converting to OF implies the lengthening of the crop rotation and the consequent reduction of the area devoted to rice, in order to avoid weed pressure (Delmotte et al. 2011). Not all soils in the Camargue are equally suitable for this conversion since shallow soils might present salinity problems after a few years of rainfed crops. Since different farm types have different soil type distribution, it is possible to identify the most probable soil types and farm types to be converted to OF in the future by the intensity of rice within the crop rotation.

To understand land use and land-use change, a geo-referenced dataset from La Tour de Valat[3] was used. This database, covering large areas of the Camargue, contains fields under rice production from 1998 to 2008 (except for 2007 that was missing). All together, 9,130 fields are described in these databases. Crossing these databases with the soil map[4] and the spatial distribution of farm types, we calculated two indices to identify past trends in terms of rice intensity: (i) the change in the proportion of rice production at the farm scale for each farm type; and (ii) the frequency of rice over the 11 years for each field in relation to the soil type.

Colour plate 25 Fig. 2a shows the evolution of the proportion of surface area devoted to rice production per soil type. At the sub-regional scale, it can be seen that rice production does not occur with the same frequency on the different soils. Fields with shallow soils are the ones that are the most frequently cultivated with rice since between 45 to 55 % of the area is cultivated with rice each year. About 37 % of the alluvial hydromorphic fields were cultivated with rice in 1998. However, this figure increased up to nearly 48 % in 2008, a level close to the one of shallow clay loamy soils. It can be seen that the deep loamy clay soils had a slightly lower frequency of rice in 2008 than shallow clay loamy soils and salty and hydromorphic soils, while deep sandy soils were always managed with lower rice frequency.

The conversion to OF of fields with shallow clay loamy soils and salty and hydromorphic soils would be less probable, which may be explained by the need to desalinate these soils since they are the most prone to salinity problems. Farmers with a high proportion of these types of soils will also probably face greater problems converting to OF than farmers with higher proportions of deep soils.

Colour plate 25 Fig. 2b shows the evolution of the percentage of fields cultivated on rice for the nine farm types. First of all, it should be observed that the typology, which was done on an independent dataset, is validated by this figure: livestock breeders have a lower proportion of rice on their farms than diversified crop producers (either organic or conventional) and rice producers or specialised rice producers. Both systems that are totally in organic (livestock breeders and diversified crop producers) have a stable proportion of rice that is always lower than 0.5. However, partially organic rice producers have a proportion of rice that is not different than

[3] http://www.tourduvalat.org/.

[4] http://www.gissol.fr/programme/bdgsf/bdgsf.php.

that of non-organic rice producers. This corresponds to the partial conversion (e.g., 20 %) of the farm area into organic farming.

Conversion to OF therefore seems to be possible in two ways: attaining a lower rice proportion in land use, the situation of diversified crop producers, or converting only a part of the farm to OF as shown by the partially organic rice growers that maintain a high proportion of the farm in conventional management.

Colour plate 25 Fig. 2c represents all farms on the basis of the past (1998–2005) standard deviation of the percentage of rice (X axis), to identify the historical variability of rice production on farms in relationship to the current proportion of rice production represented by the deviation between the average of 1998–2005 and the average of 2006 and 2008 (considering that 2007 data was missing) (Y axis).

Colour plate 25 Fig. 2c can be interpreted by grouping the farms. The "stable" group includes farms that have a low variation of area devoted to rice (low standard deviation), meaning that the production is quite stable. The "variable" group includes farms that have neither a clear decrease nor increase in surface area under rice, but where a high standard deviation exists (the difference between 2006–2008 and 1998–2005 averages is small). Finally, two groups reveal interesting trajectories in terms of conversion to OF: the farms that belong to the "increase" group are characterised by a positive difference between 1998–2005 and 2006–2008. These farms mainly belong to the "specialised rice producer" and "rice producer" categories. Their trend towards increasing the proportion of rice on the farm does not seem to be favourable for future conversion to OF because, as said before, OF requires a lower proportion of rice within the rotation to avoid herbicide use. The farms contained in the "decrease" group have a tendency to reduce their rice production area. These farms belong to the "rice producer" and the "diversified crop producer" groups that have been decreasing the proportion of rice production in the last 10 years and, therefore, seem more prone to be converted to OF.

This LUC analysis using a farm typology and a retrospective analysis of rice frequency per soil type shows that farms have different possibilities of converting to OF in relation to the proportion of different soil types present on the farm.

- Diversified crop producers and livestock breeders, with their current low rice intensity, have favourable conditions for the conversion to OF. However, at the regional scale, if all these farms were totally converted to OF, it would represent less than 15 % of the area.
- Other farm types with a high frequency of rice and less prone to conversion represent more than 50 % of the conventionally-managed agricultural area today. The partial conversion to OF of these farms on favourable soil might represent a possibility for increasing the proportion of the region under OF.

25.3.2 Plausible Scenarios for OF in the Camargue (BEM)

Two BEMs were developed that allocate optimum agricultural activities in relation to different indicators: one model that optimises land use at the farm scale and another one that optimises land use at the regional scale. Twelve indicators representing the economic, social and environmental criteria related to agricultural issues were included in the models. At the farm level, these indicators included, among others, the gross margin of the farm and of the different agricultural activities, labour required, production costs and pesticides. At the regional level, indicators included the value of agricultural production, the employment generated, the level of subsidies, water and pesticide use, among others. In both models, all indicators can be either maximised or minimised depending on the objectives of stakeholders at different scales. Indicators may also serve as constraints for optimisation (see Delmotte (2011) and Delmotte et al. (2013b) for details about the model).

The results presented here are related to the maximisation of economic indicators (i.e., gross margin and value of agricultural production for the farm and the regional level, respectively) with either conventional activities, organic or both.

Colour plate 25 Fig. 3a presents a radar graph with six major indicators for a specialised rice producer farm under the current situation and when the gross margin is maximised with either conventional or organic activities. Values presented are normalised in relation to the best value for each indicator (the outer circle of the radar). Organic and conventional activities provide similar gross margin, costs, subsidies and labour. Water used in the current situation is similar to that under organic production (in both cases, less water is used than in the conventional optimisation). In terms of pesticide use, it can be seen that the reduction of pesticides used can be achieved without considerably decreasing the gross margin and can even improve it.

Compared to the current situation, it can be seen that the gross margin can be nearly doubled with both conventional and organic forms of production. However, it has to be taken into account that the MGLP optimises a single year and allocates most land under irrigated rice and rainfed cereal production (sorghum or maize or wheat) with a large proportion under rice. In the agricultural activities used by the model, only the preceding crop is taken into account, whereas in the current situation, longer rotations are used, including alfalfa, and the level of economic productivity may therefore decrease for the previous and/or following years.

In Colour plate 25 Fig. 3b, two contrasted scenarios maximise the value of agricultural production at the regional scale with either conventional or organic production. While the conventional scenario shows marginally better values of agricultural production and employment, the organic scenario results in lower levels of subsidies, pesticide use and fuel consumption. In the conventional scenarios, rice is chosen as the main crop (67 %), which implies a high level of mechanisation of agriculture and a consequent increase in labour demand (employment), as well as higher fuel consumption.

Water use is less than half in the organic scenario than in the conventional one. However, it cannot be said that the less water used, the better, since it plays a crucial role in the ecological functioning of natural wetlands. Too little of it might result in an increase in the salt concentration and too much of it might, besides decreasing the salt concentration in the water, increase the water level of the central Vaccarès Lagoon.

25.3.3 Conversion Pathways to OF in the Camargue (ABM)

We developed an ABM for assessment of scenarios by interactive simulation with farmers and the analysis of their adaptation strategies (Delmotte 2011; Delmotte et al. 2013a). The ABM was developed under the Cormas® platform (Bousquet et al. 1998) and is based on individual interfaces for each farm type. Each participant represented a farmer of a defined farm type having specific resources in terms of farm size, soil type distribution and initial condition of cropping system.

The scenarios evaluated with farmers included issues related to subsidy suppression following the PAC reform (Delmotte et al. 2013a). Results presented here related to the conversion towards OF were obtained during interactive simulation sessions with nine MSc students from Sustainable Crop Production in 2010 (SupAgro Montpellier), each representing one farm type. Each participant decided which agricultural activity to allocate in each field. Their decision concerned the choice of crops, style of production (conventional or organic) and the level of inputs for each field. Participants also had to consider the total area of each crop on a given year at the farm scale, and the preceding/following crop couple at the field scale, in order to maintain coherent rotations. The interactive simulation was conducted for seven time steps, a time step corresponding to a year. Students were given the objective to partially or totally convert their farm to OF while maintaining their gross margin and the labour demand at the farm scale as much as possible. During the two-year conversion period to organic agriculture, inputs and outputs for organic activities were used while conventional production prices were maintained. Once choices were made by each player for a time step, indicators were calculated at the farm and regional scales. After each time step, discussions were held among participants about their individual performance and its effects at the regional scale.

In Colour plate 25 Fig. 4a, the evolution of the gross margin and the progressive conversion to OF is shown for two different farm types, a middle-size rice producer and a livestock breeder. It can be seen that for the livestock breeder, the conversion to OF has little effect, with the gross margin quite stable throughout the simulation. For the specialised large-size rice producer, the diversification of production and, therefore, a reduction in the surface area devoted to rice are implied. At the end of the simulation, this type of farm will possibly have a higher gross margin value since once the transition period has ended, the prices of organic products and subsidies for maintaining OF are used for the calculation of the gross margin. Figure 4b illustrates the evolution of different indicators for the rice producer. While water use and pesticide use are greatly reduced due to conversion to OF, the proportion of subsidies remains similar and working time increases, demonstrating the multiple criteria, beyond the gross margin, that a farmer must take into account during the conversion to OF.

Applying this ABM made it possible to determine that different farm types have different capabilities for converting to OF. The conversion of the specialised large-size rice producer implies a reduction of the gross margin in the first year of conversion, even if the conversion is supported by a subsidy of € 150 per hectare, as was the case in the simulation. There is less impact on the livestock breeder because he is less dependent on rice and has a greater crop diversity on the farm.

25.4 Regional Conversion to OF: Probable Changes, Plausible Futures and Possible Trajectories

In this chapter, we have shown three approaches that contribute to assessing scenarios of agricultural development and their application to the regional conversion to OF in the Camargue in southern France. These approaches address different and complementary issues related to the regional conversion to OF.

The LUC model allowed us to identify the *most probable spots for conversion to OF in the Camargue*. By analysing the past trends of farming systems in the Camargue, livestock breeders with a high proportion of shallow clay loamy soils and salty and hydromorphic soils (mainly under pasture) and diversified cereal producers are the farm types the most prone to switching to organic cropping systems due to a suitable combination of soil types and farming orientation to lengthen the rice rotation. Specialised and non-specialised rice producers might restrict themselves to a partial conversion to OF in specific fields with suitable soil types.

The main conversion trend in the near future will most likely take place in the deep soil of farms with a low proportion of rice in their cropping systems. However, this expected change concerns only 20 % of the arable land of the Camargue. Other farmers such as the specialised rice producers and rice producers with a fair proportion of shallow clay loamy soils and salty and hydromorphic soils, representing 45 % of the arable land of the region, would see conversion as a much more difficult task. In any case, it seems improbable that we will see a spontaneous (*vs.* assisted) conversion to OF in the near future.

The BEM model allowed us to *explore options* in what could be called *plausible futures* (van Ittersum et al. 1998). It also allowed us to calculate several indicators for multi-criteria analysis of scenarios revealing some of the trade-offs among indicators in the event of total conversion to OF. Total conversion of the region to OF is plausible. At current prices, the region would not lose in terms of economic productivity by converting to OF and, at the same time, it would protect the environment from the potential harmful effects of pesticides (Comoretto et al. 2007; Höhener et al. 2010). However, the required extensification of cropping systems (i.e., less rice in the rotation) will plausibly have negative effects on employment generation in the sector. Moreover, the volume of fresh water that is pumped into the delta might decrease and have possible effects on the level and salinity of the central lake and on the conservation of the wild habitat of fresh water species.

With the BEM, trade-off curves can be quantitatively described by maximising one indicator while setting another one as the constraint and progressively relaxing it (Lu and Van Ittersum 2004; Lopez Ridaura 2005). Such a curve might help to better understand the trade-off between productivity and the volume of water used in relation to the regional conversion of the Camargue to OF and to identify an optimal solution where both objectives are simultaneously satisfied. However, non-linearities related to the spatial distribution of fields under irrigated rice in the region and their interaction, which governs the volume of water that actually enters the lake, might not be captured by this approach, and other types of modelling would therefore be needed, e.g., agent-based modelling.

The ABM model presented here was developed for interactive simulation to elucidate *the possible trajectories towards conversion to* OF. In the participative exercise presented here, it can be seen that the transition to OF by farms specialised in rice production is much more difficult than for other types of farms, confirming what was seen in the LUC model. However, depending on the conversion trajectory, it can be seen that after several years, the profit for these farmers can almost be recovered (confirming the results of the BEM at the farm scale). These results suggest that conversion to OF for these types of farms may not be spontaneous and that greater support might be needed, at least during the conversion phase. Our current actions related to the development and use of the ABM are directed towards the organisation of participatory sessions of interactive simulation with farmers concerning the conversion to OF, in order to have a more realistic picture of the trajectories they would take, and towards the inclusion of a spatiallyexplicit water balance calculation to capture the non-linearities mentioned above.

The application of the three approaches for assessing the extension of OF in the Camargue provided the following lessons: (1) the regional conversion to OF in the Camargue is plausible; (2) the most probable spots for change in the near future are fields with enough drainage to avoid salinisation problems from livestock breeders and diversified cereal producers; and (3) the possible conversion trajectories suggest that certain farmers (specialised in rice production) might need greater assistance to ensure such a conversion to OF since their economic performance may be hampered during that period.

Methodologically speaking, the simultaneous application of these three approaches to explore the same scenario in one region revealed their complementarities for tackling a complex issue such as the regional conversion to OF from different angles: the LUC model provided information about the most probable spots of change, the BEM model made it possible to explore different futures and to evaluate their plausibility, while the ABM model focused on possible trajectories to attain a given objective. The use of these tools in a participatory manner with local stakeholders might certainly contribute to the common reflection and possibly to the development of joint actions to encourage such conversion to OF and ensure an economically-efficient agriculture while reducing its possible environmental impacts.

Acknowledgments We would like to acknowledge the financial support of ADEME (the French Environment and Energy Management Agency) and the CIAB (Internal Committee for Organic Agriculture) of INRA (the French National Agronomic Research Institute) that allowed us to conduct this research, the students of Montpellier SupAgro and ENSAIA Nancy who participated in the interactive simulation and, finally, Servane Penvern for her useful comments on an early draft of this manuscript.

References

Acs S, Berentsen PBM, Huirne RBM (2007) Conversion to organic arable farming in The Netherlands: a dynamic linear programming analysis. Agric Syst 94:405–415

Bousquet F, Bakam I, Proton H, Le Page C (1998) Cormas: common-pool resources and multi-agent Systems. Lecture Notes in Artificial Intelligence 1416:826–838

Bousquet F, Le Page C (2004) Multi-agent simulations and ecosystem management: a review. Ecol Model 176:313–332

Comoretto L, Arfib B, Chiron S (2007) Pesticides in the Rhône river delta (France): Basic data for a field-based exposure assessment. Sci Total Environ 380:124–132

Comoretto L, Arfib B, Talva R, Chauvelon P, Pichaud M, Chiron S, Höhener P (2008) Runoff of pesticides from rice fields in the Ile de Camargue (Rhône river delta, France): Field study and modeling. Environ Pollut 151:486–493

Darnhofer I, Lindenthal T, Bartel-Kratochvil R, Zollitsch W (2010) Conventionalisation of organic farming practices: from structural criteria towards an assessment based on organic principles. A review Agron Sustain Dev 30(1):67–81

de Koning GHJ, Verburg PH, Veldkamp A, Fresco LO (1999) Multi-scale modelling of land use change dynamics in Ecuador. Agric Syst 61:77–93

Delmotte S (2011) Participatory assessment of scenarios: perspectives for agricultural systems in Camargue, South of France. SupAgro Montpellier, PhD thesis, pp 380. http://tel.archives-ouvertes.fr/tel-00664304

Delmotte S, Barbier J-M, Mouret J-C, Le Page C, Wery J, Chauvelon P, Sandoz A, Lopez-Ridaura S (2013a) Participatory integrated assessment of agricultural systems at different scales in Camargue, South of France. Submitted to Ag Sys

Delmotte S, Lopez-Ridaura S, Barbier J-M, Wery J (2013b) Prospective and participatory integrated assessment of agricultural systems from farm to regional scales: Comparison of three modeling approaches. J Environ Manage 129:493–502

Delmotte S, Tittonell P, Mouret J-C, Hammond R, Lopez Ridaura S (2011) On farm assessment of rice yield variability and productivity gaps between organic and conventional cropping systems under Mediterranean climate. Europ J Agron 35(4):223–236

Ferber J (2006) Concepts et méthodologies multi-agents. In: Amblard F, Phan D (eds) Modélisation et simulation multi-agents, applications pour les Sciences de l'Homme et de la Société. Hermes-Sciences & Lavoisier, Londres

Hengsdijk H, Bouman BAM, Nieuwenhuyse A, Jansen HGP (1999) Quantification of land use systems using technical coefficient generators: a case study for the Northern Atlantic zone of Costa Rica. Agric Syst 61:109–121

Höhener P, Comoretto L, Al Housari F, Chauvelon P, Pichaud M, Cherain Y, Chiron S (2010) Modelling anthropogenic substances in coastal wetlands: application to herbicides in the Camargue (France). Environ Model Softw 25:1837–1844

Janssen S, van Ittersum MK (2007) Assessing farm innovations and responses to policies: A review of bio-economic farm models. Agric Syst 94:622–636

Laborte AG, Van Ittersum MK, Van den Berg MM (2007) Multi-scale analysis of agricultural development: a modelling approach for Ilocos Norte Philippines. Agric Syst 94:862–873

LeQuere L (2010) Création d'une base de données pour le calcul de coefficients techniques culturaux des systèmes agricoles camarguais. Rapport de stage de fin d'étude, INRA UMR Innovation AgroParisTech

Ligtenberg A, Wachowicz M, Bregt AK, Beulens A, Kettenis DL (2004) A design and application of a multi-agent system for simulation of multi-actor spatial planning. J Environ Manage 72:43–55

Lopez Ridaura S (2005) Multi-scale sustainability evaluation of natural resource management systems: quantifying indicators for different scales of analysis and their trade-offs using linear programming. Int J Sust Dev World 12:81–97

Lopez Ridaura S, Mouret JC, Hammond R (2010) Accompanying farmers in the search of technical alternatives for Organic Rice Production in Camargue, south of France In Proceedings of "Agro2010 the XIth ESA congress", Montpellier, August 29th–September 3rd, 2010. ISBN: 978-2-909613-01-7. p 2

Lu CH, Van Ittersum MK (2004) A trade-off analysis of policy objectives for Ansai, the Loess Plateau of China. Agric Ecosyst Environ 102:235–246

Mailly F, Delmotte S, Schaller N, Mouret J-C, Lopez-Ridaura S, Barbier J- M (2013) A cropping plan decision model in conventional and organic rice cropping systems to predict land use changes under various scenarios: The case of the Camargue Delta, Southern France. Cah Agric 22:424–431

Mathevet R, Bousquet F, Le Page C, Antona M (2003) Agent-based simulations of interactions between duck population, farming decisions and leasing of hunting rights in the Camargue (Southern France). Ecol Model 165:107–126

Monticino M, Acevedo M, Callicott B, Cogdill T, Lindquist C (2007) Coupled human and natural systems: a multi-agent-based approach. Environ Model Softw 22:656–663

Mouret JC (2010) L'évolution des pratiques agricoles en relation avec la mise en oeuvre d'une riziculture durable. 40ème anniversaire du Parc Naturel Régional de Camargue, Arles

Pontius RG, Cornell, J.D., Hall CAS (2001) Modeling the spatial pattern of land-use change with GEOMOD2: application and validation for Costa Rica. Agric Ecosyst Environ 85:191–203

van Ittersum MK, Rabbinge R, van Latesteijn HC (1998) Exploratory land use studies and their role in strategic policy making. Agric Syst 58:309–330

van Keulen H (1990) A multiple goal programming basis for analysing agricultural research and development. In: Rabbinge R, Goudriaan J, Keulen Hv, Vries FWTPd, Laar HHv (eds.) Theoretical Production Ecology: Reflections and Prospects, Simulation Monograph 34, Pudoc, Wageningen (1990), pp 265–276

Veldkamp A, Fresco LO (1996) CLUE-CR: An integrated multi-scale model to simulate land use change scenarios in Costa Rica. Ecol Model 91:231–248

Verburg PH, Schot PP, Dijst, M.J., Veldkamp A (2004) Land use change modelling: current practice and research priorities. GeoJournal 61:309–324

Verburg PH, Soepboer W, Veldkamp A, Limpiada R, Espaldon V, Mastura SSA (2002) Modeling the Spatial Dynamics of Regional Land Use: The CLUE-S Model. Environ Manage 30:391–405

White R, Engelen G (2000) High-resolution integrated modelling of the spatial dynamics of urban and regional systems. Comput Environ Urban Syst 24:383–400

Wilensky U (1999) NetLogo. Evanston, IL: center for connected learning and computer-based modeling, Northwestern University.http://ccl.northwestern.edu/netlogo

Colour Plate 4 (Chap. 4, Speiser et al.; Chap. 5, Simon et al.): Views of the 'Self-Regulating' Orchard

View of the 'self-regulating' orchard in the 2nd year. In the centre, diverse alleyways with flowering wild carrots and white clover in the foreground; hail protection net. For preliminary results, see Chap. 5 (Simon et al.).

Mulching of alleyways. The alleyways are mulched in a 'sandwich-system' to ensure high diversity of structures and species.

Tree strips. Tree strips are tilled in a 'sandwich-system' to limit weed competition and to provide additional floral diversity.

S. Bellon, S. Penvern (eds.), *Organic Farming, Prototype for Sustainable Agricultures*,
DOI 10.1007/978-94-007-7927-3, © Springer Science+Business Media Dordrecht 2014

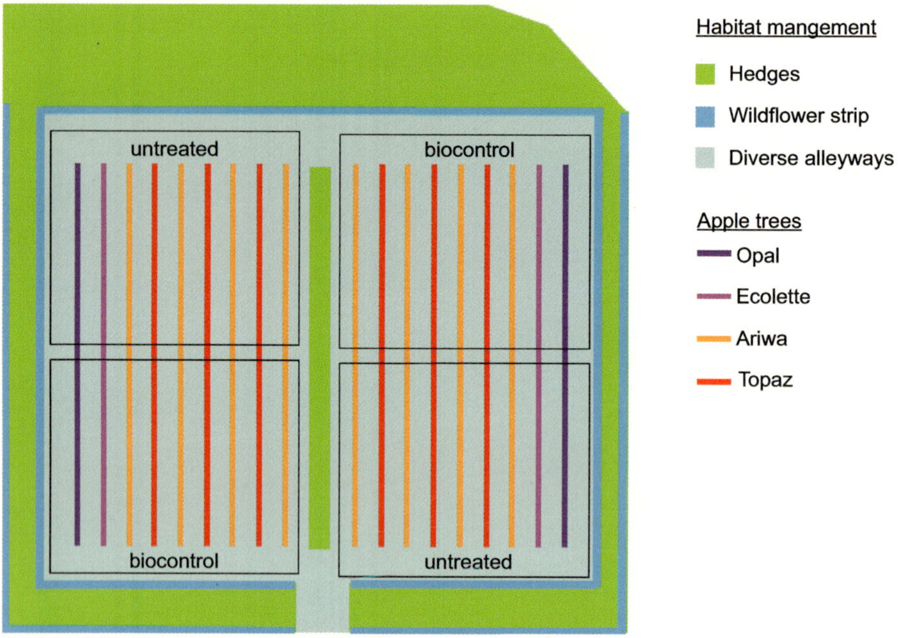

Design of the 'self-regulating' orchard

Nesting house for birds

Colour Plate 5 (Chap. 5, Simon et al.)

Pollen beetle (Meligethes aeneus) and the parasitoid Tersilochus heterocerus (Thomson, 1889) (on the right) on an oilseed rape petal (Photos: Copyright, Rothamsted Research Ltd.).

Illustrations of various landscape contexts (from simple to complex landscapes) in northwestern France.

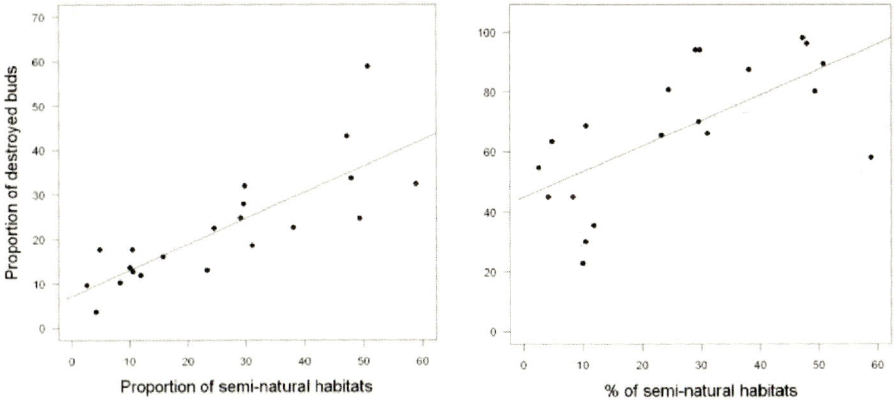

Relationships between the proportions of semi-natural habitats (computed within a 2000-m radius) and destroyed buds or parasitism rates in northwestern France.

Colour Plate 6 (Chap. 6, Deguine and Penvern): Agroecological Crop Protection Applied to the Management of Cucurbit Fly Populations in Organic Farming in Reunion Island

Habitat management. On the left, corn borders (trap plants) around a courgette crop. On the right, natural tree border (trap plants) around a pumpkin crop.

Sanitation. On the left, an augmentorium in a courgette crop. On the right, two adults of *B. cucurbitae* sequestered by the net, and an adult of *Phyttalia fletcheri* escaped from the net.

Male Annihilation technique. On the left, a trap without insecticide with a parapheromone inside. On the right, two adult males of *B. cucurbitae* entering the trap.

Conservation Biological Control. On the left, an adult of the spider *Nephila nigra* catching an adult of *D. ciliatus* in a chayotte crop. On the right, an adult of the syrphid *Ischiodon aegyptius* preying on aphids.

Colour Plate 8 (Chap. 8, Hoste et al.)

Figures 8a and 8b. The use of outdoor pastures or runs recommended by OF is usually associated with increased risks to parasites (e.g., helminth infections in chicken) as well as to some opportunities for solutions from the environment (e.g., caprine browsing behaviour favours avoidance of the infectious stage of helminths). **NB: Fig. 8a:** photo by V. Maurer (FiBL); **Fig. 8b:** photo by S. Sotiraki (NAGREF).

Colour Plate 21 (Chap. 21, Chable et al.): Evaluation of Genetic Resources and Participatory Plant Breeding for Vegetable Species in the Pays de Loire Region (France)

For six years, farmers, members of the association, Bio Loire Ocean, have collectively assessed and evaluated heirloom vegetable varieties within a participatory framework. The scientific partners' tools, methodology and knowledge are progressively added to the farmers' know-how.

What are the farmers' main motivations?

- To have varieties adapted to organic farming, to the different markets (short circuit/long circuit) and to the environmental conditions of their land, at their disposal. Jean-Michel Potiron, a participating farmer, would like "*to try to rediscover varieties adapted to our land and to break with the adjustable pattern of commercial varieties currently available*".
- To encourage and develop cultivated biodiversity. Farmers are aware that hybrids are unable to evolve with the change in environmental pressure and their cropping techniques.

Diversity in carrot From left to right: (1) 'Anthocyanée'; (2) 'Paris Market 2'; (3) 'Blanche demi-longue des Vosges'; (4) 'Jaune de Lobberich'; (5) 'Senator'; (6) 'Chantenay à coeur rouge'; (7) 'Rodelika'

'Rodelika', a biodynamic modern variety was used as a control.

How to meet their expectations?

A four-step approach was developed as the participants' involvement grew.

1. Screening of varieties mainly taken from conservatories *ex situ*: observation of their behaviour in organic farming.
2. Selection of varieties judged interesting in the field: meetings between farmers/ scientific and technical stakeholders focused on testing new varieties and defining criteria for selection together.
3. Multiplication of selected varieties by voluntary farmers: maintaining the specific traits of the variety or directing the selection to favour the expression of a specific trait (example of parsnip).

4. Evaluations over time and in various pedo-climatic environments: monitoring of the expression of traits and adaptation to new environmental pressures.

First progeny of a parsnip selection initiated by the producers and favouring a round shape. Patrick Gauthier, part of the parsnip selection said: "*This variety was cultivated thirty years ago but its seeds are untraceable today*."

Encouraging Results

Species with a pattern of selection taking shape	Parsnip Carrot Spinach Beans	Selection towards a rounded morphotype (2nd year of selection) Selection of a tasty variety for preservation Selection of a population (3rd year of selection) Observation of descendants of variants
Species being screened	Radish Lettuce Tomatoes Cabbage	The observations of varieties consistent with farmers' requests are continuing. Additional varieties are added to the trials
Species introduced in 2011–2012	Onions Peas …	At farmers' requests, additional species are added to the programme. Nevertheless, priority is given to the previously planted species

As of today, the results of the experiments are partial and are worth being confirmed through additional observations in different environments. However, some patrimonial populations created by previous generations of farmers and since forgotten have already been cultivated again.

Above and beyond the results, the trials are the time and place for the exchange of technical knowledge and know-how between producers as well as scientists. These meetings make it possible to pool the knowledge of all those involved. Working with royalty-free varieties offers the prospect of seed autonomy in committed farmers' groups. We also have to bear in mind that a regulation framework needs to be set up in cooperation with other seed agents.

"The scientists know the origins and traits of different types of vegetables that we don't have," said Nicolas Oran, producer. *"It's an opportunity to bring heirloom varieties back to life and to meet organic producers as well,"* added Emmanuel Geoffriau, responsible for the carrot and other Daucus species network for genetic resources.

The participation in European programmes that promote advancement in relation to seed regulation has given confidence to the farmers. It offers a framework that has been extended to other regions and European countries to improve methods of evaluation and protocols of selection. In parallel, the network created in the Pays de Loire region is expanding with new farmers and new varieties by making the most of the current dynamism for participatory plant breeding.

Colour Plate 25 (Chap. 25, Lopez-Ridaura et al.)

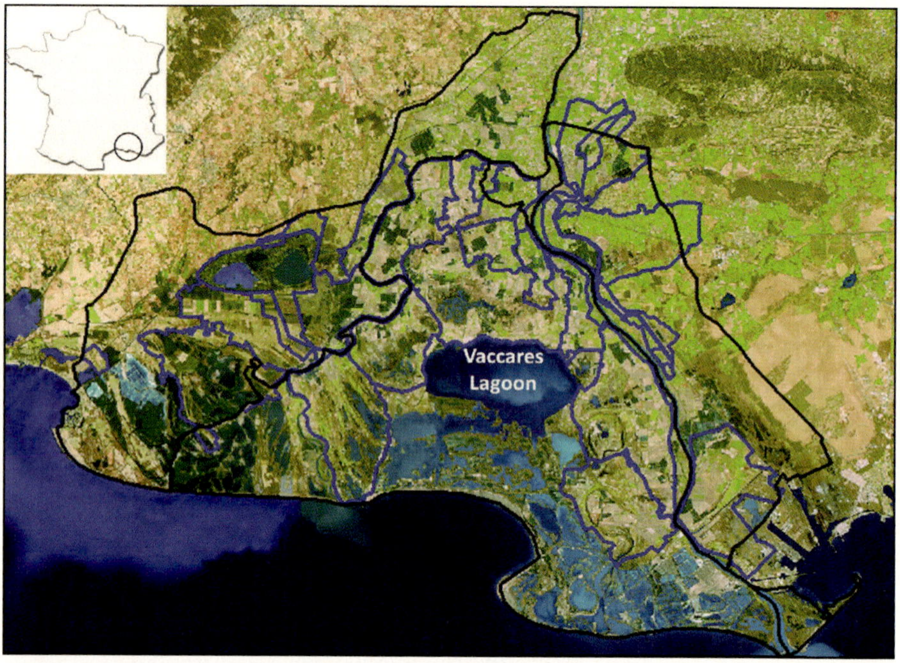

Fig. 1. The Camargue Region: the limits of the Biosphere Reserve in black, and the drainage perimeters in grey (data from the Parc Naturel Regional de Camargue, the Syndicat Mixte de la Camargue Gardoise and the Syndicat Mixte de Gestion des Associations Syndicales Autorisées).

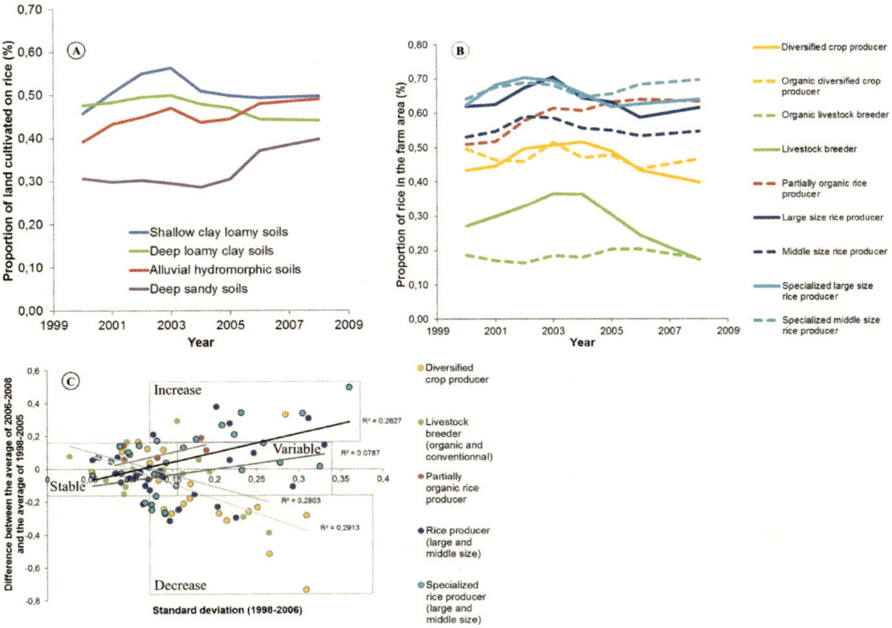

Fig. 2. A. Three-year average of the proportion of rice fields per soil type at the regional scale. **B.** Three-year average of the proportion of rice fields for the nine types of farm. **C.** Identification of four main types of trajectories of land-use evolution at the farm scale. The reader can refer to the text for more details.

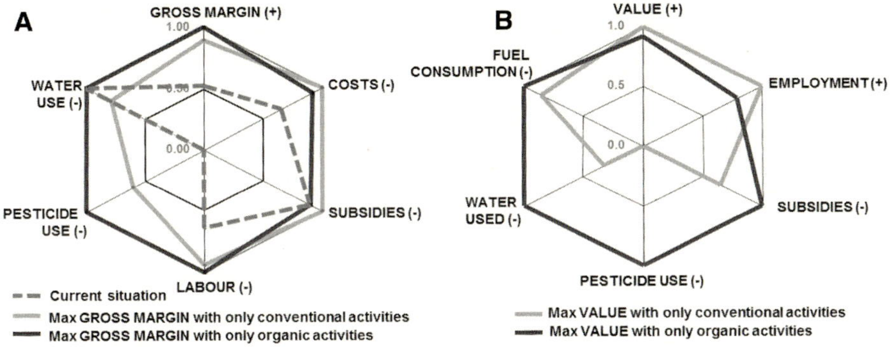

Fig. 3. A. Current situation for a middle-size specialised rice producer and under scenarios of maximisation of the gross margin with conventional and organic activity scenarios. **B.** Scenarios of maximisation of the production value in Camargue with organic and conventional activities.

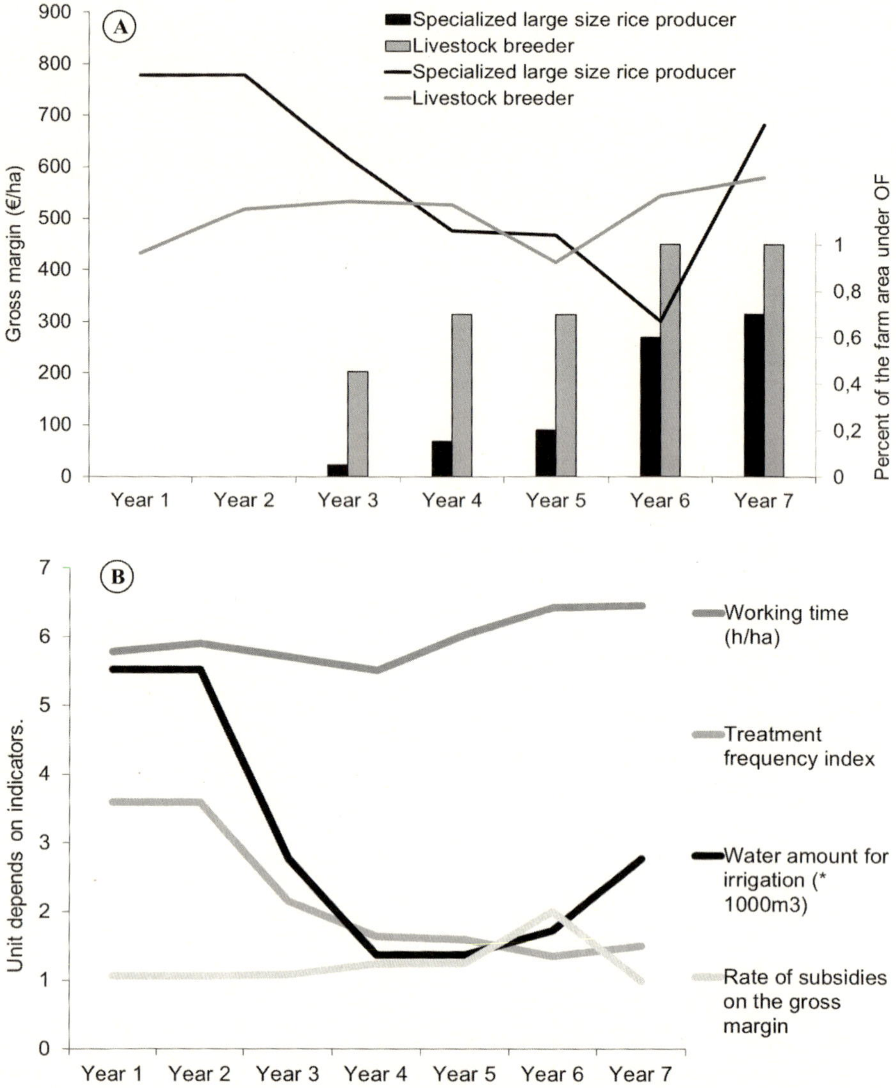

Fig. 4. A. Simulation of the evolution of the gross margin for two different farm types and the proportion of land under OF. **B**. Evolution of different indicators for a middle-size rice producer.

Index

A

Abiotic, 53, 98, 113, 115, 132, 134, 136, 144, 165

Acidification, 225, *see also* Life cycle—assessment

Adaptation

(context) adapted, 99, 79, 38, 27, 91, 132, 368, 380, 385–394, 429

adaptability (capacity), 12, 30, 116, 125, 132–144, 164, 165, 211, 213, 274, 387, 389, 442, 445

Adaptive systems, 211

Agent-based models (ABM), 455, 456, 462, 464

Agro-chemicals, 301, 342, *see also* Fungicides; Insecticides; Pesticide(s)

Agroecological crop protection, 113, 114

Agroecology, 3, 4, 6, 84, 85, 113–115, 124, 379, 380, 402, 414, 429

Agroecosystem, 11, 86–88, 90, 97, 99, 108, 115, 116, 118, 124

design, 403, 404, 406, 407, 414

functioning, 125

phosphorus management principles in, 27, 28

Alternative

type of agriculture, agricultural model or system, management or farming system, 4–8, 91, 108, 112, 281, 297, 327, 333, 352, 424, 433, 441, 456

pest management, control strategy, 114, 415, 424

plant protection products or solution, 73, 75, 79, 97, 108, 111

animal therapy, medicine, treatments, 5, 160, 161, 172, 189, 207, 208, 268, 269

methods, 3, 69, 161, 163, 213, 222, 235

breeding strategy, 135, 368, 380, 385

to organic farming, 280, 335–341

(agro-)food system, 447–449

Analytic

approach, 9

methods, 71, 298, 304, 319, 208–213

Animal, *see also* Livestock

behavior, 272, 274

capabilities, 265, 266, 272

health, 6, 151, 153, 162, 172

husbandry, 243, 249, 262, 266, 269, 282

liberty, 262, 263

monogastric, 158, 244, 249

production, 10, 27, 153, 282, *see also* Livestock—production

strategy, 172, 174

welfare, 10, 69, 150, 151, 158, 164, 248, 250, 261, 262, 264–266, 272, 279–281, 287–289, 308, 349, 350, 353, 354, 356, 361, 448

Anthropocentric approach, 202, 212, 214, 215

Antioxidant, 162, 300, 304–306, 315, 317

Apple, 3, 65, 75-79, 93, 95-97, 132–144, 222–236, *see also* Orchard(s)

breeding, 137

Arable crops, 94, 96, 388, 391, 392, 427

Architectural types, 139

Arthropods, 69, 70

Assessment

life cycle, 255

multi-criteria, 463

Association, 274, 353, 385, 386, 389, 391, 394–396, 444, 446, 448

agriculture–breeding, 326

farmers', 389, 394

learning, 353

mixed crop system, 209

S. Bellon, S. Penvern (eds.), *Organic Farming, Prototype for Sustainable Agricultures*,
DOI 10.1007/978-94-007-7927-3, © Springer Science+Business Media Dordrecht 2014

organic agriculture, 385, 446
organic animal husbandry, 273
plant breeding activities, 383
seeds, *see* Seed(s), association
Attract and kill technique, 71, 121
Authentication, 305, 314, 319, 320, 322
Authorization criteria, 74–76
Autonomy, 283, 376, 378, 379
farm, 384
farmer's, 388, 389

B
Behavioural evaluation, 172
Bifurcation (argument), 328, 339, 402
Biocontrol, 84, 86, 87, *see also* Conservation biological control
Biodiversity, *see also* Functional biodiversity; Habitat management
agents, 158
conservation, 5, 109, 391, 393
management, 117, 118
Bio-economic models (BEM), 454, 455
Bodily integrity, 266, 268, 271, 272, *see also* Animal—welfare
Breeding
on-farm, 385, 389, 395, 396
catalogue, 368, 369
participatory plant (PPB), 378, 385–390, 392
plant, 374–376, 378, 380
practices, 206, 388
programmes, 135–137, 144, 271, 373, 374, 378, 388, 390
Breeds, 165, 184, 187, 197, 267, 272, 283–287
livestock breeds, 159
Budget
nutrient farm-gate, 33, 38, 40
nutrient field budget, 33

C
Capital
natural, 411
personal, 410
social, 410
Carbon, *see also* Climate change; Global warming; Life cycle—analysis
dioxide, 225
footprint, 314, 318, 322
sequestration, 241, 250–253
Cereal
crop production, 187, 192, 193, 195
grain legume intercrops, 31, 48, 49, 51, 58
quality, 49, 51
Certification, *see* Organic—certification
Change
agents, 415
progressive, 402, 407, 409, 415–417
Chain, *see* Food chain
Civic agriculture, 348
Civil commons, 13, 349, 350, 359–361
Clay minerals, 25, 75
Climate change, 240, 241, *see also* Global warming
Co-construction, 446
Co-evolutionary, 410, 446
Coexistence, 5, 335, 336, 341
Co-grazing, 156–158
Companion plant strips, 78, 89, 93, 414
Competition, 53, 54
market, 11
for resources, 48–60
Complementary interactions (for resources), 57
Concentrate, 38, 119, 182, 185, 187, 189, 190, 192–198, 248–250, 252, 254, 300, 314–319, 321, 322, 454
input costs, 192, 193, 197
Confined animal feeding operations (CAFOs), 353
Conservation
agriculture, 97
biological control, 84
of genetic resources, 394
Consumers, 9, 11, 13, 76, 184, 268, 273, 281, 284–291, 306, 308, 313, 314, 320–322, 327, 328, 332–338, 341, 354, 358, 361, 380, 389, 393–396, 411, 439
European, 349
of fuel, 210
health of, 286, 308
of organic food, 302, 313, 318, 443
organic farming, 454
North American, 349
sustainibility in, 449
Contaminants, 296
Control
biological, 84
parasite, 152, 163, 189
Conventionalisation, 12, 328, 329, 425, 440, 441
Conversion, 5, 6, 36, 112, 150, 431, 454, 458–460, 462–464, *see also* Transition
Co-operacy, 416
Copper fungicides, 75, 79, *see also* Fungicides
Corn, 119, 120
Crop
ideotype, 134
production, 24, 32, 34, 37, 55, 86, 113, 192, 193, 195, 249, 355, 356, 462

protection, 109–111, 113, 114, 116–118, 124–126, 428, *see also* Agroecological crop protection
pest and disease management, 90, 133, 138, 144
rotation(s), 58, 67, 89, 204, 209, 245, 374, 375
Cultivars, 132, 135, 136, 139, 406
Cultural practices, 87, 89, 94, 96, 97, 99, 125, 138–143
Cycles
biological, 125, 133
crop, 113
nitrogen, 54
nutrient, 249, 444

D
Dairy
farming, 306
mixed crop systems (MCDS), 7, 203–205, 209, 211, 214
systems, 204
Deep organic, *see* Organic
Decision-making, 89, 118, 173, 204, 205, 207, 262, 333, 351, 352, 361, 410, 415, 444, 456
decision tool, 112
decision makers, 415
Demand
for different types of varieties, 375–379
for energy, 225, 234
for organic products, 11, 318, 358
Depletion
of nitrogen, 187
of phosphorus, 24, 27, 30, 38, 40
of soil, 34
of soil organic carbon, 355
Design, *see also* Orchard(s); Redesign
agroecosystem, 6, 113–116, 403, 407, 413, 414
animal husbandry, 291
designing alternative management systems, 8
of agroecosystems, 113
of farming activity, 202, 210
of intercrop management system, 56–58
plant breeding system, 379
step-by-step approach, 203–215
Diets, 209, 280, 316, 317, 320, 444, 448, 449
Diffusion, 11, 26, 27, 108, 152, 159, 165, 380, 396
Disease(s), *see also* Animal—welfare; Drugs; Pathocoenosis; Veterinary—drugs
coccidian infections, 152, 155
of crops, 49, 77, 78, 94, 136, 140, 143
fire blight, 76, 77, 136, 141
management, 227
prevention, 164, 172, 177, 262, 268, 271
scab, apple scab, 76, 77
Diversification
plant, varietal diversification, 87, 93-94, 96, 99, 389
of farming, cropping systems, 340, 373-374, 431
of the market, 379, 431
of the products, production, 379, 425, 428, 430, 431, 462
Diversity
diversified agroecosystem, 113, 124
intra-varietal, 389
of organic farming, 10, 12, 13, 440, 448, 449
Drugs
chemical, 152, 159, 161, 172
herbal, 161
DUS (Distinctness, Uniformity, Stability), 368, 374, 376, 379, 385, *see also* Breeding; Selection

E
Earthworms, 125, 153, 356
Eco-functional intensification, 48–60
Ecolabelling, *see* Label
Ecologically-based agricultures, 7
Economic
benefit, 58, 59
indicators, 458, 461
profitability, 198, 444
sustainability, 433, 445
viability, 444, 445
Ecosystems, 24, 225, 307, 359, 407, 408, 415, *see also* Agroecosystem
Education, 73, 114, 273, 274, 360
Efficiency
Efficiency-Substitution-Redesign (ESR), 4, 402, 403, 427, 428
pesticide use, 91, 92
Emotional dimension, 286
Emissions, 234, 235, 245–255, 376
aerial, 225
ammonia, 232
carbon dioxide, 230, 246, 250, 251, 443
direct field, 224, 225, 231
farming systems, 222
fertilisers, 230
fuel combustions, 234
greenhouse gas (GHG), 240, 241, 445
methane, 240, 247–250
nitrous oxide, 232, 242, 245–247

pesticides, 226, 229
phosphorus, 232
Empowerment, 207, 214, 269, 376, 378, 379
Energy, *see also* Environmental—impacts; Life cycle—assessment
non-renewable (NRE), 225, 229, 230, *see also* Fuel
use, 229, 240, 241, 244, 348, 355
Enteric fermentation, 240, 247, 248, 249
Environmental
benefits, 348, 349, 440, 443
impacts, 143, 223, 235, 440, 442, 443
performance, 5, 222, 223, 233–235, 250
resource use, 48, 51
Ethics, 159, 264, 322, 337, *see also* Animal—welfare; Organic—value
Eutrophication, 223, 225, 231–234, *see also* Life cycle—assessment
Evolution, *see also* Co-evolutionary
capability, 12, 13
evolving practices, 441
Ewe productivity (EP), 182, 185–187, 189, 190, 193, 197, 198
Exchange
of experiences, 390
of genetic resources, 390
of seed, 390, 393–395
Experiences, 11, 175, 203, 206–211
Experimenter's learning, 203

F

Farm, *see also* Mixed—farms; Organic—farms; Transition—farms
animals, 151, 274, 275
design, 348, 357
gate budget, 33, 38, 40
income, 190
inflow and outflow, 33
management, 34, 92
structure, 183, 185, 186
typology, 458, 460
Farming practices, 24, 38, 67, 254, 434, 440, 441, 444, 445, 449
Farming systems
low-input, 3, 6, 14, 59, 133, 138, 141, 143, 306, 314, 320
mixed, 37, 38, 249, 254
traditional, 394
Fatty acid (FA), 163, 248, 296, 300, 305, 306, 315, 319, 321
Feeding, 173, 182, 188
Fertilisers, *see also* Input substitution; Nitrogen; Rock phosphates
application, 229, 231, 232, 246, 247
chemical, 2, 6, 197, 251, 256, 286, 297, 306
fertilisation strategy, 138, 141
organic, 27, 34, 36, 39, 226, 227, 232, 234, 247, 249, 251–253, 301, 306
synthetic, 240, 245–247, 251, 268, 318, 322, 327
use, 37, 138, 252–254
Fertility, *see* Soil(s)—fertility
Field
nutrient budget, 28, 33, 34, 36–38, 40
potentialities, 204
Fire blight, 76, 77, 136, 141
Five freedoms, 262, 263, *see also* Animal—welfare
Food
agro-food chain, 442, 443, 446, 448
consumption, 303, 306
chain, 302, 425, 427, 430, 431, 433, 441
demand, 11, 24, 374–379
human, 297
labelling market, 328
miles, 443
prices, 34, 58, 198, 458, 462
processing, 99, 297, 307
production, 3, 4, 7, 14, 241, 255, 305–308
products, 304, 318, 319, 322
quality, 8, 298, 304, 305
retailing, 442
safety, 150, 297
security, 4, 12, 241, 255, 307, 308
Forage, 184, 185, 194
self-sufficiency, 185, 187, 192, 193
Fruit and vegetable, 425–427, 443
Fuel
consumption, 3, 210
fossil, 250, 461
Functional biodiversity, 84, 98, 113, 114, 125, 144, *see also* Biodiversity; Habitat management
Fungicides, 30, 135, 142, 227, 228, 302, 306, 327
copper fungicides, 75, 79, *see also* Clay minerals

G

Gender equality, 350, 351, 360
Genetic, *see also* Breeding; Catalogue; Ideotype(s)
modification, 27
resources, 85, 377, 380, 388, 390, 393, 394
Genotype, 30, 40, 41, 164, 187, 197, 283, 374, 378, 387, 396, *see also* Breeding; Catalogue; Ideotype(s)

Gift theory, 281, 285
Globalisation, 10, 11
Global warming, 240, 255, *see also* Carbon—sequestration; Climate change; Fossil fuels; Greenhouse gas
potential (GWP), 222, 223, 225, 230, 231, 335
Goods
public, 360
social, ecological and economic goods and services (SEEGS), 348, 350–358
Grassland management, 185, 206
Grazing
management, 155
system (GS), 39, 204, 205
Greenhouse gas, 240, 242, 255, *see also* Carbon—sequestration; Climate change; Fossil fuels; Global warming
Gross margin, 189, 190, 192–194, 196, 197, 461, 462
Ground cover, 88, 93, 95, 142

H
Habitat, *see also* Biodiversity; Functional biodiversity
management, 94, 99
manipulations, 109
semi-natural, 88, 89, 91, 92, 94, 95, 99
Health, *see also* Animal—health; Crop—protection; Disease(s); Human—health
and welfare planning, 268, 270, 270
effects, 302–304, 307
public, 302, 444, 449
Helminths, 150–152, 154, 157, 159, 161, 165
infection, 150, 153, 156, 158
Herbal drugs, *see* Drugs
Holistic
approach, 115, 116, 125, 223, 359, 361, 376
system management, 133, 348, 414
understanding, 440
vision, 7
Homeopathy, 151, 160, 164, 165, 207, 269
Host
resilience, resistance, 151, 152
Human
health, 108, 150, 225, 297, 306, 359, 444
responsibility, 264
well-being, 444
Husbandry, *see* Livestock
Hygiene, 151, 153–158

I
Ideotype(s), 6, 7, 132–134, 137–140, 144, 373, 374
concept, 134, 135
definition, 136, 137
hardy, 142
Immune
prophylaxis, 151
responses, 152, 296, 303, 304
Impact assessment, 225, 226, *see also* Environmental—impacts
Income, 58, 108, 135, 183, 185, 190, 194, 358, 359
Industrialisation, 261, 280, 281, 290, 403, 441
of pig farming, 283
Innovation, 5, 11–14, 124–126, 211, 288, 337, 352, 380, 389, 447–449
Input substitution, 6, 111, 124, 403, 440, 441, *see also* Efficiency—Efficiency-Substitution-Redesign (ESR)
Insecticides, *see also* Agro-chemicals
granulosis virus, 74, 75, 78
spinosad, 75, 77, 125
Institutionalisation, 10, 11, 331
Integrated
approach, 111, 151, 152, 356, 415
fruit production (IFP), 133, 135–138
pest management (IPM), 86, 89, 92, 95, 109–113, 116
production, 5, 133, 226, 235
Integration, 5, 58, 109–111, 115, 377
of crop and livestock production, 356
Intensification, 186, 306
agricultural, 48, 112, 116, 454
cropping systems, 36
eco-functional, *see* Eco-functional intensification
forage area, 186
pastures, 320
pigs, 153
Intercrop(s), 53–55, 57, 60, *see also* Cereal—grain legume intercrops
economic benefits, 58, 59
intercropping, 31, 48, 49, 51, 53, 56, 57, 94
International Federation of Organic Agriculture Movements (IFOAM), 6, 12, 24, 27, 37, 70, 113, 329, 351
In vitro meat, 280, 288

K
Knowledge, 10, 96, 98, 125, 132, 144, 172–178, 235, 264, 275, 281, 289, 319, 340, 360, *see also* Research
agroecological, 116
and management, 108

base, 8
bioecology of fruit fly, 119
ecological, 6, 116, 413
farmers', 9, 114, 269, 353
gaps, 84
generic programming of, 215
helminth infection, 153
integration, 60
intensive techniques, 114, 125
microclimate history, 142
organic husbandry practices, 269, 273
organic lamb meat quality, 313–315
rhizosphere, 32
scientific, 139, 209, 213, 262
SEEGS, 349–352
soil tillage, 210
system self-sufficiency, 208

L

Label, *see also* Organic—standard
erosion, 12, 114, 326, 327
labelling, 10, 70, 328, 358
regulation, 327
Labour, 135, 227, 234, 297, 321, 349, 360, 429, 431, 435, 436
demand, 457, 461, 462
Lamb
lambing rate, 186–188, 192, 193
nutritional quality of meat, 314, 315
out-of-season lambing, 185–187, 192, 195
sensory quality of meat, 315–318
Landrace, 30, 135, 388, 390, 393, 394
Landscape, 5, 85, 88, 91–97, 99, 113–115, 117, 183, 185, 224, 235, 354, 357, 359, 378, 396, 413, 434
management, 99, 354
scale, 88, 89, 95, 117
Land use change
analysis, 458, 460
model (LUC), 455, 458–460
Learning, 203, 204, 208, 212–215, 269, 273, 274, 352, 425, 432, 435, 436, 448, 449
opportunities, 352, 353
Life cycle
analysis, 10, 222, 250, 254
assessments, 255
inventory, 223
methodology (LCA), 222, 223, 235
Light interception, 53
Livestock
breeds, breeding, 159, 457
farming, 39, 264, 282, 319, 356
husbandry, 271, 273–275
production, 7, 12, 271, 281, 282, 356, 414
Long-term experiments, 7, 8, 99, 204, 208, 211, 212
Low-input
farming, agriculture, or system, 3, 14, 28-31, 32, 48-49, 51, 59, 92, 132-138, 141-144, 214, 306, 308, 314, 320, 368, 378-379, 389, 395
technique, practice, 113, 124, 210, 432-433

M

Male annihilation technique (MAT), 120, 121
Malfunction, sources of, 206
Management
practices, 117
systems, 56–58, 89, 98
Manure management, 240, 249, 250
Market, 10, 13, 34, 58, 72, 132, 268, 286, 327, 337, 374, 377, 389, 395, 424, 448, 455
analysis study, 353
apple scab, 76, 136
cereal crop, 191, 194
domestic, 441
economy, 328, 334, 335
efficiency, 11
export, 362
farm animals, 262
farmers', 351, 445
gardeners', 427, 429
global food, 2
intercrops, 59
international, 385
local, 425
open air, 430
organic products, 11, 34, 289, 304, 335, 369, 441, 443, 447
organic sheep-for-meat, 182, 185
outlets, 429, 431, 433, 435
marketing channels, 79, 349, 426, 430–432
networks, 429, 431, 433
pragmatic farmers, 424
risks, 196
seed market, 373, 379, 385, 393, 394
strategies, 431, 433
wines, 334
Maturity, 49, 58, 59
of organic systems, 8
Meat
colour, 317, 318
flavour, 315, 316
Mechanical weeding, 5, 58, 91, 230
Metapopulation, *see* Population
Methane, 225, 249
emissions, 247–250

Microorganisms, micro-organisms, 69, 71, 72, 74, 141, 172, 296
pathogenic, 296, 301
rumen, 319
soil, 26, 31, 68
Milk production, *see* Dairy—systems
Mitigation, 240, 241
greenhouse gas emission, 241
performance, 253, 254
potential, 242–244
Mixed crop, 6, 7, 12, 31, 37, 94, 158, 184, 209, 252, 254, 273, 301, 436
dairy systems, 201–215
farming systems, 38, 249, 254
grazing, 273
livestock farming, 356
Mixed farms, 33
Mixity, 5, 6, 9
Monitoring, 8, 71, 116, 120, 121, 123, 144, 177, 332, 334, 338–340, 410
Monogastric(s), 151, 156, 158–160, 165, 249
Mortality, 122, 267
Multicriteria, multi-criteria, 118
analysis, 463
evaluation, 8
Multidimensionality, 9
Multifunctional agriculture, 234, 256, 352

N

Natural enemies, 7, 66, 67, 69, 85, 87, 88, *see also* Biocontrol; Functional biodiversity; Habitat management
Naturality, 177
Networks
farmers, 391–395
local, 393
Newcomers, 11, 12
Niche, 446, 449
development, 3, 13, 14, 334, 335, 446, 449
ecological, 31, 86, 412
market, 197, 360, 377
Nitrogen
atmospheric, 48, 54, 56
availability, 54–57, 306
biological fixation of, 48
efficiency, 31
input, 48, 49, 51, 245–247
leaching, 214, 225, 234, 235, 356
oxide emissions, 245–247
Nitrous oxide, 225, 232, 240, 245–247
Nutraceutical, 162, 165
Nutrient(s), *see also* Cycles; Depletion; Nitrogen; Phosphorus; Recycling; Surpluses; Budget
cycling, 354, 439
loading, 356
management, 37
Nutrition, 162
challenge, 297, 298
nutritional quality, 314, 315
value, 298–301

O

Orchard(s)
apple orchard(s), 76–78
design, 226, 227, 234
eco-design, 144
Orchard system experiment, 223, 234, 235
Organic
certification, 328
crop protection (OCP), 109–115, 121, 124
deep organic, 402, 403, 407
definition of, 329
farmers, 9, 72, 99, 120, 123, 314, 358
farms, 40, 110, 154, 271, 348, 431, 443, 445
fruit production (OFP), 133, 135–138
organics, 2–13, 326–341, 402, 403
plant breeding, 376
plant protection, 66, 67, 71, 72, 76–79
regulation, 66, 67, *see also* Organic—principles, *see* Principles
shallow organic, 30, 402, 403
standards, 6–12, 70–79, 110, 114, 262, 264, 327, 333, 335, 337-341, 354, 387, 441, 446
values, 402
Orphan species, 368, 369
Outdoor management, 151, 155, 156

P

Paradigm
shift, 10, 417
change of, 202, 379
pest managment, 113, 116
technical, 214
new, 202, 287
Parasite control, 152, 163, 189, *see also* Animal—health
Parasitic risks, 150–152, 164
Participatory, 144, 333, 338, 339, 390, 391, 473, 475
approaches, 13, 99, 378
certification, 388
control, 443
guarantee system, 338
interactive simulation with farmers, 464
plant breeding, 378, 386–389

research, 352, 385
tools, 464
Pasture
feeding, 314–317, 319, 321, 322
rotations, 156
Pathocoenosis, 7, 173–175, 178, 273
Performance, 1, 8, 9, 13, 57, 60, 192, 141, 144, 158, 160, 162, 164, 186, 189, 197–199, 213, 222, 223, 233–235, 250, 253, 255, 256, 303, 349, 390, 396, 462, 464
assessment, 9, 112
criteria, 13
environmental, 5, 143, 222, 223, 233–235, 250
reproduction, 206, 207
Permaculture, 413, 445
Pesticide(s), 2, 6, 14, 66–73, 75, 77, 78, 86, 91, 96, 108–112, 114, 116–118, 122, 125, 126, 132, 133, 222, 224, 225, 251, 263, 268, 286, 297, 300, 301, 306, 307, 318, 322, 327, 360, 388, 405, 406, 415, 425, 426, 427, 442, 456, 457, 461, 463, *see also* Agro-chemicals; Crop—protection; Disease(s); Fungicides; Insecticides; Pest(s)
registration, 67, 71–73
substitution, 66, 112, *see also* Input substitution
use efficiency, 112, *see also* Efficiency
Pest(s), 7, 49, 53, 57, 66, 69, 71, 73, 76–79, 84, 85, 87, 89, 91, 93–99, 108–113, 118, 119, 125, 135–138, 140, 141, 143, 144, 327, 374, 375, 406, 414, 424, 425, 429, 432, *see also* Disease(s); Weeds
and diseases, 143
control, 86, 87, 405
integrated management, 189
management, 113
prevention, 407, *see also* Biocontrol; Crop—protection
Pheromones, 69, 71, 72, 74, 77
Phosphorus, *see also* Soil(s)
availability, 26, 28, 29–31, 37, 40
budget, 28, 33, 34, 36–41
depletion, 24, 27, 30, 38, 40
Phytotherapy, 151, 160, 161, 164, 165, 269, *see also* Alternative—medicines
Pigs, 150–154
Plant
breeding, 13, 376–378, 380, 384–388, 390, 391, 396
diversity, 89, 98, 99
extracts, 71, 72, 74, 161
growth, 54, 55
health, 66, 73, 117, 119
oils, 69, 74
strengtheners, 74
Planting densities, 226
Pluri-disciplinarity, 115
Population
metapopulation concept, 394
plant, 384, 395, 396
Poultry, 150–152, 154
Predator, 93, 94, 122
prey interactions, 112
Prevention
preventive measures, 66, 76, 109, 116, 119, 120, 302
preventive processes, 125
Price, 11, 34, 58, 196, 291, 463
Principles
of organic farming, 69, 185, 332, 441, 447
of IFOAM, 12, 24, 27, 37
Proactive living, 408, 416
Processes
bioecological, 115, 124, 125
biotechnical, 3
diffusion, 11
ecological, 3, 24, 85, 94, 97, 98, 115, 124, 125, 203, 355, 407
natural, 5, 411, 457
regulation, 93, 108
Productivity, 3, 6, 8, 248, 250, 272, 282, 285, 461, 463, *see also* Performance; Yields
Profits
profitability, 8, 133, 138, 182, 184, 198, 274, 339, 444, 445
profitable, 49, 184, 262, 285, 328, 358, 360, 373, 375, 444
Protein content, 209, 299, 305, 316, 388
Protection
crop, 109–118, 124–126, 428
plant, 67, 70, 71, 108, 125, 227, 230
Prototype, 2–14, 93, 108, 113, 305, 350, 442, 443, 446, 448
Prototyping, 4, 8, 9, 211
Protozoa, 150, 151, 153, 159, 165
Pruning strategies, 138
Psychology, 407–409
Psychosocial, 402, 417

Q

Quality
food, 298
label, 187, 287
product, of the product, 8, 10, 132, 135, 153, 337, 358, 389

R
Recycling, 24, 28, 222, 224, 337, 348, *see also* Nutrient(s)
of animal manure, 37
of nitrogen, 255
of nutrients, 37, 85, 113, 245
of organic matters, 37
of phosphorus, 255
Redesign, 206–208, *see also* Design
approach, strategy, 403, 404, 429–432
ecological redesign, 402
of the (marketing or farming) system, 99, 208, 213, 348, 430, 433–436
Redesign agricultural activity, 201
Regulations, 27
evolution of, 379, 425
framework, 67, 71, 79, 327
limits of, 66, 67
natural regulatory processes, 71, 79, 114, 116, 327
Reproduction, 78, 87, 188
performances, 206, 207
Research, 132, 352, 353, 385
approaches, 116
methods, 2, 7, 9
Resilience, 4, 136, 296, 307, 308, 340, 349, 368, 403, 410, 412, 415, 445
agriculture, 408
definition of, 296
ecosystem, 85, 86
farm animals, 151
host, 151, 152
natural pest control, 88
parasites, 159
selection programmes on, 160
Resistance, *see also* Cultivars; Pest(s)—management
durable, 135, 136
host plant, 151, 152
partial, 136, 137, 140, 142, 143
Resource, *see also* Energy—use
conservation, 8
natural, 69, 133, 202, 407
use, 48, 51, 56
Rhizosphere, 24, 28–32, 40, 41
Rice, 249, 457
Risks, 150–152
Rock phosphates, 24, 27
Roots, 25–27, 29, 30, 40, 41, 329, 415, 416
Rotation(s)
crop, 374, 375
pasture, 156, 157
Roughage, 154, 248–250
Ruminants, 150–153, 155–158, 161–163, 165, 247, 249
Rural
community, 350, 351

S
Sanitation practices, 140, 141, 227
practices, 140, 227
Scenario assessment, 458–462
Seed(s)
association, 386, 391, 392, 396
community banks, 395
production, 385, 388, 390, 391, 393, 395
regulation, 377, 385, 391–395
variety, 368–374
Selection
genetic, 151, 164, 271
natural, 159, 378, 395
Self-regulation, 7, 78, 403, 408, *see also* Regulation
self-regulating process, 109
self-sufficient system, 203, 204, 212–215
Self-sufficiency, 5–8, 38, 190, 198, 204–206, 210, 214, 215
improvement of, 208
in organic farming, 6, 40
of feed, 181, 184, 185, 194, 206, 318, 321
of forage, 185, 187, 188, 192–194, 199
of nutrients, 40
thresholds, 198
transition, 211
Services
ecological, 113, 357, 358
ecosystem, 85, 354
social, ecological and economic goods (SEEGS), 348–358
Sensory
characteristics, 315, 316
problems, 320, 321
Shallow organic, *see* Organic
Sheep
farms, 176, 182, 193, 194, 197,198
Sequestration (soil carbon), 241–253
Slaughter
age, slaughtering age, 283, 320
slaughterhouse, 284, 285
Social
ecology, 410, 411
equity, 443, 444
gender equality, 351, 352
innovations, 447–449
justice, 4, 409, 444, 449
learning opportunities, 352, 353
responsibility, 445
Sodium bicarbonate, 74, 75
Soil(s)
acidification, 30

biodiversity, 116
erosion, 359
fertility, 2, 6, 8, 58, 117, 124, 144, 251, 253, 255, 444
management, 242, 354, 429
microbial biomass, 26, 31, 356
microorganisms, 26, 31, 32
mineral nitrogen, 55
nutrients, 24, 28, 29, 32, 354, 444
organic carbon (SOC), 241, 355, 357
organic matter, 249, 251, 354, 374
phosphorus, 25
tillage, 29, 210
Specialisation, 89, 98, 248, 430–432, 435, 454
of organic farms, 12, 431
of pests and natural enemies, 89, 98
crop, 429
Species, 369–371, 374, 375
dominant, 57
Spinosad, 75, 77, 121
Standards, 6–12, 70–79, 110, 114, 262, 264, 327, 333, 335, 337-341, 354, 387, 441, 446, *see also* Label—labeling; Organic—certification; Principles; Regulations
standardisation, 9, 97, 161, 307, 334, 374
Stocking rate, 150, 156, 183, 196, 197, 248, 254
Stockless, 27, 34, 36, 38–40
Stress, 3, 114, 115, 155, 267, 301
environmental, 306
water, 37, 58
nutrient, 136
Substitution
input, 6, 111, 124, 403, 440
pesticide, 66, 112
Supply, 11, 60, 91, 144, 226, 227, 318
short food supply chains, 425
Surpluses, 356
phosphorus, 38
Sustainability
organic orchards, 76, 78, 96, 135, 233
sustainable agriculture, 3, 4, 5, 86, 126, 201, 246, 255, 256, 306
System(ic)
approach, 9, 241, 253–255, 439, 442–445, 448, 449
definition and boundaries, 224, 225
experimentation research, 353

T

Territory, 204, 205, 209, 327, 379, 458
territorial level, 13
Tillage, 91
no-tillage systems, 94, 242
reduction, 95, 96, 242
Toxicity and ecotoxicity, 228, 229
Trade-offs, 250, 463
Traditional farming, 394
Training strategies, 138
Trajectories, 4, 6, 9, 415, 425–427, 430, 435, 436, 460, 463, 464
Transdisciplinarity, 9, 14
Transformation, 327, 329, 337, 446–449
Transition, 2, 5, 6, 9, 13, 111–113, 116, 211, 403, 411, 424–426, 429–432, 434–436, 440, 446–449, 462, 464
farms, 425
pathways, 10–13, 431
to sustainability, 440, 446–448
trajectories, 9
Transport, 245, 255, *see also* Food—miles; Non-renewable energy
Trap plant systems, 119
Treatments, 160–163, 177
organic farming, 30, 31
affecting parasite biology
herbal drugs, 161, 162
nutraceuticals, 162, 163
targeted selective treatment (TST),163, 177
Tree architecture manipulation, 140

U

Utilitarianism, 281, 286

V

Vaccination, 159, 268
Values for Cultivation and Use (VCU), 368, 379, 385
Variety (of crops)
intra-varietal diversity, 389
locally adapted, 388, 390, 395
Veterinary
drugs, 318, 322
expenses, 189
profession, 269
sociology, 171, 172
Viticulture, 331, 342

W

Water
catchment, 14
watershed approach, 359
Weeds
control, 12, 48, 49, 142, 230, 387
management, 117, 249, *see also* Mechanical weeding

Welfare
  animal, 261, 262, 265, 287, 353, 354
  human well-being, 444
Well-being, 262–264, 281, 359, 403, 410, 411, 417, 418
Work, 9, 13, 30, 95, 96, 126, 142, 170, 174, 190, 193, 207, 210, 212, 215, 230, 235, 262, 263, 265, 280–291, 297, 306, 351, 352, 375, 389, 390, 395, 402, 408, 410, 411, 415, 417, 429, 431, 446

Y

Yeoman's approaches, 413
Yields, 8, 24, 32, 36, 41, 48, 49, 58, 60, 124, 132, 234, 255, 305, 327, 358, 375, 424, 432, 433, 458, *see also* Productivity

Printed by Printforce, the Netherlands